ESO ASTROPHYSICS SYMPOSIA
European Southern Observatory

Series Editor: Bruno Leibundgut

A. Kaufer   F. Kerber   (Eds.)

The 2007 ESO Instrument Calibration Workshop

Proceedings of the ESO Workshop
held in Garching, Germany,
23-26 January 2007

 Springer

Volume Editors

Andreas Kaufer
European Southern
Observatory (ESO)
Alonso de Cordova 3107
Santiago 19
Vitacura
Chile
akaufer@eso.org

Florian Kerber
European Southern
Observatory (ESO)
Karl-Schwarzschild-Str. 2
85748 Garching
Germany
fkerber@eso.org

Series Editors

Bruno Leibundgut
European Southern Observatory
Karl-Schwarzschild-Str. 2
85748 Garching
Germany

ISBN: 978-3-540-76962-0   e-ISBN: 978-3-540-76963-7

ESO Astrophysics Symposia ISSN: 1431-2433

Library of Congress Control Number: 2008923198

© 2008 Springer-Verlag Berlin Heidelberg

This work is subject to copyright. All rights are reserved, whether the whole or part of the material is concerned, specifically the rights of translation, reprinting, reuse of illustrations, recitation, broadcasting, reproduction on microfilm or in any other way, and storage in data banks. Duplication of this publication or parts thereof is permitted only under the provisions of the German Copyright Law of September 9, 1965, in its current version, and permission for use must always be obtained from Springer. Violations are liable to prosecution under the German Copyright Law.

The use of general descriptive names, registered names, trademarks, etc. in this publication does not imply, even in the absence of a specific statement, that such names are exempt from the relevant protective laws and regulations and therefore free for general use.

Cover design: WMXDesign GmbH, Heidelberg

Printed on acid-free paper

9 8 7 6 5 4 3 2 1

springer.com

Preface

The first ESO Instrument Calibration Workshop took place from 23–26 January 2007 at the ESO headquarters in Garching, Germany. It attracted more than 120 participants with a good representation of the diverse ESO user community and ESO's operations groups.

The La Silla Paranal Observatory is currently operating 19 optical, NIR, and MIR instruments (9 VLT, 2 VLTI, 8 La Silla). Successful scientific operation of such an instrument suite is a complex task. To monitor and calibrate both the performance of each of these instruments and the quality of the data they deliver, ESO executes dedicated calibration plans. The calibration plans describe – for each instrument – systematic measurements that are routinely performed in order to aid in the calibration of data from science programs, at least to specified levels of accuracy.

The first ESO/ST-ECF workshop on calibrating and understanding HST and ESO instruments was held in 1995 to review the calibration strategies of HST and ESO La Silla instruments and to prepare for the start of operation of the VLT.

We felt that after more than seven years of science operation with the innovative, complex, and still growing instrumentation suite at the VLT, it was timely to review the achievements and limitations of the established instrument calibration plans together with the ESO user community.

Hence the goals of this workshop can be summarized as:

- to foster the sharing of information, experience and techniques between observers, instrument developers, and instrument operation teams,
- to review the actual precisions and limitations of the applied instrument calibration plans, and
- to collect the current and future requirements by the ESO users.

The workshop tried to cover a large variety of aspects through a series of overview talks given by the Instrument Operations Team (IOT), invited talks by expert users, contributed talks and posters.

A total of 11 sessions high-lighted the various instruments: Optical Spectro-Imagers, Optical Multi-object Spectrographs, NIR and MIR Spectro-Imagers, High-Resolution Spectrographs, Integral Field Spectrographs, Adaptive Optics Instruments, Polarimetric Instruments, Wide-field Imagers,

Interferometric Instruments as well as other crucial aspects such as data flow, quality control and data reduction software and atmospheric effects.

These present proceedings collect the majority of the workshop contributions and document in large detail the status quo of instrument calibration at ESO. *Calibration is a life-long learning process* as Gianni Marconi put it during the workshop. In this sense these workshop proceedings will be a reference point for all future efforts to improve our instrument calibration procedures.

We would like to thank the members of the organizing committee R. Hanuschik, F. Patat, M. Peron, M. Romaniello, M. Sterzik, and L.E. Tacconi-Garman for their efforts in the preparation of the workshop. We would further like to thank K. Boutsia, G. Dremel and his crew from General Services, E. Janssen, S. Lowery, M. Lyubenova, S. Mieske, S. Pedicelli, F. Saitta, E. Siml, and B. Sjoeberg. Special thanks go to C. Stoffer for her exquisite support during the preparation and execution of the workshop. We also thank P. Bristow, I. Bronnert and S. Strunk for their help during the preparation of these proceedings.

Santiago, Garching December 2007

Andreas Kaufer
Florian Kerber

Contents

Part I Introduction

Instruments Calibration at the La Silla Paranal Observatory
G. Marconi 3

The Detector Monitoring Project
P. Amico, P. Ballester, W. Hummel, G. LoCurto, L. Lundin,
A. Modigliani, P. Sinclaire, L. Vanzi 11

Part II Session 1: Optical Spectro-Imagers

IOT Overview: Optical Spectro-Imagers
F. Patat 25

Photometry with FORS
W. Freudling, P. Møller, F. Patat, S. Moehler, M. Romaniello,
E. Jehin, K. O'Brien, C. Izzo, E. Pompei 35

Radial Velocity Measurements Based on FORS2 and FLAMES Spectra
M. Monelli, M. Nonino, G. Bono, P. François, F. Thévenin,
R. Buonanno, F. Caputo, C. E. Corsi, L. Pulone, H. A. Smith,
P. B. Stetson, A. R. Walker..................................... 41

The VLT-FLAMES Survey of Massive Stars: Instrumental Stability and Detection of Massive Binaries
C. J. Evans, D. J. Lennon, I. Hunter, T. Augusteijn, S. J. Smartt 49

VIMOS Total Transmission Profiles for Broad-Band Filters
S. Mieske, M. Rejkuba, S. Bagnulo, C. Izzo, G. Marconi 53

Laboratory Measurements of Calibration Sources for X-Shooter
F. Saitta, F. Kerber, P. Bristow, H. Dekker, S. D'Odorico, C. Dupuy,
J.-L. Lizon, A. Norup Sorensen, J. Vernet 57

Modelling the Fringing of the FORS2 CCD
J. R. Walsh, H. Kuntschner, E. Jehin, A. Kaufer, K. O'Brien,
M. Riquelme, A. Smette 63

Master Response Curves for Flux Calibration of VIMOS Spectroscopy
B. Wolff 69

Part III Session 2: Optical Multi-object Spectrographs

IOT Overview: Optical Multi-Object Spectrographs
L. Schmidtobreick, S. Bagnulo, E. Jehin, G. Marconi, K. O'Brien,
E. Pompei, I. Saviane 75

Quantitative Spectroscopy with FLAMES
P. Bonifacio 81

Good News for MOS, MXU & Co. – The New Spectroscopic Pipeline for the FORSes
S. Moehler 87

Calibrating the VIMOS Redshift Survey Data
M. Scodeggio, P. Franzetti, B. Garilli, M. Fumana, L. Paioro,
and A. Zanichelli 95

Part IV Session 3: Near-IR and mid-IR Spectro-Imagers

IOT Overview: IR Instruments
E. Mason 109

Some (Little) Thing(s) about VISIR
E. Pantin, L. Vanzi, U. Weilenmann 119

Infrared Spectrograph Calibration Issues: Using CRIRES, the High Resolution Infrared Spectrograph for ESO's VLT as an Example
H. U. Käufl, The CRIRES Team 131

Calibrating Mid-Infrared Standard Stars
D. Dobrzycka, L. Vanzi 139

Stellar Calibrators for the Medium Resolution Spectrometer of the Mid Infra-Red Instrument on Board the JWST
E. Bauwens, L. Decin, J. A. D. L. Blommaert 145

ISAAC LW Telluric Line Atlas
C. Papadaki, L. Schmidtobreick, W. Hummel, A. Smette 149

Building-Up a Database of Spectro-Photometric Standard Stars from the Ultraviolet to the Near-Infrared
J. Vernet, F. Kerber, S. D'Odorico, R. Bohlin, V. Ivanov, C. Lidman,
E. Mason, T. Rauch, F. Saitta, A. Smette, J. Walsh, R. Fosbury,
P. Goldoni, P. Groot, F. Hammer, M. Horrobin, L. Kaper,
P. Kjaergaard-Rasmussen, R. Pallavicini, F. Royer 153

Part V Session 4: Data Flow and Data Reduction Software

The VLT Data Flow System
M. Peron 159

ESO Reflex: A Graphical Workflow Engine for Running Recipes
R. N. Hook, M. Romaniello, M. Ullgrén, P. Järveläinen, S. Maisala,
T. Oittinen, V. Savolainen, O. Solin, J. Tyynelä, M. Peron, C. Izzo,
T. Licha 169

Astronomical Data Reduction Pipelines at NASA: 30 Years and Counting
D. Lindler 177

The End-to-End Pipeline for HST Slitless Spectra PHLAG
M. Kümmel, R. Albrecht, R. Fosbury, W. Freudling, J. Haase,
R. N. Hook, H. Kuntschner, A. Micol, M. R. Rosa, J. R. Walsh 185

A Bottom-Up Approach to Spectroscopic Data Reduction
C. Izzo, Y. Jung, P. Ballester 191

Model Based Instrument Calibration
P. Bristow, F. Kerber, M.R. Rosa 199

From Predictive Calibration to Forward Analysis – Preparing for the ELT Era
M. R. Rosa 207

Efits: A New Efficient and Flexible FITS Library
A. Grado, M. Pavlov, L. Limatola 217

Error Propagation in the X-shooter Pipeline
M. Horrobin, P. Goldoni, F. Royer, P. François, G. Blanc, J. Vernet,
A. Modigliani, J. Larsen 221

The CRIRES Data Reduction Challenges
Y. Jung, P. Bristow 225

THELI - A Pipeline for UV- to Mid-IR Imaging Data Reduction
M. Schirmer, T. Erben 229

MUSE: Design and Status of the Data Reduction Pipeline
P. Weilbacher, J. Gerssen, M. M. Roth, P. Böhm, The MUSE Team . 233

Part VI Session 5: Adaptive Optics Instruments

IOT Overview: Adaptive Optics
N. Ageorges, C. Lidman, C. Dumas 239

AO Assisted Spectroscopy with SINFONI: PSF, Background, and Interpolation
R. Davies 249

Imaging and Treatment of the PSF in AO Instruments: Application to NACO
Y. Clénet, E. Gendron, G. Rousset, T. Fusco, C. Lidman,
M. Kasper, N. Ageorges, O. Marco 259

Probing μ-arcsec Astrometry with NACO
A. Seifahrt, T. Röll, R. Neuhäuser 271

Part VII Session 6: Integral Field Spectroscopy

IOT Overview: Integral Field Spectroscopy
P. Amico, S. Bagnulo, C. Dumas, C. Izzo, G. Marconi, C. Melo 279

Integral Field Spectrographs: A User's View
E. Emsellem 289

Specsim: A Software Simulator for Integral Field Unit Spectrometers
N. P. F. Lorente, A. C. H. Glasse, G. S. Wright, S. K. Ramsay,
and C. J. Evans 295

Integral Field Spectroscopy with VIMOS
M. M. Roth, A. Monreal-Ibero, L. Christensen, T. Becker, J. Gerssen,
P. Weilbacher 301

KMOS: Design Overview and Calibration Requirements
R. M. Sharples, S. K. Ramsay, R. Davies, M. Lehnert 311

Calibration of the KMOS Multi-Field Imaging Spectrometer
S. K. Ramsay, S. Rolt, R. M. Sharples, R. Davies 319

MUSE: A Second-Generation Integral-Field Spectrograph for the VLT
R. M. McDermid, R. Bacon, S. Bauer, P. Boehm, D. Boudon, S. Brau-Nogué, P. Caillier, L. Capoani, C. M. Carollo, N. Champavert, T. Contini, E. Daguisé, B. Delabre, J. Devriendt, S. Dreizler, J. Dubois, M. Dupieux, J. P. Dupin, E. Emsellem, P. Ferruit, M. Franx, G. Gallou, J. Gerssen, B. Guiderdoni, T. Hahn, D. Hofmann, A. Jarno, A. Kelz, C. Koehler, W. Kollatschny, J. Kosmalski, F. Laurent, S. J. Lilly, J. L. Lizon, M. Loupias, A. Manescau, C. Monstein, H. Nicklas, L. Parès, L. Pasquini, A. Pécontal-Rousset, E. Pécontal, R. Pello, C. Petit, J.-P. Picat, E. Popow, A. Quirrenbach, R. Reiss, E. Renault, M. Roth, J. Schaye, G. Soucail, M. Steinmetz, S. Stroebele, R. Stuik, P. Weilbacher, L. Wisotzki, H. Wozniak, P. T. de Zeeuw 325

SPHERE-IFS: A Tool for Direct Detection of Giant Planets
R. U. Claudi, J. Antichi, R. G. Gratton, S. Desidera, A. Berton, D. Mesa, M. Turatto, M. Feldt, K. Dohlen, J.-L. Beuzit, D. Mouillet, P. Puget 337

The Origin of Fringing in the VIMOS IFU
E. Jullo, L. Christensen, A. Smette, S. Bagnulo, C. Izzo, G. Marconi .. 343

A User's View of VIMOS-IFU Calibrations
H. Kuntschner 347

Part VIII Session 7: High Resolution Spectroscopy

IOT Overview: High Resolution Spectrographs
G. Lo Curto, P. Francois, C. Ledoux, C. Melo, E. Pompei, A. Smette, L. Vanzi 353

Quantitative Spectroscopy with UVES
P. E. Nissen 365

High Resolution and High Precision-Spectroscopy with HARPS
F. Pepe, C. Lovis 375

Laser Comb: A Novel Calibration System for High Resolution Spectrographs

C. Araujo-Hauck, L. Pasquini, A. Manescau, Th. Udem,
T. W. Hänsch, R. Holzwarth, A. Sizmann, H. Dekker, S. D'Odorico,
M. T. Murphy 385

Instrumental Line Shape Function for High Resolution Fourier Transform Molecular and Atmospheric Spectroscopy
M. Badaoui, F. Schreier, G. Wagner, M. Birk 391

Calibration Sources for CRIRES
F. Kerber, A. Seifahrt, P. Bristow, G. Nave, H. U. Käufl, M. R. Rosa, C. J. Sansonetti, R. Siebenmorgen, A. Smette 397

Recent Developments in the Optimal Extraction of UVES Spectra
J. M. Larsen, P. Ballester, V. D'Odorico, C. Ledoux, A. Modigliani, J. Pritchard, H. Sana, A. Smette 403

A new Generation of Spectrometer Calibration Techniques Based on Optical Frequency Combs
P. O. Schmidt, S. Kimeswenger, H. U. Käufl 409

Part IX Session 8: Atmospheric Effects

Photometric Calibration of LSST Data
D. L. Burke, The LSST Collaboration 415

The GTC Photometric Calibration Programme
J. M. Rodríguez Espinosa, A. Di Césare, P. L. Hammersley 425

Measuring the Amount of Precipitable Water Vapour with VISIR
A. Smette, H. Horst, J. Navarrete 433

NIR Extinction Coefficients in Paranal
E. Mason, G. Lombardi, C. Lidman, A. O. Jaunsen 439

Improvements in the Residual OH Emission Removal in SINFONI Pipeline Spectra
A. Modigliani, R. Davies, C. Dumas, M. Neeser 443

Calibration of the Relationship Between Precipitable Water Vapor and 225 GHz Atmospheric Opacity via Optical Echelle Spectroscopy at Las Campanas Observatory
J. Thomas-Osip, A. McWilliam, M. Phillips, D. Osip 447

Part X Session 9: Interferometric Instrumentation

IOT Overview: Calibrations of the VLTI Instruments (MIDI and AMBER)
S. Morel, F. Rantakyrö, T. Rivinius, S. Stefl, C. Hummel, S. Brillant,
M. Schöller, I. Percheron, M. Wittkowski, A. Richichi, P. Ballester ... 451

AMBER on the VLTI: Data Processing and Calibration Issues
F. Millour, R. Petrov, F. Malbet, E. Tatulli, G. Duvert, G. Zins,
E. Altariba, M. Vannier, O. Hernandez, G. Li Causi 461

QC and Analysis of MIDI Data Using *mymidigui* and OYSTER
C. A. Hummel.. 471

Calibration of AMBER Visibilities at Low Spectral Resolution
P. Cruzalèbes, A. Spang, S. Sacuto 479

VLTI Instruments: From J to N Band Instrumental Calibrations, from Short to Long Baseline Astronomical Calibrations
I. Percheron... 483

Part XI Session 10: Polarimetry

IOT Overview: Polarimetry
N. Ageorges 489

Polarimetry with ESO Instruments
H. M. Schmid 499

Faint NIR Polarimetric Standards
R. Barrena, J. A. Acosta-Pulido, A. Manchado 511

Developmental Aspects of a Multi-Slit Spectro-Polarimeter
K. George, K. Sankarasubramanian, R. Bayanna, H. Lin,
and P. Venkatakrishnan 515

Current and Future Instruments Providing a Spectro-Polarimetric Mode
S. Hubrig, I. Ilyin.. 519

Circular Polarization Observations at ESO Using UT2 and FORS1
S. Katajainen, V. Piirola, H. Lehto, A. Berdyugin 523

Part XII Session 11: Wide Field Imagers

IOT Overview: Wide-Field Imaging
F. J. Selman 529

Processing Wide Field Imaging Data
M. J. Irwin 541

The Garching-Bonn Deep Survey (GaBoDS) Wide-Field-Imaging Reduction Pipeline
H. Hildebrandt, T. Erben, M. Schirmer, J. P. Dietrich,
and P. Schneider 553

Calibrating VISTA Data
J. Emerson, S. Hodgkin, P. Bunclark, M. Irwin, J. Lewis 559

The VISTA Data Flow System
J. Lewis, M. Irwin, P. Bunclark, S. Hodgkin...................... 565

SkyMapper and the Southern Sky Survey a Resource for the Southern Sky
S. C. Keller, B. P. Schmidt, M. S. Bessell 573

Quality Control Monitoring for WFCAM
M. Riello, M. Irwin 581

Relative and Absolute Calibration for Multi-Band Data Collected with the 2.2 m ESO/MPI and 1.54 m Danish Telescopes
A. Calamida, C. E. Corsi, G. Bono, P. B. Stetson, L. M. Freyhammer,
R. Buonanno 589

ALBUM: A Tool for the Analysis of Slitless Spectra and its Application to ESO WFI Data
C. Martayan, D. Baade, A.-M. Hubert, M. Floquet, J. Fabregat,
and E. Bertin 595

Part XIII Workshop Summary

The 2007 ESO Instrument Calibration Workshop – A Personal Summary
D. Baade 601

Index 611

List of Contributors

Jose Acosta-Pulido Instituto de
Astrofisica de Canarias
Vía Láctea s/n E-382005 La Laguna
Canary Islands Spain
jap@iac.es

Nancy Ageorges
ESO
Alonso de Cordóva 3107, Vitacura
casilla 19001
Santiago 19
Chile
nageorge@eso.org

Carlos Alvarez Iglesias
Grantecan S.A.
Cuesta de San Jose s/n
38712 Brena Baja
Spain
carlos.alvarez@gtc.iac.es

Catarina Alves de Oliveira
ESO
Karl-Schwarzschild-Strasse 2
85748 Garching
Germany
coliveir@eso.org

Paola Amico
ESO
Alonso de Cordóva 3107, Vitacura
Santiago 19
Chile
pamico@eso.org

Ebrahim Amini Biparva
University of Padua
Vicolo dell'Osservatorio 2
35122 Padova
Italy
amini_biparva@yahoo.com

Gloria Andreuzzi
Telescopio Nazionale Galileo
Calle Alvarez de Abreu, 70
38700 Santa Cruz de la Palma
Spain
andreuzzi@tng.iac.es

Constanza Auraujo-Hauck
ESO
Karl-Schwarzschild-Strasse 2
85748 Garching
Germany
caraujo@eso.org

Dietrich Baade
ESO
Karl-Schwarzschild-Strasse 2
85748 Garching
Germany
dbaade@eso.org

Mohammed Badaoui
IAV Hassan II, Physics Unit
B.P. 6240 Madinat al Irfane
10101 Rabat
Morocco
nouralhakim@yahoo.com

Pascal Ballester
ESO
Karl-Schwarzschild-Strasse 2
D-85748 Garching
Germany
pballest@eso.org

Rafael Barrena
Instituto de Astrofísica de Canarias
Vía Láctea s/n
E-382005 La Laguna
Canary Islands
Spain
rbarrena@iac.es

Eva Bauwens
Instituut voor Sterrenkunde
K.U. Leuven
Celestijnenlaan 200D
B-3001 Heverlee
Leuven
Belgium
Eva.Bauwens@ster.kuleuven.be

Konstantina Boutsia
ESO
Karl-Schwarzschild-Strasse 2
85748 Garching
Germany
kboutsia@eso.org

Piercarlo Bonifacio
GEPI
Observatoire de Paris, CNRS
Université Paris Diderot
5 Place Jules Janssen
92190 Meudon
France
Piercarlo.Bonifacio@obspm.fr

Paul Bristow
ESO
Karl-Schwarzschild-Strasse 2
85748 Garching
Germany
bristowp@eso.org

Peter Bunclark
Cambridge Astronomical
Survey Unit
Institute of Astronomy
University of Cambridge
Madingley Road
CB3 0HA Cambridge
UK
psb@ast.cam.ac.uk

D. L. Burke
Kavli Institute for Particle
Astrophysics and
Cosmology Stanford Linear
Accelerator
Center Stanford University
Stanford
CA 94309
USA
daveb@slac.stanford.edu

Antonio Luis Cabrera Lavers
GTC Project Office
c/Via Lactea s/n
38200 La Laguna
Spain
antonio.cabrera@gtc.iac.es

Annalisa Calamida
INAF – Osservatorio Astronomico di
Roma
Via Frascati 33
00040 Monte Porzio Catone
Rome
Italy
calamida@mporzio.astro.it

Massimo Capaccioli
INAF – Osservatorio Astronomico di
Capodimonte
Via Moiariello 16
80131 Naples
Italy
capaccioli@na.astro.it

Mark Casali
ESO
Karl-Schwarzschild-Strasse 2
85748 Garching
Germany
mcasali@eso.org

Hector Castaneda
Instituto de Astrofisica de Canarias
c/Via Lactea s/n
38200 La Laguna
Spain
hcastane@iac.es

Sandra Castro
ESO
Karl-Schwarzschild-Strasse 2
85748 Garching
Germany
scastro@eso.org

Yann Clénet
LESIA
Observatoire de Paris
5 place Jules Janssen
92195 Meudon Cedex
France
yann.clenet@obspm.fr

Pierre Cruzalèbes
Observatoire de la Côte d'Azur
Laboratoire Gemini
UMR-CNRS 6203, av. Copernic
F-06130 Grasse
France
pierre.cruzalebes@obs-azur.fr

Richard Davies
Max Planck Institute for Extraterrestrial Physics
Postfach 1312
85741 Garching
Germany
davies@mpe.mpg.de

Silvano Desidera
INAF – Osservatorio Astronomico di Padova
Via dell'Osservatorio 5
35122 Padova
Italy
silvano.desidera@oapd.inaf.it

Luca Di Fabrizio
Telescopio Nazionale Galileo
Calle Alvarez de Abreu, 70
38700 Santa Cruz de la Palma
Spain
difabrizio@tng.iac.es

Danuta Dobrzycka
ESO
Karl-Schwarzschild-Strasse 2
85748 Garching
Germany
ddobrzyc@eso.org

Sandro D'Odorico
ESO
Karl-Schwarzschild Strasse 2
85748 Garching
Germany
sdodoric@eso.org

Jim Emerson
Astronomy Unit
Queen Mary University of London
Mile End Road
E1 4NS London
UK
j.p.emerson@qmul.ac.uk

Eric Emsellem
Université de Lyon 1
Centre de Recherche Astrophysique de Lyon
Observatoire de Lyon
9 avenue Charles André
69230 Saint-Genis Laval Cedex
France

and
CNRS
UMR 5574
ENS de Lyon
Lyon
France
emsellem@obs.univ-lyon1.fr

Chris J. Evans
UK Astronomy Technology Centre
Royal Observatory
Blackford Hill
EH9 3HJ Edinburgh
UK
cje@roe.ac.uk

Wolfram Freudling
ESO
Space Telescope European Coordinating Facility
Karl-Schwarzschild-Strasse 2
85748 Garching
Germany
wfreudli@eso.org

Koshy George
ISRO Satellite Centre
Bangalore
India
koshyastro@gmail.com

Paolo Goldoni
APC/UMR 7164
Paris
France
and
Service d'Astrophysique
Centre d'Etudes de Saclay
F-91190 Gif-sur-Yvette Cedex
France
goldoni@apc.univ-paris7.fr

Aniello Grado
INAF – Osservatorio Astronomico
di Capodimonte
Via Moiariello 16
80131 Naples
Italy
agrado@na.astro.it

Reinhard Hanuschik
ESO
Karl-Schwarzschild-Strasse 2
85748 Garching
Germany
rhanusch@eso.org

Hendrik Hildebrandt
Argelander-Institut füg Astronomie
Universität Bonn
Auf dem Hügel 71
53121 Bonn
Germany
hendrik@astro.uni-bonn.de

Richard N. Hook
ESO
Space Telescope European Coordinating Facility
Karl-Schwarzschild-Strasse 2
85748 Garching
Germany
rhook@eso.org

Ulrich Hopp
University Observatory Munich and MPE
Scheiner Strasse 1
81679 Munich
Germany
hopp@usm.uni-muenchen.de

Matthew Horrobin
University Amsterdam
Postbus 19268
1000 GG Amsterdam
The Netherlands
horrobin@science.uva.nl

Swetlana Hubrig
ESO
Alonso de Cordova 3107, Vitacura
Santiago
Chile
shubrig@eso.org

C. A. Hummel
ESO
Alonso de Cordóva 3107, Vitacura
Santiago
Chile
chummel@eso.org

Wolfgang Hummel
ESO
Karl-Schwarzschild-Strasse 2
85748 Garching
Germany
whummel@eso.org

Mike Irwin
University of Cambridge
Madingley Road
CB30HA Cambridge
United Kingdom
mike@ast.cam.ac.uk

Carlo Izzo
ESO
Karl-Schwarzschild-Strasse 2
85748 Garching
Germany
cizzo@eso.org

Yves Jung
ESO
Karl-Schwarzschild-Strasse 2
85748 Garching
Germany
yjung@eso.org

Seppo Katajainen
Tuorla Observatory
University of Turku
Vaisalantie 20
FI-21500 Piikkio
Finland
sekataja@utu.fi

Andreas Kaufer
ESO
Alonso de Cordóva 3107, Vitacura
Santiago
Chile
akaufer@eso.org

H. U. Käufl
ESO
Karl-Schwarzschild-Strasse 2
85748 Garching
Germany
hukaufl@eso.org

Stefan Keller
Research School of Astronomy and
Astrophysics
Mt. Stromlo Observatory
Cotter Rd.
2611 Weston
Australia
stefan@mso.anu.edu.au

Florian Kerber
ESO
Karl-Schwarzschild-Strasse 2
85748 Garching
Germany
fkerber@eso.org

Stefan Kimeswenger
Institute of Astro- and Particle
Physics
University Innsbruck
Technikerstrasse 25
A-6020 Innsbruck
Austria
Stefan.Kimeswenger@uibk.ac.at

List of Contributors

Martin Kümmel
Space Telescope European
Coordinating Facility
Karl-Schwarzschild-Strasse 2
85748 Garching
Germany
mkuemmel@eso.org

Harald Kuntschner
Space Telescope European Coordinating Facility
Karl-Schwarzschild-Strasse 2
85748 Garching
Germany
hkuntsch@eso.org

Jonas Moeller Larsen
ESO
Karl-Schwarzschild-Strasse 2
85748 Garching
Germany
jmlarsen@eso.org

James Lewis
Cambridge Astronomical Survey Unit
Institute of Astronomy
University of Cambridge
Madingley Road
CB3 0HA Cambridge
United Kingdom
jrl@ast.cam.ac.uk

Don Lindler
Sigma Space Coperation
4801 Forbes Boulevard
9801 Greenbelt Road
Lanham, MD 20706
USA
don.lindler@gsfc.nasa.gov

Gaspare Lo Curto
ESO
Alonso de Cordova 3107, Vitacura
Santiago 19
Chile
glocurto@eso.org

Luis Lopez-Martin
Instituto de Astrofisica de Canarias
c/ Via Lactea s/n
382005 La Laguna
Spain
luislm@ll.iac.es

Nuria Lorente
UK Astronomy Technology Centre
Royal Observatory Edinburgh
Blackford Hill
EH9 3HJ Edinburgh
United Kingdom
npfl@roe.ac.uk

Lars Lundin
ESO
Karl-Schwarzschild-Strasse 2
D-85748 Garching
Germany
llundin@eso.org

Mariya Lyubenova
ESO
Karl-Schwarzschild-Strasse 2
D-85748 Garching
Germany
mlybeno@eso.org

Patrizia Manzato
INAF – OATs
via G.B. Tiepolo, 11
34143 Trieste
Italy
manzato@oats.inaf.it

Gianni Marconi
ESO
Alonso de Cordóva 3107, Vitacura
Casilla 19001
Santiago 19
Chile
gmarconi@eso.org

Christophe Martayan
GEPI-Observatoire de Paris-Meudon
5, place Jules Janssen
92195 Meudon
France
christophe.martayan@obspm.fr

Elena Mason
ESO
Alonso de Cordóva 3107, Vitacura
Santiago
Chile
emason@eso.org

Richard McDermid
Leiden University
Niels Bohrweg 2
2333 CA Leiden
The Netherlands
mcdermid@strw.leidenuniv.nl

Sabine Mengel
ESO
Karl-Schwarzschild-Strasse 2
85748 Garching
Germany
smengel@eso.org

Steffen Mieske
ESO
Karl-Schwarzschild-Strasse 2
85748 Garching
Germany
smieske@eso.org

Florentin Millour
LAOG/LUAN
Parc Valrose
6000 Nice
France
florentin.millour@obs.ujf-grenoble.fr
and
Max-Planck-Institut für Radioastronomie
Auf dem Hügel 69
Germany
fmillour@mpifr-bonn.mpg.de

Andrea Modigliani
ESO
Karl-Schwarzschild-Strasse 2
85748 Garching
Germany
amodigli@eso.org

Sabine Moehler
ESO
Karl-Schwarzschild-Strasse 2
85748 Garching
Germany
smoehler@eso.org

Palle Moller
ESO
Karl-Schwarzschild-Strasse 2
85748 Garching
Germany
pmoller@eso.org

Matteo Monelli
Instituto de Astrofisica de Canarias
c/ Via Lactea s/n
382005 La Laguna
Spain
monelli@iac.es

Sebastien Morel
ESO
Alonso de Cordova 3107, Vitacura
Santiago
Chile
smorel@eso.org

Poul Erik Nissen
Department of Physics and Astronomy
Aarhus University
Ny Munkegade
8000 Aarhus C
Denmark
pen@phys.au.dk

David Osip
Observatories of the Carnegie
Institution of Washington
Las Campanas Observatory
Colina El Pino
Casilla 601 La Serena
Chile
dosip@lco.cl

Ralf Palsa
ESO
Karl-Schwarzschild-Strasse 2
85748 Garching
Germany
rpalsa@eso.org

Eric Pantin
CE Saclay
CEA/Centre de Saclay
91191 Gif-sur-Yvette
France
epantin@cea.fr

Ferdinando Patat
ESO
Karl-Schwarzschild-Strasse 2
85748 Garching
Germany
fpatat@eso.org

Silvia Pedicelli
ESO
Karl-Schwatzschild-Strasse 2
85748 Garching
Germany
spedicel@eso.org

Francesco Pepe
Observatoire Astronomique de l'
Universitéve Geneve
51, ch. des Maillettes
CH-1290 Sauverny
Switzerland
Francesco.Pepe@obs.unige.ch

Isabelle Percheron
ESO
Karl-Schwarzschild-Strasse 2
85748 Garching bei München
Germany
ipercher@eso.org

Michele Peron
ESO
Karl-Schwarzschild-Strasse 2
85748 Garching
Germany
mperon@eso.org

Monika Petr-Gotzens
ESO
Karl-Schwarzschild-Strasse 2
85748 Garching
Germany
mpetr@eso.org

Francesca Primas
ESO
Karl-Schwarzschild-Strasse 2
85748 Garching
Germany
fprimas@eso.org

John Pritchard
ESO
Karl-Schwarzschild-Strasse 2
85748 Garching
Germany
j.pritchard@eso.org

Piero Rafanelli
University of Padua
Vicolo dell'Osservatorio 3
35122 Padova
Italy
piero.rafanelli@unipd.it

S. K. Ramsay
UK Astronomical Technology Centre
Royal Observatory
Blackford Hill
EH9 3HJ Edinburgh
UK
skr@roe.ac.uk

Marina Rejkuba
ESO
Karl-Schwarzschild-Strasse 2
85748 Garching
Germany
mrejkuba@eso.org

Helena Relke
MPE Garching and USM
Scheinerstrasse 1
81679 Muenchen
Germany
relke@usm.uni-muenchen.de

Andrea Richichi
ESO
Karl-Schwarzschild-Strasse 2
85748 Garching
Germany
arichich@eso.org

Marco Riello
Cambridge Astronomy Survey Unit
Institute of Astronomy
Cambridge University
Madingley Road
Cambridge CB3 0HA
UK
mriello@ast.cam.ac.uk

Jose Miguel Rodríguez Espinosa
Instituto de Astrofísica de Canarias
La Laguna
Tenerife
Spain
jmr.espinosa@iac.es

Martino Romaniello
ESO
Karl-Schwarzschild-Strasse 2
85748 Garching
Germany
mromanie@eso.org

Michael R. Rosa
Space Telescope European
Coordinating Facility
Karl-Schwarzschild-Strasse 2
85748 Garching
Germany
mrosa@stecf.org

Piero Rosati
ESO
Karl-Schwarzschild-Strasse 2
85748 Garching
Germany
prosati@eso.org

Martin M. Roth
Astrophysikalisches Institut
Potsdam
An der Sternwarte 16
D-14482 Potsdam
Germany
mmroth@aip.de

Frederic Royer
GEPI
Observatoire de Paris
5 place Jules Janssen
92195 Meudon Cedex
France
frederic.royer@obspm.fr

Gero Rupprecht
ESO
Karl-Schwarzschild-Strasse 2
85748 Garching
Germany
grupprec@eso.org

Francesco Saitta
ESO
Karl-Schwarzschild-Strasse 2
85748 Garching
Germany
fsaitta@eso.org

Mischa Schirmer
Isaac Newton Group of Telescopes
Apartado de correos 321
38700 Santa Cruz de La Palma
Tenerife
Spain
mischa@ing.iac.es

Hans Martin Schmid
Institute of Astronomy
ETH Zurich
CH-8092 Zurich
Switzerland
schmid@astro.phys.ethz.ch

Linda Schmidtobreick
ESO
Alonso de Cordova 3107,
Vitacura
Santiago
Chile
lschmidt@eso.org

Marco Scodeggio
INAF – IASF Milano
Via Bassini 15
20133 Milano
Italy
marcos@lambrate.inaf.it

Andreas Seifahrt
ESO
Karl-Schwarzschild-Strasse 2
85748 Garching
Germany
aseifahr@eso.org

Fernando J. Selman
ESO
Alonso de Cordóva 3107,
Vitacura
Santiago
Chile
fselman@eso.org

Ray M. Sharples
Centre for Advanced Instrumentation
Durham University
NetPark
Sedge-field TS21 3FB
UK
r.m.sharples@durham.a

Alain Smette
ESO
Alonso de Cordóva 3107,
Vitacura
Santiago
Chile
asmette@eso.org

Stan Stefl
ESO
Alonso de Cordóva 3107,
Vitacura
Santiago
Chile
sstefl@eso.org

Michael Sterzik
ESO
Karl-Schwarzschild-Strasse 2
85748 Garching
Germany
msterzik@eso.org

Lowell Tacconi-Garman
ESO
Karl-Schwarzschild-Strasse 2
85748 Garching
Germany
ltacconi@eso.org

Bruno Valat
LUAN
Parc Valrose
6100 Nice
France
bruno.valat@unice.fr

Marc Vallbe Mumbru
Instituto de Astrofisica de Canarias
c/ Via Lactea s/n
382005 La Laguna
Spain
marc.vallbe@iac.es

Mario van den Ancker
ESO
Karl-Schwarzschild-Strasse 2
85748 Garching
Germany
mvandena@eso.org

Joel Vernet
ESO
Karl-Schwarzschild-Strasse 2
85748 Garching
Germany
jvernet@eso.org

Jeremy R. Walsh
ESA
Space Telescope European Coordinating Facility
Karl-Schwarzschild-Strasse 2
85748 Garching
Germany
jwalsh@eso.org

Peter Weilbacher
Astrophysikalisches Institut
Potsdam
An der Sternwarte 16
D-14482 Potsdam
Germany
pweilbacher@aip.de

Markus Wittkowski
ESO
Karl-Schwarzschild-Strasse 2
85748 Garching
Germany
mwittkow@eso.org

Burkhard Wolff
ESO
Karl-Schwarzschild-Strasse 2
85748 Garching
Germany
bwolff@eso.org

The participants of the 2007 ESO Instrument Calibration Workshop held on 23–26 January 2007 at the ESO headquarters in Garching bei München, Germany (Photo: ESO – Louis Calcada)

Part I

Introduction

Instruments Calibration at the La Silla Paranal Observatory

G. Marconi

ESO, Alonso de Cordóva 3107, Vitacura, Casilla 19001, Santiago 19, Chile;
gmarconi@eso.org

1 La Silla Paranal Observatory

Almost seven years ago the first night of scientific operations started at VLT-UT1; after seven years at the Paranal observatory there are 4 main UTs telescopes and 4 auxiliary ATs telescopes fully in operations. Together with the 3 telescopes still in operation at La Silla Observatory the La Silla Paranal Observatory (LPO) enumerates a total of 11 optical telescopes in activity. The total number of instruments in operations is currently 12 (ISAAC, 2 FORSes, CRIRES, FLAMES, UVES, VISIR, VIMOS, SINFONI, NACO, MIDI, AMBER), and 9 (HARPS, EFOSC2, CES, SOFI, EMMI, SUSI2, WFI, FEROS, GROND) for the VLT and La Silla respectively.

1.1 Scientific Operations: Statistic

What means scientific operations for the La Silla Paranal observatory? It means that all the telescope time not used in commissioning new instruments or spent for engineering purposes, is devoted to science. In Fig. 1 the details of how and how much time has been dedicated to the various activities in P78 (April–September 2006), is shown. Up to more than 90% of the total telescope time has been available for science, while only a very small fraction of it (less than 3%) has been lost for technical reasons.

The efficiency of the VLT telescopes is quite evident, but, are the instruments behaving as well as the UTs? The answer is summarized in Fig. 2 where the efficiencies (total shutter time over total observing time) are reported for all the instruments in operations. Excluding the VLTI instruments, for which a large amount of time is still spent in engineering activities, the average shutter efficiencies, in the same period above, has been of 67.1% for the VLT and 61.4% for La Silla instruments respectively. As for the telescope, the technical downtime due to the instruments, have been contained within few percent (less than 3%) of the total available time.

Finally, is the scientific productivity following the remarkable performances of the telescopes+instruments systems? It is evident in the histogram of Fig. 3, that, if before the 1999 (start of the VLT operations), La Silla was contributing with an average of a publication per day, after the advent of

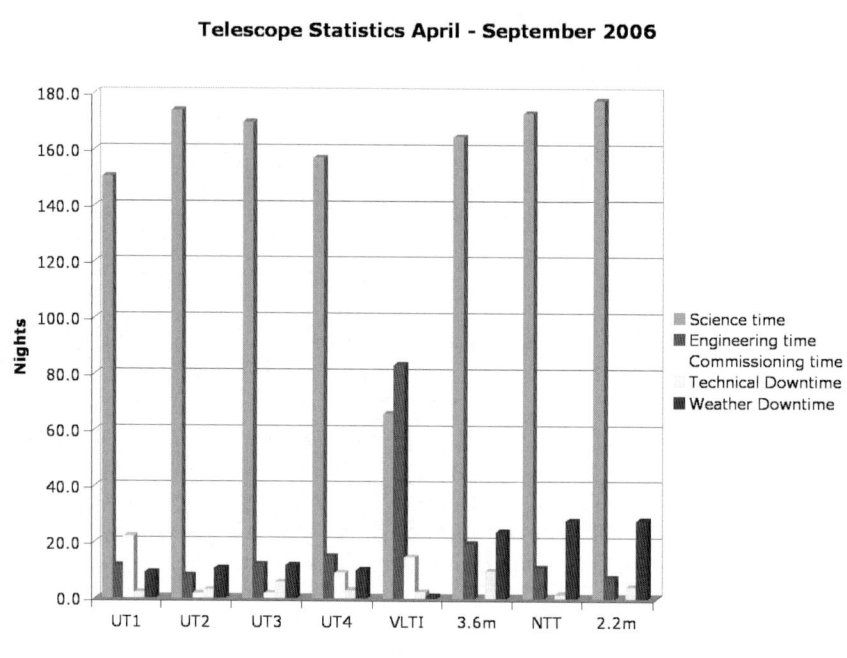

Fig. 1. Telescopes statistic April–September 2006

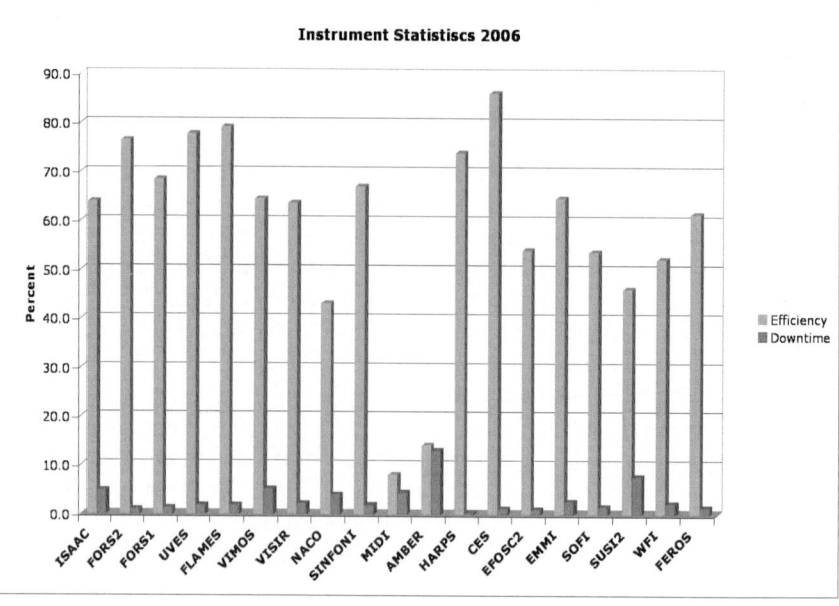

Fig. 2. Instruments statistic 2006

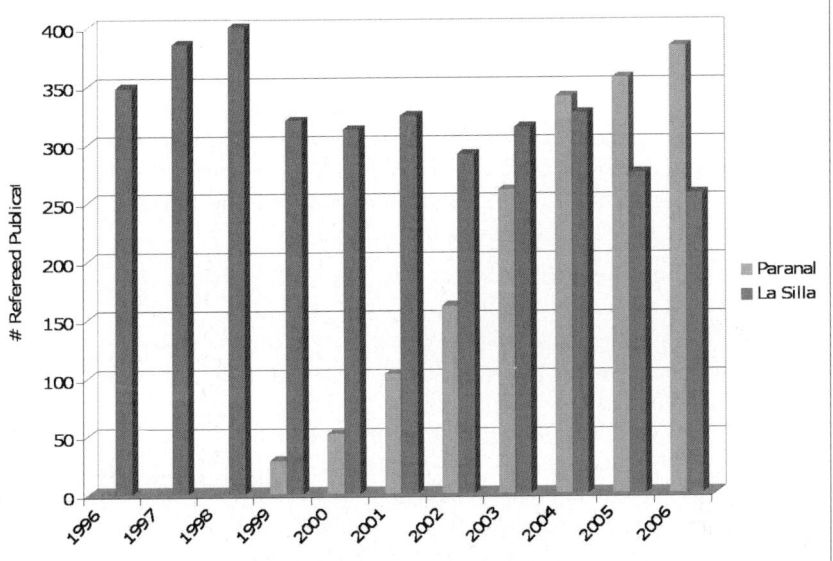

Fig. 3. Publications rate over the last 10 years

the VLT, the situation did not change substantially. In the last 10 years the publication rate for La Silla only slightly decreased, while the VLT publication rate grew rapidly to reach in 2004 the average publication rate of the La Silla pre-VLT era. For La Silla the decrease could be considered as a psychological effect related first of all to the decommissioning of several telescopes and related instruments, and second to the fact that the scientific interest moved rapidly to the better performances of the VLT instruments. On the other hand, the rise in the scientific productivity of the VLT is an obvious effect of the increasing number of the instruments entering in operation. What is instead worth noticing is the fact that, moving from La Silla scheme for operations to the VLT scheme, has been a big step in terms of both philosophy and practical aspects of the operations. In the La Silla scheme the users themselves (visitor mode) were (are) taking care of their calibration needs; in the VLT hybrid scheme, were a copicuous fraction of observations is conducted in service mode, the observatory is in charge to provide to the users all the necessary instrument calibrations to grant the usefulness of the scientific data. The smooth transition between the two schemes, as evidentiated in Fig. 3 , is somehow testifying the success of the adopted VLT *modus operandi*. As a matter of fact, Paranal and La Silla observatories, are together currently producing 2 referred papers per day, of which almost 40% are still based on La Silla telescopes observations.

2 From the Theory (VLT White Book) to the Practice (Operations)

The origin of the evident outcame of the "VLT model" originated well before the beginning of the VLT operations in 1999, and derived from a clear philosophical vision, coupled with the well defined guidelines to achieve it. Both of them are summarized in the *VLT White Book* [1]. Some of the requirements stated in [1] constitute the pillars on which stand the operations and the calibrations of the VLT instruments.

- The VLT must be operated in a way that will result in *the greatest scientific return*
- The VLT instruments should be *continuously calibrated and monitored* to ensure the accuracy of the calibrations and the long term performances of the instruments
- For all supported instrument modes, an associated *calibration plan* will be developed. This plan will be executed automatically and produce the calibration data that will ensure the long term usefulness of the raw science data. Calibration plan data will enable ESO to monitor instrument performances and long term trends in instruments and telescopes behaviour.
- VLT operations will produce *quality controlled science data* combining raw and calibration frames into final data products which contains physical as well as instrumental units.

2.1 The Calibration Plan

The calibration plan is intrinsically a *continuous learning process*. For every VLT instrument, it is born with the instrument, evolves with it, and, to be effective, needs continuous feedback from the Instrument Scientists as well as from the expert users.

Purposes of the Calibration Plan

The final absolute goal of the calibration plan is the usefulness of the scientific data; in other words, it means to define and provide to the users all the necessary tools to fully exploit the science data. At the same time it should provide to the expert users (instrument scientists in particular), the tools to keep the instruments performances under control. It consist of:

- Fully characterize/monitor the performances of an instruments; in order to create/mantain software simulator tools (i.e. Exposure Time Calculator)
- Define which calibrations are needed to remove instrument signatures from science data
- Define which calibrations are needed to project back into the physical parameters space
- Define which calibrations are needed for routine daily instruments health check and/or medium/long term trend analysis

Who Executes the Calibration Plan

The night calibrations are executed by the night astronomers (NAs) accordingly to the requirements of the executed science. The day calibrations are automatically created by a tool *CalobBuild* according to the requirements of the calibration plan, and for all the setups modes used during an observing night. The day calibrations are executed and certified by the day astronomers (DAs).

2.2 The Quality Control (QC)

The Quality Control of the data as it is currently conceived [2, 3], implies the continuous control of the following aspects:

- the quality of the raw and reduced scientific data
- the quality of the raw and master calibration data
- the quality of the products and of the product creation process
- the performances of the instruments

Given the importance QC has in the exploitation of the scientific data, the full process is for the observatory a shared responsability between Data Flow Operations (DFO), QC Garching, and La Silla-Paranal Science Operations (LPSO). There are indeed always QC issues which require immediate reaction and intervention; these issues can only be properly handled on site at the Observatory. This naturally leads to the concept above of **shared QC** which means that part of the QC tasks are done at the Observatory (in real time, by nighttime and daytime astronomers), and part in Garching (off-line, by QC Scientist and by the Instrument Scientist). In practice we could distinguish between:

On-site QC. Basic quality checks on the science and on the calibrations data are performed by the astronomers at the observatory. Once the data have been taken the online pipeline processed them into the calibration products; both raw data and products are visually inspected and certified by the astronomers. The online pipelines also produce a set of essential QC basic parameters (health check) relating to fundamental instrument properties, which, in case of failure, would jeopardize the usefulness of the scientific data (i.e. the grism alignment).

Off-line QC. Anything which is not time critical, but requires in-depth analysis, is later analysed in Garching. The full set of quality checks is applied to the QC database there, as well as the complex trending analysis which requires extended data sets and/or more sophisticated algorithms (i.e. color and extinctions terms, long term instrument efficiencies etc.).

In addition to the complete QC database in Garching, the relevant QC monitoring parameters are also stored in the local database "automatic report" (autrep) at the observatory which also contains the full telemetry of the telescopes as well as environmental quantities (i.e. seeing). This database represents in practice one of the most powerful tools to check/monitor in real time the behaviour of any instrumental parameter against any possible variable of the global system telescope-instrument-environment.

2.3 The Instrument Operation Team (IOT)

The most fundamental step in the QC process is the exchange of information between the two sites (Garching QC and Observatory) and in particular between the instrument experts of both sites. The main room for that are the *Instrument Operation Teams IOT*. These structures are composed by a delegate of each team participating in the operations of a specific instrument. The Instrument Scientist (normally from LPSO), the Instrument Responsible and the Software Responsible (both from Engineering Dept), on the observatory side; the User Support Scientist (User Support Dept, USD), the Quality Control Scientist (Data Flow Operations and Quality Control Group), the Pipeline Development Responsible (Pipeline Group), the Instrument Scientist and the Instrument Responsible (both from Instrumentation Division) on Garching side. The IOT is led by the Instrument Scientist and is the platform where all the expertise about a VLT instrument are focused and where most efficiently the loop between QC results and monitor/improvement of instrument performances is closed. Is it enough to close the loop in this way? The answer is clearly no and the reason is simple: despite the wide overview proper of these teams, them would never fulfill the complete range of expertise and requirements coming from the large ESO users community. That is the reason why the feedback from the external users is not only a wish, but a real need in order to further improve the performances of any instrument. Is it possible for the users to interact with the IOT? The answer is clearly yes and such possibility is already in place as a part of the VLT operations scheme. The users are indeed advised step by step from the preparations of the observations (USD), through the execution of them (USD, LPSO) down to the certification and release of the data (QC, DFO). All the main actors of this process are advertised on the WEB pages of the instruments and are reachable any time their expertise is required. Could these interactions be improved? The answer is once more yes and in particular is highly desirable to encourage the users to provide their feedback and suggestions; such a feedback is particularly important in the very first phases of the life of every instrument when the *continuous learning process rate* is faster and any suggestions could easily improve the scientific performances of an instrument.

Acknowledgement. All the IOTs of the La Silla Paranal Observatory instruments.

References

1. ESO: *The VLT White Book*, ed by ESO, Garching 1998
2. P. Quinn: The Messenger **84**, 30 (1996)
3. R. Hanuschik, D. Silva: The Messenger **108**, 4 (2002)

Discussion

U. Hopp: You mentioned that one of the first steps of QC is visual inspection, which can be done for existing instruments. What about future instruments like VISTA and OmegaCAM?

G. Marconi: For large array mosaic like scans, we should indeed figure out an automatic way to inspect the files. The visual inspection is out of discussion excluding a small sample of them.

P. Rosati: It is useful to remind that when ESO talks about "too many calibration data", or "overcalibration", they are referring to detector calibrations, i.e., removal of instrumental signatures. However, when dealing with spectro-photometric calibrations, the situation is quite different. There is a large room for improvement in this case (i.e. more spectrophotometric standards, illumination correction maps) are often needed to reach adequate photometric calibration.

M. Scodeggio: In my experience, the main reason for people to complain about the calibration data is the fact that too many calibration frames are provided to the user, and he/she gets lost in the process of finding which ones to use.

G. Marconi: Indeed we should move in the direction to release to the users only the subsample of calibration data really necessary to reduce their data and/or improve the information on the data themselves.

R. Hanuschik: The service mode data packages contain detailed information (reports) which calibrations have been used for calibrating/processing of which science data. In that way, we (the Quality Control Group) try to prevent confusion about too many calibrations.

The Detector Monitoring Project

P. Amico[1], P. Ballester[2], W. Hummel[2], G. LoCurto[1], L. Lundin[2], A. Modigliani[2], P. Sinclaire[1], and L. Vanzi[1]

[1] ESO, Alonso de Cordóva 3107, Vitacura, Casilla 19001, Santiago 19, Chile; pamico@eso.org
[2] ESO, Karl-Schwarzschild-Strasse 2, 85748 Garching, Germany

Abstract. Many detectors, optical CCDs and IR arrays, are currently in operation onboard ESO instruments at the La Silla Paranal Observatory. A unified scheme for optical detector characterization has been adopted since several years in La Silla, and it is used by the Science Operation team to monitor the 18 CCDs belonging to the eight instruments operated by ESO at the Observatory. This scheme has been proven successful in ensuring a high quality performance of the detectors along the years. In Paranal the science operation team and QC Garching monitor the performance of the detectors using instrument-specific data reduction pipelines.

Understanding the performance limits and the calibration requirements of an instrument is fundamental in the operational scheme followed at the observatory and crucially depends on our knowledge of the nature of the detector arrays, their key performance parameters and the way these are defined and measured. Recently two issues were addressed: (1) despite the many commonalities among detectors, different ways to check their performance are in use and (2) the characterization matrix is often incomplete, i.e. not all crucial parameters are measured for all detector systems.

The detector monitoring project arises from the desire to improve the efficiency of the data flow, simplify Quality Control (QC) operations and promote standardization of testing procedures among detectors in use at ESO. The goals of the project are: (a) designing a detector monitoring plan that covers all the detectors in use; (b) reviewing the current reduction procedures and associated pipeline recipes; (c) standardizing the test procedures whenever applicable; (d) merging the test procedures for IR and optical detectors when possible and describe the differences in all the other cases; (e) consolidate the measurement procedures and the use of data reduction recipe and algorithms. The ultimate goal is to provide the observatory and the instrument operation teams (IOTs) with a complete and homogeneous detector-monitoring scheme.

1 Introduction

Modern detectors for astronomy are almost perfect devices when compared to the expectations of astronomers:

1. The ability to detect 100% of the incident light is achieved with detectors with quantum efficiency up to 99%.

2. The availability of progressively larger focal plane arrays is accomplished in modern instruments with few hundred million of pixels (e.g. Megacam at CFHT, Omegacam at ESO).
3. Very low noise performance for faint object astronomy is reached by modern detectors with noise as low as $2\,e^-$ rms at reasonable readout speed (e.g. 255 kbps)

On the other hand, when dealing with performance of modern detectors in detail, the astronomers are confronted with many issues that can significantly affect their science projects. For example, when looking at the signal in a detector, it is important to know the signal gain variation and stability, the QE spatial variations (intrinsic or as a result of external contamination), the effects of flux, bias voltages and temperature on linearity, the conditions under which signal persistence becomes significant, just to name a few. When looking at the noise, possible sources include the bias spatial variations, the dark current, the amplifier glow, electronic ghosts, readout noise, microphonic noise, pickup-up noise, odd-even column effects, etc. All of these effects have to be recognized, characterized and measured routinely for monitoring purposes.

At ESO, a lot of attention is given to detector systems: in Garching the Infra-Red [1] and Optical detector groups [2] use state-of-the-art facilities [3] to characterize in great details the qualities and the weaknesses of detectors to be used onboard ESO instruments and to set parameters for best performance on sky. At the Observatories, engineers and astronomers (Instrument Scientists in particular) monitor the behaviour of detector systems and characterize the quantities that may depend on the changing ambient conditions (e.g. electric interferences).

The current scheme is among the most advanced in the astronomical world and serves well the ESO scientific community. Is it possible to improve it? Is there something missing?

This chapter describes an effort initiated recently at ESO to implement a more thorough and modern plan of performance monitoring for those detectors that are being used onboard scientific instruments and relevant auxiliary systems (e.g. wave-front sensors) at the La Silla Paranal Observatory (LPO). Sections 2 and 3 describe the *status quo* at the Observatories, Sect. 4 explains the rationale of the project, and gives some details of the new monitoring plan and its current status. Section 5 gives some details on the Common Pipeline Library, which provides the basic set of tools for the practical implementation of the plan.

2 Status at the La Silla Observatory

The optical instruments in La Silla make use of a total of 1 IR array and 18 CCD detectors, of several types (EEV, MIT/LL, LLes/UV flooded, TK1024)

Table 1. Summary of available detector in La Silla

Instrument	Detector	Size	Pixel (μm)	Format	Wavelength Range (nm)
SUSI2	E2V 44-82	2048×4096	15	M of 2	300–1000
EMMI blue	Tektronix 1024 AB	1024×1024	24	S	300–500
EMMi red	MIT/LL CID-20	2048×4096	15	M of 2	400–1000
WFI	E2V 44-82	2046×4098	15	M of 4x2	350–1100
FEROS	E2V 44-82	2048×4102	15	S	350–920
HARPS	E2V 44-82	2048×4096	15	M of 2	380–530 530–690
EFOSC2	Loral/Lesser MPP Thin	2048×2048	15	S	300–1100
CES	E2V 44-82	2048×4096	15	S	350–1000
SOFI	Rockwell Hawaii HgCdTe	1024×1024	18	S	900–2500

Note: Format: S – single detector, M – Mosaic

and with different characteristics, covering an integrated wavelength range from 340 to 1000 nm. Table 1 lists the detectors currently in use at the observatory.

To monitor and characterize (if needed) all these detectors in a comprehensive and uniform way, common test procedures had to be developed over the years. CCD tests are performed weekly by personnel of the Science Operations department, and generally involve use of the "beta-light". The beta-light consists of a tritium radioactive pellet encapsulated inside a glass whose interior is covered by fluorescent material. The electrons from the decay of the tritium excite the fluorescent material generating a steady glow. The flux emitted by these sources is extremely stable since it follows a known law of decay with time with a time scale set by the half life of the tritium (12.4 years).

A different source, but the same strategy, is used for both HARPS and FEROS. The HARPS CCD is not accessible, and beta light tests are therefore not feasible. Similarly, for FEROS, insertion of the beta light is inconvenient, as it has an impact on the instrument stability. Instead LEDs, placed just above the CCDs, are used. This technique is proven to be fully satisfactory comparing the data from the FEROS LEDs (high stability power source, no feedback) with the beta light data purposely acquired for the comparison. This is achieved thanks to the use of a mechanical shutter so that the LED can be let to stabilise before the images are taken [4].

The tests are performed via graphic user interfaces (GUI) with a similar "look and feel" for the several detectors. The test strategy and the reduction algorithm are the same for all detectors. While the test is running, feedback is displayed on the screen. Once (If) certified, the test results are automatically saved in the database and published on the web.

The extracted values are: bias level, average count rate, gain, readout noise, linearity and shutter delay. The analysis method used is the traditional photon transfer method, documented, among others, by [5]. The final result is a homogeneous database spanning over several years to be used for trending, troubleshooting and health checks of the detectors.

3 Status at the Paranal Observatory

The instruments in operation at the Paranal Observatory include both optical CCDs and infrared arrays. Table 2 and 3 list all the optical and infrared detectors currently in use plus the one foreseen for future instruments.

The monitoring of the performance of these detectors is a collaboration of the Science Operations and Engineering/Instrumentation departments of the Paranal observatory and the Quality Control Group in Garching: test data are routinely acquired at the observatory, primarily during the day by means of calibration technical templates and the main detector parameters are checked and trend analysis is performed offline. According to the calibration plan of the observatory, all instruments and detectors should be

Table 2. Optical detectors datasheet for present and (near) future instruments at the LPO

Instrument	Detector	Size	Pixel (μm)	Format	Wavelength Range (nm)
FORS1	Tek 2048EB4-1	2048×2048	24	S	330–1100
FORS2	MIT/LL CCID-20	2048×4096	15	M of 2	330–1100
FLAMES (Giraffe)	E2V 44-82	2048×4102	15	S	370–950
UVES Blue	E2V 44-82	2048×4096	15	S	300–500
UVES Red	E2V 44-82 MIT/LL CCID-20	2048×4096	15	M of 2	420–1100
VIMOS	E2V 44-82	2048×2048	15	M of 4	360–1100
NAOS OPT	E2V CCD50	128×128	24	S	450–1100
Future Instruments					
OMEGACAM	E2V 44-82	2048×4096	15	M of 32	330–1000
GALACSI/ GRAAL	E2V L3 CCD220	240×240		M of 5	450–950
MAD	E2V CCD 39	80×80	24	S	450–950
VISTA OPT	E2V CCD 42-40	2048×2048	15	M of 4	850–1100
X-Shooter UVB	E2V 44-82	2048×4096	15	S	320–500
X-Shooter VIS	MIT/LL CID-20	2048×4096	15	S	500–1100

Table 3. Near-IR detectors datasheet for present and (near) future instruments at the LPO

Instrument	Detector	Size	Pixel (μm)	Format	Wavelength Range (μm)
ISAAC	Rockwell Hawaii	1024×1024	18	S	0.9–2.5
	HgCdTe SBRC Aladdin	1024×1024	27	S	1–5
VISIR	DRS BIB Si:As	256×256	50	M of 2	5–25
CONICA	SBRC Aladdin 3 InSb	1024×1024	27	S	1–5
NAOS	Rockwell Hawaii HgCdTe	1024×1024	18	S	0.9–2.5
SPIFFI	Rockwell Hawaii HgCdTe	2040×2048	18	S	0.9–2.5
MIDI	Raytheon IBC SI:As	320×240	50	S	5.0–25
AMBER	Rockwell Hawaii HgCdTe	1024×1024	18	S	0.9–2.5
CRIRES science	Raytheon Aladdin III	1024×1024	27	M of 4	0.95–5.2
CRIRES SV	Raytheon Aladdin III	1024×1024	27	S	1–2.2 (5)
Future Instruments					
X-Shooter NIR	Rockwell Hawaii 2RG	2048×2048	18	S	1.1–2.4
Hawk-I	Rockwell Hawaii 2RG	2048×2048	18	M of 4	1.1–2.4

monitored in the same way, but still some differences exist among instruments: for example, for most instruments, basic parameters such as bias level and noise are measured daily but the conversion factor is not monitored for all and other important measurements (e.g. linearity, fringing, contamination, etc) are not yet monitored. Additionally, measurements of the same parameter are made differently for different instruments, typically because they have different data reduction pipelines and recipe implementation. One example of this is the readout noise: we find cases in which it is measured as the sigma value in one raw file within a specific X×Y pixels window, corrected for fixed-pattern contribution but also measured on each whole single raw frame, with no corrections. Although both methods are acceptable and give consistent results, more uniformity would be desirable. This situation is exemplified in Table 4, which shows for various instruments in Paranal which measurements are checked and/or monitored for longterm variability (trend analysis).

The quality control web pages in Garching [6] provide links to the data measured for each instrument, and also allow queries to the parameters database (trending plot archive) for specific periods of time, which is a powerful tool for troubleshooting and for use of archival data mining.

Table 4. Detector parameters currently monitored

Optical Instrument	RON	Bias Level	Inverse Gain	Linearity	Bad Pixels
FORS1/2	Ch/Tr	Ch/Tr	Ch/Tr	–	–
UVES	Ch/Tr	Ch/Tr	–	–	–
VIMOS	Ch/Tr	Ch/Tr	Tr	Ch/Tr	–
GIRAFFE	Ch/Tr	Ch/Tr	Tr	Tr	–

IR Instrument	RON	Dark Current	Inverse Gain	Linearity	Bad Pixels
ISAAC	Ch	Ch	–	Tr	Tr
CONICA	Ch	Ch	Tr	–	Tr
SINFONI	Ch	Ch	Tr	Tr	Tr
CRIRES	–	–	–	–	–
AMBER	–	–	–	–	–
MIDI	Ch/Tr	–	–	Ch/Tr	–
VISIR	Ch	Ch	–	–	Tr

Note: Format: Ch – checked, Tr – Trend analysis

4 The Detector Monitoring Plan

The establishment of a monitoring plan for IR and optical detectors at the Observatory comes from the need to expand and unify as much as possible the detector testing strategy, with the aim of gaining in efficiency and quality of the measurements. The monitoring plan is mostly complementary and does not supersede the characterization and testing plans implemented in Garching by the infrared and optical detector teams (IDG and ODT): as such, it is limited at considering those parameters that can be measured at the telescope by means of day-time and night-time calibrations. All other parameters are already characterized before the instrument is commissioned at the observatory or are included in the manufacturers' specifications documents. It is however possible that, as a result of an upgrade (e.g. the implementation of a new read-out mode), or some other structural change (e.g. change of the operating temperature of a camera) a new characterization has to be performed at the observatory. Additional testing procedures are also designed and carried out by the instrumentation group at the observatory, like for example the measurement and monitoring of variable environmental effects such as electrical interferences. In the interest of economy of efforts and of ESO-wide standardization, the plan includes the relevant test procedures devised in the labs in Garching and tries to make use of proven recipes, algorithms and procedures already in place, for example at the La Silla Observatory. For the same reasons, the Common Pipeline Library [7] was the chosen tool for the practical implementation (data reduction recipes, see Sect. 4 for details). This plan is meant to achieve the following broad goals:

- Standardize test procedures whenever applicable (e.g. define a unique way to measure linearity).
- Unify test procedures for IR and Optical detectors whenever applicable and describe the differences in all the other cases (e.g. define unique ways to acquire data sets).
- Unify the measurement procedures and the use of data reduction recipes and algorithms.
- Utilize available resources, such as data taking OBs, pipeline recipes, existing reporting tools (e.g. QC web pages).

Some additional specific goals:

- Generate a list of all the parameters, the required accuracy and their measurement frequencies (Table 4).
- Provide a description of each parameter, its impact on science and its dependencies on operating conditions, as provided at the observatory.
- Set the requirements for data acquisition and the operating conditions for measurements.
- Give a description of the algorithms for measuring each parameter and the methods for data analysis.

The monitoring plan also provides useful by-products: (1) a higher level implementation roadmap to serve as guide for the pipeline developers who have the task of providing the data reduction recipes, (2) quantitative measurements to plan for interventions and to monitor the health of the instruments for the instrument scientist and the engineers. In this way, the detector monitoring plan interests a broader audience:

- The night astronomer, who operates the instruments and is interested in optimizing their scientific output.
- The day astronomer who needs guidelines to validate the calibration data.
- The user support astronomer, who needs the nitty-gritty specifics of the detectors to advise the users when setting their observational tactics and techniques.
- The quality control scientist, who is in charge of the back-end instrument health checks.
- The astronomer (user), who is interested in planning a strategy for his observations and achieving specific scientific goals.

Not all of the parameters listed in Table 5 affect performance at a system level. Therefore the monitoring plan also lists the parameters, their importance at a system level and their dependencies on operating conditions. It is important to understand the dependencies on operating conditions in order to set test frequencies and to use predictive tools.

For example, the readout noise has high importance, since it affects all aspects of astronomical applications and it is a good indicator of instrument

Table 5. Detector parameters in alphabetical order and IR/OPT compatibility

Parameters	OPT	IR	MIR	Accuracy (unit)	Method	F	P
Amplifier Glow	X	X	X	Signal (ADU)	Statistics: Bias/Dark	N	14
Bias Level	X			<1	Bias Statistics	D	1
	X			1 digit (ADUs)			
Bad pixels, Cosmetic Quality	X	X	X	N/A	Bias Frames (OPT) Dark Frames (IR) Flat frames (ALL)	Y	2
Contamination	X				UV Flats analysis	Y	16
Conversion Factor	X	X	X	2 digits [e^-/ADU]	Transfer Curve TDI images	BA M	3
Cosmic Ray Sensitivity	X				Long darks statistics	N	17
Crosstalk (multiple ports)	X	X	X	1/300,000	Bright star sequence	Y	8
CTE (V/H)	X			10^{-6}	EPER (overscan); Signal variance;	O	6
Dark Current	X	X	X	2 digits [e^-/px/s] –[e^-/px/hr]	Dark Frames Statistics	D	1
Dark Signal Non uniformity variable fixed (DSNU +FPN)	X			<10%		BA	5
Full Well Capacity	X			1%	Linearity curve	BA	4
Linearity	X	X	X	<1%	Transfer Function	BA	3
Persistence (Remanence)	X	X	X	% of original	Dark current measurements after illumination	Y	13
Readout Noise (RON)	X	X	X	2 digits [e- RMS]	Bias Frames (OPT) Dark Frames (IR)	D	1
50Hz Pick up Noise	X	X	X	2 digits (ADU)	FFT analysis	Y/N	7
Microphonic Noise		X				BA	8
N-pixel correlated Noise		X				M	9
Odd-Even column effect	X					D	2
Shutter Pattern/Error	X			% of maximum signal	Linearity curve Shutter procedure	Y/N	11
Spatial Uniformity, Response non-Uniformity (PRNU) - Fringing	X	X	X	<2%	Analysis of flats	M/N	10
Stability		X	X	N/A	Visual inspection bias frames	D	15
Stray light	X				Flat frames	W	18
Temperature stabilization	X	X	X		% error	Y	12

Notes: **F**=Frequency, **P**=Priority, D=Daily, W=weekly, M=monthly, BA=biannual, Y=yearly, O=once, N=when needed.

electronic problems. It can change with temperature, it may be influenced by detector electronics and by external electromagnetic interference (pick up noise). The crosstalk has medium importance, since it affects spatial resolution. It is important at design level and it is usually stable, independent of pixel location, normally proportional to signal. The pick up noise is considered of low importance, it depends on the detector integration time, readout speed, environment conditions, detector mounting and dismounting and other nearby equipment that is altered.

The monitoring plan is being developed in phases, currently in different stages of completion:

1. Requirement analysis and definitions: completed and recorded in [8].
2. Analysis of current status of detector monitoring at ESO: completed.
3. Implementation of pipeline recipes: on-going.
4. Implementation of observing templates for data collection: to be initiated.

The plan will become gradually operative, following the development status of the pipeline recipes and the observing templates. After an initial phase of testing and cross checks with independent tools it will become part of the routine operations at the LPO.

5 The Common Pipeline Library

The ESO Common Pipeline Library (CPL, [7]) has been developed to unify, share and centralize the pipelines code, reduce their maintenance costs. The CPL is a self-contained ISO-C library, designed for use in a C/C++ environment. CPL is used for the development of all ESO pipelines, both for internal developments and for developments by the external consortia.

The CPL itself is split into four major components, reflecting the algorithmic level of the software contained therein (Fig. 1):

- The CPLCORE library provides the fundamental CPL data types (to support/define images, tables, vectors, matrix) the operations defined on these data types, and elementary utility functions.
- The CPLUI (UI for User Interface) provides services defining the standard interface for recipes and provides more complex data reduction related utilities and services. In particular, it contains the necessary types to handle the plugin interface, the parameters used to handle command-line options and the (set of) frames used to store the input list of files.
- The CPLDRS (DRS for Data Reduction System) uses the CPLCORE data types and functions to implement higher level data processing algorithms. CPLDRS is itself organized in sub-modules corresponding to different stages of data reduction: detector calibration, geometrical calibration, aperture definition, and photometry.

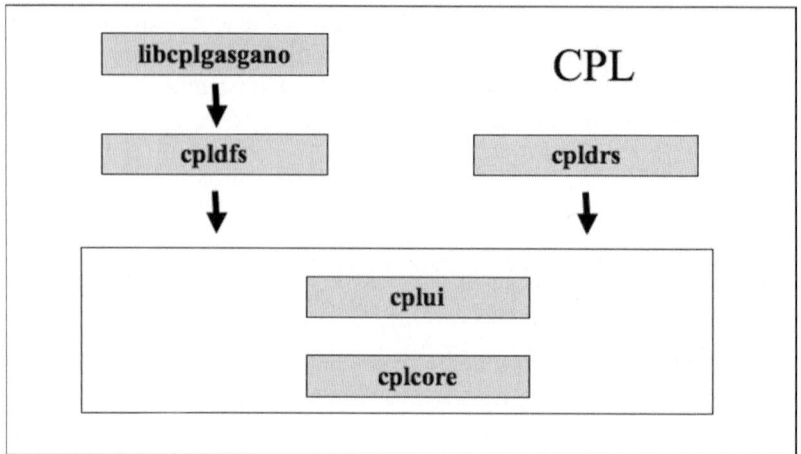

Fig. 1. Software layers of the ESO Common Pipeline Library

– Finally, the most recent addition to CPL, the CPLDFS module (DFS for Data Flow System) provides DFS related utilities for the creation of DFS compliant data products, or quality control logs.

The routines of the detector monitor project will be provided as part of the detector correction module in the CPLDRS layer. All pipelines will be able to invoke the same function, e.g. the detector linearity algorithm. Each pipeline will call these functions from the instrument specific recipes, thus providing the flexibility required to cope with instrument-specific features, keywords, or data structures.

Acknowledgement. The authors would like to thank Nicolas Haddad, Pedro Mardones, Emanuela Pompei and Martin Roth for the fruitful discussions and comments. James Brewer and Thomas Agustejin for having pioneered the work for a common algorithm for detector monitoring in La Silla. Emilio Barrios and Paul Le Saux who developed the GUI and are maintaining the whole software.

References

1. IR detector department web pages http://www.eso.org/projects/iridt/
2. ESO Optical Detector Team (ODT) web pages: http://www.eso.org/projects/odt/ODTnew/index.html
3. P. Amico, T. Boehm, T.: ESO's New CCD Testbench, Optical Detectors for Astronomy. Proceedings of an ESO CCD workshop held in Garching, Germany, October 8–10, 1996. Edited by James W. Beletic and Paola Amico. Kluwer Academic Publishers, Boston, MA, 1998 (Astrophysics and Space Science library), Vol. 228, p. 95

4. J. Alonso, ESO Doc. No.: 3P6-DSD-ESO-60400-0007
5. M. Downing, D. Baade, P. Sinclaire, S. Deiries, F. Christen: CCD riddle: (a) signal vs time: linear; (b) signal vs variance: non-linear, High Energy, Optical, and Infrared Detectors for Astronomy II. Edited by Dorn, David A. and Holland, Andrew D. Proceedings of the SPIE, Vol. 6276 (2006)
6. Quality control pages: http://www.eso.org/qc
7. The Common Pipeline Library, http://www.eso.org/cpl
8. The Detector Monitoring Plan document. VLT-MAN-ESO series, 1st release, Oct. 2006

Discussion

D. Baade: This looks like a very meritorious project. But how comes that the two detector departments, which have provided nearly all of the detector systems, do not know about it?

D. Baade: Were the many scary anonymous examples of defective detectors taken from present ESO detector systems?

P. Amico: The examples are not specific to ESO's systems.

P. Bonifacio: I would like to know with detail the effects of cosmic rays on detectors. Therefore, I would like its monitoring to be put at least at a medium importance level. Long exposures, necessary to beat the RON on very faint objects. The experience gained can be very important for the design and construction of future instruments.

P. Amico: The cosmic rays rate is mainly dependent on the detector's environment (typically the surrounding material and the location/altitude). This quantity is characterized in Garching. It is checked for stability at the Observatory, typically during commissioning. We plan to monitor the cosmic rays rate, but since it is not expected to vary significantly, this measurement has lower priority than others.

D. Osip: How many CCDs do you operate with a Nod & Shuffle mode? Do you have different characterization and monitoring plans for this mode?

P. Amico: There are no instruments at the Observatory, which offer Nod & Shuffle.

Part II

Session 1: Optical Spectro-Imagers

IOT Overview: Optical Spectro-Imagers

F. Patat

ESO, Karl-Schwarzschild-Strasse 2, 85748 Garching, Germany; fpatat@eso.org

Abstract. Taking the FORS instruments as a representative case, I review the Calibration Plan for optical spectro-imagers currently offered at ESO, discussing various aspects related both to the scientific outcome and the instrument/site monitoring. I also describe ongoing and future calibration projects planned by the Instrument Operations Teams, trying to give an objective view on the limitations of the Calibration Plans currently implemented at ESO for this class of instruments.

1 The Long Tradition of Optical Spectro-Imagers at ESO

The history of optical spectro-imagers at ESO starts in April 1985, when the ESO Faint Object Spectrograph and Camera (EFOSC1) is offered at the 3.6 m telescope [1]. After that, many others followed both at ESO (DFOSC, EMMI, EFOSC2, FORS1&2, VIMOS) and in many other observatories around the world (**A**FOSC, **B**FOSC, **C**AFOS just to cite a few).

Currently, five optical spectro-imagers are offered at ESO (see Table 1). As time went by, the complexity has been steadily growing. From the compactness of EFOSC1 and 2, which has indeed been the key of their success, optical spectro-imagers have evolved into significantly more complex machines (FORS1 and 2), offering relatively large field of views and enhanced multiplexing capabilities (VIMOS).

Due to the high number of modes featured by these instruments, the calibration plans are quite articulated, as the case of the FORSes, which I will take as a representative case, clearly show (see Table 2). For this reason, there are unfortunately a number of calibrations which are not supported as part of the observatory calibration plan, but that can still be obtained upon request:[1]

- twilight flats and standard stars for non-standard CCD settings;
- any specific standard star to correct for telluric absorption bands;
- radial velocity standards;

---

[1] For additional night time calibrations, however, the user is charged in terms of execution time.

Table 1. Optical spectro-imagers currently offered at ESO and their modes

	EFOSC2	EMMI	FORS1	FORS2	VIMOS
IMG	×	×	×	×	×
IMG/OCC	×		×	×	
LSS	×	×	×	×	
MOS	×	×	×	×	×
LPOL	×		×		
CPOL	*		×		

Notes: IMG=Imaging, IMG/OCC=Imaging with occulting masks/bars, LSS=Long Slit Spectroscopy, MOS=Multi Object Spectroscopy, LPOL=Linear Polarization, CPOL=Circular Polarization
*This mode is being implemented and tested at the time these proceedings are edited

Table 2. FORS1/2 calibration plan

Calibration Mode	Collimator	Frequency	Number	Time	Results	Accuracy
Bias	–	weekly	5	Day	bias lev., RON	RON/2
Darks	–	monthly	3	Day	dark curr.	–
Screen Fl. UBVRI	SR	weekly	2	Day	CCD check	–
Astrometry	SR/HR	annually	1	Night	dist., scale	1 px
Img. Sky Flats	SR	weekly	4	Twi.	norm. flat	2%
	HR	as needed	4	Twi.	norm. flat	2%
UBVRI Ph. Std.	SR	nightly	1	Night	zero points	5%
	HR	as needed	1	Night	zero points	5%
Gunn & other	SR/HR	as needed	1	Night	response	10%
airm>1.6 UBVRI	SR/HR	weekly	1	Night	ext. coeff.	5%
Scr. Flats LSS, MOS,MXU,ECH	SR/HR	as needed	5	Day	norm. flat	5%
Scr. Arcs LSS, MOS, MXU, ECH	SR/HR	as needed	1	Day	disp. coeff.	0.3 px
Specphot. Std.	SR/HR	as needed	1	Night	response	10%
IPOL Pol. Std.	SR	as needed	1	Night	zero angle (L)	1 deg
IPOL Unp. Std	SR	annually	1	Night	instr. pol. (L)	–
IPOL Unp. Std.	SR	annually	1	Night	instr. pol. (C)	–
PMOS Arcs	SR	as needed	1	Day	disp. coeff.	0.3 px
PMOS Fl. (45 deg)	SR	as needed	5	Day	norm. flat	5%
PMOS Pol. Std.	SR	as needed	1	Night	zero angle (L)	1 deg
PMOS Unp. Std.	SR	annually	1	Night	instr. pol. (L)	–
PMOS Unp. Std.	SR	annually	1	Night	instr. pol. (C)	–

Source: Adapted from [2], Table 4.1.

- spectro-photometric standards for 2nd order spectroscopy;
- any day or night calibrations for slitless spectroscopy;
- any day or night calibrations for spectroscopy with non-standard order sorting filters;
- any day or night time polarimetric calibrations for retarder plate angles different from 0,22.5,45,67.5 (linear) and −45,45 (circular);
- any PMOS screen flats at retarder plate angles different from 45 degrees;
- any IPOL screen flats;
- any IPOL day or night time calibrations with the HR collimator.

In general, the accuracy promised by the Calibration Plan can be definitely achieved by the calibrations that it provides. The problem arises when the user needs to reach a higher precision, especially when this means pushing the instrument to its limits.

2 Open Issues

A crucial issue with focal-reducers (as most optical-spectro imagers are) is the proper flat-fielding of the data. As an extensive study has shown [3, 4], FORS1 shows a feature, also known as The Big One, whose position depends on the adapter rotator angle (see also the contribution by W. Freudling, these proceedings). This and the light concentration, typical of this kind of instruments, require a proper set of calibrations and a special reduction procedure [5], which altogether make 1% photometry across the whole field of view a non trivial task.

Another potential problem is represented by the lack of suitable standard stars for 8 m-class telescopes, a fact that certainly limits the photometric accuracy one can reach. Even though steps in this direction are being considered (see the contributions by W. Freudling and J. M. Rodriguez Espinosa, these proceedings), a final solution to this issue is currently not available, even though it is probably just a matter of time.

Indeed, with modern instruments (at least those discussed here) it seems like we have reached a level where calibrating the instrument is relatively easy and what dominates the whole process is *calibrating the atmosphere*. In this respect, a crucial issue is the monitoring of absolute extinction. As the photometric standard stars observations are currently implemented in the Calibration Plan (see Table 2), extinction coefficients can be derived only on a monthly basis [6], using data taken on different photometric nights, as in the example shown here in Fig. 1. This procedure (see reference [7] for a detailed description) is currently maintained by the Quality Control Group,[2] which provides online monthly full photometric solutions (zeropoints, color terms and extinction coefficients). Nightly zeropoints are also provided whenever a

---
[2] http://www.eso.org/qc

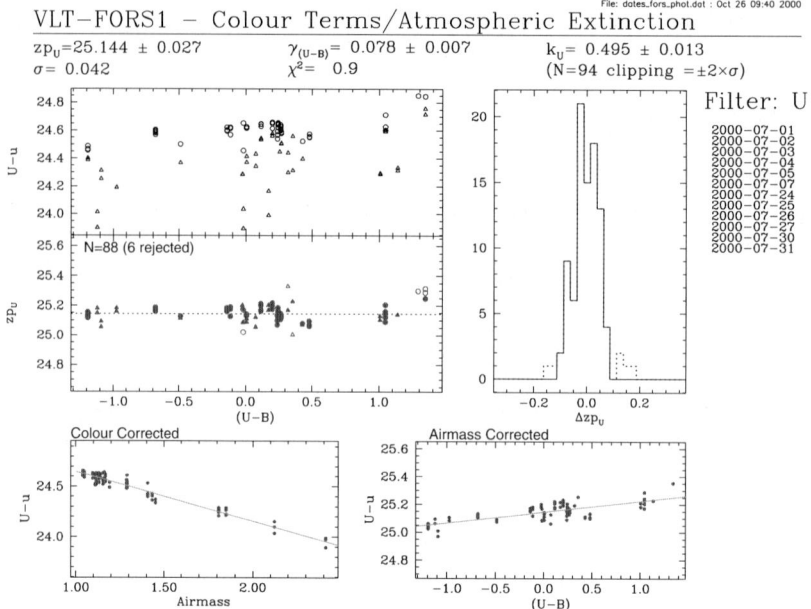

Fig. 1. Example of monthly, photometric solution for FORS1, July 2000, U passband. Courtesy of the Quality Control Group. *Upper left*: observed zeropoints (*top*) and color-, airmass-corrected zeropoints (*bottom*). *Upper right*: deviations from the observed zeropoints of all (*dotted*) and clipped (*solid*) data. *Lower left*: airmass dependency of color-corrected data. *Lower right*: color dependency of airmass-corrected data

photometric standard field is observed but, in these cases, the extinction is assumed and not determined. While this is sufficient for instrument trending purposes, it is probably not optimal, at least for some scientific cases. Also, the lack of a direct estimate of prevailing transparency conditions poses some operational problems during the night, when the astronomer has to take a decision about the execution of a given observation requiring some specific transparency conditions.

Sky transparency variations are indeed monitored at Paranal by LOSSAM,[3] but absolute and calibrated extinction coefficients are not yet available. Nevertheless, ESO is investigating this issue, with the final aim of providing online two-dimensional atmospheric extinction maps at different times during any given night.

Another pending problem is the removal of fringing in the red portion of the spectral range, which severely affects some instruments like EFOSC2, VIMOS and FORS1 (after its blue upgrade, April 2007. See next section). Even

---

[3] http://archive.eso.org/asm/ambient-server

though some solutions are being proposed (see for example the contribution by H. Kuntschner et al., these proceedings), in this kind of issues one is indeed confronted with something we cannot control, i.e. the random fluctuations of night sky emission lines.

3 Instrument Operations Team (IOT) Parallel Activities

Besides taking care of the routine operations, each IOT carries out a series of parallel activities, aimed at characterizing both the site and the instruments. Examples are the studies of the optical night sky brightness at Paranal [6, 8, 9] (see Figs. 2 and 3), the effect of scattered moonlight on optical imaging and low resolution spectroscopy [10], the determination of optical extinction curve [6, 8, 11] and the study of twilight brightness [12] and the characterization

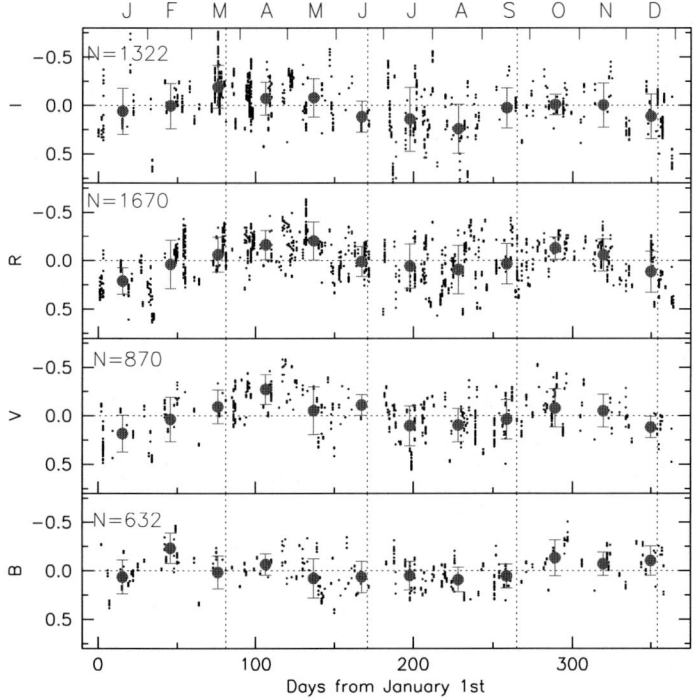

Fig. 2. Seasonal night sky brightness fluctuations derived from 6 years of FORS1 BVRI imaging data. The data have been corrected for differential zodiacal light contribution and solar flux dependency. The *small* symbols mark single observations, while *larger* symbols indicate monthly averages

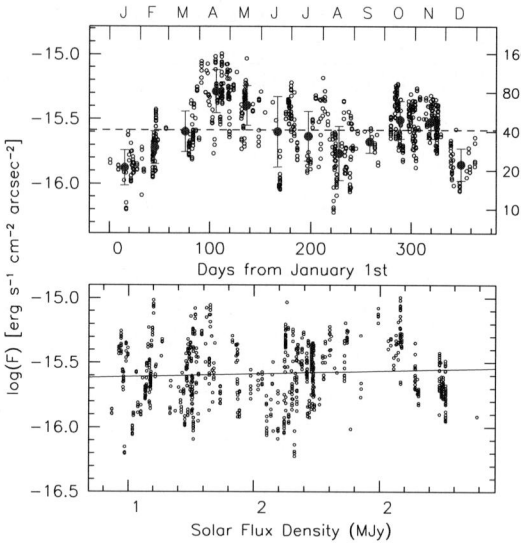

Fig. 3. Dark time Na I D line flux as a function of solar density flux (*lower panel*) and time from the beginning of the year (*upper panel*). The large symbols are monthly averages and the error bars indicate the RMS deviations

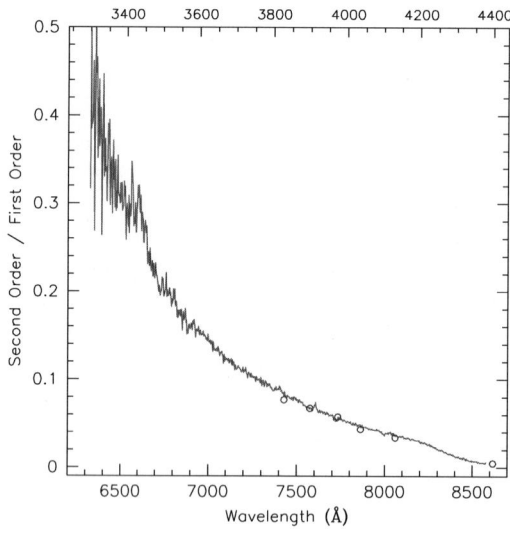

Fig. 4. Second order contamination for FORS1/2 Grism G300V derived from spectrophotometric standard star observations (*line*) and He I lamp exposures (*empty circles*). The plot shows the ratio between the second order spectrum (*lower wavelength scale*) and the first order (*upper wavelength scale*)

of VIMOS filter efficiency (see the contribution by S. Mieske et al., these proceedings).

Very recently, the FORS IOT has addressed the problem of second order contamination in those cases where the user is allowed to request a grism without order sorting filter (see Fig. 4 for an example). This action was actually triggered by a user and this is a clear example of constructive interaction between the user community and the Observatory.

Other ongoing and planned projects are listed here:

- Blue CCD upgrade for FORS1;
- Derive flexure models for both FORSes;
- Derive offsets for the various filters;
- Determine off-axis response for Volume Phase Holographic (VPH) Grisms;
- Implement high time resolution (HIT) mode Calibration Plan;
- Accurate definition of FORS photometric system.

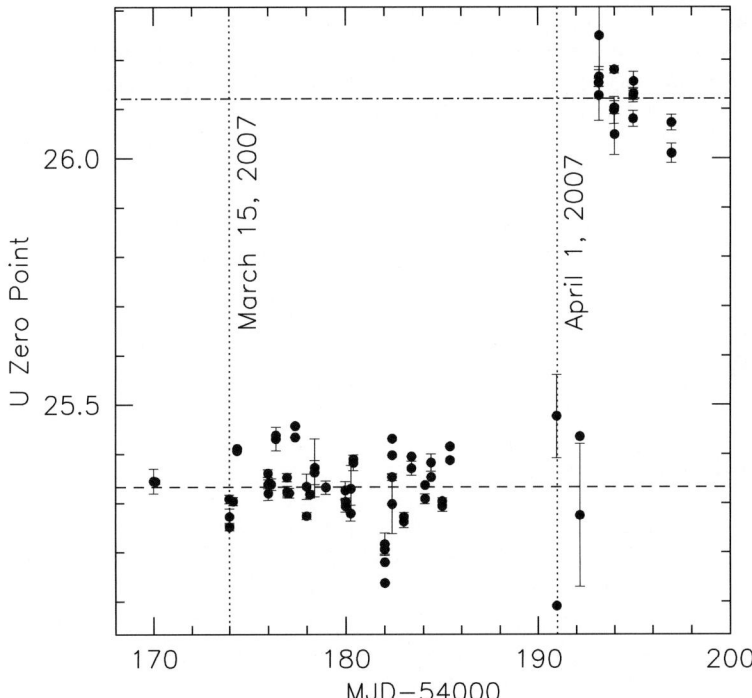

Fig. 5. Effect of FORS1 Blue CCD Upgrade. The horizontal lines mark the average values before (*dashed*) and after the intervention (*dotted-dashed*). Data courtesy of the Quality Control Group (http://www.eso.org/qc)

The FORS1 Blue CCD upgrade has been carried out in two phases, in February (while this meeting was taking place) and in early April 2007. Starting with ESO Period 79, FORS1 is offered with a new 2 CCD mosaic, which features a much higher sensitivity in the blue. The data, processed by the Quality Control group in Garching, show an increase of about 0.75 mag in the U band (see Fig. 5).

References

1. J. Melnick, H. Dekker, S. D'Odorico: *ESO Operating Manual No. 4*, ESO, Garching Vol. 2 (1989)
2. E. Jehin, K. O'Brien: *FORS User Manual*, VLT-MAN-ESO-13100-1543, Issue 79.0 (2006)
3. W. Freudling, M. Romaniello, F. Patat, P. Møller, E. Jehin, K. O'Brien: *The FORS Absolute Photometry Project*, VLT-TRE-ESO-13100-4006 (2006)
4. W. Freudling, M. Romaniello, F. Patat, P. Møller, E. Jehin, K. O'Brien: in *The Future of Photometric, Spectrophotometric, and Polarimetric Standardization*, Photometry with FORS at the ESO VLT ed. C. Sterken, ASP, Provo, UT, Vol. 364, p. 113 (2007)
5. P. Møller, et al.: *FORS: An Assessment of Obtainable Photometric Accuracy and Outline of Strategy for Improvement*, VLT-TRE-ESO-13100-3808 (2005)
6. F. Patat: Astron. Astrophys. **400**, 1183 (2003)
7. F. Patat: *Photometric Solutions via One Step Multilinear Least Squares Fit*; http://www.eso.org/observing/dfo/quality/FORS1/qc/photcoeff/photcoeff.ps (2002)
8. F. Patat: Messenger **115**, 18 (2004)
9. F. Patat: Astron. Astrophys., in press, ArXiv 0801.2270 (2008)
10. F. Patat: Messenger **118**, 11 (2004)
11. I. Gavignaud, et al.: Astron. Astrophys., in preparation
12. F. Patat, O.S. Ugolnikov, O.V. Postylyakov: Astron. Astrophys. **455**, 385 (2006)

Discussion

P. Bonifacio: The SDSS has been a very successful project with respect to the quality of the photometry. Their philosophy was to have two telescopes. The main survey telescope and a small photometric telescope. The photometric telescope was used to observe standard stars and calibrate extinction. Do you think such an approach would be valuable also for VLT instrumentation?

F. Patat: We first need to say that probably most of VLT science cases do not need a photometric accuracy much better than what one can achieve with the current calibration plan. It is certainly true, though, that more demanding science cases would definitely profit from a more systematic and properly calibrated extinction monitor.

M. Capaccioli: Does the seasonal pattern of the night sky brightness repeat identical over the years?

F. Patat: Yes, in general this repeats almost unchanged from year to year. Of course, the change in the sun ionizing flux produces an overall variation in the night sky brightness.

Photometry with FORS

W. Freudling[1], P. Møller[1], F. Patat[1], S. Moehler[1], M. Romaniello[1], E. Jehin[2], K. O'Brien[2], C. Izzo[1], and E. Pompei[2]

[1] ESO, Karl-Schwarzschild-Strasse 2, 85748 Garching, Germany; wfreudli@eso.org
[2] ESO, Alonso de Cordóva 3107, Vitacura, Santiago, Chile

Abstract. Photometric calibration observations are routinely carried out with all ESO imaging cameras in every clear night. The nightly zeropoints derived from these observations are accurate to about 10%. Recently, we have started the *FORS Absolute Photometry Project (FAP)* to investigate, if and how percent-level absolute photometric accuracy can be achieved with FORS1, and how such photometric calibration can be offered to observers. We found that there are significant differences between the sky-flats and the true photometric response of the instrument which partially depend on the rotator angle. A second order correction to the sky-flat significantly improves the relative photometry within the field. We demonstrate the feasibility of percent level photometry and describe the calibrations necessary to achieve that level of accuracy.

1 Introduction

The two FORS instruments FORS1 and FORS2 are, among their multiple functionalities, optical imaging cameras at the VLT. They are all-dioptric instruments for the wavelength range from 330 to 1100 nm and provide an image scale of 0".2/pixel, or 0".1/pixel with the high resolution collimator. Since the FORSes are not direct imaging cameras, they suffer from sky concentration and other effects which complicate accurate photometry. The goal of the *FORS Absolute Photometry (FAP)* project is to characterize the photometric performance of the FORS instruments and investigate if and how the routine calibration of the instruments can be changed to offer improved photometric zero points. Our work has so far concentrated on the FORS1 camera, but most of the findings will equally apply for FORS2.

2 Data

To characterize the photometric response, we analyzed archival data taken with FORS1, and obtained new observations of the Stetson fields Mark A, L92, L113 and PG1633. To allow investigation of the flatfield response, data were taken with a dither pattern shown in Fig. 1.

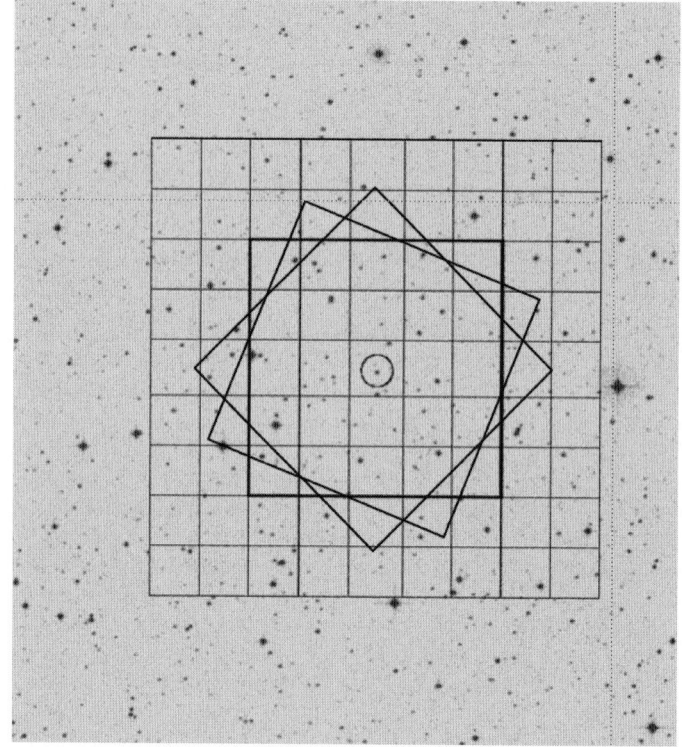

Fig. 1. Dither pattern of the *FAP* observations on the Mark A field. The pattern includes two rotated frames. The pattern was chosen so that an illumination correction to the flatfield can be derived from all stars within the FORS1 field-of-view which were observed more than once

3 Results

A detailed description of the results of *FAP* is given in [1]. Here, we give a brief summary of the most important findings.

3.1 Twilight Flats

The master flats routinely produced by the pipeline from twilight flats differ from night to night or even within a night by up to 5%. Some of the pattern seen in the flats rotate with the adaptor angle, which is shown in Figs. 2 and 3. For highly accurate photometry, it is necessary to remove this rotating pattern from the flatfields.

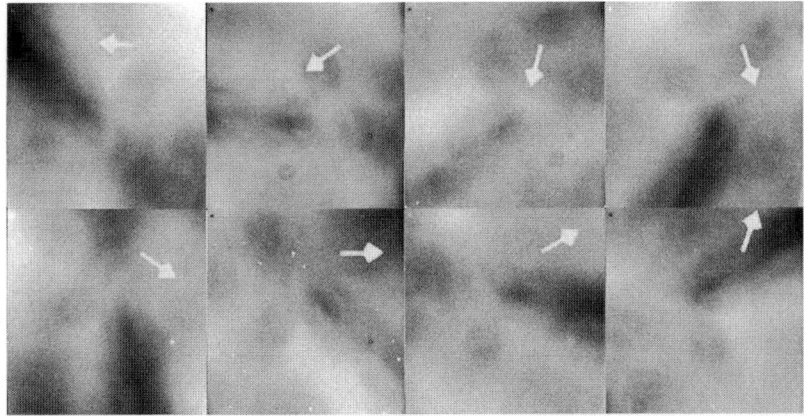

Fig. 2. A sequence of FORS1 B sky-flats, divided by the mean of the flats. The rotator adapter angles $(-105°, -73°, -35°, 0°, +30°,$ and $+70°)$ are indicated by arrows in the upper right corner. It can be seen that the dark band across the detector rotates with the adapter. The intensity scale range is 3%

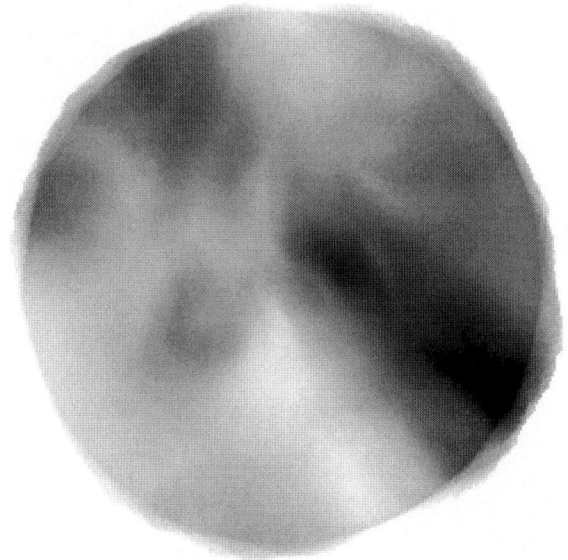

Fig. 3. The sky-flats shown in Fig. 2 counter-rotated by the adaptor angle and averaged. The visible structure is the rotating part of the flat fields. The intensity scale range is 1%

3.2 Response

After the rotating and fluctuating pattern has been removed from flatfields, they still can be used to correct science observations for variations in the throughput on scales of a small fraction of the detector. The flatfield response on larger scales has to be determined independently. Correcting flatfields for a large scale illumination pattern will also remove any effect of the sky concentration. This response on large scales can accurately be derived from the dithered observations of standard fields. As an example, the flatfield response of the R-band filter for FORS1 is shown in Fig. 4.

3.3 Error Budget

The main contribution to the photometric error are the magnitude measurement errors, fluctuation in the extinction, uncertainties in the standard magnitudes and remaining flatfielding errors. A careful analysis of the error budget (see Fig. 5) shows that when the commonly used master flats are corrected as described above, the remaining flatfielding error are significantly less than 1%.

Fig. 4. An illumination pattern of a R-band flat derived from standard fields dithered on a 5×5 grid, with a step size of 20% of the field of view. Such illumination correction should be applied to all FORS flats if percent-level photometric accuracy is required. The intensity scale range is 2%

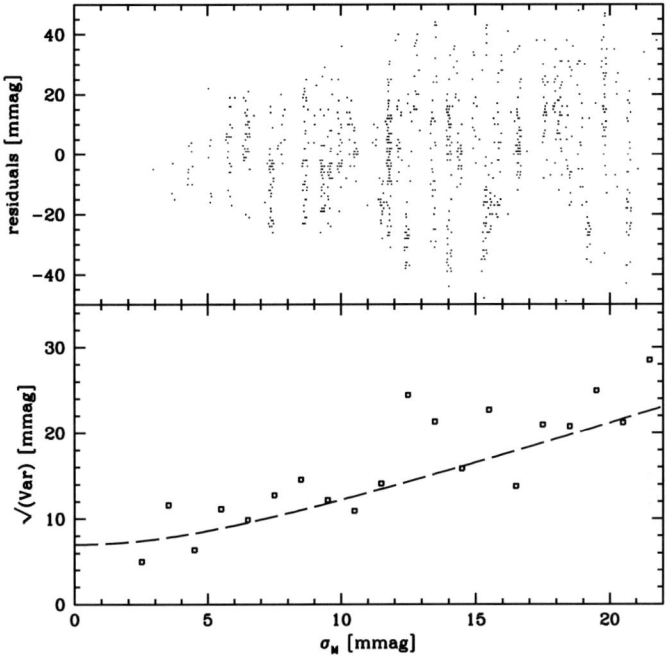

Fig. 5. *Upper panel*: Residuals from extinction fits as a function of the combination of magnitude measurement errors and errors in the listed standard magnitudes. *Lower panel*: the variance of the residuals. The *dashed line* is the error model which includes statistical errors as well as extinction fluctuations. The error model does not include any residual flatfielding errors. A detailed analysis suggests an upper limit on residual flatfielding and other sources of errors of less than 10 mmag

3.4 Calibrating Nights with Two and Three Standard Fields

Stetson standard star fields [2, 3] contain a large number of photometric standard stars suitable for 8m-class telescopes. Appropriate regions selected from within the Stetson fields contain typically 20–30 standard stars within the FORS field of view. With such a large number of standard magnitudes, the main source of errors in the determination of the extinction is the short term fluctuations in the extinction. The photometric zero point can be determined with an accuracy of about 10 mmag with three standard observations in perfectly photometric nights (see Fig. 6) This demonstrates that percent-level photometric calibration is feasible with the FORSes when appropriate standard fields are used.

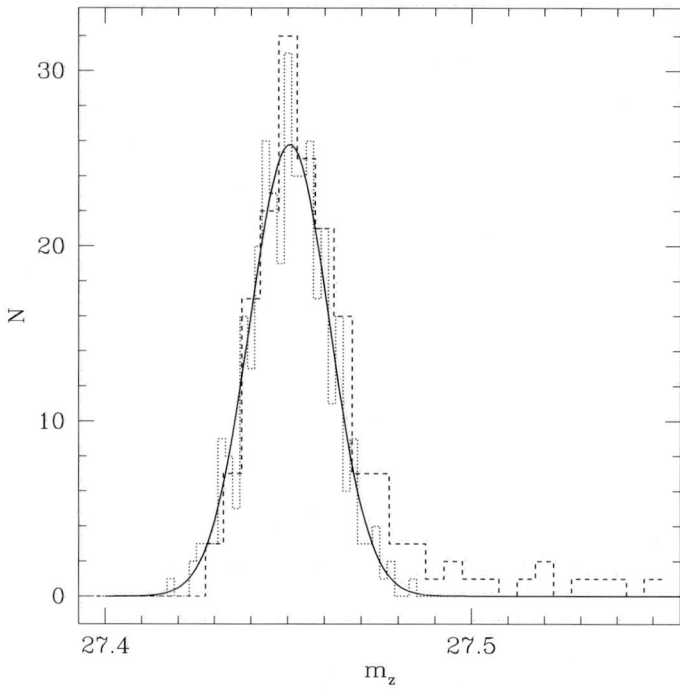

Fig. 6. Distribution of the residuals from extinction fits when two (*dashed histogram*) or three (*dotted histogram*) standard fields are used. The *solid line* is a Normal distribution with σ of 10 mmag. The plot demonstrates that under ideal conditions, three photometric standard fields are sufficient to achieve percent-level accuracy

Summary and Outlook

FAP has shown that it is possible to achieve percent-level photometry with FORS1 with moderate effort. Over the next year, we plan to prepare new photometric standard fields suitable to obtain more accurate photometric solutions and simultaneously derive flatfield corrections and to incorporate new flatfielding algorithms into the FORS pipeline.

References

1. W. Freudling, M. Romaniello, F. Patat, P. Møller, E. Jehin, K. O'Brien: in *"The Future of Photometric, Spectrophotometric, and Polarimetric Standardization"*, ed. C. Sterken, ASP, 364, 2007, pp. 113
2. P.B Stetson: PASP **112**, 925 (2000)
3. P.B. Stetson: 2006, fields listed at http://cadcwww.dao.nrc.ca/cadcbin/wdbi.cgi/astrocat/stetson/query

Radial Velocity Measurements Based on FORS2 and FLAMES Spectra

M. Monelli[1], M. Nonino[2], G. Bono[3], P. François[4], F. Thévenin[5],
R. Buonanno[6], F. Caputo[3], C. E. Corsi[3], L. Pulone[3], H. A. Smith[7],
P. B. Stetson[8], and A. R. Walker[9]

[1] Instituto de Astrofísica de Canarias, C/ Via Lactea, 38205 La Laguna, Tenerife, Spain; monelli@iac.es
[2] INAF–Osservatorio Astronomico di Trieste, Via Tiepolo 11, 34131 Trieste, Italy
[3] INAF–Osservatorio Astronomico di Roma, Via Frascati 33, 00040 Monte Porzio Catone, Rome, Italy
[4] GEPI–Observatoire de Paris-Meudon, 92125 Meudon Cedex, France
[5] Dépt. Cassiopée, UMR 6202, Obs. de la Côte d'Azur, BP 4229, 06304 Nice Cedex 4, France
[6] Dip. di Fisica, Univ. di Roma Tor Vergata, Via della Ricerca Scientifica 1, 00133 Rome, Italy
[7] Dept. of Physics and Astronomy, Michigan State University, East Lansing, MI 48824
[8] Dominion Astrophysical Obs., HIA-NRC, 5071 West Saanich Road, Victoria, BC V9E 2E7, Canada
[9] Cerro Tololo Inter-American Observatory, NOAO, Casilla 603, La Serena, Chile

Abstract. We present an extensive spectroscopic investigation on evolved old- and intermediate-age stellar tracers of the Carina dSph galaxy. We discuss the approach we devised to calibrate the two multi-object spectrographs (FLAMES, FORS2) available at the European Southern Observatory. Pros and cons of the two instruments are also mentioned. Finally, we present preliminary results concerning the radial velocity measurements and the comparison with data from the literature

1 Introduction

The use of large field of view mosaic CCD cameras provide the unique opportunity to investigate in detail the stellar populations of low-density stellar systems such as the dwarf spheroidal galaxies in the Local Group (LG). These new instruments also played a crucial role in tracing stars in different evolutionary phases over a substantial fraction of the body of large globular clusters [11, 1]. The spectroscopy, due to the complexity of the hardware, typically lags both in limiting magnitude and in multiplicity when compared with photometry. However, during the last few years the introduction of multi-objects spectrographs such as FLAMES@VLT and FORS@VLT at ESO; OCTOPUS@4m at CTIO; AAOmega at the AAT; GMOS at Gemini; and DEIMOS at Keck partially filled this gap. The large sample of low- and medium-resolution spectra that have been collected provided fundamental insights not only on abundance peculiarities in evolved cluster stars [2, 12], but also on the radial velocity distribution of evolved stars in several dwarf galaxies of the LG [4, 3, 5].

In this context, the dSph Carina is a cornerstone because of its vicinity to the Galaxy, the discontinuous star formation history and the mix of stellar populations hosted. Moreover, the wide sky area covered and the low central density make it ideal target to fully exploit the multiplicity capabilities of multi-object spectrographs.

A few years ago we undertook a long-term photometric and spectroscopic project on Carina. As far as the spectroscopy is concerned we have already collected a sizable sample (880) of low-resolution spectra in good seeing conditions with FORS2@VLT. The targets are distributed across the Carina center. Together with these data we also retrieved a large sample of medium-resolution spectra collected with FLAMES and available in the ESO archive. In the following we discuss pros and cons of these two spectrographs, and in particular several thorny problems that might affect not only the accuracy but also the precision.

2 Data Samples

Five fields were collected using GIRAFFE with the grism LR8 centered on the Calcium triplet (CaT) and ranging from 8200 to 9400 Å. These data have already been independently reduced and discussed by [4] and by [8]. The FORS2 spectra were secured with the blue grism 1400 V in the range from 4560 to 5860 Å. With this instrument we also adopted to collect a few sets of spectra with the grism 1028z centered on CaT and ranging from 7730 to 9300 Å [9]. The medium resolution GIRAFFE spectra cover a substantial fraction of the body of Carina, while the low-resolution FORS2 spectra have been collected in the very center and across the core radius.

Data plotted in Fig. 1 show that the bulk of FORS2 targets are either intermediate-mass red clump (RC) stars or low-mass horizontal branch (HB) stars. Moreover, we also collected spectra for a limited sample of hot stars located along the blue plume identified by [7]. In order to compare both radial velocities and abundances based on low-resolution spectra with medium and high resolution spectra a few relatively bright red giants have also been included in the target list.

The intrinsic accuracy of both radial velocity and abundance measurements of this data set is crucial to accomplish the scientific goals of this experiment. This means that the observational and the data reduction strategies were carefully planned. The FORS2 spectra were reduced using standard Iraf *apextract* and *longslit* tasks, and in particular using day time associated bias frames, flat field and calibration lamps. The entire set of spectra were interactively extracted using the *apall* task, during which the sky was also interactively set.

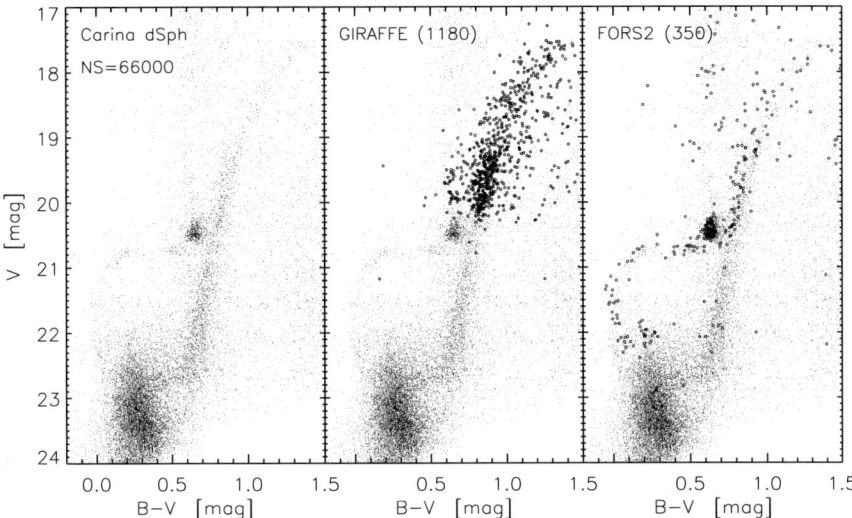

Fig. 1. *Left*: Optical $(V, B - V)$ Color-Magnitude Diagram of Carina. *Middle*: distribution in the CMD of the GIRAFFE targets (1180). *Right*: distribution in the CMD of FORS2 targets (350)

The wavelength calibration was performed using daytime lamp frames. However, we found systematic shifts of the order of 1 pixel (≈ 0.6 Å) between the position of the night-time and in the day-time mask. The systematic shift in wavelength was corrected using the spectrum extracted by the *apall* task, but without performing the sky subtraction. This procedure gives a spectrum with the skyline at 5577.347 Å and all the spectra were shifted in wavelength until this line attained the correct wavelength. We found that the shifts in wavelength were quite similar across the two chips and hence spectra of individual pointings were corrected by adopting a linear correction, which was different for every mask and exposure.

The formal solution in the wavelength fit was ≤ 0.02 Å, corresponding to an intrinsic accuracy of individual spectra of $\approx 1.5 \, \mathrm{km \, s^{-1}}$. Therefore, the precision of the Radial Velocity (RV) measurements based on the corrected co-added spectra is typically better than $2 \, \mathrm{km \, s^{-1}}$. On the other hand, individual uncorrected spectra provided RV measurements affected by a spread of the order of 20–$30 \, \mathrm{km \, s^{-1}}$, due to the small shifts in the mask positioning between the night- and the day-time masks [9]. Once this effect is properly taken into account for each individual mask, the lines in the coadded spectrum are sharper and centrally peaked when compared with the uncorrected spectra.

2.1 FLAMES Spectra

The Giraffe raw data were retrieved from the ESO Science Archive. These data were also reduced using Iraf. After the standard bias and flat correction, the night *attached* flats were used to extract the traces with *apall*. These traces were then used to extract the science spectra. The wavelength calibration was obtained from daily calibration lamps: the formal solution rms was ≤ 0.02 Å, corresponding to intrinsic accuracy in RV better than $1\,\mathrm{km\,s^{-1}}$. All the spectra were visually inspected in order to check skies and remove bad spectra. After an initial wavelength calibration, for each setup/exposure, all skies were cross-correlated, using the Iraf package *rvsao 2.4.3*, with the first one, and shifted accordingly. We selected from [10] 13 skylines, in the wavelength range 8300–8800 Å, i.e. the range we decide to optimize in the reduction of the GIRAFFE spectra. Using the task *continuum*, we subtract a fitted continuum to all spectra. This was done to allow the subsequent automatic (*splot*) evaluation with a gaussian fit of the integrated intensity of the selected sky lines for both the spectra and the skies. Cross correlation among the co-added skies showed that the relative accuracy in wavelength calibration in RV measurements is better than $1\,\mathrm{km\,s^{-1}}$.

Figure 2 shows the coadded spectra of three RGB stars spanning the whole magnitude range of the sample. In order of decreasing magnitude the total exposure times are 20, 10 and 11 hr, and the S/N at $\lambda \simeq 8600$ Å are $\approx$150, 50 and 20, respectively. Figure 3 shows the comparison between the FLAMES (top panel) and FORS spectra (bottom) for a bright star. The S/N ratio is high in both cases, $\approx$100 and 70, respectively, but it is worth noticing that

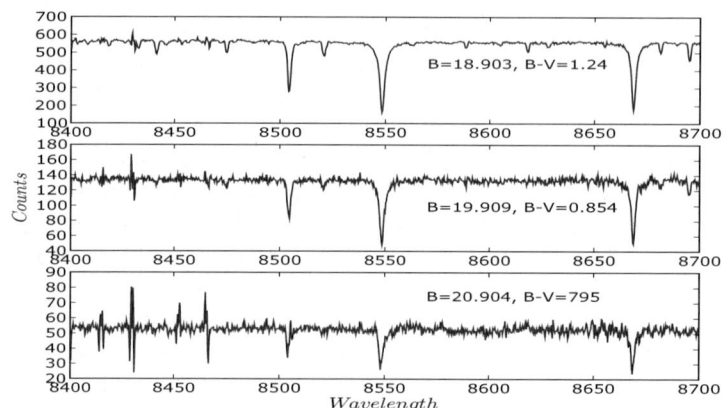

Fig. 2. FLAMES spectra of 3 three RGB Carina stars ranging from the region around the Tip (*top panel*) to the faintest stars of the sample (*bottom*), with magnitude slightly brighter than the intermediate-age RC ($V \approx 21$) stars. The S/N are $\approx$ 150, 50 and 20, respectively

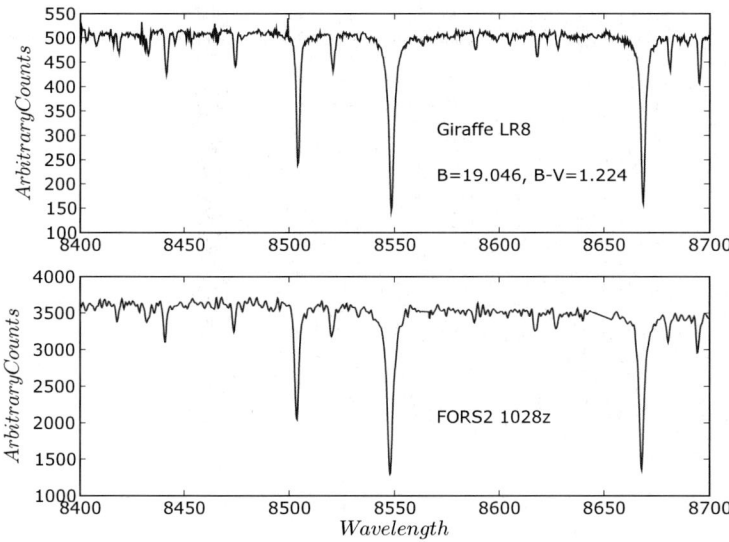

Fig. 3. Comparison between FLAMES and FORS2 spectra for a bright RGB star. The S/N ratios obtained with the two instruments are similar, but the exposure times are considerably longer for FLAMES

the exposure time are drastically different: 15 hr for the FLAMES spectrum, 0.4 hr for the FORS2 one.

To verify the accuracy of our measurements we performed a comparison with data available from the literature. Figure 4 shows the difference between both our dataset with data published by [6] and [8]. Radial velocities measurements for 23 bright RGB stars by [6] were based on high resolution spectra collected with the 2.5 m DuPont telescope, while [8] measurements rely on both medium-resolution MIKE spectra and the same FLAMES archive data. Interestingly enough, both FORS2 and FLAMES measurements are in excellent agreement with radial velocity measurements by [6] (within 2%), but there is evidence of a small systematic difference with [8] ($\approx 3\%$).

3 Conclusions

We have presented preliminary results concerning an extensive radial velocity investigation on Red Giant and central He-burning structures (RC, HB) in the Carina dSph. Large samples of low- and medium- resolution spectra have been collected with FLAMES@VLT and with FORS2@VLT. In order to limit subtle systematic uncertainties in the radial velocity measurements we devised specific reduction strategies for the two different spectrographs.

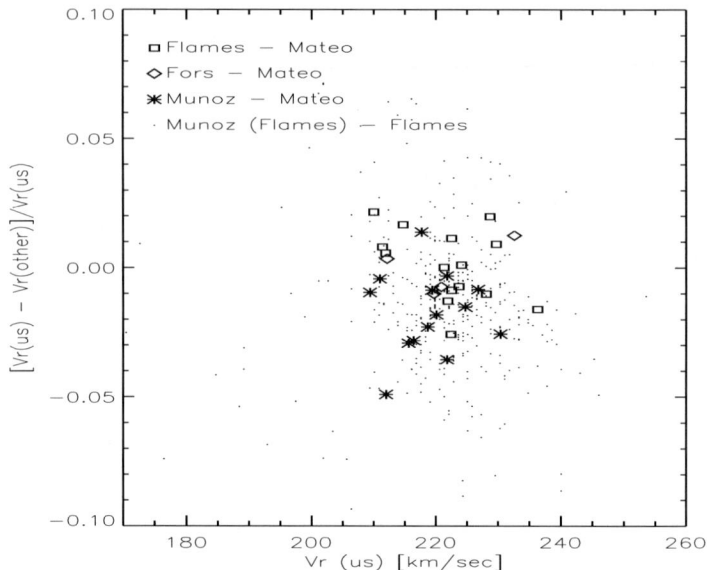

Fig. 4. The comparison with data from the literature discloses good agreement with [6], but there is a small shift with the measurements by [8]

As a whole, we measured the radial velocity for more than 1450 stars and the intrinsic accuracy of individual radial velocity measurements is on average of the order of 1–2 km s^{-1}, and in good agreement with data from the literature.

References

1. V. Castellani, M. Castellani, P. Prada Moroni: A&A **457**, 569 (2006)
2. E. Carretta, et al.: AJ **131**, 1766 (2006)
3. M.G. Coleman, G.S. Da Costa, J. Bland-Hawthorn: AJ **130**, 1065 (2005)
4. A. Koch, et al.: AJ **131**, 895 (2006)
5. S. Majewski, et al.: AJ **130**, 2677 (2005)
6. M. Mateo, et al.: AJ **105**, 510 (1993)
7. M. Monelli, et al.: AJ **126**, 218 (2003)
8. R.R. Munoz, et al.: ApJ **649**, 201 (2006)
9. M. Nonino, et al.: ASP **364**, 295 (2007)
10. D.E. Osterbrock, et al.: PASP **108**, 2770 (1996)
11. K. Pollard, et al.: ApJ **628**, 729 (2005)
12. P. Shetrone, et al.: ApJ **585**, 45 (2003)

Discussion

A. Smette: Have you checked the accuracy of the wavelength calibration using sky lines?

M. Monelli: No, I haven't.

A. Smette: Sky spectra are used more and more in the IR. It can also be used in the optical. We are checking with CRIRES how stable the velocity of sky lines is. Certainly better than 100 m/s, which is sufficient for your purpose.

G. Marconi: Did you try to apply the sky line correction to the common stars in the 2 samples of FORS2/FLAMES?

M. Monelli: Not, yet.

The VLT-FLAMES Survey of Massive Stars: Instrumental Stability and Detection of Massive Binaries

C. J. Evans[1], D. J. Lennon[2,3], I. Hunter[2,4], T. Augusteijn[5], and S. J. Smartt[4]

[1] UK ATC, Royal Observatory, Blackford Hill, EH9 3HJ Edinburgh, UK; cje@roe.ac.uk
[2] Isaac Newton Group, Apart. de Correos 321, E-38700, S/Cruz de La Palma, Canary Islands, Spain
[3] Dept. de Astrofísica, Universidad de La Laguna, Avda. Astrofísico Francisco Sánchez, s/n, E-38071 La Laguna, Spain
[4] Dept. of Physics & Astronomy, Queen's University Belfast, Belfast BT7 1NN, Northern Ireland, UK
[5] Nordic Optical Telescope, Apart. de Correos 474, E-38700, S/Cruz de La Palma, Canary Islands, Spain

Abstract. We introduce our VLT-FLAMES Large Programme that has observed ∼800 early-type stars in the Galaxy and the Magellanic Clouds. The service-mode data sample a wide range of observational epochs, revealing a large number of massive binaries. In the course of disentangling the kinematics of these systems, cross-correlation techniques reveal the excellent wavelength-stability of the FLAMES spectra (± 2 km s^{-1}). This stability enables well-constrained orbital parameters to be found.

1 Introduction

Massive stars are the principal source of chemical enrichment and kinetic energy in star-forming galaxies, both via their stellar winds and as supernova explosions. The VLT-FLAMES Survey of Massive Stars is an ESO Large Programme to quantify the effects of metallicity and environment on the evolution of massive, luminous stars [1, 2]. The survey comprises 7 FLAMES fields, centered on open clusters in the Milky Way and Magellanic Clouds, as summarised in Table 1.

The observations were undertaken in service mode, using six of the standard high-resolution Giraffe settings. To yield a useful signal-to-noise ratio in the final spectra, each of the wavelength settings was observed a minimum of six times for the LMC and SMC fields. Although not one of the primary objectives of the programme, this provided reasonable time-sampling for the detection of massive binaries, finding lower limits to the binary fraction in the young cluster fields of ∼30%.

Table 1. Summary of VLT-FLAMES fields

	'Young' Clusters (<5 Myrs)	'Old' Clusters (10–20 Myr)
Milky Way	NGC 6611	NGC 3293 & 4755
LMC	N11	NGC 2004
SMC	NGC 346	NGC 330

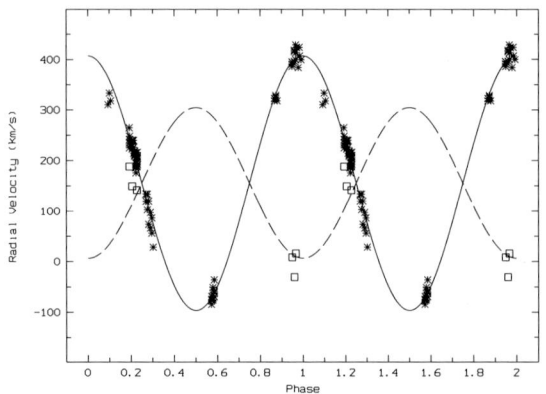

Fig. 1. Phase plot for star NGC346-013. This object is a peculiar double-lined binary with an approximately B1-type primary. The period is 4.2 days

2 Binary Parameters

The high-resolution FLAMES spectra are of sufficient quality to investigate the nature of many of the newly discovered systems. Orbital periods and mass ratios are found for both single- and double-lined systems, e.g. the two components in NGC 346-013 can be disentangled as shown in Fig. 1. Specific systems will be the subject of forthcoming papers, and we are seeking further monitoring of the fields to better constrain the binary fraction.

3 Instrumental Stability

The interstellar Calcium K line allows a check on the wavelength stability of the Giraffe spectra within each field, and with time. Figure 2 shows the relative velocities of Ca K for the targets in our NGC 2004 FLAMES field. Figure 3 shows the relative Ca K velocities for a selected star from each field. The precision obtained in both cases is typically ± 2 km s^{-1}. A more thorough examination of this aspect of the data (such as the contribution to the uncertainties by noise) is now underway.

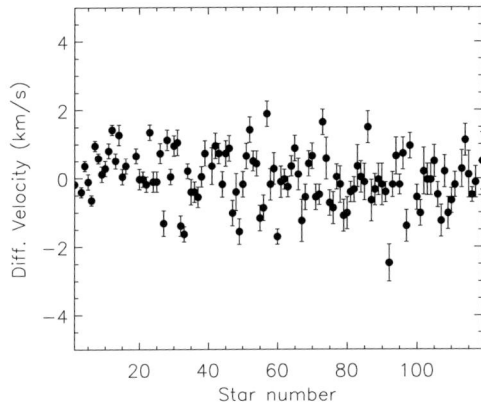

Fig. 2. Relative velocities of the Ca K line for the FLAMES targets in the NGC 2004 field

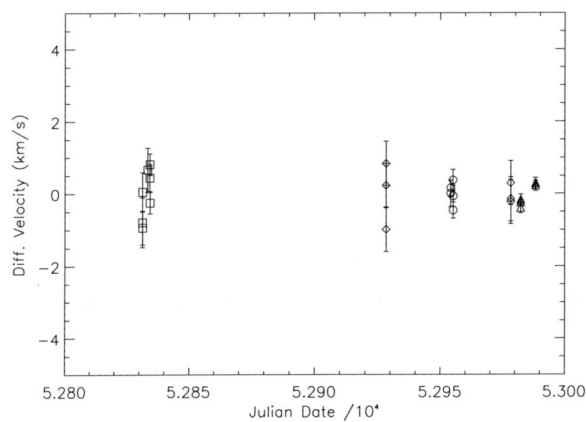

Fig. 3. Relative offsets from mean Ca K velocities for one star in each of our fields: N11 (*diamonds*), NGC 2004 (*triangles*), NGC 330 (*squares*), NGC 346 (*circles*)

References

1. C.J. Evans, S.J. Smartt, J.-K. Lee, et al.: A&A **437**, 467 (2005)
2. C.J. Evans, D.J. Lennon, S.J. Smartt, C. Trundle: A&A **456**, 623 (2006)

VIMOS Total Transmission Profiles for Broad-Band Filters

S. Mieske[1], M. Rejkuba[1], S. Bagnulo[2], C. Izzo[1], and G. Marconi[2]

[1] ESO, Karl-Schwarzschild-Strasse 2, 85748 Garching, Germany;
smieske@eso.org
[2] ESO, Alonso de Córdova 3107, Vitacura, Casilla 19001, Santiago 19, Chile

Abstract. VIMOS (http://www.eso.org/instruments/vimos) is a wide-field imager and spectrograph mounted on UT3 at the VLT, whose FOV consists of four 7'x8' quadrants. Here we present the measurements of total transmission profiles – i.e. the throughput of telescope + instrument – for the broad band filters U, B, V, R, I, and z for each of its four quadrants. The results are compared with estimates from the VIMOS consortium.

1 Transmission Profiles of Telescope + Instrument

What has been available to ESO up to now are efficiency curves provided by the VIMOS consortium that are averaged over the four quadrants and were determined with laboratory measurements, not by directly observing standard stars. Prompted by user requests to have separate efficiency estimates for each quadrant, two spectro-photometric standard stars were observed in 2006: Hiltner 600 for all filters, and in addition LTT7379 for the U and B band. These observations were performed with the broadest possible slit (5") and the grisms LR-blue (for U, B, and V) and LR-red (for R, I, and z). Under the assumption that the total star-light is contained in the slit, the efficiency curve is derived from the measured flux. The required input consists of the tabulated standard star flux, the extinction curve, and the grism transmission as provided by the VIMOS consortium. The resulting curves are shown in Fig. 1 (they are available on the VIMOS web-page).

There is good agreement between the newly derived efficiency estimates and those from the consortium. The substantial sensitivity drop in quadrant 3 for the U-band is found independently for both LTT 7379 and Hiltner 600 (note that the measurement for Hiltner 600 was not used for the efficiency estimate in the U-band since it suffered from flux loss in all four quadrants, due to a slight misalignment of the standard within the slit). Since the U-band photometric zero-points for Q3 do not show such a strong drop, a centering problem due to instrument flexure is a possible reason for this lack of measured flux. In addition, the consortium's estimate of the grism transmission in Q3 for $\lambda <3800$ Å may be too high. We suggest that the efficiency estimates especially in the U-band be always re-scaled to the (integrated) photometric zero-points.

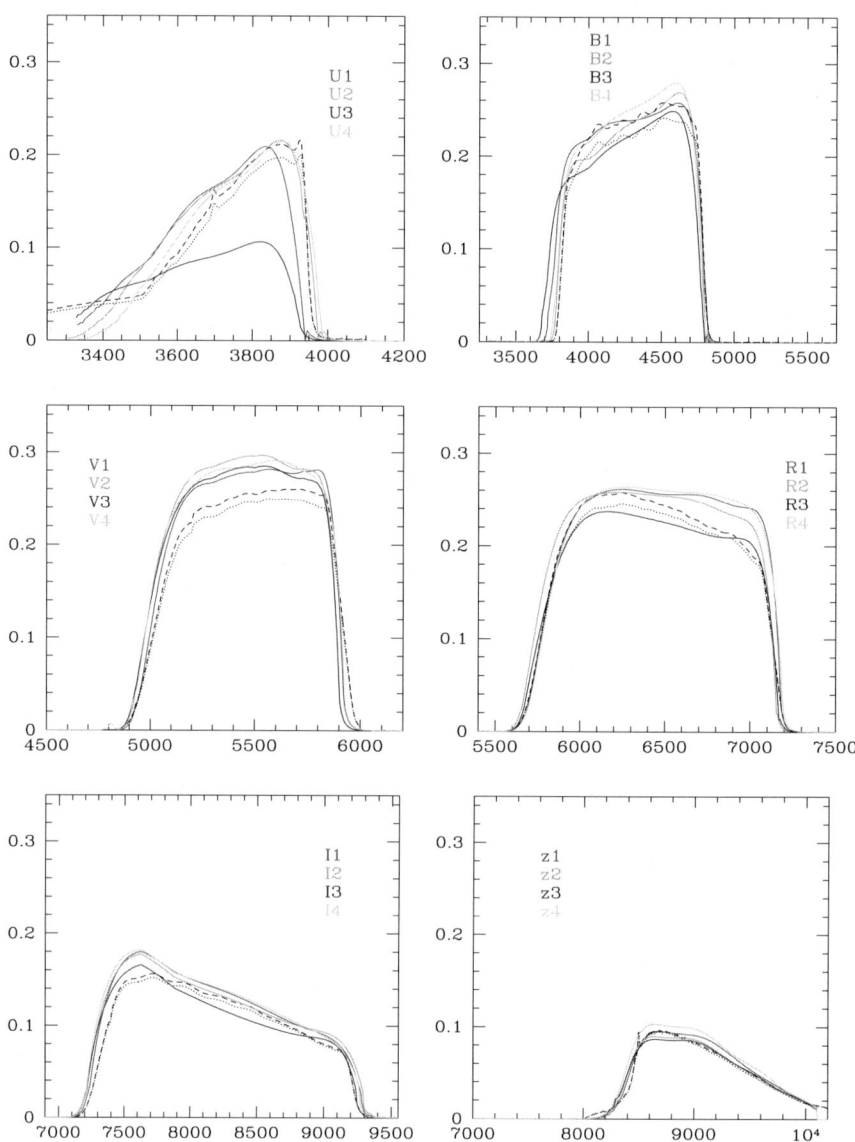

Fig. 1. Total VIMOS telescope+instrument efficiency curves for the broad band filters U, B, V, R, I, and z. X-axis is in units of Ångstrom. The *dotted* and *dashed lines* are the efficiency curves available from the VIMOS Exposure Time Calculator on the web, assuming the La Silla (*dotted*) and CTIO (*dashed*) extinction curve. The U-band estimate is based on the star LTT 7379. The B-band estimate is the mean of the efficiency derived from Hiltner 600 and LTT 7379. The VRIz estimates are based on observations of Hiltner 600

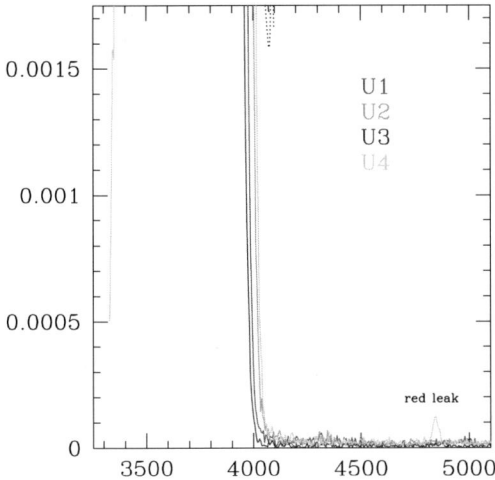

Fig. 2. Detection of a very small "red leak" in the U-band efficiency curve of Q4. x-axis is in units of Ångstrom U1 to U4 denote different quadrants. Note that the y-axis has been zoomed in by a factor of 200 compared to Fig. 1. The peak of the red leak in U4 at 4850Å is about 0.1% of the efficiency peak

2 Transmission Profiles of Broad-Band Filters

In addition to the total instrument+telescope efficiency estimates, we used the continuum lamp screen flats taken with and without inserted filter to measure the filter transmissions for each quadrant. The results (dotted lines) are shown in Fig. 3 and compared with the consortium estimates (solid lines). There is very good agreement for the V,R,I, and z filters. For the U-band, the consortium estimates are about 20% above those derived from the screen flats. Also for the B-band, the consortium estimates are about 5–10% higher.

Unlike in the consortium filter transmission curves from Fig. 3, we do *not* detect a pronounced red leak of the U-band filter around 4200 Å. Only in quadrant 4, there is a very minor leak at about 4850 Å (see Fig. 2).

3 Conclusions

The measured telescope+instrument efficiency curves for the VIMOS broad band filters agree very well with the consortium estimates. A discrepancy in the U-band is observed for one quadrant, possibly due to flexure and low grism transmission at short wavelengths. The filter transmissions also agree very well, except for the B and especially U-band, where the newly measured transmission is about 20% lower.

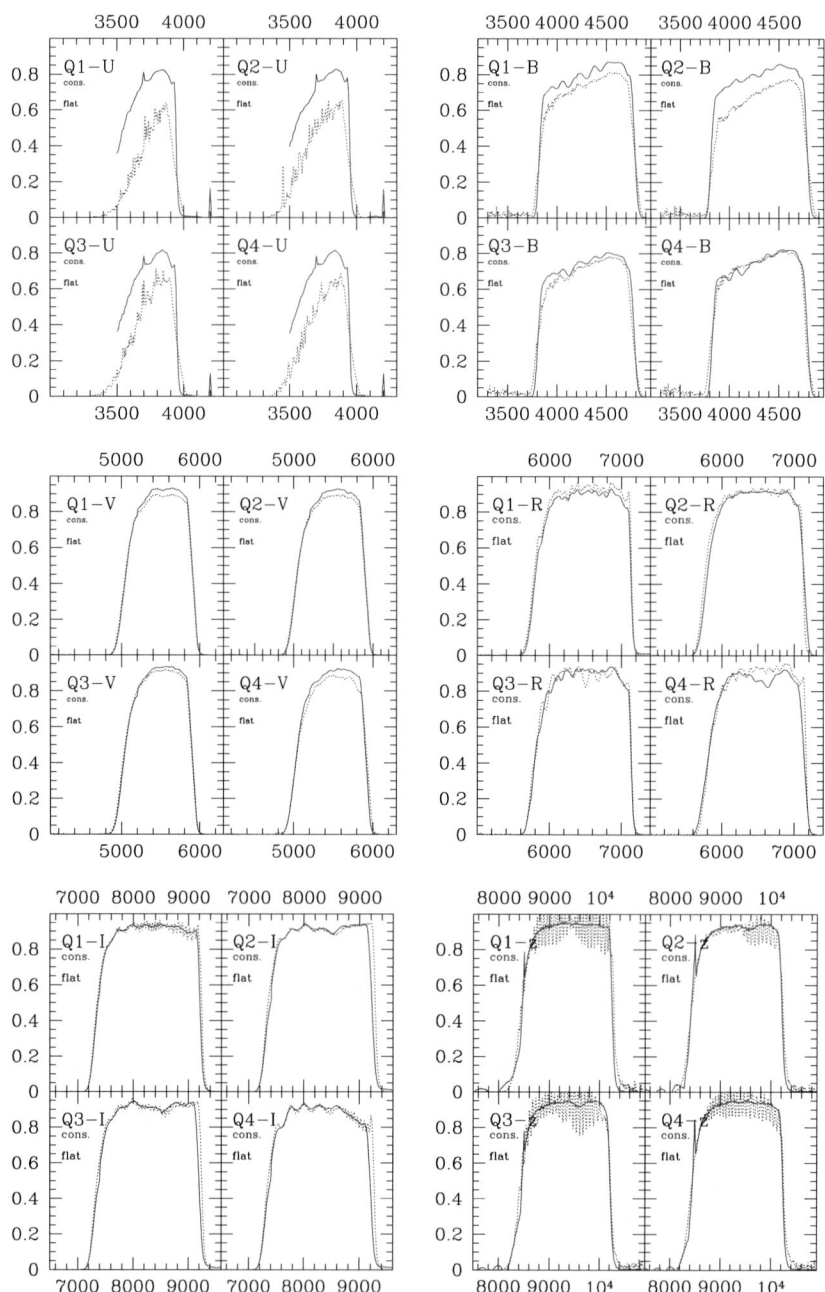

Fig. 3. *Dotted* (blue) lines indicate filter transmission profiles for the VIMOS broad band filters as derived from continuum lamp exposures taken with and without the filters inserted. *Solid* (red) lines indicate the transmission estimates provided by the VIMOS consortium. X-axis is in units of Ångstrom

Laboratory Measurements of Calibration Sources for X-Shooter

F. Saitta[1], F. Kerber[1], P. Bristow[1], H. Dekker[1], S. D'Odorico[1], C. Dupuy[1], J.-L. Lizon[1], A. Norup Sorensen[2], and J. Vernet[1]

[1] ESO, Karl-Schwarzschild-Strasse 2, 85748 Garching, Germany;
fsaitta@eso.org
[2] University Observatory, Juliane Maries Vej 30, DK-2100 Copenhagen, Denmark

1 Introduction – X-Shooter

X-shooter [1] is a single target spectrograph for the Cassegrain focus of one of the unit telescopes of the Very Large Telescope (VLT) covering in a single exposure the spectral range from the UV to the K band (320–2500 nm). It is designed to maximize the sensitivity in this spectral range by splitting incoming light in three arms (UVB, VIS and NIR) with optimized optics, coatings, dispersive elements and detectors. X-shooter will be a unique instrument on 8 m class telescopes in that it is capable of recording – over such a large wavelength range – the spectrum of an astronomical target in a single exposure. The instrument is currently under construction and commissioning is scheduled for 2008. The main characteristics of X-shooter are listed below:

Spectral format	Prism cross-dispersed echelle (order separation $\geq$ 12")
Wavelength range	320–2500 nm, split in 3 arms by dichroics
Resolution	4000–7000 for 1" slit
Slit configuration	long slit ($\sim$ 12"); widths: 1" (standard), 0.6" (high resolution), 5" (flux calibration); IFU 1.8" x 4" input area
Detectors	2k x 4k CCDs (UVB and VIS arms), 2k x 2k (1k x 2k used), Hawaii LPE MCT (NIR)
Auxiliary functions	Calibration Unit; acquisition and guide unit with 1' x 1' field and filter set; ADC for the UVB and VIS arms.

2 Wavelength Calibration for X-Shooter

Excellent wavelength calibration across all three arms is essential in order to realize the scientific potential of X-shooter. For the UVB and VIS arms, Th–Ar hollow cathode lamps (HCLs) have been chosen as calibration sources following the successful operations of such lamps in e.g. FEROS, SINFONI, UVES and HARPS. For the NIR arm several options are under study: a

Th–Ar HCL or a combination of pen ray lamps with a noble fill gas: Ne, Ar, Kr, or Xe. In the X-shooter calibration unit the individual lamps are going to be used in parallel and their light is going to be combined by means of an integrating sphere. For X-shooter we are using ESO's laboratory Fourier Transform Spectrometer (FTS), see Fig. 1, to study the spectra and other relevant properties of these lamps to then select the best combination of calibration sources for the near-IR wavelength region which traditionally has relied on atmospheric features for wavelength calibration. Recently, ESO has gained significant experience with NIR wavelength standards as part of the Cryogenic IR Echelle Spectrometer (CRIRES) [4] project (see Kerber et al., these proceedings).

2.1 Calibration Sources

Pen ray lamps are called "Pencil" lamps because of their size and shape. They are made of double bore quartz tubing with two electrodes at one end sealed into a handle. The lamps produce narrow, intense lines from the excitation of various rare gases and metal vapors. They are commercial products widely used for wavelength calibration of spectroscopic instruments such as monochromators, spectrographs, and spectral radiometers.

Modern commercial **hollow cathode lamps** are sealed-off glass tubes that contain a metal cathode, a metal anode and a fill gas at a defined pressure. They emit a rich spectrum of narrow emission lines from both the gas and metal atoms and ions in the plasma. The Th spectrum was studied from 278 nm to about 1000 nm at high resolution more than 20 years ago [6]. Two valuable studies [3, 2] of the Th–Ar spectrum in the near IR have

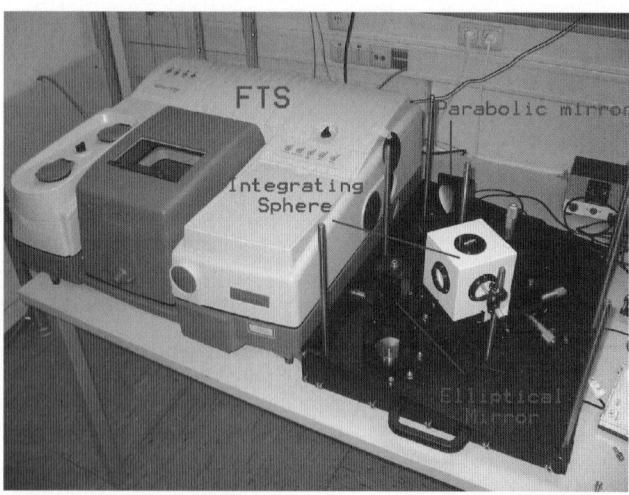

Fig. 1. FTS and permanent set-up built to measure external light sources

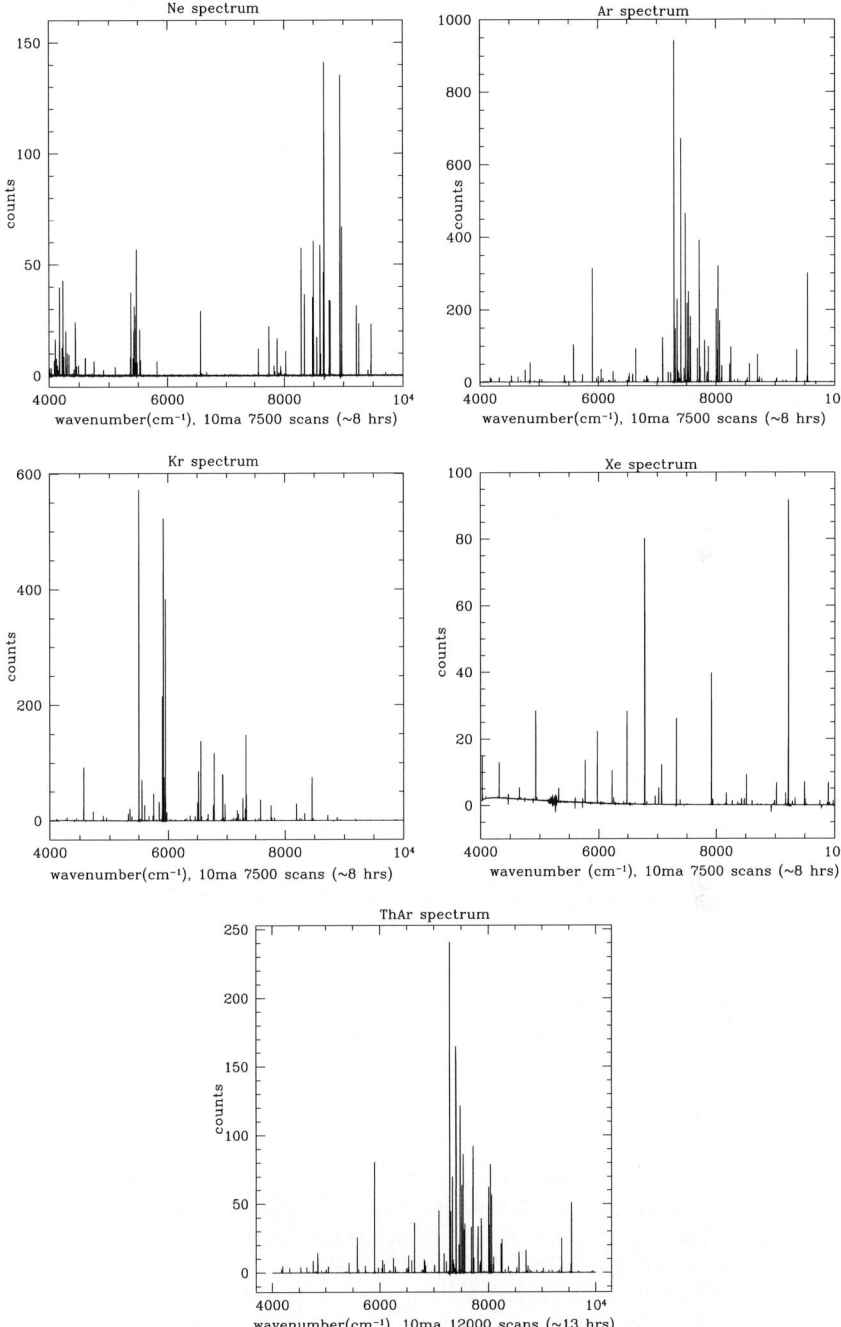

Fig. 2. Deep spectra of four pen ray lamps and a Th–Ar HCL obtained with ESO's FTS. Spectral range (1000–2500 nm) and resolution correspond to the X-shooter NIR arm

recently been published, but neither is directly applicable to the operation of X-shooter. Low current Th–Ar HCLs have extensively been studied by a collaboration [5] of ESO and the US Institute of Standards and Technology (NIST) for CRIRES (Kerber et al., these proceedings). X-shooter will benefit directly from this expertise.

3 First Results and Next Steps

Our FTS has a permanent set-up which replicates part of its optical train allowing us to analyze light from external sources (Fig. 1). One main goal of the study is to find out what combination of the four pen ray lamps will give a composite spectrum best suited for calibration of X-shooter in terms of line density and dynamic range. For this we need to establish relative intensities of the lines and the lamps relative to each other.

For each lamp we have taken a very deep exposure of about 8 h at an operating current of 10 mA in order to obtain a high S/N spectrum (Fig. 2). Using literature data we have identified the following number of lines in the range 4000–10,000 cm^{-1} (1000–2500 nm) in the deep spectra: Ne 80 lines, Ar 155 lines, Kr 130 lines, Xe 65 lines; while the overall relative output of Ne/Ar/Kr/Xe has intensity ratios of 2/3.5/1.7/1. A preliminary analysis suggests that a combination of Ne, Ar and Kr will provide a similar number of lines ($\approx$ 350) as the Th–Ar HCL while providing a comparable or better coverage for a given dynamic range. In the immediate future we will combine the laboratory data with simulations done with a physical instrument model (Bristow, et al., these proceedings). With this refined analysis we will get quantitative results for the line density and line ratios for the Th–Ar and pen ray lamps per echelle order. After a detailed investigation of line blending we will then establish the best combination of calibration lamps for X-shooter.

Acknowledgement. We thank G. Nave and C. Sansonetti (NIST) for providing line data prior to publication. We gratefully acknowledge the support from the X-shooter and CRIRES projects.

References

1. S. D'Odorico, H. Dekker, R. Mazzoleni, et al.: X-shooter, UV- to K band intermediate-resolution high-efficiency spectrograph for the VLT: status report at the final design review. In: *Proceedings of the SPIE 6269*, part two, ed by I.S. McLean, M. Iye, SPIE (2006) pp 33-1–33-10
2. R. Engleman Jr., K.H. Hinkle, L. Wallace: JQSRT **78**, 1 (2003)
3. K.H. Hinkle, R.R Joyce, A. Hedden, et al.: PASP **113**, 548 (2001)
4. H.-U. Käufl, P. Ballester, P. Biereichel, et al.: CRIRES: a high-resolution infrared spectrograph for ESO's VLT. In: *Proceedings of the SPIE 5492*, ed by A.F.M. Moorwood, M. Iye, SPIE (2004) pp 1218–1227

5. F. Kerber, G. Nave, C.J. Sansonetti, et al.: The spectrum of Th-Ar hollow cathode lamps in the 900–4500 nm region: establishing wavelength standards for the calibration of VLT Spectrographs. In: *Proceedings of the SPIE 6269*, ed by I.S. McLean, M. Iye, SPIE (2006) pp 2O-1–2O-11
6. B.A. Palmer, R. Engleman Jr.: Los Alamos Report, LA 9615, USA (1983)

Modelling the Fringing of the FORS2 CCD

J. R. Walsh[1], H. Kuntschner[1], E. Jehin[2], A. Kaufer[2], K. O'Brien[2], M. Riquelme[2], and A. Smette[2]

[1] ESA, Space Telescope European Co-ordinating Facility,
 Karl-Schwarzschild-Strasse 2, 85748 Garching, Germany; jwalsh@eso.org,
[2] ESO, Alonso de Córdova 3107, Vitacura, Santiago, Chile

Abstract. Thinned CCD detectors display fringing which arises from the interference of multiply reflected light in the layers of the CCD. If the layer construction – the thicknesses and refractive indexes of the layers – is known, then the observed fringing can be accurately modelled and used to correct imaging and spectroscopic data for its effects. In practice the specifications on the actual deposited layer thicknesses may not be known to sufficient accuracy to predict the fringe behaviour. Thus calibration data, in the form of monochromatic flat fields, is required and can be modelled using the technique outlined by Malamuth et al. 2003, which has been applied to ACS CCDs.

Initial tests to model the observed fringing of the FORS2 MIT CCD are described. A set of six monochromatic flat fields was generated by shining light from a tunable monochromator into the FORS2 instrument calibration unit. Peak-to-peak fringe amplitude of 7–8% was measured at a wavelength of 970 nm. In order to model the CCD structure, a series of monochromatic flats at closely spaced wavelengths is required, which would have been very time-consuming to acquire with the monochromator. An alternative strategy is outlined. The fringing characteristics of CCDs should be systematically calibrated while still in the test laboratory.

1 Observational Setup

A commercial monochromator was used to feed monochromatic light into the FORS2 calibration unit using a fibre. The slit of the monochromator was set to 250 µm, which produced a beam of width slightly less than 1.0 nm. Given the wavelength period of the fringing of around 3 nm, this provides satisfactory sampling of the fringes. Table 1 provides brief details of the observations.

In order to determine the actual wavelength of the monochromatic light illuminating the detector, each monochromatic image was accompanied by an exposure with the 600z+23 grism, with OG590 order sorting filter, and a 0.3 arcsec slit. Wavelengths were simply determined by linear interpolation, using a linear dispersion calculated from the positions of the 885.387 and 865.438 nm lines on an arc lamp exposure (dispersion determined as 0.0803 nm/pix).

Table 1. Observational details

Meas. λ (nm)	Meas. FWHM (nm)	Exp. Time (s)	Fringe Pk-to-Pk (%)
774.0	0.84	60	
876.0	0.75	1800	2.2
906.0	0.65	1800	3.0
926.1	0.60	1800	5.1
956.1	0.75	1800	7.0
986.0	0.75	1800	7.5

2 Reduction

The monochromatic flats were simply reduced by subtracting a constant value of 185 for the bias value and normalised by dividing by a 2D Gaussian smoothed (σ=30 pix) version of the bias-subtracted image. The smoothing σ was determined by examining the rms of a selected region of the chip. Too low values of σ fit the fringe scale (around 40 pixels peak-to-peak for the closest spaced fringes), while large values of σ would not remove the large-scale illumination pattern, which was peaked to the lower right of Chip 2 (slave). Figure 1 shows four images, normalised to unity by the smoothed version and hereafter referred to as fringe flats. The flat at 774 nm had low signal-to-noise and no fringes were visible, so it is not shown.

3 Results

An indication of the peak-to-peak amplitude of the fringing is listed in Table 1, measured in a region where the fringing is highest (around pixel 1300,1300 on the master chip). Whilst the fringing increases considerably from 900 to 960 nm, the increase levels off to higher wavelength. Figure 2 shows the fringe contrast in two locations on the master and slave chips, again chosen to have high fringe amplitude. The crosses in Fig. 2 derive from the positions selected to have a high value in the 956 nm fringe map, whilst the triangles are from a nearby position (in the adjacent valley on the 956.0 nm image).

Shown in Fig. 2 (right) is a possible set of fringes. These have been derived by taking the wavelength period of the fringing to be 2.9 nm, as measured from some 300I flats taken in the ESO GOODS programme, and approximating the fringing by a simple sine function. An envelope of the fringe amplitude with wavelength, which would be produced by a thinner layer deposited on the Silicon, is approximated by a $\sin^2$ function. This is purely illustrative; there is not sufficient data at closely spaced wavelengths to determine if the period of the fringing varies across the detectors or if the envelope of the fringe amplitude is more complex (see for example Malamuth [1] for STIS and Walsh et al. [2] for ACS detectors).

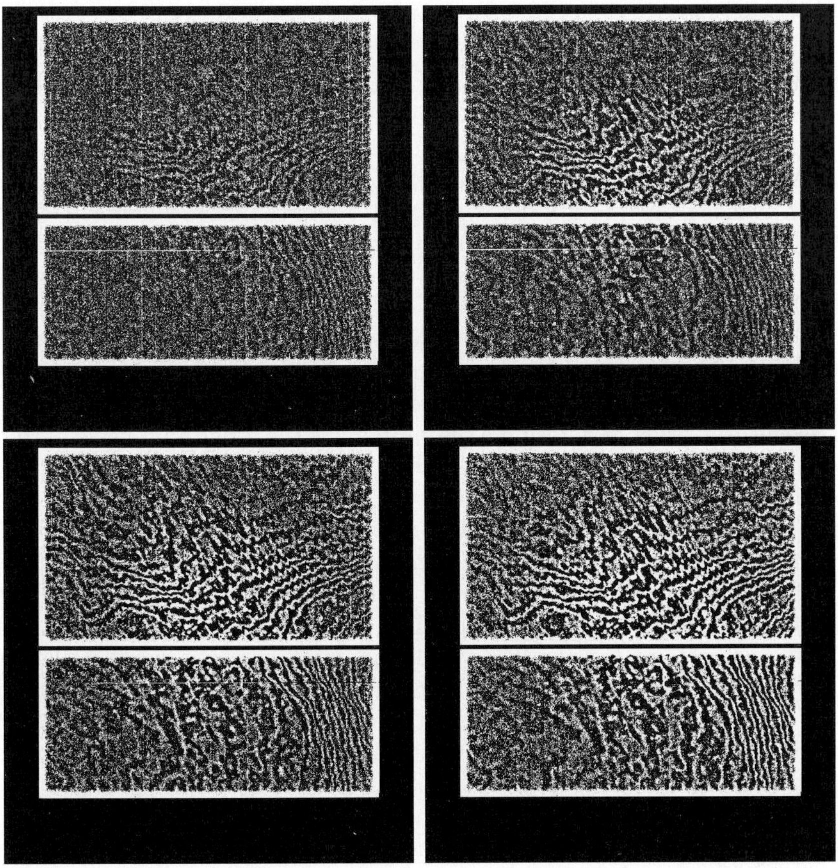

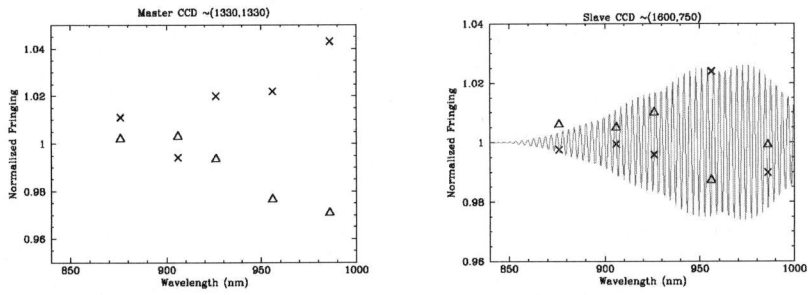

Fig. 1. Set of FORS2 fringe flats at wavelengths of 876, 906, 926, 956 nm

Fig. 2. Fringe contrast at two positions on the FORS master chip around pixel (1330, 133) are shown (*left*). Fringe contrast at two positions on the slave chip around pixel (1600, 1700) are shown (*right*). A possible set of fringes is shown for illustration (not a model)

4 Prospects

The fringe flats with the monochromator provide excellent data for modelling the fringing through fitting of the CCD layer thicknesses, given a knowledge of CCD composition and the refractive index of the layers (Malamuth et al. [1]; Walsh et al. [2]). Walsh et al. [2] modelled the layer structure of the HST Advanced Camera for Surveys (ACS) High Resolution Channel (HRC) CCD, manufactured by SiTe, using multiple monochromatic full field flats. In order to perform such modelling, it is necessary to have a closely-spaced sampling of wavelengths. For the case of the FORS2 CCD then in order to sample the fringe period adequately, monochromatic sampling about every 0.7 nm would be required, at least at a moderate number of wavelengths. This closely-spaced sampling can be supplemented by fringe flats at other isolated wavelengths in order to determine the envelope of the fringe amplitude. Given the typical exposures of 30 min which were required to reach counts of 5000–15000 electrons per unbinned pixel (lower to redder wavelengths) with the monochromator and FORS calibration unit, then another approach needs to be considered for detectors mounted on operational instruments.

The suggested approach would be to use a grism and the MOS single long slit mode spanning all the multi-slits, placing the slit at many positions over the detector to provide flat field illumination at many wavelengths for each pixel. The 600z grism would be well suited, or the 300I. The dome lamp could be used and a high signal could be quickly reached, allowing dense sampling of wavelengths over a few regions and sparser sampling of other wavelength regions for each pixel, in a relatively short time. With a choice of grisms, coverage of the wavelength region with highest fringing amplitude (950–1000 nm) could be achieved. A concern here is that the rail between the slitlets obscures some pixels. The obscuration between slitlets is at maximum 3 (2×2 binned) pixels. In the worst case no useful fringe information is available across these regions. When modelled, the detector layer thickness could be interpolated across these regions; probably the CCD layer thicknesses vary fairly smoothly, so the loss in fidelity may not be important.

A few full-field deep monochromatic flats could be taken in addition to the scanned slit data as a minimum check. There must be some concern that the different beam, entering the telescope from dome flats as compared to the astronomical objects (and sky), may alter the measured fringing behaviour. Some on-sky data should also be analysed. An out of focus image of a bright star could be trailed along the MOS slit at a number of slit positions as a check. Similar procedures for determining the fringing of the VIMOS CCD's could be considered using aligned slitlets or specially designed masks.

Ideally the set-ups to measure fringing could be performed in the CCD test laboratory away from the limitations of an operational observatory environment. Important here would be to simulate the input beam on the detector appropriate for the particular spectrometer.

References

1. E.M. Malamuth, R.S. Hill, T. Gull, et al.: Publ. Astron. Soc. Pacific **115**, 218 (2003)
2. J.R. Walsh, W. Freudling, N. Pirzkal, A. Pasquali: Modelling the fringing of the ACS WFC and HRC chips. ACS ISR 03-12 (2003)

Master Response Curves for Flux Calibration of VIMOS Spectroscopy

B. Wolff

ESO, Karl-Schwarzschild-Strasse 2, 85748 Garching, Germany; bwolff@eso.org

1 VIMOS Spectroscopic Modes

The VIMOS instrument [1] at the Nasmyth B focus of the VLT-Melipal telescope has four identical optical arms (quadrants); each is equipped with 6 different grisms of spectral resolution between 200 and 2500 (for 1″ slits). Two spectroscopic modes are offered: Multi Object Spectroscopy (MOS) using masks with up to 200 slits per quadrant and Integral Field Unit (IFU) spectroscopy with 6400 fibres (80×80).

Spectro-photometric standard stars are measured on a regular basis in both spectroscopic modes. They are usually observed near in time to science observations with the same grism setting. Standard star observations are used for monitoring of the spectral instrument efficiency and are also intended for providing response curves for flux calibration of scientific data. Limitations for the usage of response curves arise since standards are not always observed under optimum weather conditions (and not in the same conditions as the science data) and telescope scheduling requirements may prevent timely observation of standards.

2 Master Response Curves

The dependency on individual standard star measurements can be relaxed by using averages of response curves from a dedicated time interval for relative flux calibration. These averages are called master response curves.

Individual response curves $R(\lambda)$ have been derived with recipes of the VIMOS pipeline:

$$R(\lambda) = F_{\text{std}}(\lambda)/f_{\text{std}}(\lambda) \qquad (1)$$

where $F_{\text{std}}(\lambda)$ is the tabulated standard star flux in erg cm^{-2} s^{-1} Å^{-1} and $f_{\text{std}}(\lambda)$ is the observed flux corrected for airmass and extinction, i.e.

$$f_{\text{std}}(\lambda) = f_{\text{ins}}(\lambda) 10^{0.4\mu \Delta m(\lambda)} \qquad (2)$$

with $f_{\text{ins}}(\lambda)$ in e^- s^{-1} Å^{-1}, extinction $\Delta m(\lambda)$ in magnitudes and airmass μ.

Master response curves have to be created for each spectroscopic setting in MOS and IFU for each quadrant separately. Individual response curves measured during a dedicated period have been inspected, clearly deviating curves have been de-selected, and the remaining ones have been averaged to create a final master $R_{\mathrm{mst}}(\lambda)$ for this setting and quadrant.

3 Verification

The resulting master response curves have been verified by selecting a random set of standard star observations, applying the response $R_{\mathrm{mst}}(\lambda)$ to the extracted spectrum $f_{\mathrm{ins}}(\lambda)$

$$F_{\mathrm{obs}} = R_{\mathrm{mst}}(\lambda) f_{\mathrm{ins}}(\lambda) 10^{0.4 \mu \Delta m(\lambda)} \; , \qquad (3)$$

and comparison to the tabulated flux. The typical accuracy is 10% for relative flux calibration within an individual spectrum.

Figure 1 shows an example for MOS. The top panel illustrates the tabulated flux for the standard star CD $-32°9927$. The lower panel shows the pipeline-extracted flux from an observation on 31 March 2006 with the HR_blue grism. A flux calibration has not yet been applied. The extracted spectra in quadrants 1 and 3 differ significantly over a major part of the wavelength range.

Figure 2 shows the same observation with master response curves for MOS HR_blue applied. All four quadrants have now similar flux and spectral shape. The difference between quadrants 1 and 3 has disappeared.

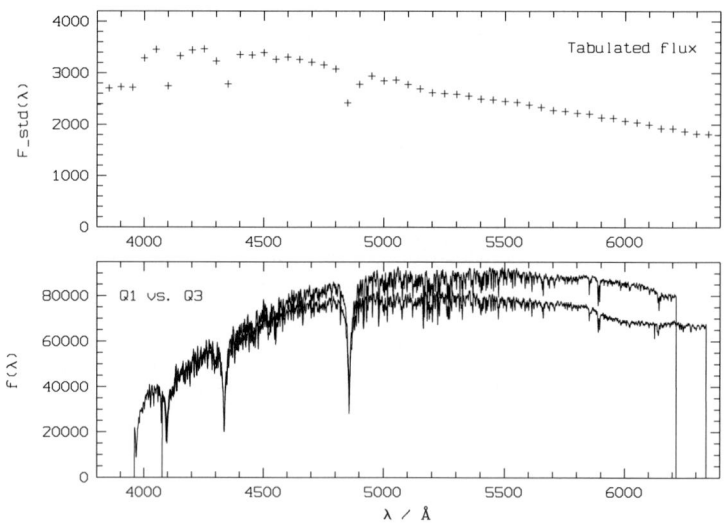

Fig. 1. Tabulated flux (in 10^{-16} erg cm^{-2} s^{-1} Å^{-1}) of the standard star CD $-32°9927$ (*top*) and pipeline-extracted spectra without flux calibration of a MOS observation with HR_blue grism from 31 March 2006 (*bottom*)

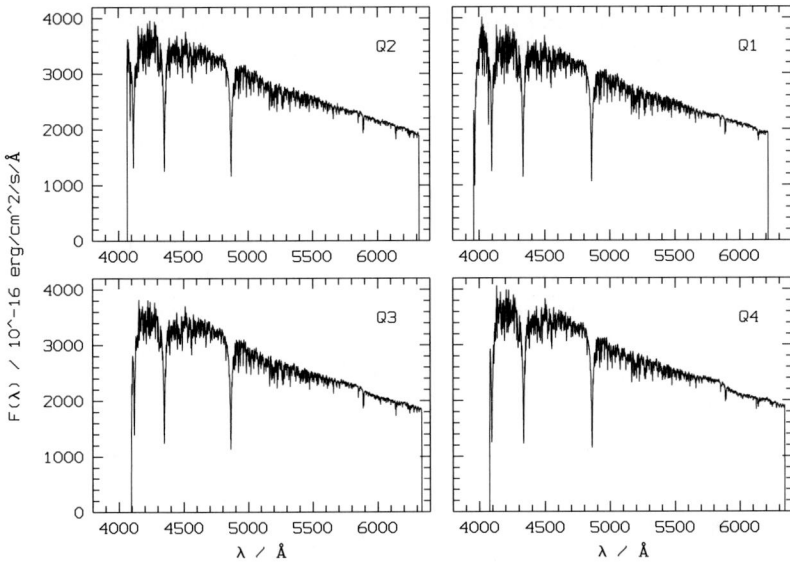

Fig. 2. Flux-calibrated observation of CD −32°9927: master response curves have been applied in all four quadrants

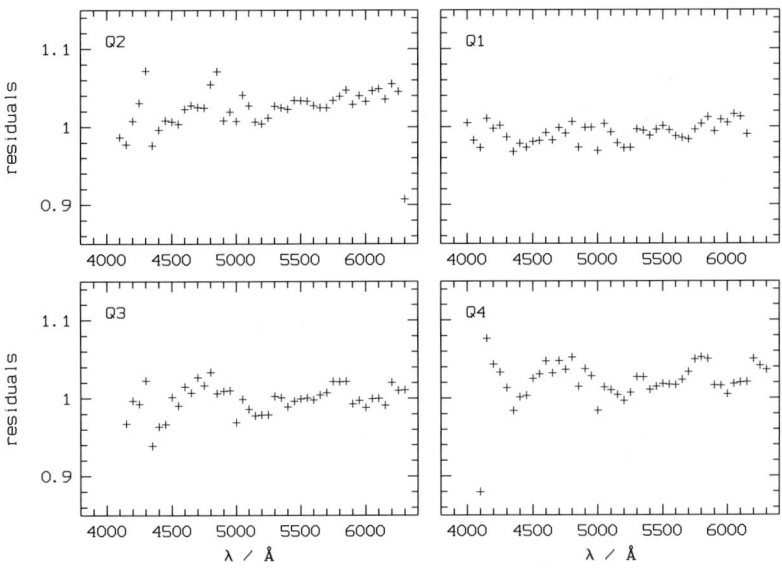

Fig. 3. Residuals: flux-calibrated spectra of CD −32°9927 divided by tabulated flux

Finally, Fig. 3 shows the residuals of the flux calibration: flux-calibrated spectra divided by tabulated flux. The deviations are within 10% of the actual flux. Some differences in the Balmer lines occur because of different sampling of the tabulated flux and the observation.

IFU flux-calibrated spectra have also been verified. Here, the pipeline product is a 3D data cube with spatial and spectral information. The flux calibration is applied per fibre. The 2D image of a standard star is usually much larger than the spatial resolution of each fibre. For a comparison with the tabulated flux, the spectra from all fibres in a quadrant have been added. The background has been estimated from fibres without signal and subtracted. The results show an agreement similar to the MOS case.

4 Limitations and Availability

Master response curves have been created with data from September 2005 to November 2006. The number of available observations and therefore the accuracy of the masters depend on the setting. Some general intrumental and observing limitations also have an influence. For MOS, slit losses can occur and are dependent on wavelength because of atmospheric dispersion. In case of IFU, the spectral response varies over the field for each quadrant (see contribution by M. Roth, these proceedings). This is not completely corrected by throughput correction from flat fields.

Changes of the global efficiency with time cannot be accounted for with this method. Applying the curve to an arbitrary observation will result in a small offset with respect to the absolute flux. However, regular standard star measurements show that the relative response within each setting is in general stable. Master response curves have been created for almost all VIMOS spectroscopic settings. They can be downloaded from http://www.eso.org/qc/VIMOS/qc/response.html. They will be used for pipeline reduction of VIMOS Service Mode data.

Reference

1. O. LeFevre, M. Saisse, D. Mancini, et al.: Commisioning and performances of the VLT-VIMOS instrument. In: *Instrument Design and Performance of Optical/Infrared Ground-based Telescopes*, ed by M. Iye, A. Moorwood (Proc. SPIE, vol 4841, 2003) pp 1670–1681

Part III

Session 2: Optical Multi-object Spectrographs

IOT Overview: Optical Multi-Object Spectrographs

L. Schmidtobreick, S. Bagnulo, E. Jehin, G. Marconi, K. O'Brien, E. Pompei, and I. Saviane

ESO, Alonso de Cordóva 3107, Vitacura Casilla 19001, Santiago 19, Chile; lschmidt@eso.org

Abstract. We give an introduction to the several instruments that ESO operates and which are able to perform optical multi-object spectroscopy. We point out the standard ways of reducing these spectra, the problems that occur, and the way we deal with them. A short introduction is given on how the quality control is performed.

1 What is Multi-Object Spectroscopy?

Quite self-explanatory, multi-object spectroscopy is an instrument mode in which spectra are taken of several objects at the same time. In general, this is achieved by placing individual slits on the targets (see Fig. 1 for an illustration). Pre-images are used to define these slits, and depending on the instrument specifics, masks are then manufactured according to these definitions and placed in the instrument. During the acquisition, a best match has to be achieved between the targets and the slitlets. A grating or grism is then used to disperse the light of all targets at the same time.

2 ESO's Optical MOS Instruments

ESO offers six optical instruments that can be operated in MOS mode. On La Silla, these are EFOSC2 at the 3.6 m telescope and EMMI at the NTT; on Paranal these are the two FORSes, VIMOS and FLAMES. In the following we will give a short overview on these instruments and their MOS capabilities.

2.1 EFOSC2

The ESO Faint Object Spectrograph and Camera v.2 (EFOSC2) [4] is a focal reducer instrument, allowing imaging and low resolution spectroscopy. The instrument has multi mode capability including normal/polarimetric imaging/spectroscopy (several submodes in each), multi-object spectroscopy and coronography. It is mounted in the Cassegrain focus of the 3.6 m telescope on La Silla. In general, MOS is executed in visitor mode. The MOS plates are individually punched, the possible slit widths are 1.15, 1.35, and 1.75 arcsec.

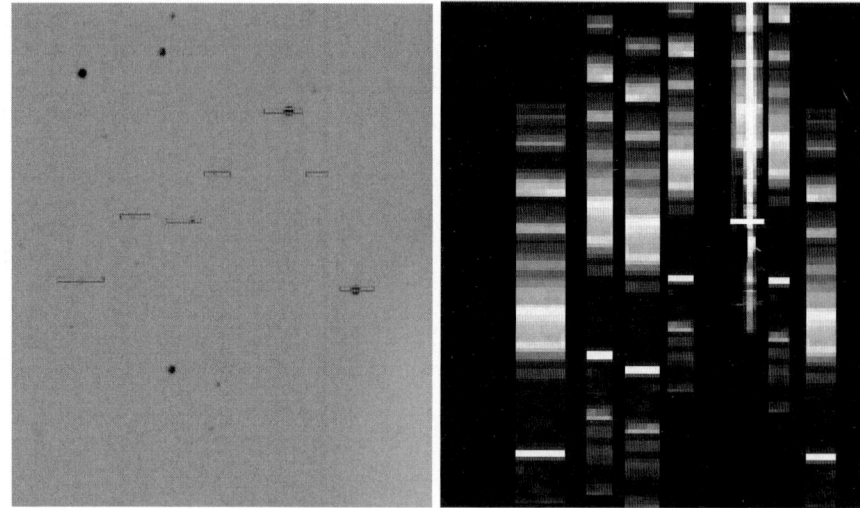

Fig. 1. An illustration of multi–object spectroscopy. Small, individual slits are placed on the objects (*left*) to get a spectrum (*right*) for each of them

Up to 5 masks can be loaded at the same time, and an exchange of masks is also possible during the night. However, as this results in a time-loss of about 30 min it is recommended to avoid such changes if possible. EFOSC2 offers several grisms which provide FWHM resolutions between 0.6 and 6 nm and cover the range from 320 up to 1100 nm.

2.2 EMMI

The ESO Multi Mode Instrument (EMMI) is a versatile instrument, allowing a wide range of observations like imaging and low, medium, and high resolution spectroscopy while using either the red or the blue arm. EMMI is mounted in one of the Nasmyth foci of the NTT. Again, the MOS observations are normally executed in visitor mode. The MOS plates are individually punched, the possible slit widths are 0.8, 1.02, 1.34, and 1.87 arcsec. Up to four masks can be loaded at the same time, but no exchange of the masks is possible during the night. The MOS mode of EMMI is only offered on the red arm in low resolution, thus constraining the spectral range between 400 and 1000 nm, and the resolution between 0.4 and 4.5 nm. EMMI's MOS field of view is 8×5 arcmin.

2.3 FORS1

The FOcal Reducer Spectrograph, v.1 (FORS1) [1] is designed as an all-dioptric optical instrument. It allows imaging and spectroscopy and has the capability for polarimetry and spectro–polarimetry that can be combined with the MOS mode. FORS1 is mounted in the Cassegrain focus of Kueyen (UT 2) on Paranal. The slits of its MOS mode are built by pairs of slitlets which are continuously mounted and are individually driven. This allows a maximum of 19 slits to be built. Note that for the polarimetric mode the amount of slits is limited to 9, to allow the space for the dispersion of the Wollaston prism and the separation of the two polarisation beams. MOS observations on FORS1 are executed in visitor or service mode. Since no physical mask has to be manufactured for these MOS runs it is also possible to use it for Target of Opportunity (ToO) observations. The possible spectral range lies between 330 and 1100 nm, the FWHM resolution ranges between 0.1 and 6 nm. The field of view is 6.8 × 6.8 arcmin.

2.4 FORS2

FORS2 is similar designed as FORS1 and is mounted in the Cassegrain focus of Antu (UT 1) on Paranal. Its capabilities are different from FORS1. FORS2 offers no polarimetric mode, but has instead a larger MOS capability. For once, it provides the same MOS mode as FORS1, building the slits via pairs of slitlets and thus allowing the use of this mode for Targets of Opportunity (ToO) as well as classical visitor and service observations. In addition it contains a mask exchange unit (MXU) which can accommodate up to 10 exchangeable slit masks for MOS of approximately 80 objects each. The individual masks are cut with a dedicated laser cutting machine, slits can be placed in all positions and all orientations.

2.5 VIMOS

VIMOS [5] is a wide field imager and multi object spectrograph mounted in the Nasmyth focus of Melipal (UT 3) on Paranal. The instrument consists of four identical arms, each one has a FOV of 7 × 8 arcmin. The gap between each quadrant is about 2 arcmin. The MOS mode is carried out using masks (one per quadrant) which are cut using the laser cutting mask manufacturing unit. Depending on the grisms used, the spectral resolution varies from 200 to 2500, and the observable range is from 360 to 1000 nm. The maximum number of slits per mask (quadrant) varies from 40 at R=2500 to 150–200 at R=200. VIMOS observations are carried out in visitor and service mode.

2.6 FLAMES

FLAMES [7] is a multi-object, intermediate and high resolution spectrograph mounted in one of the Nasmyth focus of Kueyen (UT 2). Its FOV is round with a diameter of 25 arcmin. Different from the other MOS instruments, the objects are not selected via slits, but fibres are placed on the object positions which then feed the GIRAFFE spectrograph. Up to 130 targets can thus be selected, the resolution is between R=5600–46000 and covers the range from 370 to 950 nm. FLAMES observations are carried out in visitor and service mode.

3 Science that Is Done Using MOS Modes

Typical science done using MOS are studies of cluster members in either stellar clusters or galaxy clusters.

One such example is the REFLEX survey [3], an X-ray flux-limited sample of 447 galaxy clusters. The redshifts have been determined using MOS on EFOSC2 and thus the distance to the clusters could be determined. This revealed superstructures in the spatial distribution of the clusters. Furthermore, constraints could be made on the cosmic density of matter including both, dark matter Ω_m and dark energy Ω_Λ. Together with the constraints from supernovae and the cosmic microwave background, the values for the universe are determined around $\Omega_m \approx 0.3$ and $\Omega_\Lambda \approx 0.7$.

Another such project is the VIMOS VLT Deep Survey (VVDS), which aims at understanding the major steps in the evolution of galaxies [6]. For 1599 objects with $I_{AB} \leq 24$ around the Chandra Deep Field South, the redshift has been determined and released to the community.

In terms of stellar astrophysics, a search has been made for links between magnetic fields and stellar evolution by studying the magnetic fields in open cluster A- and B-type stars. While mass, age, metallicity are determined from the cluster membership, the magnetic field of 235 early-type stars was measured from circular polarisation in several lines using the PMOS mode of FORS 1 [2].

4 Calibration of MOS Data

Calibration of MOS data is in general similar to that of long–slit spectroscopy. The main steps include:

- bias and overscan
- flatfield with the mask
- wavelength calibration: arc lamps with mask
- spectrophotometric calibration: standard star with longslit

However, there are some caveats. The MOS slitlets are not necessarily aligned along the central column, hence different slitlets cover different spectral ranges. The spectrophotometry is even more difficult than for long–slit spectroscopy, as the slit loss is basically not known and depends strongly on the quality of the mask and acquisition. Hence, only relative spectro–photometry, i.e. the correction for instrument function, can be done without major efforts. In terms of operation it is important that all calibration data are taken before the masks are moved.

4.1 Special Calibration Issues: VIMOS

VIMOS has two special instrumental features that have to be considered for the observations or in the calibration: the instrument flexure and the lack of an Atmospheric Dispersion Corrector due to its enormous size.

The instrument flexure is a flexing of the whole instrument depending on its position angle. Folding-mirrors may to some extent compensate for such flexures. The effect is quantified by measuring the image position of a pinhole mask at various instrument position angles. The results are summarised at: http://www.eso.org/instruments/vimos/inst/flexures.html. It affects the fit of the mask to the CCD and results in a high uncertainty concerning the slit loss. In severe cases, the objects might not even fall on the slit anymore. Furthermore, it affects the wavelength calibration, i.e. calibrations taken at day-time at a fixed position angle do not necessarily represent the actual situation during the science observations. For this reason, all data which have not sufficient sky lines for the wavelength calibration, i.e. the blue modes, need to be calibrated during night-time and wavelength calibration data are taken just after the science exposures.

The fact that VIMOS contains no Atmospheric Dispersion Corrector results in a field differential refraction and a chromatic dispersion (see http://www.eso.org/instruments/vimos/inst/atmos.html for details). To minimise the effect, MOS observations as well as pre-imaging data are generally taken at a field orientation of 90° with the slits oriented in north-south direction. The data is usually taken only within 2 h of the meridian to avoid the strong chromatic dispersion. Motivated exceptions from this strategy can be required through a waiver.

5 Monitoring the Instrument Performance

Again, most of the performance monitoring that is done for the MOS mode is similar or even the same as done for long–slit spectroscopy. These include the lamp–efficiencies and the detector parameters. However, some special monitoring is done for the wavelength calibration. As an example, the RMS of the dispersion solution is plotted for VIMOS in Fig. 2. All arc lamp measurements taken to calibrate MOS science observations are continuously reduced by the pipeline and the results shown in such trending plots, which are checked on a daily basis. These trending plots can be found at http://www.eso.org/observing/dfo/quality/ALL/daily_qc1.html for all Paranal instruments.

VIMOS: rms of IDS, MOS MR (last 90 days)
date range: 2006-10-23 ... 2006-12-18; last Paranal data: 2007-01-20

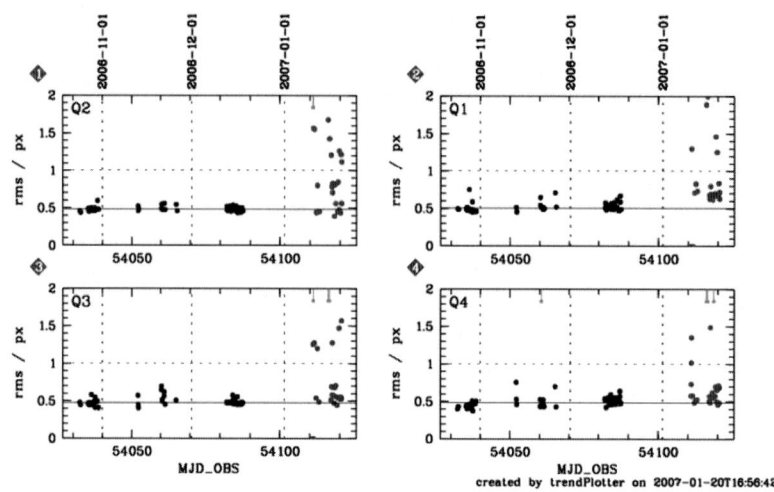

Fig. 2. The RMS of the dispersion solution is plotted for the four VIMOS quadrants

References

1. I. Appenzeller, et al.: The Messenger **94**, 1 (1998)
2. S. Bagnulo, J.D. Landstreet, E. Mason: A&A **450**, 777 (2006)
3. H. Böhringer, P. Schuecker, L. Guzzo, et al.: A&A **425**, 367 (2004)
4. B. Buzzoni, et al.: ESO Messenger **38**, 9 (1984)
5. O. Le Fevre, et al.: SPIE **4841**, 1670 (2003)
6. O. Le Fevre, G. Vettolani, S. Paltani, et al.: A&A **428**, 1043 (2004)
7. L. Pasquini, et al.: The Messenger **110**, 1 (2002)

Quantitative Spectroscopy with FLAMES

P. Bonifacio[1,2,3]

[1] CIFIST Marie Curie Excellence Team; Piercarlo.Bonifacio@obspm.fr
[2] GEPI, Observatoire de Paris, CNRS, Université Paris Diderot, 5 Place Jules Janssen 92190, Meudon, France
[3] INAF – Osservatorio Astronomico di Trieste, Via Tiepolo 11, I-34131 Trieste, Italy

Abstract. Starting from the experience gained in these years with FLAMES observations I shall summarise some of the lessons learnt in preparing and conducting observations with this facility. Finally I will talk about my approach to the analysis of FLAMES data focusing on aspects which are peculiar to this facility with respect to other spectrographs.

1 Preparing the FLAMES Observations

1.1 Fibre Positioning

The multi-fibre facility FLAMES [1] allows to obtain high resolution spectra of multiple objects over a field of 20′ diameter. The Medusa mode can deploy 132 fibres feeding the Giraffe spectrograph, 8 fibres feeding the UVES spectrograph and 4 "Fiducial Acquisition Bundles" (FACB) which serve to check the field alignment. There is no target acquisition system, which means that the fibres must be placed on the basis of predefined coordinates for all the objects. This requires to provide an object list with (relative) positional accuracy of 0.2″. The astronomers who are knowledgeable in analysis of stellar spectra are, typically, not equally knowledgeable in wide field photometry and astrometry. As a result I notice that an instrument like FLAMES has fostered very fruitful collaborations between spectroscopists and photometrists. The definition of FLAMES observations is usually a challenging collaborative task. With this regard I must acknowledge the meritory work done by the EIS Pre-FLAMES Survey [2, 3] which has acquired wide field images of most of the interesting targets for FLAMES in the southern sky.

Once you have available your astrometrised photometric catalogue, you can select the targets which are of interest to you and start the process of fibre allocation. You do this with the FPOSS software[1], which works remarkably well. Here are a couple of tips about the use of FPOSS, which have come from personal experience:

- make sure that your catalogue has a rather large number of "fiducial" stars of suitable brightness for the FACB;

---

[1] http://www.eso.org/observing/p2pp/OSS/FPOSS/FPOSS-tool.html

- it is often useful to place manually the FACB and the UVES fibres, and then let FPOSS work its way with the other allocations;
- if there are a few targets which you want to observe at all costs, place the fibres on them manually, in spite of the use of priorities it is difficult to guarantee that FPOSS will place a fibre on them;
- make sure your catalogue has a large number of suitable "sky" positions marked. By large I mean at least a factor of 5 larger than the actual number of sky fibres you want to place.

The last point brings me to what I think is a crucial point for preparing successful observations: the measurement of the background sky. To make an accurate measurement of equivalent widths and chemical abundances one should be able to subtract reliably the sky background. Since the science targets for FLAMES are, typically, faint and the instrument is used often in grey or even bright time, it is not unusual that your scientific target gives you only 20% of the counts provided by the sky background. To estimate the latter you must make use of the signal measured by the "sky" fibres. One should not be too greedy and place too many fibres on the scientific targets and too few on the sky. As a rule of thumb one should place 1 sky fibre every four or five objects. It is generally unwise to allocate fewer sky fibres than these. This is fine in a situation in which the sky background is fairly smooth and slowly changing across the FLAMES field of view. If instead there is a considerable gradient in the sky background one must place as many fibres as necessary to proper sample this gradient and *place them in the right positions*. Note that in some situations a background with a rather steep gradient is of "astronomical" nature (e.g. in a compact Globular Cluster the inner regions have a higher sky background than the outer regions).

In crowded regions it may be impossible to find enough sky positions and one has to be content with whatever is available. Typically one writes some sort of iterative procedure, which explores the photometric catalogue, searching for regions devoid of stars, or, at least devoid of stars which are brighter than 3 magnitudes *more* than your science target.

Once your "sky positions" are found it is highly advisable to check these positions with a good quality image of your field (the EIS Pre-FLAMES images are great for this) and verify that these positions actually correspond to what you would call as "sky". The reason is that a position may be "empty" in your photometric catalogue for reasons which have nothing to do with "emptiness" on the sky. Typical encountered examples are: background galaxies, which are absent from the photometric catalogue of *stellar* objects; bright stars near to the position, these have caused saturation in the surrounding region in the images used to obtain the photometry, thus creating a *hole* in the catalogue; defects in the image used to determine the photometry (e.g. a gap in a mosaic or a bad CCD row); stars which are not so faint, but fainter than the completeness limit of the catalogue. The last point clearly encourages to use photometric catalogues which are complete down to, at least, three magnitudes fainter than the faintest science targets.

1.2 Spectrograph Settings

FLAMES feeds only the red arm of UVES, for which you can choose only three settings, centred at 520 nm, 580 nm and 860 nm. Note that with the 520 nm setting only 6 fibres can be deployed, given the smaller inter-order separation.

Giraffe is a considerably more complex instrument and leaves to choose among 8 low resolution settings and 22 high resolution settings. By browsing the ESO archive you may realise that, although the low resolution mode is most often used by the extra-galactic community, the resolution is high enough that some very interesting work can be done on stellar abundances and radial velocities. The high resolution settings provide a range of spectral resolutions and coverages, so that a detailed investigation based on synthetic spectra or real spectra, of stars similar to the scientific targets, is necessary to select a good combination. One has to resist the temptation to observe the same targets in many ($n \geq 5$) settings in order to obtain a large spectral coverage, because this largely defeats the multiplexing advantage of FLAMES. Therefore the objective is to define the minimum number of settings to be used in order to achieve the science goals. One should contrive as much as possible to fully exploit the spectral range of each setting. Often in a Giraffe setting there are lines of some element, which are not typically used in high resolution abundance analysis. The reasons are often: poor atomic data, line blending, saturation. It is often really worth to devote extra attention to these lines, perhaps they can be usefully used, e.g. using a solar or astrophysical gf value; blends can often be satisfactorily treated by spectrum synthesis techniques; saturated lines can provide meaningful abundances, provided we know how to fix the microturbulent velocity or that we use 3D hydro simulations. Many lines which are "left out" in the analysis of a traditional echelle spectrum, just because there are "better" ones of the same element, can become precious in the limited coverage of a few Giraffe settings.

As obvious from the above considerations the planning of a Giraffe observation can take up a considerable work, however the pay back can be very large.

2 Analysis of Giraffe Data

In this section I will briefly discuss some issues on the analysis of Giraffe data, FLAMES-UVES spectra can be analysed in pretty much the same way as SLIT-UVES spectra and will not spend any time on them.

2.1 Sky Subtraction

The first step is of course to inspect your "sky" spectra and verify if all of them have been correctly placed (some may have landed on a star or a galaxy..). Next one should assess if there is any significant variation in the sky background or if it may be assumed to be more or less constant across the field. N.B. this assumes that you have chosen the sky position in a way suitable to sample the field. If, for example, in the case of a Globular Cluster, you have placed all your sky fibres in the outer regions of the cluster, they may appear more or less similar, although when you move to the inner regions the background is effectively increasing in the inner regions. If the background is varying you must choose a recipe to assign the "appropriate" sky to be used for each object fibre, e.g. the nearest. You may devise more sophisticated schemes by which you interpolate a sky spectrum for each object, using the available sky spectra. The interpolation scheme must be clearly two-dimensional. If the background is mainly astronomical in nature (as opposed to lunar illumination), you may define the shape of the background by using the image you used for your photometry/astrometry and use that shape (e.g. a 2D polynomial or a spline) to suitable scale the available sky spectra at the positions of the object fibres.

If the background is fairly constant you may simply median the spectra from all the available fibres to obtain a median sky spectrum.

Whichever the case you will find that in most situations your sky spectra are of very poor S/N ratio. In this case it would be a very bad idea to simply subtract your sky spectrum from your science spectrum, because all the noise in the sky will be added to your science spectrum significantly degrading its quality. At very minimum you should heavily smooth your sky spectrum in order to boost its S/N ratio. The fact that then the sky spectrum will have effectively a lower resolution than the science spectrum is generally inconsequential, especially if the sky spectrum does not display prominent absorption or emission lines. One should always be weary of this though, and often a simulation can help to assess if you are likely to introduce artifacts in your science spectrum. The simplest simulation is to take a synthetic spectrum, degraded at the Giraffe resolution, add to it the real sky and then subtract the smoothed sky. The comparison between the noiseless synthetic spectrum ad the sky subtracted spectrum will show you if significant artifacts are created.

Often in the wavelength range of a Giraffe setting the sky does not show large variations with wavelength so that it may be sufficient to take the median number of counts of the sky and subtract that number from the object spectrum. In an only slightly more complex situation the sky can be modelled by a straight line and you may subtract the straight line to the spectrum. Whenever acceptable such approach should be preferred since you add no noise to your scientific spectrum.

2.2 Abundance Analysis from Giraffe Spectra

Of course one may perform abundance analysis on Giraffe spectra pretty much in the traditional way: star by star, line by line. However one must take into account three peculiarities of the Giraffe spectra:

1. the resolution of Giraffe, even in the high-resolution mode, is slightly lower than what is typically used for abundance analysis, with the consequence, that, especially for the cooler and not so metal-poor stars, there are precious few "unblended" lines, for which we can measure equivalent widths, if any;
2. the objects which are the main targets of Giraffe are often quite faint, so that the final S/N ratio is in the range 15–30, i.e. many stars have a lower S/N ratio than you would typically like for abundance analysis;
3. for each plate you acquire of the order of 100 spectra, so you would better build an automatic or semi-automatic procedure, unless you want to pass the rest of your career analysing Giraffe spectra.

For these reasons we have developed an automatic procedure [4] for the abundance analysis of high resolution spectra. In that paper we illustrated the general principles underlying the procedure, however the implementation was specifically tailored for the analysis of giants in the Sgr dSpH from Giraffe spectra in setting HR09 and HR14. The key points that make it useful for the analysis of Giraffe spectra are the capability of using blended lines, through line fitting, and the fact that a large fraction of each setting is used, so that even if the S/N ratio is low, and many features have to be discarded because of a too low goodness-of-fit, the average metallicity is fairly accurate and robust.

Acknowledgement. It is a pleasure to acknowledge the help of the friends and colleagues who have shared with me these exciting years of work on FLAMES data. Therefore thanks very much to (in strict alphabetical order): E. Caffau, P. Di Marcantonio, L. Girardi, G. Marconi, P. Molaro, L. Monaco, S. Monai, L. Sbordone, S. Zaggia.

References

1. L. Pasquini, et al.: The Messenger **110**, 1 (2002)
2. Y. Momany, et al.: Astronomy and Astrophysics **379**, 436 (2001)
3. S. Zaggia, et al.: The Messenger **105**, 25 (2001)
4. P. Bonifacio and E. Caffau: Astronomy and Astrophysics **399**, 1183 (2003)

Discussion

P. E. Nissen: How well is the flat fielding of the GIRAFFE spectra? What is the accuracy in percent?
P. Bonifacio: It is generally of the order of a few percent (1–5%).

M. Rosa: Are some of your stars observed in more than one fiber or on another instrument? What experience do you have therefore on repeatability or cross-calibration issues?

P. Bonifacio: UVES-FLAMES spectra compare very well to the UVES-slit spectra, although they are generally of lower S/N ratio. Reproducibility is of the order of 1/(S/N), as can be expected theoretically. The comparison UVES/GIRAFFE has to take into account the different resolution of the two spectrographs. In the S/N ratio, we work (S/N = 15–30) I would not trust any line measured with GIRAFFE with an equivalent width smaller than about 2.5 pm. This explains why we cannot measure weak Fe lines. This value is in good agreement with the estimated error on EW derived from the Cayrel formula (Cayrel 1988 IAU symposium 132 p. 345): $\sigma(EW) \sim 1.5(FWHM \times PIXELSIZE)^{1/2}(S/N)^{-1}$.

D. Osip: You mentioned "large fiber losses", could you quantify? Do you see additional losses for fibers placed off-axis?

P. Bonifacio: UVES fiber link loses about 50% of the light, due to the length of the fibers. We do not see any extra losses for off-axis fiber; however, an accurate flux calibration should be performed in order to definitely assess this issue. We only work on normalized spectra for abundance analysis, therefore we never flux-calibrate the spectra.

Good News for MOS, MXU & Co. – The New Spectroscopic Pipeline for the FORSes

S. Moehler

ESO, Karl-Schwarzschild-Strasse 2, 85748 Garching, Germany; smoehler@eso.org

Abstract. Since October 1, 2006, spectroscopic data from the two FORS instruments have been reduced with a new pipeline, which is based on a bottom-up calibration approach. I give a short description of the pipeline and discuss first experiences with automatic data reduction using this software, which has significantly increased the percentage of processed data for both instruments. I will also describe possible new options for Quality Control.

1 How Does the Pipeline Work?

The spectroscopic pipeline for the FORS instruments can handle long-slit and multi-object spectroscopic data using slitlets. In order to be as flexible as possible most information is obtained directly from the observed calibration data, minimizing the need for assumptions (for details of the underlying principles see the contribution by Izzo et al., these proceedings).

1.1 Calibration Data

A first guess of the positions and lengths of the slitlets (or long slit) is obtained from an arc lamp frame (see Fig. 1). The positions of arc lamp spectra on the CCD are determined by a pattern matching technique applied to the detected emission lines. For that an estimate of the linear dispersion as well as a line catalog (which may both be provided by the user) are required. Adjacent slitlets with no offsets in dispersion direction are not distinguished and will form one longer slitlet.

The spatial curvature is determined by tracing the edges of the corresponding flat field spectra (cf. Fig. 2) when possible. The curvature is fit by default with a second order polynomial. The final wavelength calibration is obtained taking into account the spatial curvature of the spectra.

The pipeline provides the following output calibration files:

- slit positions on the CCD at the central wavelength of the grism
- coefficients for the dispersion relation and the spatial curvature
- master bias
- master flat field (normalized and not-normalized)

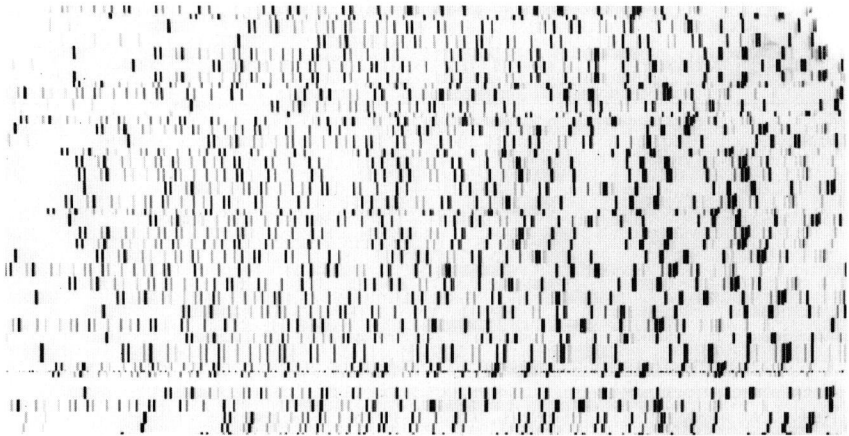

Fig. 1. Example of a raw arc frame for FORS2 MXU

- wavelength map and spatial map
- reduced arc lamp frame
- spectral resolution and line widths

The wavelength map specifies at the position of each original CCD pixel its corresponding wavelength. Thus the user can avoid the re-sampling of

Fig. 2. Example of spatial curvature for a raw FORS2 MXU flat field

the spectra usually done after the wavelength calibration. The spatial map provides in the same manner the position of the CCD pixel within its associated slitlet. The reduced arc lamp frame allows to judge the quality of the calibration (straight arc lines, no wriggles, no offsets). If successful (see Sect. 2) the pipeline achieves an accuracy of 0.1 pixel both in spatial and dispersion direction.

1.2 Science Data

The default reduction of spectroscopic science data is as follows: The data are first corrected for bias. Then the extraction mask derived above (i.e. the positions of the slitlets and the curvatures of the resulting spectra) is applied to the science data; they are flat-fielded and re-mapped eliminating the optical distortions. Afterwards they are rebinned to constant wavelength steps. The wavelength calibration can be adjusted using sky emission lines. This allows to correct for shifts between night-time science and day-time calibration data. Such shifts (of the order of 1 pixel) are known to happen due to flexure for MXU observations. Finally object spectra are detected and extracted, together with the corresponding error spectra and sky spectra. For long-slit spectroscopy the sky background is determined using a median. This is a valid approach if not more than half of the pixels contain flux from objects *and* there is no spatial gradient in the sky background. For multi-object spectroscopy instead the sky is subtracted before remapping, i.e. when the spectra are still in the original CCD coordinate system. The sky is determined with a robust linear fitting, which allows for a linear spatial gradient in the background. Also in this case, however, not more than half of the pixels may contain flux from the object(s).

The output science files are:

- positions and widths of object spectra
- extracted object spectra, error spectra, and sky spectra
- unmapped (corrected for spatial distortions, not rebinned) science frames
- mapped (i.e. corrected for spatial distortions and rebinned) science frames
- adjusted dispersion coefficients and wavelength map

2 Strengths and Weaknesses

The major strength of the new pipeline is its flexibility and robustness. It requires only very limited input information (mainly dispersion estimate and line catalog) and can therefore be applied to a large variety of instrument configurations. Thus it is now possible for the first time to *automatically* reduce long-slit (LSS) and multi-object spectroscopic data (MOS/MXU) from the two FORS instruments for all grisms and both collimators (standard and high-resolution). To verify its flexibility the pipeline was applied to data from

Fig. 3. Example of missed slitlets in FORS2 MOS data (*left*: averaged screen flat, *right*: normalized screen flat)

the Low-Resolution Spectrograph of the Hobby Eberly Telescope, which it handled without any problems.

The pipeline does have problems if there are too few arc lines, e.g. for slitlets with large offsets. In such cases it sometimes fails to correctly identify the slitlet. This can be most easily recognized by comparing the not-normalized and the normalized screen flat (cf. Fig. 3 – one slitlet is partly missing, another one was not processed at all). Large gaps between arc lines can be problematic as non-linear dispersion terms can become important, while the pipeline currently uses only linear dispersion estimates for the pattern matching.

The pipeline will fail in case of regularly spaced arc lines, like those created by a laser comb, as its pattern matching (see contribution by Izzo et al.) does not work for such data.

3 Quality Control

In order to check the quality of observational data a pipeline is of great importance. Otherwise it is difficult to distinguish between well-known and correctable instrument effects and real problems. Quality control shall ensure that the data observed with the FORSes can be calibrated. In addition the trending of certain parameters allows to have an eye on the instrument health (see http://www.eso.org/dfo/quality for more details). Therefore the resolution, central wavelength, and number of identified arc lines are regularly monitored. A change in the number of identified lines can indicate reduced flux of the arc lamp and possibly impending failure of the lamp. A change in resolution could indicate focus problems.

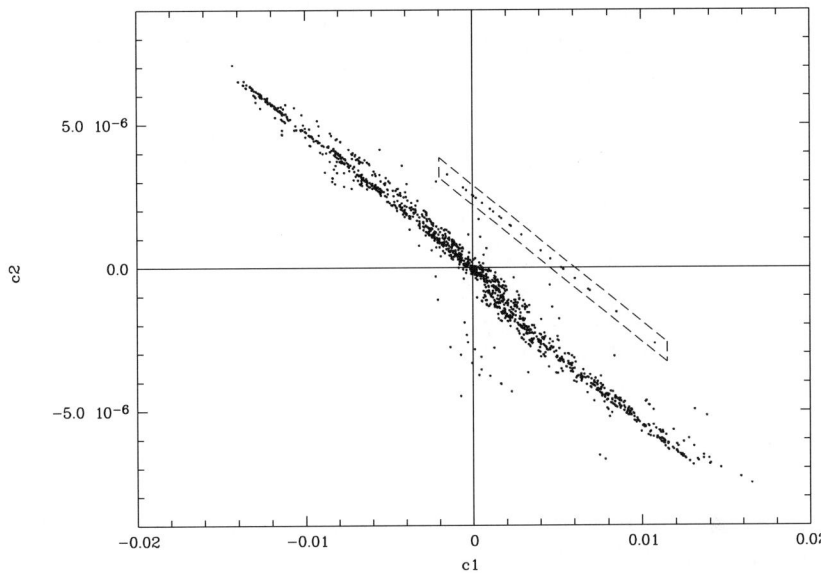

Fig. 4. Curvature coefficients for FORS2 MXU data for the 600z grism. The lower branch resulting from chip 2 data indicates that chip 2 is mounted at an angle of $0.079° \pm 0.007°$ with respect to chip 1 (the known value being $0.083°$). The reduced arc lamp image corresponding to the coefficients in the dashed box is shown in Fig. 5

The situation for the quality control of FORS data has vastly improved with the new pipeline, as now almost all data can be reduced (with polarimetric data being the only exception). Moreover, the new pipeline also provides additional information on the instrument. For example, the combination of the curvature coefficients can be used to monitor grism alignment: The coefficients $c1$ and $c2$ describe the slope and the curvature of a slitlet, respectively. If the curvature is 0, then the slope should be 0 as well – otherwise this indicates that the grism is not well aligned. Figure 4 shows the coefficients obtained for the 600z FORS2 grism. Obviously the correlation for the majority of solutions passes through zero.

Some solutions, however, show significant offsets from the general trend. Figure 5 shows a reduced arc lamp image for such a deviating case (marked by the dashed box in Fig. 4). Obviously the wavelength calibration obtained for several slitlets is rather bad (wriggles instead of straight lines). Thus diagrams like Fig. 4 allow to look for bad solutions efficiently in case of large data volume, which may prohibit checking all solutions individually.

Other new parameters become available if there are more than 12 slitlets distributed across the CCD. In this case the pipeline calculates a global

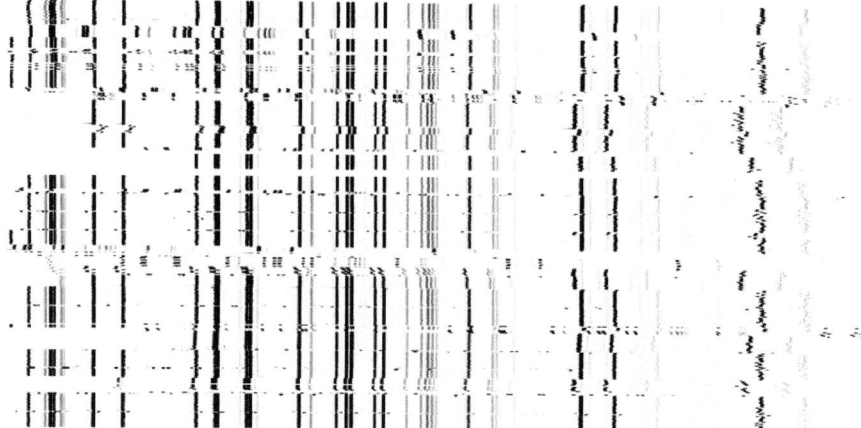

Fig. 5. Reduced arc lamp frame for the deviating points marked in Fig. 4

distortion model, which contains good indicators for the monitoring of the instrument's health, like the instrument scale.

4 Concluding Remarks

The new spectroscopic pipeline for the FORS instruments has been a great success in that it allows to reduce *automatically* long-slit and multi-object spectroscopic data for all FORS grisms. The fact that it does not need a first guess from an instrument model makes it a promising candidate for the reduction of spectroscopic data from a variety of instruments. In addition this approach is extremely helpful in the presence of instrument instabilities as it allows to reduce data whose configurations deviate from the expected one (e.g. due to instrument aging or earthquakes).

Acknowledgement. I would like to thank C. Izzo and U. Hopp for their valuable comments on this manuscript.

Discussion

A. Smette: You mentioned the pipeline has problems when there are too few arc lines. Are there some set-ups of FORS for which this problem occurs?
S. Moehler: Yes, settings: 600B, 1400 V for large offsets from center of the CCD.
A. Smette: Should we consider the use of other arc lamps?
S. Moehler: Yes, if the problem remains after further checks of the pipelines parameters and the line catalogs.

D. Baade: The much improved robustness of the pipeline in the case of NOT well-behaved input data is very good news indeed. Do you also find any gain in the qualify of reduced well-behaved input data?

S. Moehler: Yes, the residuals of the wavelength calibration are significantly smaller (about a factor of 2) e.g. for 300 V.

S. Mieske: Will the new FORS pipeline also do spectrum extraction?

S. Moehler: Yes, it applies the calibration (i.e. dispersion solution, correction for spatial curvature) to the science data; does a correction for the sky background and an optimum extraction of all detected object spectra including error propagation.

S. Stefl: What is the main problem with the spectropolarimetry? Why is the spectropolarimetric mode not included in the pipeline?

C. Izzo: There are no problems, as far as the extraction geometry determination is concerned. However, further data reduction steps (such as sky subtraction, object detection and combination, etc.) would require to treat the ordinary and extraordinary beams separately: this is not yet implemented in the FORS pipeline.

Calibrating the VIMOS Redshift Survey Data

M. Scodeggio[1], P. Franzetti[1], B. Garilli[1], M. Fumana[1], L. Paioro[1], and A. Zanichelli[2]

[1] INAF – IASF Milano, Milano, Italy; `marcos@lambrate.inaf.it`
[2] INAF – IRA Bologna, Bologna, Italy

Abstract. We discuss the accuracy of redshift measurements obtained using VIMOS spectroscopic observations, on the basis of analysis of more than 50,000 spectra, obtained as part of the VVDS and zCOSMOS surveys. At the core of the data processing pipeline we have VIPGI, the VIMOS Interactive Pipeline and Graphical Interface, which provides a very efficient set of tools for the reduction and analysis of VIMOS spectroscopic data.

1 The VIMOS Spectrograph and the Redshift Survey Data

VIMOS is an imaging spectrograph, installed on the Unit Telescope no. 3 (Melipal) of the VLT (see [1, 2] for a more detailed description). It was designed to cover as much as possible the unvignetted part of the focal plane at the VLT Nasmyth focus (a circular area with a diameter of 22 arcmin on the sky). Since this large area corresponds to a large linear scale (almost one meter), it was decided to achieve the coverage by splitting the instrument into four identical optical channels. The field of view covered by each channel (generally referred to as a VIMOS quadrant) is approximately 7×8 arcmin, with a pixel scale of 0.205 arcsec/pixel, and a gap approximately 2 arcmin wide is present between quadrants. For spectroscopic observations six different grisms provide a spectral resolution ranging from R~200 to R~2500. Order sorting filters are used to avoid an overlap between first and second grating orders. Laser-cut masks, one per quadrant, are used for multi-object spectroscopic (MOS) observations. The number of slits that can be placed on each mask varies from approximately 40 at high spectral resolution up to approximately 250 at low spectral resolution.

Two large redshift surveys are being carried out trying to take full advantage of VIMOS large multiplexing capabilities: the VIMOS VLT Deep Survey (VVDS), and the zCOSMOS survey. The VVDS (see [3]) was designed to use in the best possible way the Guaranteed Time allocated to the VIRMOS Consortium, and its original goal was to obtain a redshift for some 150,000 galaxies. The galaxy sample targeted by the survey was actually divided into three parts: a wide survey of 100,000 galaxies, covering four separate 2×2 deg. wide sky areas, and reaching a limiting magnitude $I_{AB} \leq 22.5$, a deep survey of 50,000 galaxies, covering a 1×1 deg. wide area inside one of the four wide areas, reaching a limiting magnitude $I_{AB} \leq 24.0$, and an ultra-deep survey of some 1500 galaxies inside the deep area, reaching a limiting magnitude $I_{AB} \leq 25.0$. Because of a significant cut in the Guaranteed Time allocation decided by ESO after the delivery of VIMOS, only approximately 30% of the VVDS wide and deep observations have been carried out so far, for a total of some 40,000 spectra, while the ultra-deep part has just been started. zCOSMOS (see [4] for details) has been designed to provide redshift coverage for a subset of the galaxies in the COSMOS field (see [5]). Like the VVDS, also zCOSMOS is divided in different parts: a wide survey of 20,000 galaxies, covering the whole 2 sq. deg. COSMOS field, and reaching a limiting magnitude $I_{AB} \leq 22.5$, and a deep survey of some 10,000 galaxies, covering the central 1 sq. deg of the COSMOS field, color selected to lie in the redshift interval $1.4 \leq z \leq 3.0$. So far, some 10,000 spectra for the wide part, and 1000 spectra for the deep part have been obtained.

2 The VIMOS Data Reduction Pipeline

Among the last-generation spectrographs VIMOS is perhaps the most challenging in terms of data production, being currently the instrument with the highest multiplexing capabilities available to any astronomer. For this reason the VIRMOS Consortium was contracted by ESO to deliver all the elements necessary to build a VIMOS-specific automatic data reduction pipeline. In collaboration with the ESO Data Management Division, we have developed a C library of data reduction procedures, the VIMOS Data Reduction Software (DRS), which is now at the core of both the on-line VIMOS data reduction pipeline at ESO, and the VIRMOS Consortium semi-automatic data reduction pipeline, the VIMOS Interactive Pipeline and Graphical Interface (VIPGI; see [6] for details).

The VIMOS DRS routines have been coded using the C language (for a total of some 150,000 lines of code) to obtain the maximum speed for these computationally intensive tasks. The routines range from the basic "opening a file and reading its content" to wavelength calibration, spectral extraction and IFU 3-dimensional data cube reconstruction, and are organized in a relatively small number of reduction recipes. Generic tasks like the handling of FITS files, or of the World Coordinate System are devoted to special-purpose external software packages (the CFITSIO and WCSTools libraries, respectively) The global data reduction scheme implemented by the DRS is a fairly traditional one, following broadly the one implemented by the IRAF longslit package. However a special effort was made to achieve a very high efficiency during the repeated application of this scheme to the large set of VVDS data, by tailoring all aspects of the data reduction scheme to the specific characteristics of VIMOS. Still, the various reduction functions are general enough that they could be adapted for the reduction of data produced by any MOS spectrograph with a minimal effort (an experimental adaptation for the FORS2 data is under way).

The ESO on-line VIMOS pipeline was then implemented using the DRS to implement fully automated data reduction recipes, taking advantage of the general facilities provided by ESO Gasgano or ESORex software packages. With VIPGI, (Fig. 1) instead, we have kept the DRS capability for a very fast data reduction process, but we also have added many interactive data reduction quality control points, a user friendly graphical user interface, and a simple but effective VIMOS-specific method to organize the raw and reduced data. VIPGI is implemented mostly in Python (some 20,000 lines of code), with the exception of a few tasks devoted to the handling of complex FITS files, that have been written in C.

The performance of the VIPGI pipeline depends on the computer hardware being used to run it, and on the kind of data being reduced. As an example, the reduction of a typical VVDS deep pointing (4 masks for a total of some 550 slits, 10 jittered exposures to be combined), on a Linux PC with a 1 Gb of physical memory takes approximately 25 CPU seconds for the complete wavelength calibration of the 4 quadrants, and 1170 CPU seconds to complete the reduction of the science data. From the end user point of view, it takes approximately 1 hour to go from the raw data to a set of fully reduced

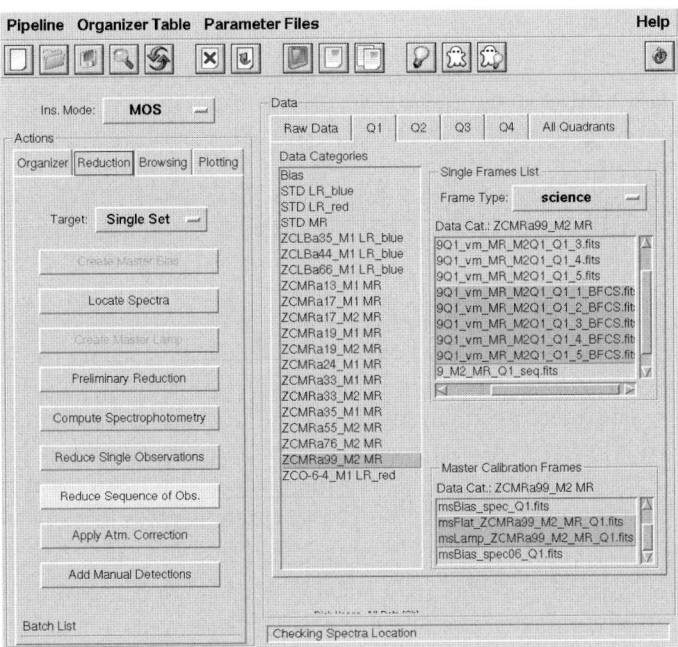

Fig. 1. The main VIPGI window, showing on the left the commands buttons area, and on the right the data organizer area

spectra, including the CPU time required to run the reduction recipes, the time spent selecting the appropriate input files for the recipes, and the visual inspection and verification of the instrument calibrations.

3 The Calibration of VIMOS Spectroscopic Data

The high efficiency achieved with the VIPGI data reduction process is mostly due to a significant simplification of the calibration process for VIMOS data. VIPGI recipes require an already existing calibration of the instrument properties as part of their input, to be used as first guesses which will be subsequently refined by the recipes themselves, and such a calibration is always provided to the astronomer, stored in the VIMOS raw data FITS headers. This operational choice saves the astronomer the time normally devoted to the task of deriving instrument calibrations, and it also results in a more homogeneous data reduction process, since all astronomers effectively start the reduction from the same calibration data, an important fact for a survey instrument like VIMOS. A fundamental component of the whole pipeline architecture is therefore the VIMOS instrument model, which analytically describes the main calibration relations required for the extraction of object spectra from VIMOS spectroscopic observations, via a fully bi-dimensional mapping of the VIMOS focal plane. The model is separated into three different components:

(1) the Optical Distortion Model, which provides a mapping between positions on the VIMOS focal plane and pixel coordinates on the CCD frame. This mapping is obtained for an arbitrarily fixed wavelength, and is described by two independent polynomial relations.

(2) the Curvature Model, which provides a description of the geometrical shape of each spectrum on the CCD, to allow for its tracing and extraction. As optical distortions in the VIMOS focal plane are very small, this model is obtained using a low order (≤ 2) polynomial.

(3) the Inverse Dispersion Solution, which provides the mapping between wavelength and pixel coordinates along the geometrical shape traced by the Curvature Model. It is measured using offsets with respect to the reference wavelength and pixel position defined by the Optical Distortion Model, to allow for multiple spectra on a given CCD column, as required to achieve the large VIMOS multiplexing. Also this mapping is described by a relatively low order polynomial, with the order depending on the grism being used (but generally limited to ≤ 4).

The first two components together define what we call spectra location, while the third component effectively provides the wavelength calibration for the VIMOS spectroscopic data.

As described in Sect. 2, VIPGI recipes are designed to be run in a semi-automated way, and the astronomer using them has available interactive tools

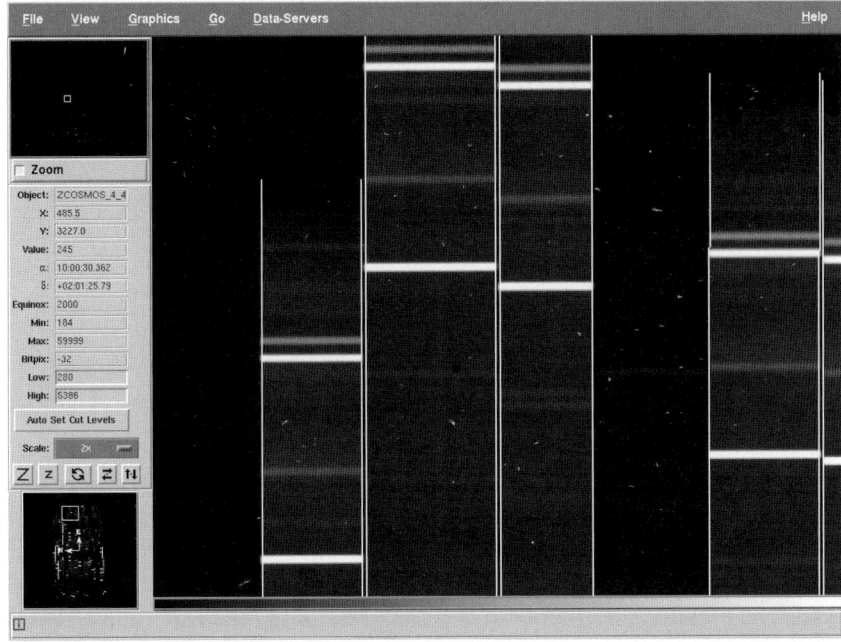

Fig. 2. The output of the spectra location verification tool. The white vertical lines superposed to the original VIMOS exposure frame identify the left and right edges of the spectra location for each slit in a VIMOS MOS mask. Only a small portion of the full frame is shown

to adjust their behavior and to verify the quality of the various data reduction steps. In particular, experience has shown that it is very important to be able to adjust the instrument calibration first guesses, because of the relatively large flexures affecting VIMOS, and therefore the stability of its calibrations. Figure 2 provides an example of the output of one quality control tool, designed to verify the quality of the automated spectra location on the CCD. Should this or similar other tools show the need for an adjustment of the calibration first guesses, an interactive visual procedure has been devised, taking advantage of the image display and external communication capabilities of the DS9 display tool. Using the so-called DS9 regions, and the first guess calibration stored in the FITS file header, VIPGI can overlay to a VIMOS arc lamp exposure frame the expected location of the arc lamp emission lines. The user can then shift the entire set of DS9 regions (dragging around the DS9 regions with the mouse), simulating a shift in the first guesses, until the expected location of the arc lamp lines visually matches the real position in the exposure. Finally, when the match is considered satisfactory, the first guess calibrations are automatically recomputed on the basis of the new location of the DS9 regions on the image. An example of two steps of such an interactive adjustment is provided in Figs. 3 and 4.

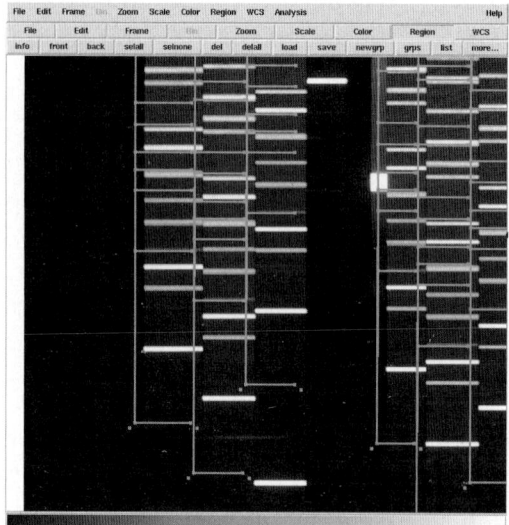

Fig. 3. The starting point of the first guesses adjustment process. The thin horizontal lines, connected by vertical thin lines, visualize the expected location of the arc lamp emission lines. They are superposed to the image of an arc lamp exposure, with the real emission lines appearing as the thick horizontal lines. Only a portion of the lamp frame is shown

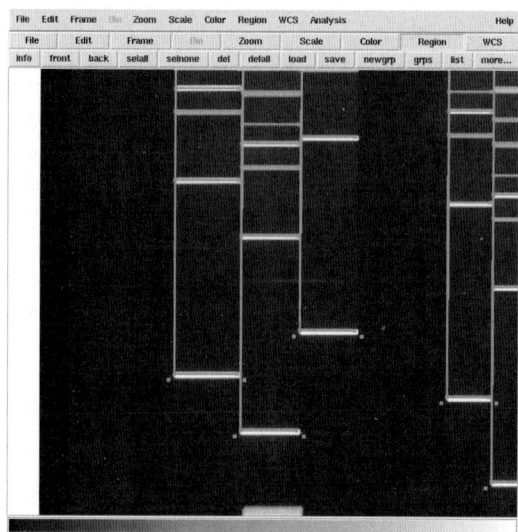

Fig. 4. A slightly smaller part of the arc lamp frame as in Fig. 3, except that now the expected location of the arc lamp lines has been shifted to match the observed position of the lines

4 The VIMOS Data Wavelength Calibration

The Inverse Dispersion Solution, which provides the wavelength calibration for VIMOS spectra, is obtained for each MOS slit by measuring the position on the CCD of a number of strong emission lines. This operation can be carried out using either a calibration arc lamp exposure (a combination of He, Ne and Ar lamps can be used for this purpose) or the night-sky lines present in a science exposure. The list of lines to be measured is provided by a pre-defined Line Catalog. The position of each line is derived computing the barycenter of the light distribution within a pre-defined extraction window, and measured positions are fitted against the known line wavelengths using a low order polynomial function via an iterative sigma clipping procedure.

The accuracy of the wavelength calibration changes slightly from grism to grism, but the *rms* residuals around the best fitting relation typically amount to one fifth of a pixel, when the calibration is carried out using a lamp exposure. The accuracy which can be achieved using night-sky lines is typically ten times worse, due to the combination of a weaker signal for each line with lines crowding in the red part of the spectrum, or with sheer lack of lines in the blue part of it. For the VVDS data obtained with the Low Resolution Red grism, which produces spectra with a linear dispersion of 7.14 Å/pixel, a fit with a third degree polynomial to calibration lamp exposures results in a wavelength calibration with a median *rms* residual of 1.005 Å (measured over a total of 20,000 slits). Although the distribution shows the presence of a tail extending to high rms residual values, only approximately 0.9 percent of the slits show a wavelength calibration with rms residuals larger than 3.0 Å. No systematic variation in the calibration accuracy is observed as a function of slit position within the MOS masks, or as a function of VIMOS optical quadrant.

Still, the accuracy of the arc lamp calibration fit in itself does not guarantee a comparable quality in the wavelength calibration of the science exposures used to derive the redshift measurements, mostly because of the instrument flexures inside VIMOS. Therefore one very important step of the data reduction process is the refinement of the wavelength calibration that is carried out on the science exposures themselves, using some of the brightest night sky lines. For each individual slit and for each individual exposure (typical deep VIMOS exposures are split in a jitter sequence of 5–10 individual exposures) an offset is applied to the wavelength solution obtained from the arc lamp data, to obtain a match between the observed and the expected wavelength of the sky lines. As a final quality check on the accuracy of the wavelength calibration, we extract the night sky spectrum for each MOS slit, and resample it to a linear wavelength scale, using exactly the same procedure used to extract and resample the spectra for the astronomical targets in the slits. We can therefore compare the observed and expected position of the night sky lines in the final data, and quantify very precisely and directly the

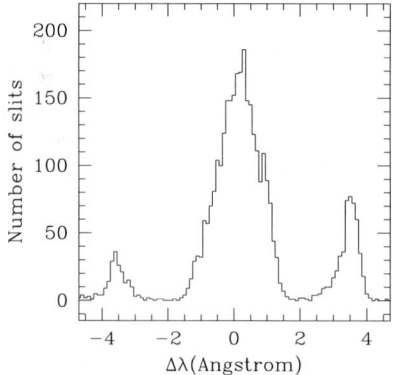

Fig. 5. The distribution of the differences between observed and expected wavelength for one night sky line for 4,200 slits over 32 VIMOS MOS masks. The data are obtained with the Low Resolution Red grism, with a dispersion of 7.14 Å/pixel. The median offset is 0.27 Å, while the two quartiles of the distribution are at −0.41 Å and at 1.16 Å, respectively. The effect of the data resampling is visible in the two small peaks offset by approximately one half of the dispersion value

effect of wavelength calibration uncertainties on the redshift measurements. One example of such a comparison is presented in Fig. 5.

5 Redshift Measurements

The most time-consuming step in the reduction and analysis of VIMOS spectra is, by far, the redshift measurement. A large effort has been devoted to develop a software package that could automatize to a large degree the procedure, and also interact with VIPGI and use its output files, so that the astronomer would see the data reduction and analysis as steps of one global procedure, carried out within a single software environment. The result of this effort is Easy-Z (EZ), which is a software package developed using the C language for the computationally intensive tasks, and Python for the user interface. EZ can be used as a stand-alone tool, but it is also capable of working in tight integration with VIPGI, and it can be run entirely from within the VIPGI graphical user interface.

A typical redshift measurement run with EZ is carried out in two steps. First, EZ is run in a blind, fully automated way over all spectra extracted from a VIMOS mask. In the second step, the astronomer examines interactively each single spectrum, and decides which measurements from the first step are correct, and which are to be repeated. During this phase also a quality flag, which is used to quantify the reliability of the redshift, is assigned to each measurement. The human control over each single redshift determination is made necessary by at least three factors. First, there is the intrinsic difficulty in deriving spectra of very faint galaxies, with a surface brightness only a few percent of the sky one, at the red edge of the optical spectrum, a wavelength domain contaminated by many strong sky emission features. Second, there is a limitation to the data quality introduced by the relatively strong fringing affecting the VIMOS CCDs. The data reduction procedure is designed to remove to a certain degree the fringing residuals present in the spectra for $\lambda \geq 8500\,\text{Å}$, but once again instrument flexures limit quite severely the effectiveness of this removal. Third, there is the problem of the contamination between slits. To obtain a very high multiplexing VIMOS spectra are packed quite tightly on the CCD, and in particular more than one spectrum can be located on one CCD column (up to four spectra with the low resolution grisms, and up to two with the medium resolution one). It is therefore possible for the zero order spectrum produced by a given slit to overlap the spectrum produced by a slit below it in the MOS mask, and/or for the second order spectrum to overlap the spectrum produced by a slit above it in the MOS mask. Since each MOS slit is reduced independently with the VIPGI recipes, there is no way at the moment to take automatically into consideration these overlaps and the contamination they might produce in a given spectrum.

In the case of both the VVDS and zCOSMOS, redshift measurements are actually being carried out for each VIMOS pointing by two astronomers independently, and their measurements are then compared and discussed to resolve all possible discrepancies. At the end of such a labor intensive procedure, the measurements can be demonstrated to have a very high quality. Such a demonstration is obtained from two sets of comparisons: either an internal comparison from repeated measurements of a small fraction of objects, or an external comparison with redshift measurements obtained with other instruments for the same objects. Internal comparisons have been obtained both for the VVDS and for the zCOSMOS data. In the VVDS case, repeated observations of 150 galaxies, carried out with the Low Resolution Red grism (resolution 210), yield an rms scatter in the redshift differences of 0.001, equivalent to a radial velocity uncertainty of $175\,\text{km}\,\text{s}^{-1}$ at the median redshift of the VVDS sample ($<z>=0.76$). In the zCOSMOS case, repeated observations of 91 galaxies have been obtained so far, carried out with the Medium Resolution grism (resolution 580). They yield an rms scatter in the radial velocities of $72\,\text{km}\,\text{s}^{-1}$. The distribution of these differences is shown in Fig. 6. An external comparison is available for the VVDS sample with the FORS2 observations in the GOODS south field, from both the K20 and the GOODS surveys (41 and 27 galaxies, respectively). This comparison shows an rms scatter in the radial velocities of $283\,\text{km}\,\text{s}^{-1}$ (see Fig. 6 in [6]).

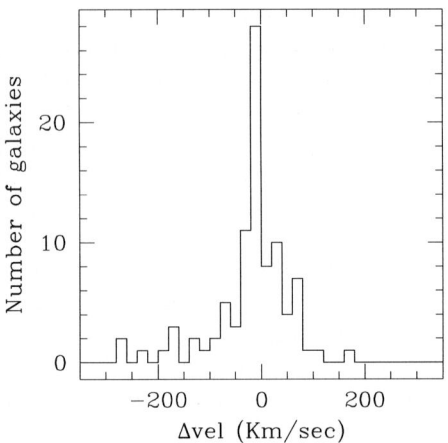

Fig. 6. Differences in redshift measurement (converted into radial velocity differences) from repeated observations of 91 galaxies in the zCOSMOS sample. Observations were obtained with the VIMOS Medium Resolution grism, which produces spectra with a resolution of 580 and a dispersion of 2.5 Å/pixel. The rms scatter in the velocity differences is $72\,\mathrm{km\,s^{-1}}$

6 Conclusions

Using the combination of VIPGI and EZ it is possible to obtain redshift measurements with VIMOS of a quality that correspond to the design specifications for the instrument. The price to pay for the achievement of such high a quality is a relatively large degree of interactivity in both the data reduction and in the redshift measurement processes, which still makes the development of redshift surveys with VIMOS a very demanding task in terms of human resources. While this is still a manageable task with the current instrumentation, it is quite clear that the current operational scheme will not be capable to withstand the impact of the next generation of instruments and surveys, where one could foresee the goal of measuring one million redshifts.

References

1. O. Le Fèvre, et al.: SPIE **4008**, 546 (2000)
2. O. Le Fèvre, et al.: The Messenger **109**, 21 (2002)
3. O. Le Fèvre, et al.: A&A **439**, 845 (2005)
4. S. Lilly, et al.: ApJS **172**, 70 (2007)
5. N. Scoville, et al.: ApJS **172**, 1 (2007)
6. M. Scodeggio, et al.: PASP **117**, 1284 (2005)

Discussion

L. Guzzo: The level of effort that you have to add in measuring z's with the VIMOS pipeline strongly depends on how strongly you are pushing the limit of the instrument. If you go to $I_{AB} \sim 24$ like VVDS-Deep, with 4 to 5 hour exposures, you are working at the limit of VIMOS and therefore you need a significant effort to measure z's for the faintest objects. With more relaxed observations (less aggressive strategy) and better SNRs, we can end up measuring virtually all z's automatically. Obviously, eliminating the fringing would improve significantly this issue.

M. Rejkuba: By how much are the velocity determinations affected by the ripples introduced by OS/red (OS/blue) filters that are used together with low resolution grisms?

M. Scodeggio: For normal galaxies they are not an issue, because real spectral features are much narrower. They can be an issue for broad lined AGNs, as they can mimic faint broad features.

F. Patat: I happen to have here one of your VIMOS masks. May I circulate it in the audience, so that people get to see one of your problem's sources?

M. Scodeggio: Mask problems are, of course, a cause of technical downtime, but from the pipeline point of view, we only get to see the results of successful mask insertion.

D. Baade: Has it been tested whether flatfield observations obtained at night can alleviate the fringing problems?

M. Scodeggio: For the red grisms it looks like no improvement can be obtained. The flatfield lamp and the night sky are just too different to produce comparable fringing patterns.

U. Hopp: You emphasized that every step in the data reduction should be controlled by hand which is fine for smaller to medium sized samples. Given the projects in cosmology aiming for several million redshifts of galaxies, what are the bottlenecks of your pipeline and when and how can it be improved?

M. Scodeggio: At the moment with VIPGI, it takes between one and two hours to fully reduce one VIMOS pointing. To that we need to add one to three days of work to measure redshifts for faint galaxies (especially for objects at $z > 1.4$). Thus the bottleneck is entirely in the redshift measurements. Either we improve vastly our methods for measuring redshift or we will need spectra of much higher SNR to work with.

D. Osip: Why are the night time flats such a poor fit to night spectra so that fringing cannot be removed?

M. Scodeggio: There are basically two factors that come into play. One is the different spectral energy distribution of the lamp with respect ot the night sky. The other is the fact that the flat is taken using the Nasmyth shutter screen, producing an extreme near-field illumination.

Part IV

Session 3: Near-IR and mid-IR Spectro-Imagers

IOT Overview: IR Instruments

E. Mason

ESO, Alonso de Cordóva 3107, Vitacura, Santiago, Chile; emason@eso.org

Abstract. In this instrument review chapter the calibration plans of ESO IR instruments are presented and briefly reviewed focusing, in particular, on the case of ISAAC, which has been the first IR instrument at VLT and whose calibration plan served as prototype for the coming instruments.

1 Introduction

NIR and IR instruments at ESO are:

- **ISAAC** is a spectro-imager consisting of two arms which operate in the wavelength range 1–2.5 µm (SW) and 2.5–5 µm (LW). As imager it offers a FoV of $2.5'\times2.5'$ and $0.148''$/pix in the SW arm (J, H and K) and $1.25'\times1.25'$ and $0.07''$/pix in the LW (L and M). Linear polarimetry imaging is possible in the SW arm; while, fast photometry (Burst mode) is now offered in the LW arm. As a spectrograph, ISAAC offers low ($R\sim500$) and medium ($R\sim3000$–5000) resolution spectroscopy in both arms.
- **SofI** (son of ISAAC) is a copy of ISAAC SW arm. It operates in the wavelength range 0.9–2.5 µm and offers imaging in two different FoVs and pixel scales (FoV = $2.5\times2.5'$ or $5\times5'$ with scale of 0.144 and 0.28 $''$/pix, respectively), and low resolution spectroscopy ($R\sim600$–1500). SofI, too, allows linear polarimetry imaging observations.
- **NaCo** it is a NIR spectro-imager (λ range $\sim$0.8–2.5 µm) adaptive optic assisted. As imager it offers also polarimetric and coronographic observations with 3 different FoV and pixel scales (FoV = 56, 28 and 14$'$ with pixel scales of 0.054, 0.027 and 0.013 $''$/pix, respectively). Low resolution spectroscopy (up to $R\sim1000$) is possible with NaCo.
- **Sinfoni** is an integral field spectrograph operating in the wavelength range 1.1.–2.45 µm. It offers 3 different spatial resolutions ($0.25''$, $0.10''$ and $0.025''$/pix) and FoV ($8\times8''$, $3\times3''$ and $0.8\times0.8''$, respectively).
- **VISIR** is currently the only MIR spectro-imager at ESO. It operates in the N and Q bands between 8 and 24.5 µm. Its FoV and pixels scale depend on the wavelength in use ($\sim19\times19''$ at $0.075''$/pix, and $\sim32\times32''$ at $0.127''$/pix, in the N and Q band, respectively). As a spectrograph it offers low ($R\sim200$), medium ($R\sim3000$) and high ($R\sim25000$) resolution spectroscopy.

CRIRES will join the IR instrument team starting from period 79 and will offer very high spectral resolution (up to 10^5) in the wavelength range 0.95–5.4 µm.

Adaptive optic instruments, integral field units and high resolution echelle spectrograph will be reviewed in other contributions. Therefore, here, I will present only on the basic calibrations required by IR observations, focusing in particular on ISAAC.

2 The Calibration Plan

A calibration plan consists on the set of frames taken to allow data reduction, removing all the instrumental signatures. However, it also includes monitoring of the instrument performances. Indeed, we can distinguish different types of calibrations depending on their purpose:

- **calibrations for science**: they include all the day and night time observations which are taken to allow science data reduction and removal of the instrumental signatures. Example of these are the standard star observation, flat fields, dark/bias frames, wavelength calibration frames, etc ...
- **calibrations for the instrument characterization**: they are also calibration for the removal of the instrumental signatures but they are not needed in all the science cases. They can include the field distortion, or the illumination frames for the second order correction of the detector response, etc.
- **calibrations for monitoring**: they are frames and/or checks which are not normally distributed to the user. They are meant for internal use in order to monitor the status and the performance of the instrument. However, some of the calibration for science also serve to monitor the instrument status.
- **technical calibrations**: these are mainly tests and measurements taken after an instrument intervention in order to redefine all its configuration files. They include, for example, the procedures to determine the center of rotation, the exact orientation on sky of the detector, the grating dispersion and the pixel to wavelength correspondence for the spectroscopic set ups.

These types of calibrations are common to all the instruments, regardless of their operational wavelength. However, the actual set of calibrations included in each type depend on the instrument. Namely, a calibration plan is the result of the combination of the instrument history (the longer the instrument has been in operation and the larger might be its calibration plan), the instrument specific problems (sets of calibration might be taken to monitor a misbehaving function), and the instrument usage (e.g. modes which are never requested by the user community might end up receiving minimal support in term of calibrations).

The calibration plans of the IR instruments currently offered at ESO are detailed in Table 1. Note that the table does not include the technical calibrations as they are not relevant to this workshop and chapter.

In the following sections I will present some specific aspects of both science and monitoring calibrations, comparing, in particular, the current ISAAC plan with ideal cases and the plan of the other instruments.

3 Calibrations for Science: A Few Notes

All NIR spectrographs (ISAAC, SoFI, NaCo, Sinfoni) provide darks, spectroscopic flats, and arcs to the user. These frames represent the basic calibration set to reduce IMG and SPC data. These calibrations might not be sufficient for very high S/N observations or very high precision velocity determinations. The user who aims at high accuracy photometric or spectrophotometric measurements should require extra calibrations either at night or day.

3.1 Darks

IR detectors have no shutter so there is no bias, i.e. the zero time exposure to evaluate the detector background level/noise. The corresponding calibration is the dark. As the darks level varies with the readout mode and the detector integration time (DIT), science frames are accompanied by a set of 3 darks taken with exactly the same DIT and readout.

Note that dark subtraction is not strictly necessary if we consider that NIR observations are performed by offsetting (jitter and nodding) the telescope or tilting the secondary mirror (chopping) and that the subtraction of a pair of nodded frames automatically removes the background level. However, if the user wants to scale to a common background level a set of frames before stacking the images, then the dark subtraction is critical, as it otherwise leads to incorrect estimates of the background level. This is why imagers such as ISAAC, SofI, and NaCo deliver dark frames to the user.

3.2 Flat Field

Flat fields allow to correct for different response of the pixels in the array. ISAAC and NaCo offer twilight flats for the SW IMG observations[1]. Twilight flats uses the decreasing sky brightness to sample the dynamic range of the instrument and compute a normalized gain map of the detector. However Hawaii detectors (ISAAC SW arm detector) have dark levels which

---

[1] LW IMG flats are dominated by the thermal background and can be taken any time of the night (sky flats).

Table 1. Details on the calibration plans of ESO's IR instruments

Calibration Type	Science	Instrument Characterization	Monitoring
ISAAC			
	– darks – sky/tw flats (IMG, IPOL) – screen flats (SPC) – arcs – phot std – telluric std – night flat/arc	– star trace – illumination correction – field distortion	– darks (SW, LW, all readout modes), ZP, telluric of the month, lamp efficiency – slit image – grating check – flat check – linearity
SofI			
	– darks – dome flats (IMG, SPC, IPOL) – arcs – phot std – telluric std	– star trace	– linearity
NaCo		-	
	– darks – sky/tw flats (IMG) – screen flats (IMG, IPOL, SPC) – arcs – phot std – telluric std – night flats/arc		– dark, ZP, lamp efficiency – slit image – linearity – gain
Sinfoni		-	
	– darks – screen flats – arcs – telluric std		– dark, lamp efficiency, off-lamp check, grating check, resolution
VISIR		-	
	– phot std – telluric std		– dark, sensitivity/ZP – IMG flats on both detectors

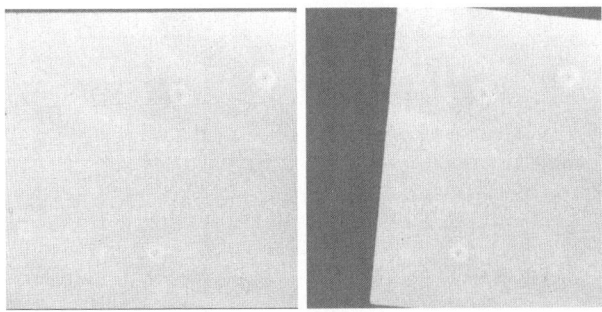

Fig. 1. Sofi dome flats. *Left*: full frame flat. *Right*: masked flat

depend on the detector illumination (total flux), being higher for stronger illuminations. The ideal case would require the subtraction of a true dark level from each frame. This is currently not possible within ISAAC. However, SofI, being equipped with dome flats and having the possibility to partly mask few columns of its detector, has developed a particular procedure to provide the most accurate master flat. Pairs of high and low count level flats are taken. Each pair consisting of a fully illuminated flat and a masked one (see Fig. 1). Within each pair, the difference between the fully illuminated frame and the exposed part of the masked frame, provides an estimate of the additional dark level in the strongly illuminated Hawaii detector. This additional term is subtracted from the full frame flats before computing the master flats.

It should be noted that Aladdin (ISAAC SW arm, NaCo) detectors are not equally sensitive to the frame illumination and the dark is much more stable, thus they do not require similar care.

Spectroscopic flats are taken during day time. They consist of three pairs of images, each pair being a lamp ON frame and a lamp OFF frame.

Care must be taken when flat fielding spectra, if there is need to remove fringing. ISAAC SW MR spectroscopy is affected by fringing at a $\sim 2\%$ level. Movements of the objective and/or of the grating produce different fringe patterns which are not perfectly removed by the flat field correction. In principle, the only way to safely remove fringing is by taking night time flats both at the end of the science observation and the telluric star. Alternatively the science and the telluric should be observed exactly at the same positions thus, when correcting for the telluric, the fringes are automatically removed too. However, ISAAC's objective is subject to flexure (Fig. 2) which we compensate for through frequent (namely, at each spectroscopy acquisition) realignments of the function, but we cannot avoid. These flexures depend on the rotator position and therefore are more noticeable whenever the rotator moves fast (i.e. for observations which cross the meridian in proximity of the Zenith)

SW spectroscopy in the LW arm of ISAAC should be attempted, to see whether high S/N, fringing free observations can be offered.

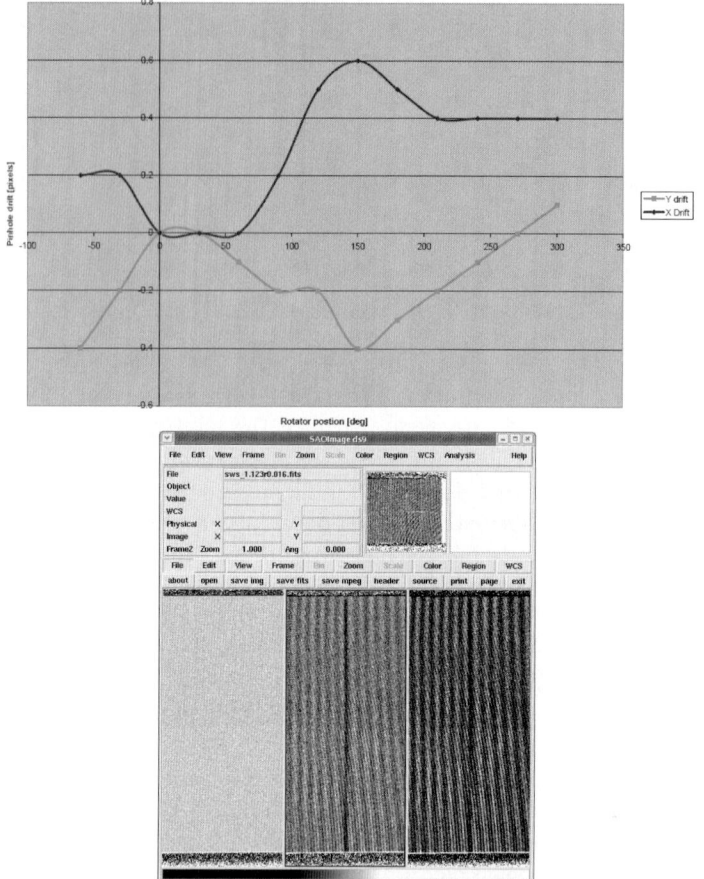

Fig. 2. *Top*: observed flexure as function of the rotator position. *Bottom*: examples of ISAAC fringing patterns caused by the misaligned objective. From *left* to *right*: ratios of flats taken with aligned (*left*) and misaligned objective (*center and right*)

Detector fringing is observed also in VISIR. In particular, the "scanner" for the fine positioning of the grating in high resolution mode suffers from flexure, too, which depend on the telescope pointing direction/declination (Fig. 3). This causes both fringing and degradation of the spectral resolution. Solutions such as *(i)* modeling the flexure for each pointing position or *(ii)* wavelength calibration of each single frame before image stacking are currently under study.

Note that detector fringing depends also on the stability of the sky conditions and not just on the stability of the instrument. Figure 4 shows two different stars observed with VISIR and exactly the same setup. The brighter one (left panel) is affected by fringing which could not be removed as expected

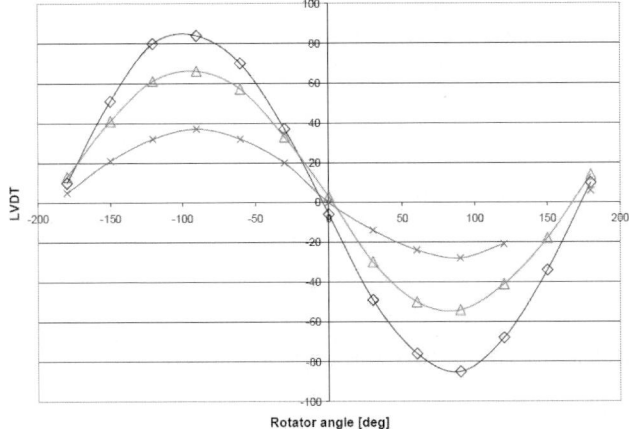

Fig. 3. Scanner flexure observed in VISIR at different declination (*diamonds*: $\delta = 45°$, *triangles*: $\delta = 60°$, *crosses*: $\delta = 75°$ above the horizon) and different rotator angles. Horizontal lines are in step of 5 pixel. Flexure of the scanner can be as large as 20 pixels, on the detector

by the combination of the chop-nodded pair. A probable solution, which still needs to be verified, will be to scale to a common background the individual frames, before image stacking.

3.3 Photometric Standards

Few words should be spent also on the photometric standard stars. Low and high airmass photometric standards are typically observed during the evening twilight for the classification of the night and as a check of the instrument status (the latter being particularly true in the case of NaCo).

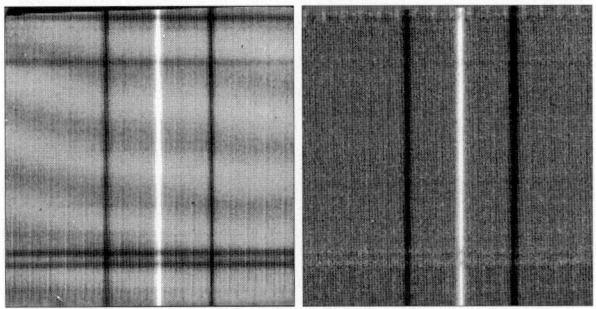

Fig. 4. An example of VISIR fringing caused by variable sky background during the exposure (*left*), compared to observations not affected by fringing (*right*)

This calibration plan might not be sufficient for accurate photometry to a few percent level, nor provides information about the stability of the night. Therefore, though the observatory is working on the improvement of its calibration plans, the users are still recommended to submit their own photometric standard whenever the observatory calibration plan is judged inadequate to their science goals. Soszynski et al. [1] by providing a set of six photometric standard stars which bracket their science targets have reached a 0.03–0.04 mag accuracy for 18–19 mag Chepheids in nearby galaxies, with ISAAC.

4 Calibrations for Monitoring: Examples

Calibration for monitoring are intended to check the status of the instrument and its health. Several of the science calibrations can be used for this purpose. For example, the darks are used to check the stability and/or variation in the dark level and read out noise (RON). The dark level might vary with the temperature; while higher RON values might be symptomatic of problems in the electronic controlling the detectors.

The zero points are another example of science calibrations used to check the instrument health. NaCo, observes, daily, a standard star at the beginning of the night in order to check the instrument and the AO alignment.

There are other calibration, however, which are just for internal use. The parameters which are monitored and the frequency of the monitoring depend on the instrument. As an example, detector linearity is checked on a semi regular basis every 6 months at ISAAC, taking a sequence of flats a different level of illumination. NaCo and Sinfoni monitor the detector linearity with about the same frequency while VISIR does not. The detector gain, on the contrary is currently monitored only at NaCo. Another example is the odd even column effect which has been observed several times, in the past in the ISAAC Hawaii detector. It consists of a count level offset between the odd and the even columns of the array and depends on both the detector illumination and the readout speed. The odd-even column effect is checked every time twilight flats are taken. Should it appear larger than 1%, the minimum DIT and the read speed are reviewed (increased).

IR instruments, more that optical ones, might suffer from reproducibility problems. The monitoring of a misbehaving function is twofold. First, it allows to check that the function motion or any parameter associated to it stays within the tolerated limits. In the case those limits have been passed, the appropriate action is taken. Second, a long term monitoring of a function helps in characterizing it, its problems and their possible causes. All the monitored parameters are in the quality control (QC) web pages. Important and/or critical parameters are posted in the "daily health check" pages and checked every day for outliers and trends. The QC1 database gathers all the quantities monitored by QC and produced by the instrument pipeline. These quantities can be plotted and/or retrieved to check long term trends, etc.

An example of a monitored function in the QC1 database is given by the ISAAC objectives which suffers of flexure and reproducibility problems. Small movements of the objective can be spotted through the slit image which appears horizontally shifted with respect to its nominal position. These shifts can occasionally be as large as ≥ 1 pixels and causes poor centering of the scientific target in the slit (see Fig. 5, top panels). Therefore we need to continuously correct for those offsets. In particular, the objective is re-aligned at each spectroscopic acquisition. The two bottom panels of Fig. 5 show the horizontal slit position vs time and the corresponding histogram. By fitting the histogram with a normal Gaussian distribution, we derive $\sigma \sim 0.3$ pixels. Paranal Science Operations (PSO) currently realigns the objective, every time the slit image has drifted by more than 0.1 pixels. 0.1 pixels

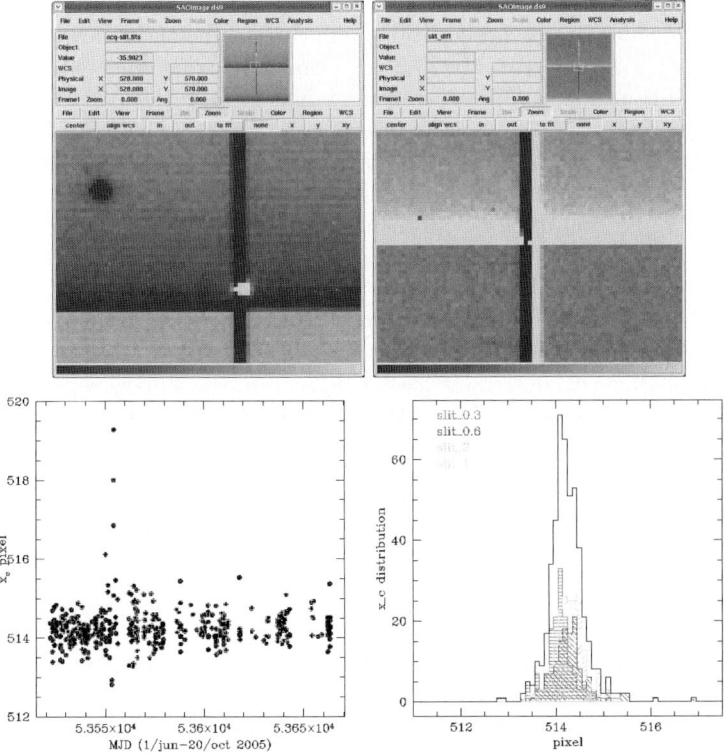

Fig. 5. *Top panels*: combined images showing (*left*) the star centered at the nominal slit position superposed with the real slit position at the time of the acquisition; (*right*) the difference between the aligned (*bright*) and the misaligned (*dark*) slit images. *Bottom panels*: x position of the slit image across time (*left*) and the corresponding histogram (*right*)

represent 5% of the 0.3" slit width, but only 1.5% of the 1" slit. Adopting a 5% tolerance for every slit, PSO would need to realign the objective in roughly 50% of the acquisitions (assuming that all the slits – but the 0.8" one – are equally popular among the users). A more automatic alignment procedure with different tolerances for different slits is currently under "construction".

5 Conclusions

In summary, I have illustrated some of the reason for which specific calibration (either for science and monitoring) are taken. In many cases instrument calibration plans are very similar to each other, as obvious. Still, the ideal calibration plan should be tailored to each instrument depending both on the science which is typically performed with that instrument and the instrument capability. For example: wide field imagers used for deep imaging surveys need second order correction of the detector response (illumination correction), which is not needed by small FoV instruments devoted to observations of a single object at a time. An instrument affected by severe reproducibility problems of its functions might require that the calibrations are taken just after the science observation and not on the following morning after the telescope has been moved and different instrument setups have been used. It is at this point that the feedback from the user community is important and reports about problems encountered during data reduction or requests for specific science cases are valuable input for the continuous improvement of the instrument performances. I conclude with the example of ISAAC Burst mode which has been offered after tests triggered by user request. Now, the use of ISAAC Burst mode for the observations of lunar occultations allows to determine stellar diameters and binary separations to the limit magnitude K$\sim$11–12 and resolution Φ $\sim$0.5–1.5 mas, in competition with VLTI (see Richichi et al. [2]).

References

1. I. Soszynski, W. Gieren, G. Pietrzynski, F. Bresolin, R. P. Kudritzki, J. Storm: ApJ **648**, 375 (2006)
2. A. Richichi, O. Fors, E. Mason, J. Stegmeier: Messenger **126**, 24 (2006)

Some (Little) Thing(s) about VISIR

E. Pantin[1], L. Vanzi[2], and U. Weilenmann[2]

[1] DSM/DAPNIA/SAp, CE Saclay, UMR 7158, France; `eric.pantin@cea.fr`
[2] ESO, Alonso de Cordóva 3107, Vitacura, Santiago, Chile

Abstract. VISIR is the VLT mid-infrared Imager and Spectrometer. It offers a comprehensive set of observing modes, imaging in N and Q bands, at the limits of the telescope diffraction, as well as spectroscopy in the same bands. In particular, VISIR provides a very high-resolution spectroscopy mode with an achieved resolution up to 30000 in N band; this mode is so far unique in the southern hemisphere. VISIR calibration is quite specific when compared to standard visible/near-infrared ones. Various dedicated methods have to be developed to remove the instrumental signatures and obtain the best scientific return.

1 The Instrument

VISIR is mounted on the VLT unit telescope Melipal at Paranal and has been commissioned in April 2004. Since April 2004, it operates routinely to deliver mid-infrared (mid-IR) images and spectra in N and Q bands (atmospheric windows centered on 10 and 20 µm respectively). The Imager and the Spectrograph are two entities physically separated, having each one their own optics and detector. Given the huge background emitted by both the atmosphere and telescope, background cancelling techniques are compulsory (see Fig. 1). The instrument itself emits a negligible background since its interior is cryogenically cooled between 50 and 70 K. VISIR Imager uses typical chopping frequencies around 0.25 Hz and nodding period around 1 min. VISIR spectrometer uses the same parameters in low and medium resolution modes, but given the small number of photons impinging onto the detector at much higher spectral dispersions, 10 times larger periods are set for the high resolution mode.

1.1 The Imager Sub-instrument

The imager is equipped with a 3-mirror anastigmat all reflective design collimator, feeding the detector with diffraction-limited performance images. It is equipped with a DRS 256×256 pixel detector designed for high-level fluxes. The imager provides 11 filters (at several wavelength of interest, such as the PAH bands at 8.6 and 11.3 µm) covering the N-band and 3 in Q-band, matched to the atmospheric sub-bands between 17.5 and 19.5 µm. The imager is equipped with calibration devices such as pupil imaging and a pinhole

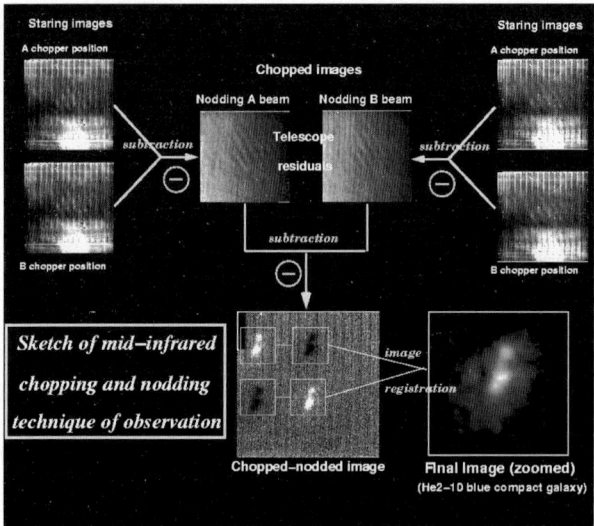

Fig. 1. Scheme of chopping and nodding technique used to suppress the very high mid-IR background. First, the elementary images are averaged (resulting in staring images) over half a chopping period. Then, the chopping is applied for each telescope nodding position. Finally, chopping residuals due to different optical paths in the two chopper positions, are canceled by subtracting the two nodding beams

mask, used for measuring the pixel scale and the optical distortion in the field.

1.2 The Spectrometer Sub-Instrument

VISIR spectrometer optics have a similar all-reflective design. Various gratings are mounted in the instrument, allowing to perform low (LR mode, $R \approx 300$ in N band) and medium resolution (MR mode) spectroscopy ($R \approx 3000$ in N band, $R \approx 1500$ in Q band). VISIR offers also a high spectral resolution mode ($R \approx 30000$ in N band, $R \approx 15000$ in Q band) thanks to a duo-echelle grating. Some order-sorting filters are provided allowing to observe spectral lines of physical interest (e.g. H_2, Ne II) with a slit length of 32.1″ on the sky (long-slit, HRS mode); a cross-dispersed mode allows to set any central wavelength onto the detector, but with a reduced slit length (about 4″, HRX mode). See [4] for a more complete description of the spectrometer sub-unit.

2 The Warm Calibration Unit (WCU)

On top of VISIR enclosure, lies the Warm Calibration Unit (WCU). This unit comprises a point source (monochromator), and extended source (Peltier plate) presenting an adjustable surface temperature (-20 to $50^\circ C$), and a telescope simulator (Offner design). This unit allows to perform daytime checks and calibrations (e.g. image quality, relative sensitivity, flat data measurements).

3 Detector Features

The two detectors that equip VISIR have to be fine tuned for the conditions we are working in. They are cooled by pulsed Sumitomo cryo-coolers. Although much efforts have been devoted to damp temperature oscillations, the detectors still feel periodic (1 Hz) temperature fluctuations inducing slight changes in the detector gain. To minimize this effect, some heaters are glued to the detector sockets, and stabilize the detectors at temperatures between 6.0 and 9.0 K. Residual temperature fluctuations (therefore detector gain variations) are canceled by keeping in phase the chopper with the cryo-coolers.The two DRS detectors have well-capacities around $2\times10^7\,e^-$ (large capacity mode) and $2\times10^6\,e^-$ (small capacity mode). Quantum efficiency is typically around 50%. For a detailed study of the detectors characteristics, see [3].

The detector's working point is nominally set around +5000 ADU on the sky (1 ADU $\approx$ 275 e^-, given a total range of $[-32000, 32000]$ ADU. The detectors start to saturate when the levels reach about 8000 ADU and more (about 2/3 of the total capacity). Some pixels have an abnormal behaviour (gain) and will trigger striping (due to multiplexer amplifiers collapse, see Fig. 2) over a usually wide range of rows on the detector. This striping is periodic, with a period of 16 columns, since one amplifier reads every 16 pixels sequentially.

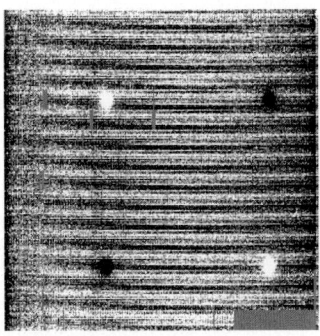

Fig. 2. Imager data showing the "striping effect". This striping is periodic (16 pixels) and depends at first order, on background fluctuations that affect "hot" pixels

4 Instrument Performances

Instrument performances are mainly monitored using standard stars from Cohen's database [1] (see also Sect. 4.2) which are systematically observed at the beginning of a VISIR observing night. The background level, the Point Spread Function (PSF) as a measure of image quality, and the point source conversion factor (between detector ADU and physical units) and sensitivity are estimated. All these values are reported in a Quality Control summary provided by the ESO's Data Processing and Quality Control Group (http://www.eso.org/observing/dfo/quality/VISIR/qc/qc1.html, and also Dobrzycka, these proceedings).

4.1 Sensitivity

Imager

VISIR imager sensitivity over the period Sep 04–Dec 04 is reported in Fig. 3. The median sensitivities are close to expected values within a factor of 2. Elements concerning effects possibly degrading these performances are given in Sect. 5. While the point source sensitivities in N band are remarkably stable in time and under various weather conditions, Q band ones are much more sensitive to the precipitable water vapor (PWV) content and airmass.

Image Quality

VISIR imager image quality achieves diffraction-limited performances in N band when the optical seeing is $\leq$ 0.8 arcsec. For instance, in PAH1 filter (8.6 μm), the spatial resolution of VISIR is PSF (FWHM) is close to 0.25 arcsec. On a 8.2 m telescope, the atmosphere turbulence starts to play a significant role, degrading the spatial resolution for standards optical seeings

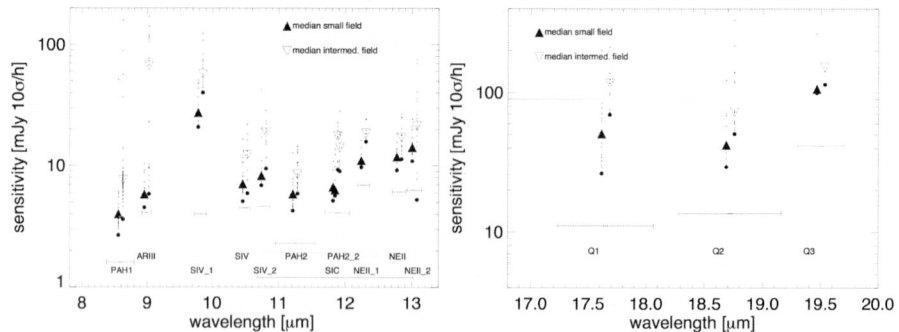

Fig. 3. Sensitivities of the VISIR Imager, for all filters, in N and Q bands. Both pixel scales small field (0.075″/pixel) and intermediate field (0.127″/pixel) are shown

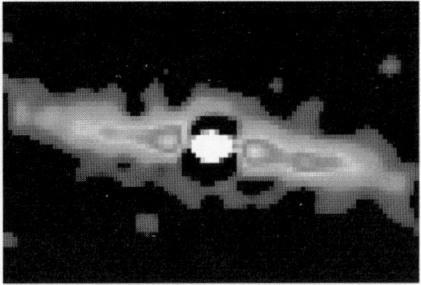

Fig. 4. VISIR image of the β Pictoris dust disk at 11.2 μm (PAH2 filter) (*left panel*) has been deconvolved using a standard star observations to estimate the PSF (*right panel*). This PSF differs slightly from the true one and creates spurious deconvolution artefacts (e.g. an inner dark ring around the star). With such data, myopic deconvolution is compulsory

(0.8 arcsec and more). In Q band, diffraction-limited PSF are routinely observed. VISIR images deconvolution is still a challenge since VISIR PSF, half affected by random seeing fluctuations, half by stable diffraction, is varying in time (see Fig. 4)

4.2 Instrument Calibration

Photometric and Spectro-Photometric Calibration

Imager photometric and Spectrometer spectro-photometric calibrations are usually performed using standard stars. These calibrators (81 in southern skies) have been chosen among Cohen's radiometric database [1] and adding a selection of A0 type stars for which the spectrum is well-known and relatively free of lines in the mid-IR. The experience over 3 years now shows that derived conversion factors (from ADU to physical Jy units) are quite stable in N-band (15% typical variations), and more variable in Q band (30% and more). In the specific case of high-resolution spectra, it is advised to use also bright asteroids observations to get a more accurate correction of the telluric absorption.

Wavelength Calibration of Spectra

VISIR data contain both sky-subtracted images and sky data. Sky spectra can be usefully compared to models of mid-IR atmospheric emission to derive a wavelength calibration. In the majority of the cases, except some very "clean" windows in N-band at medium resolution, it is possible therefore to "self-calibrate" the data.

5 Instrument Characteristics and Signatures

5.1 Detector Noise Studies

Given the level of the background, the detector dark current (smaller than 7000 e$^-$) and readout noises (1771 and 296 e$^-$ in high and low capacity modes) are negligible against the background shot noise. Rapid imaging experiments have been performed in order to study any other source of noise. These experiments show that a too low temperature of the detector (6.5 K), a 1 Hz oscillations noise generated by the cryocoolers pulse tube is observed (see Fig. 5). When slightly increasing the working temperature of the detector to 8.0 K, these thermal oscillations vanish (see Fig. 5), but at the cost of deterioration of detector cosmetics (increase of detector "striping", see Sect. 5.5). At 8.0 K also, the shot noise increases as expected (square-root of impinging number of photons), whereas its behaviour seems more erratic and less easily interpretable at 6.5 K.

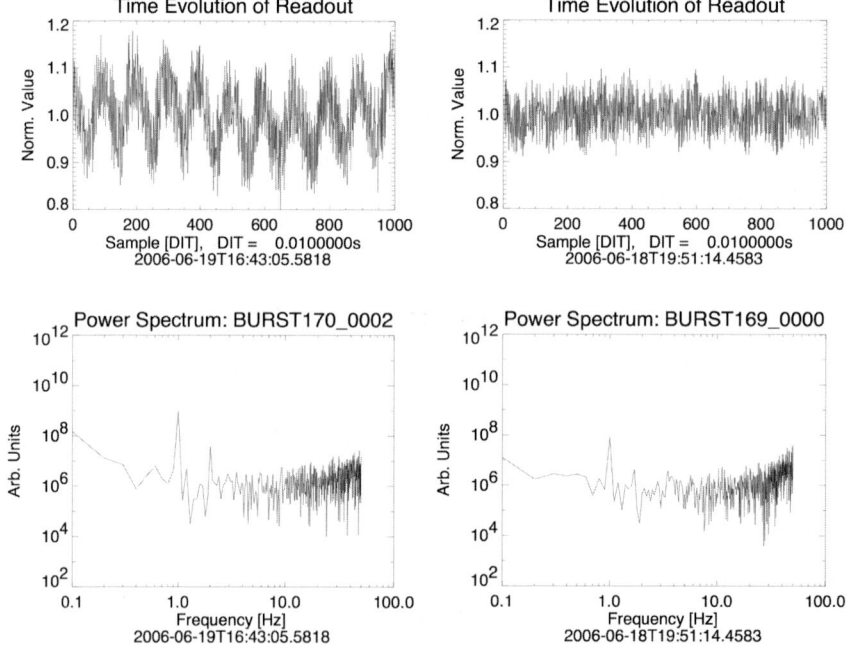

Fig. 5. Detector output power spectrum as a function of the detector temperature (*upper row* are the time signals). At 6.5 K (*left side*), a clear 1 Hz fluctuations noise is evidenced. At 8.0 K (*right side*), the 1 Hz noise is significantly reduced, and the time signal does not show a strong periodic modulation any longer

5.2 Background Noise

Once the detector is properly set at a temperature around 8.0 K, observation parameters such as the chopping frequency, number of rejected images after a chopper movement, etc, shall be adjusted. A study of the background noise (image as a whole) has been carried out. As seen in Fig. 6, for chopping frequencies larger than typically 1 Hz, the background noise is close to the expected noise level. However, for chopping frequencies lower than 1 Hz, the background noise starts to increase in a different way, according to the filter. This degree of noise increase seems to be linked to the quality of the atmospheric transmission in the given filter; the more affected by atmospheric transmission is a filter, the higher is the increase of background noise.

5.3 Background Errors and Noise

As seen in Fig. 7, VISIR background noise deviates from a pure gaussian and white noise behaviour. The excess of noise at low-frequencies degrades VISIR performances for faint extended objects making it particularly difficult to assess the object spatial extent and measuring its photometry (e.g. a faint spatially extended comet). This effect is also particularly annoying when trying to extract spectra. On the other hand, the point source performances are only weakly affected by this effect. This excess of low/medium spatial frequencies noise is probably related to the global effects described in Sect. 5.2. Some measurements of spectral characteristics of VISIR background noise as a function of observing parameters would be helpful.

5.4 How to Correct for this Background Errors?

As seen previously, this "extra noise" at medium and low spatial frequencies can be assimilated to a problem of varying background one has to deal

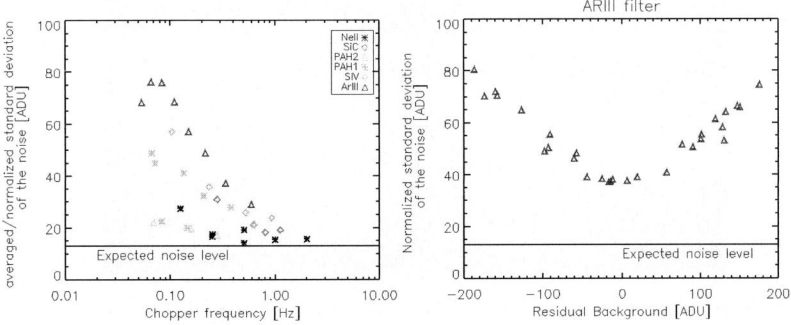

Fig. 6. *Left panel*: Background noise as a function of the chopper frequency. *Right panel*: Background noise as a function of the residual background level after chopping. As evidenced, the noise level is correlated with background fluctuations

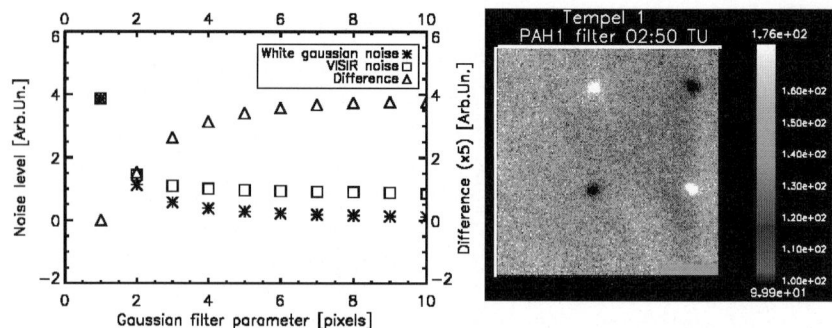

Fig. 7. Plot of the noise level as a function of a low-pass gaussian filter parameter σ. As the cut-off frequencies decrease, *the noise level gets higher than a pure gaussian noise having the same standard deviation than VISIR noise in original image*. This shows that VISIR noise has a different behaviour than a pure gaussian noise, and presents an excess of noise at lower frequencies, as seen visually in the *right panel* image

with. In order to properly assess some objects photometry, it is possible to mask the observed objects and reconstruct the underlying background. Usual methods are very poor in doing this job. However, using wavelets+curvelets reconstruction [2], it is possible to get reasonable results (see Fig. 8).

5.5 Detector Striping

As already discussed in Sect. 3, VISIR DRS detectors are affected by striping triggered by some abnormal behaviour of some pixels. Although the striping

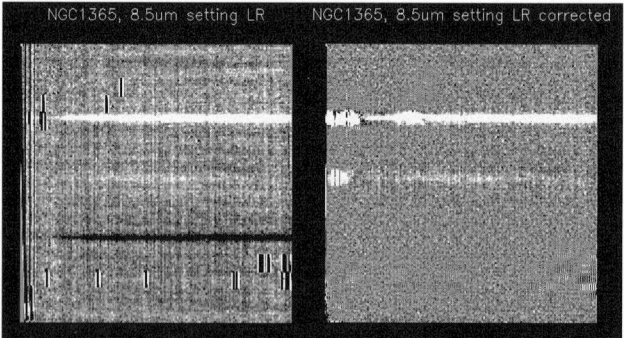

Fig. 8. *Left panel*: low-resolution VISIR spectrum showing some important background errors from the atmosphere. This extra "noise" cannot be easily filtered in Fourier space because it is spread over a wide range of frequencies. *Right panel*: the corrected image (the negative beam has not been reconstructed using the wavelets+curvelets method (see text)

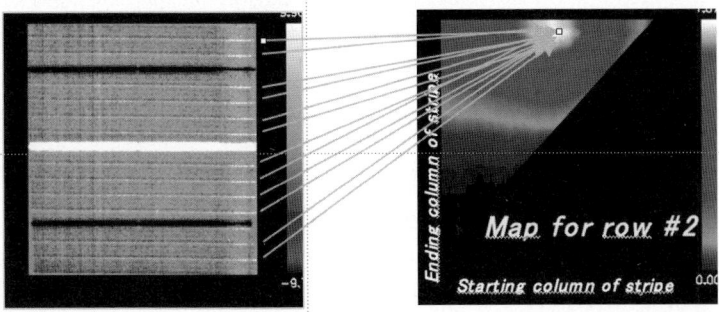

Fig. 9. Periodic stripes (*left panel*) are "concentrated" via a dedicated image transform, into a dual space (*right panel*). Once the "stripe" detected in this dual space where is appears as a single peak, the stripe is then reconstructed and subtracted from the original image. Since we usually deal with more than only one stripe, the process is iterated until no more stripes are detected (this threshold is computed based on the noise level in image space)

is sensibly reduced thanks to a "masked" read-out scheme by the sequencer, some residual periodic (16 pixels) stripes appear very often. Then, the only method is to suppress the stripes in the off-line processing. The currently adopted method consists in using an "adapted image transform towards a dual space" in which a given stripe starting at line l and row r will produce a peak concentrating the full stripe signal in the dual space (Fig. 9).

5.6 Gain Map and Flat-Fielding

Optical and near-infrared usually include flat-field correction for the data. One can wonder if this is applicable or even compulsory for mid-IR data; there is a long debate in the literature about this question. Pragmatically, experiments show typical variations of a few percent of photometric estimate on the four different beams of a standard star in a VISIR image using the most sensitive N-band filter (PAH2), the dispersion having a standard deviation of about 1%. The background shot-noise itself can contribute at most to about typically 0.5% of photometric error and therefore, cannot explain a few percents of variation peak-to-peak. The observed variations must be then attributed either to background errors (see Sect. 5.3), or to a spatially varying differential response of the detector. Using the method described above to correct for background errors, one can discard any predominance of them in the observed variations, on the other hand laboratory measurements of the detectors gain, showed peak-to-peak variations of about 1% of individual pixels gains [3]. To reach the highest photometric accuracy, it would be probably interesting to be able to correct for individual gain response. Estimating this response is not an easy task. On one hand, sky dips can be affected by instrument flexures, on the other hand, observations of the extended source of the WCU and varying the plate temperature can be corrupted by any warm element on the light path. So far, flat-fielding attemps did not improve the photometric accuracy.

6 Conclusions

VISIR offers to the ESO community the possibility to obtain valuable infrared data in N and Q bands. Although its sensitivity, limited by the background shot noise, remains two to three orders of magnitude worse than a space observatory (e.g. SPITZER), its high spatial resolution, typically 10 times better, provides mid-IR imaging with a unprecedented sharpness. In addition, the VISIR spectrometer sub-unit with spectral resolution up to 30000, opens a new "window" for both spatially **and** spectrally resolved mid-IR emission. However, on a 8.2 m telescope, the spatial resolution degradation due to the atmospheric turbulence is not negligible any longer, and usual deconvolution techniques are useless. There is then a need for simple pre-focal adaptive optics systems that would allow to recover purely diffraction limited data and stabilize the PSF.

References

1. M. Cohen, R.G. Walker, B. Carter, et al.: AJ **117**, 1864 (1999)
2. M. Elad, J-L. Starck, D. Donoho, et al.: Appl Comput Harmon Anal **19**, 340 (2005)
3. P. Galdemard, F. Garnier, and P. Mulet: Characterisation of DRS Technologies 256 × 256 mid-IR arrays for VISIR. In: *Proceedings SPIE*, vol 4841, 129 (2002)
4. VISIR users manual, http://www.eso.org/instruments/visir

Discussion

T. Mueller: Are there any airmass dependencies in the calibration process?
E. Pantin: Yes, but they still need to be characterized.
M. Sterzik: Gain map (GM) construction and its application as a viable flat field correction has – in the case of TIMMI2 – been only possible with sky background data that have been recorded simultaneously with science frames. By definition, the GM requires some dynamical range, and this is only present under variable and mediocre atmospheric conditions. Thus flatfield corrections seem to work in the case of worse conditions, and, unfortunately not for good and stable conditions.
H.U. Käufl: A VISIR intervention is planned for May 2007. The temperature stabilization for the detectors will be fundamentally improved. We then

expect that it will be possible to choose the operating temperature freely which may improve the cosmetic quality.

E. Pantin: The temperature stabilization and flexibility in choosing the working point are indeed highly valuable to efficiently reduce detector gain variations due to temperature oscillations. Working at higher temperature where, although the sensitivity and dark levels might be degraded, the detector gain oscillation can be significantly reduced. However, on should be careful that higher temperature means also higher detector striping. A compromise must be found.

C. Hummel: With the acquisition images of MIDI, we see background fluctuations faster than 1 Hz, where presumably the atmospheric fluctuations should die down. MIDI does not chop. Does this mean that we see instrumental effects faster than atmospheric ones?

E. Pantin: Hard to tell. A Fourier analysis of the signal variations should be performed, and compared to expected atmospheric variations. An instrumental origin would presumably show up at a precise frequency, whereas atmospheric effects would appear as a decreasing, continuous noise as a function of the frequency. Our measurements on VISIR indicate that the overall noise (instrument and atmosphere) is strongly correlated with some left-over background and increases fastly at chopping frequencies lower than 1 Hz.

J. Emerson: How do you decide if a night is good (dry?) enough for observing at Q-band to be productive? Do you use a water-vapor meter?

E. Pantin: We have no PWV meter at Paranal. Some weather forecasts are provided on the ESO website, predicting the PWV for the current and next nights. At the beginning of the night, some background measurements are performed, eventually in Q-band. The background level in Q-band is already a good zero-order indicator of the Q-band sensitivity. A. Smette has developed a procedure based on a medium-resolution spectroscopic observations at 19.5 micron compared with an atmospheric model that gives a fairly good estimate of the PWV. See talk by A. Smette.

Infrared Spectrograph Calibration Issues: Using CRIRES, the High Resolution Infrared Spectrograph for ESO's VLT as an Example

H. U. Käufl and The CRIRES Team

ESO, Karl-Schwarzschild-Strasse 2, 85748 Garching, Germany; hukaufl@eso.org

Abstract. CRIRES, a pre-dispersed **CR**yogenic **I**nfrared **E**chelle **S**pectrograph, provides a resolving power $\lambda/\Delta\lambda \approx 10^5$ (or $\Delta v \approx 1.5\,\mathrm{km\,s^{-1}}$ per pixel) between 1000 and 5000 nm (56–315 THz) at the 8 m ESO VLT-UT 1. CRIRES is a unique instrument, at least with respect to its spectral resolution, its detector format and the spatial resolution, due to an integrated adaptive optics system. Remote insertion of gas cells to measure high precision radial velocities is possible. Some specifics of the calibration of this instrument will be described.

1 Introduction: General Spectrograph Calibration Issues

For CRIRES,[1] as for any high-resolution long-slit spectrograph at least the following items need to be calibrated through rigorous procedures:

- spatial scale along slit
- spatial scale and distortion of slit viewer
- slit position angle on sky
- differential refraction effects
- distortion of the main spectrograph camera

All those issues are identical to similar optical spectrographs and thus not further discussed; the infrared specific issues are, that molecular rotational-vibrational lines are everywhere in the IR-spectral range. Therefore **all** IR astronomy is always high resolution spectroscopy, even filter photometry. Thus special considerations for these absorption lines drive the calibration

---

[1]The CRIRES team comprises of: Paola Amico, Pascal Ballester, Eduardo Bendek, Peter Biereichel, Paul Bristow, Mark Casali, Bernhard Delabre, Reinhold Dorn, Siegfried Eschbaumer, Raul Esteves, Enrico Fedrigo, Gert Finger, Gerhard Fischer, Gordon Gillet, Domingo Gojak, Michael Hilker, Gotthard Huster, Yves Jung, Florian Kerber, Jean-Paul Kirchbaumer, Jean-Louis Lizon, Lars Lundin, Manfred Meyer, Enrico Marchetti, Leander Mehrgan, Manfred Meyer, Alan Moorwood, Sylvain Oberti, Jérome Paufique, Jean-Francois Pirard, Eszter Pozna, Francesca Primas, Hughes Sana, Ricardo Schmutzer, Andreas Seifahrt, Ralf Siebenmorgen, Armin Silber, Alain Smette, Barbara Sokar, Jörg Stegmeier, Lowell Tacconi-Garman, Sebastien Tordo, Stefan Uttenthaler, Elena Valenti, Jakob Vinther, Ueli Weilenmann & Burkhard Wolff.

requirements for CRIRES, especially for precision spectro-photometry and absorption line spectroscopy.

2 CRIRES in a Nut Shell

Figure 1 shows the general lay-out of CRIRES. For commissioning reports and pictures see [1] and [2]. The main characteristics are summarized in Table 1. CRIRES and its operational concept are described in [3]. For the latest state see the ESO users online documentation. To the calibration unit a $ThAr$ hollow cathode lamp[2] was added to provide a calibration spectrum with sufficient line density in the 1000–2500 nm regime. The ThAr lamp is unfortunately by orders of magnitude too faint to illuminate CRIRES via the integrating sphere; provisionally, the cathode of the lamp – mounted above the table – is re-imaged on to the entrance slit by a folding mirror on a spare position of the calibration slide. Gas-cells[3] can be moved into the beam for

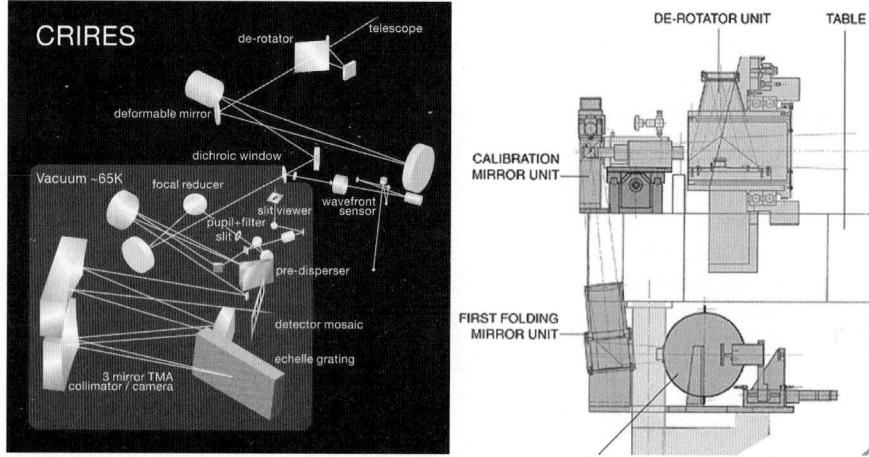

Fig. 1. *Left*, CRIRES Optical Design: The VLT Nasmyth focus (*f15*) is close to the first mirror of the de-rotator assembly. The *shaded area* is the cryogenic part. *Right*, Schematics of the Calibration Unit: A Halogen Lamp, an IR emitter and a classical Neon spectral lamp illuminate an integrating sphere. A motorized baffle allows for vernier adjustment of the illumination level. A kinematic mirror selects, whether CRIRES views the telescope or the calibration unit. Next to the de-rotator there is a motorized slide, which allows to insert pinholes (for focus and first order scale calibrations) and gas-cells as frequency standard. The gas-cells can also serve as frequency reference for very high precision radial velocity studies

---

[2] A precision infrared $ThAr$-lamp spectral atlas was created, c.f. [4].

[3] Gases used are CO, N_2O and OCS (*carbonyl sulfide*); for these gases the line positions have been connected to the Cesium time standard using heterodyning

Table 1. CRIRES main characteristics

Spectral Coverage	$\lambda \approx 950\text{–}5300\,\text{nm}$ ($\nu \approx 56\text{–}315\,\text{THz}$)
Spectral Resolution	$\lambda/\Delta\lambda \approx 10^5$ or $\Delta v \approx 3\,\text{km s}^{-1}$; 2 pixel Nyquist sampling
Array Detector Mosaic	4 x 1024 x 512 Aladdin III InSb mosaic; λ-coverage 2% pixel-scale 0.1" pixel^{-1}
Dark Current	$0.05\text{–}0.1\,\text{e}^-/\text{s}^{-1}$ per pixel
Infrared Slit Viewer	Aladdin III $InSb$ with J, H and K filters
Precision	calibration and stability (goal) $\approx 75\,\text{m s}^{-1}$ i.e. 1/20th of a pixel or 5 mas tracking error
Adaptive Optics	curvature sensing ESO-MACAO system 60 sub-apertures, R-band wave front sensor

calibration purposes and for searches for very small radial velocity changes similar to the Iodine-cell technique.[4] The de-rotator is followed by a curvature sensing adaptive optics system [5, 6] with the deformable mirror on a kinematic gimbal mount. The entrance window to the cryostat is a dichroic, separating the visible light with high efficiency for the AO wavefront control. The cryogenic optical bench is cooled by three Closed Cycle Coolers to $\approx$ 65 K. Temperatures are stabilized, in critical areas to the mK level. The slit-viewer has a pixel scale of 0.05 arcsec/pixel and an unvignetted field-of-view of $25\times50\,\text{arcsec}^2$. The main slit is continuously adjustable up to several arcsec. The pre-disperser uses a ZnSe prism in retro-reflection. The collimator mirror can be slightly tilted with a Piezo actuator for vernier adjustment to compensate for stick-slip effects in the grating and the prism drives. For order selection after the pre-disperser there is a second motorized intermediate slit (close to the small folding mirror next to the prism). The main collimator, a three-mirror anastigmat, produces a 200 mm collimated beam which illuminates a R2 Echelle grating (31.6 gr/mm).

The CRIRES 'genealogical tree' consists of a great variety of infrared spectrographs at ESO: IRSPEC [7] (1986), TIMMI [8] (1994), SOFI [9] (1998), ISAAC [10] (1998), TIMMI2 [11] (2000), CONICA [12] (2001) and VISIR [13] (2004). So to that end some experience in operating and calibrating IR spectrographs had accumulated at ESO.

---

and frequency chains. For details see the web-site of NIST http://physics.nist.gov/PhysRefData/wavenum/html/spect.html. It should be emphasized, that using these gases CRIRES is in principle calibrated absolutely in frequency and not in wavelength. Indeed one should seriously consider to move from the nm to the GHz. This would also end once and forever the equally annoying and confusing discrimination between wavelength in vacuum and air.

[4]CRIRES a classical long slit spectrograph without image slicer or scrambler is fully sensitive to tracking problems (e.g. 5 mas yield a spectral shift of $\approx 75\,\text{m s}^{-1}$).

3 Special Challenges for Calibrating CRIRES

The 'bread-and-butter' science of CRIRES is high signal to noise absorption line spectroscopy: atomic species in the 1–2.5 µm interval and molecular lines between 2–5 µm range. This in turn implies, that spectral calibration and spectral resolution need to be extremely well defined, not so much because of the scientific aspects but to be able to measure and correct for the very many narrow telluric absorption lines. Telluric linewidths are of order of the speed of sound, i.e. $\approx 500\,\mathrm{m\,s^{-1}}$. This leads to the overall requirement for stability and precision of calibration in CRIRES (c.f. Table 1). The dichotomy in the species to be observed in astrophysical sources reflects itself into the selection of calibration lamps. Atomic lines become less and less abundant with increasing wavelength, not only in stars but also in calibration lamps.

From Table 2 it is obvious, that there is not one single method for frequency calibration, but that a combination of the ThAr hollow cathode lamp, atmospheric emission lines[5] and molecular lines from the gas cell is necessary for calibration. In Fig. 2 it is shown, how in the thermal infrared, i.e. when the thermal emission of a ambient temperature black-body exceeds detector read noise and dark current, absorption lines show up as emission lines against the relatively dark sky. It should be noted, however, that for a rigorous solution for the dispersion of CRIRES a physical model [15] has been developed which in many cases is fundamental, especially in the near-infrared and for long-slit spectroscopy.

Table 2. CRIRES frequency calibration trade-offs

Source	λ-Range [nm]	Brightness	Line-Density	Operational Constraints	Absolute Precision
atmospheric absorption lines	950 –1900	na	marginal to acceptable	not for long-slit mode	$\approx 20\,\mathrm{m\,s^{-1}}$
atmospheric absorption lines	1900 –5200	acceptable to good	acceptable to good	needs telescope	$\approx 20\,\mathrm{m\,s^{-1}}$
gas cell absorption	2000 –5200	acceptable to good	very good	cumbersome operations	$\leq 1\,\mathrm{m\,s^{-1}}$ absolute (ν)
gas discharge lamps	950 –2500	very good	marginal	none	$\leq 10\,\mathrm{m\,s^{-1}}$
hollow cathode lamps	950 –2500	marginal	good to very good	illumination compromises	$\leq 10\,\mathrm{m\,s^{-1}}$

---

[5]The OH air-glow spectrum is discussed in [14] and references there-in.

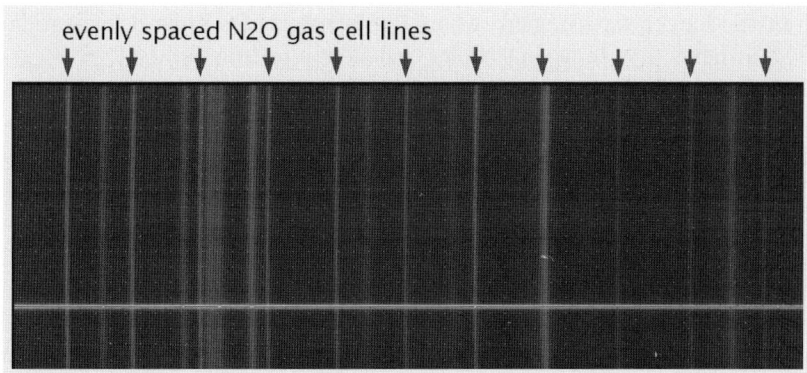

Fig. 2. CRIRES 2-D Spectrum: One of the 4 detectors is shown with a star observed at $\lambda \approx 4500$ nm through the N_2O gas-cell (raw data). In the lower part, the continuum of the star under investigation ia clearly visible (data from the science verification programme by E. Guenther et al. aimed to establish feasibility of high precision radial velocity searches). The comb of nearly equally spaced molecular lines, in emission, illuminating the slit nearly uniformly is very evident. A detailed analysis of nodded images then shows, as expected the same lines in absorption imprinted on to the stellar continuum

4 Other Tricky Calibrations

Flatfielding, both in dispersion and in slit direction is rather tricky especially as very high S/N-ratios have to be achieved. The method used here is a combination of lamp-flats taken with the calibration unit, also important to remove detector non-linearities, together with standard star spectra. So far, however, quite convincing results have been reported from science verification data [16] (a S/N-ratio of ≈ 300 was achieved in chemical abundance studies). Up to now stellar lines in the standard stars have not been a serious problem,

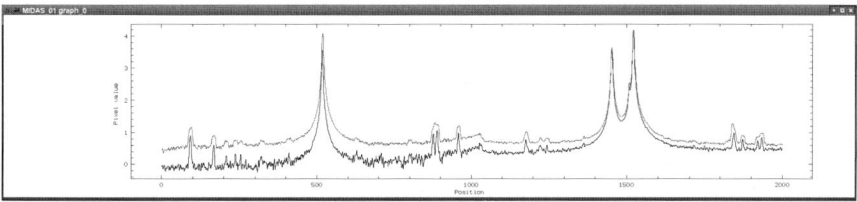

Fig. 3. CRIRES Inter Order Straylight: These spectra (two of the four detectors, extracted, in pixel coordinates) show Ne-lines observed at $\lambda \approx 2350$ nm on a logarithmic intensity scale. The *black curve* is for the intermediate slit at nominal width (0.3 mm), the *grey* one is for 0.6 mm. The ratio between line peaks and ghosts and/or straylight exceeds 10^3

even at the longer wavelengths, where one is brightness limited for standard stars, and hence has to accept cooler and slowly rotating stars.

For absorption line spectroscopy it is of paramount importance to assess and calibrate inter-order straylight. Here, however, the good news is, that laboratory tests (c.f. Fig. 3), that the level of inter-order straylight is $\leq 10^{-3}$, hence there is no need to calibrate and correct for this effect.

5 Conclusions and Outlook

CRIRES, an unique instrument allowing to access a largely unexplored parameter space, is now fully integrated into the VLT observatory. Various small improvements are envisaged to improve operations. In the next step polarimetry (all 4 Stokes parameters) will be implemented including a corresponding calibration unit. Taking advantage of this extra effort the assembly of the slide, carrying the gascells (c.f. Fig. 1, right) will be modified to allow the use of the ThAr lamp without restrictions, i.e. to freely illuminate any point of the slit or to use it with a pinhole, while maintaining uniform illumination of the spectrograph pupil. A dedicated pumping station with gas-handling equipment is being prepared for the use at the observatory to allow for gas-cell maintenance, and within limits use of other gases specified by the CRIRES user community.

References

1. H. U. Käufl, et al.: The Messenger **124**, 2 (2006)
2. H. U. Käufl, et al.: The Messenger **126**, 32 (2006)
3. H. U. Käufl, et al.: Proc. SPIE **5492**, 1218 (2004)
4. F. Kerber, et al.: These proceedings (2007)
5. R. Arsenault, et al.: Proc. SPIE **4839**, 174 (2003)
6. J. Paufique, et al.: Proc. SPIE **5490**, 216 (2004)
7. A. Moorwood, et al.: The Messenger **44**, 19 (1986)
8. H.U. Käufl: The Messenger **78**, 4 (1994)
9. A. Moorwood, et al.: The Messenger **91**, 9 (1998)
10. A. Moorwood, et al.: The Messenger **94**, 7 (1998)
11. H. U. Käufl, et al.: Proc. SPIE **4841**, 117 (2003)
12. R. Lenzen, et al.: Proc. SPIE **4841**, 944 (2003)
13. P.O. Lagage, et al.: The Messenger **117**, 12 (2004)
14. P. Rousselot, et al.: A&A **354**, 1134 (2000)
15. P. Bristow, et al.: These proceedings (2007)
16. P.E. Nissen, et al.: A&A **469**, 319 (2007)

Discussion

M. van den Ancker: Does the fact that CRIRES is an AO-fed instrument raise any particular issues with the calibration, especially as the AO correction for science target and telluric standard star will in general be different?

H. U. Käufl: We expect some problems from polarization of the light at the slit as the AO allows us to use a slit width close to the diffraction limit. This has not yet been fully assessed.

G. Lo Curto: On which time scale is the CRIRES line shift?

H. U. Käufl: 1/20 pixel in 5 min.

P. E. Nissen: Concerning the present CRIRES ETC, I think that the division with a factor of 4 of the SNR expected from the photon statistics is far too pessimistic. Are you going to update the ETC soon?

A. Smette: Indeed, the ETC is currently conservative. When it had to be released, a number of functions were not yet properly calibrated: centering of the objects (differential atmospheric refraction, slit viewer filter offsets, slit width, etc.). In addition, the blaze function had not been measured and was not taken into account – it is now since Phase 2 Proposal Preparation. It is an objective of the commissioning period in early February to provide an ETC that produces more realistic values.

U. Hopp: CRIRES will be used to study star formation regions and other objects which emit mostly in emission lines. Can the pre-images include narrow band filters in its filter wheel to precisely point to interesting objects in their lines?

H. U. Käufl: Unfortunately, this is not possible for technical reasons. The CRIRES slit viewer can be used as a quasi-imager and sensitivies are more or less the same as for NACO. In that way, at least the precise location of the slit can be recorded.

Calibrating Mid-Infrared Standard Stars

D. Dobrzycka[1] and L. Vanzi[2]

[1] ESO, Karl-Schwarzschild-Strasse 2, 85748 Garching, Germany;
ddobrzyc@eso.org
[2] ESO, Alonso de Cordova 3107, Vitacura, Santiago, Chile

Abstract. The VLT Spectrometer and Imager for the Mid-Infrared (VISIR) is a Paranal instrument dedicated to observations through the two mid-infrared (MIR) atmospheric windows: N band (8–13 µm) and Q band (16.5–24.5 µm). As part of standard operations, VISIR has been continuously observing MIR standard stars for more than a year. The derived conversion factors, sensitivities, etc. have been systematically collected in the database, allowing for statistical analysis of their temporal behavior. We analyzed long time variations of the conversion factor of selected VISIR standard stars and found that they show variability of less than 10% in the N band and less than 20% in the Q band.

1 Introduction

The VLT Spectrometer and Imager for the Mid-Infrared (VISIR) is an instrument dedicated to observations through the two mid-infrared (MIR) atmospheric windows: N band (8–13 µm) and Q band (16.5–24.5 µm). It was installed at ESO Paranal Observatory in April 2004 in the Cassegrain focus of the 3rd VLT Unit Telescope – Melipal. The first light was obtained in May 2004 and VISIR was offered to the community from beginning of April 2005. This cryogenic instrument combines diffraction limited high sensitivity imaging capabilities over a field of view of up to 51″ and long slit grating spectroscopy capabilities with a range of spectral resolutions between 150 and 30,000. So far, two pixel scales have been offered: 0.075″ (small field, SF) and 0.127″ (intermediate field, IF).

Ground observing in MIR is quite challenging. It's not only that our atmosphere absorbs the majority of MIR radiation from astronomical sources, but it also emits strong background with the spectral shape of a 253 K black body. The telescopes contribute an additional MIR background, estimated to be <15% in N band. The VISIR instrument is cooled to avoid internal thermal contamination. The detectors are kept at 5–6 K and the interior of the cryostat is kept at 33 K. Special observing techniques are applied to suppress the elevated MIR background. They include differential observations using chopping and/or telescope nodding.

As MIR observations depend strongly on the ambient conditions, such as humidity, temperature or airmass, the science observations are accompanied by relevant calibration standard star observations obtained no further than 3 h apart. As part of standard operations, VISIR has been continuously observing MIR standard stars for more than a year. The derived conversion factors, sensitivities, etc. have been systematically collected in the database. We present long time coverage of the observations of selected MIR photometric standard stars and discuss possible variability of their conversion factor.

2 The MIR Spectro-Photometric Standard Star Catalog

VISIR calibrators are selected from the MIR spectro-photometric standard star catalog of the VLT (http://www.eso.org/instruments/visir). It is based on the radiometric all-sky network of absolutely calibrated stellar spectra by Cohen et al. [1] and supplemented with the MIR standards used by TIMMI2. Zero point fluxes (in Jy) have been calculated for the VISIR filters set by taking into account the measured transmission curve, the detector efficiency and an atmosphere model.

For the project of monitoring photometric precision of VISIR observations, 81 stars have been selected from the catalog. They fulfill basic criteria: (a) non-variability (according to Hipparcos), (b) not being visual binaries (according to SIMBAD), and (c) having absolute calibration errors less than 20%. Out of these, 12 photometric standard stars were further selected for frequent observations. These targets have similar spectral types and are uniformly distributed in the Right Ascension. In addition, their flux in N band, of the order of 10 Jy, is bright enough to be observable in the Q band without reaching non-linearity levels in the N band even in non-ideal background conditions. Every effort is made to observe at least one star from this reduced catalog on each night VISIR is in use. This allows to monitor observable properties of these targets over the long time scale.

3 Data Processing and Quality Control

From the very beginning, the operation of VISIR was designed to follow a scheme common to all the VLT instruments [2]. The standard VISIR data flow operation covers all steps of data handling from initial inspection of the raw frames and quick-look products at Paranal to thorough classification, processing, quality control and delivering data to principal investigators.

The observations are obtained either in visitor or service mode. The visitor mode data are packed on-site and collected by the visiting astronomer, while all the calibration frames and service mode data are transferred to ESO in Garching for further processing and inspection by the Data Flow Operation Quality Control Group. Here, the data are classified and reduced using a pipeline. The quality control (QC) process includes assessing the quality of the raw data (also done at Paranal), quality of products created by pipelines, as well as monitoring performance of the instrument through temporal behavior of the QC parameters [3].

3.1 Handling of the Standard Star Observations

The VISIR standard star observations arriving at Garching are first classified, sorted by the instrument setup and processed with dedicated pipeline recipes: *visir_img_phot* and *visir_spc_phot*. The recipes are based on the ESO common pipeline library (CPL). Most of the algorithms have been written by the instrument consortia, while the ESO DFS Department implemented them into common ESO pipeline environment.

The pipeline products for standard stars are checked for sufficient number of input frames, proper flux level (over or under exposed), adequate background removal, unusual noise pattern, striping, etc. To assure that only the best calibrations are further considered and used for reducing science data only certified products are archived. Flagged calibrations may also indicate instrument problems and thus, are closely investigated.

The VISIR pipeline recipes are used not only to create master standard star frames, but also to extract a number of specially designed parameters.

Some of the most important QC parameters are:

- Conversion Factor: It measures conversion between ADU and Jy
  – $F_{tot.observed}/F_{model}$ [ADU/Jy], where F corresponds to flux.
- Sensitivity: The sensitivity in a given instrument setup (filter, pixel field of view) is defined as the limiting flux of a point-source detected with S/N of 10 in 1 h of on-source integration.
- Mean Background Level: It is measured from the Half-cycle frames. It's value can vary between two extremes of $-32,000$ and $+32,000$ ADU.

3.2 VISIR QC1 Database

The QC parameters calculated by the recipes are extracted from the headers of standard star pipeline products and are further stored in the QC1 database. The QC1 database is publicly available at http://archive.eso.org/bin/qc1_cgi. It contains information for all supported VLT instruments. Each database table includes not only values of the QC parameters but also values of corresponding general and instrument keywords. This enables to correlate temporal behavior of the QC parameters with e.g. instrument setup.

For VISIR there are four database tables:

- *visir_zp_img* – contains parameters from the products of imaging standard stars;
- *visir_zp_spc* – contains parameters from the products of spectroscopic standard stars;
- *visir_flat_img* – contains parameters from the master flat fields of the imaging detector;

- *visir_zp_spc* – contains parameters from the master flat fields of the spectroscopic detector.

The QC1 database web site offers two types of services – the required parameters can be printed or plotted in selected periods of time.

The values collected in the tables since the beginning of VISIR operation cover now about a year and a half. This allows a statistical analysis of the QC parameters with respect to instrumental and atmospheric conditions. In particular, we can monitor the temporal behavior of the conversion factor for chosen VISIR standard stars.

4 Results

To ensure good statistical coverage we limited our analysis to targets from the list of the 12 photometric standards selected for frequent observing.

We excluded data points taken at particularly bad observing conditions (significantly larger sensitivity, mean background level) or high airmass.

Figure 1 shows variations of the conversion factor of the standard star HD 178345 measured from the observations taken with filter PAH1 (centered at 8.6 µm) and small pixel scale (SF) [4]. The variations appear to be at the level of 10% only. Similar variations of the conversion factor for the standard star HD 26961 are shown in Fig. 2. The plotted data points correspond to

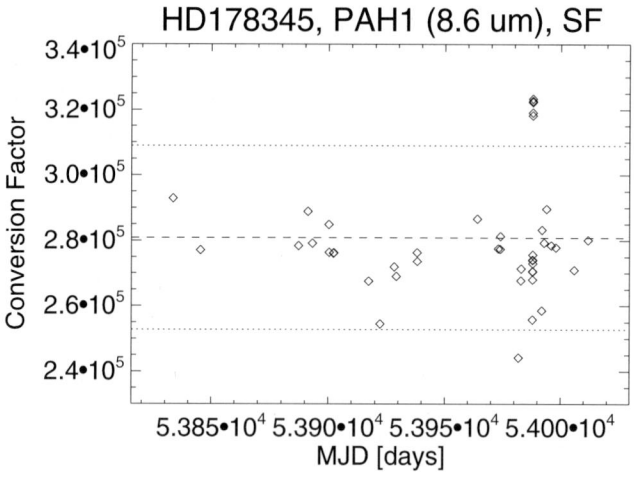

Fig. 1. Conversion factor of the VISIR standard star HD 178345 as a function of time. Only data taken with filter PAH1, centered at 8.6 µm, and small pixel scale (SF) are plotted. The *dashed line* corresponds to the average value and *dotted lines* show 10% variations

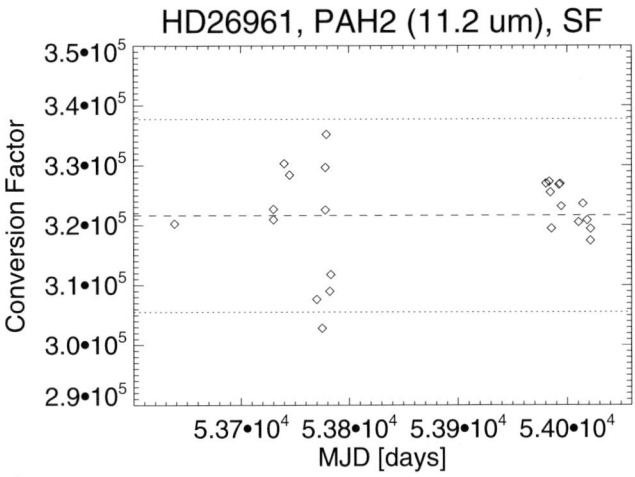

Fig. 2. Conversion factor of the VISIR standard star HD 26961 as a function of time. Only data taken with filter PAH2, centered at 11.2 μm, and small pixel scale (SF) are plotted. The *dashed line* corresponds to the average value and *dotted lines* show 5% variations

the observations taken with filter PAH2 (centered at 11.2 μm) and small pixel scale (SF). Here the variations appear to be around 5%.

Table 1 summarizes some of the results. We measured variations of the conversion factor for 10 VISIR standard stars that have been most frequently observed so far. Two filters in the N band, PAH1 and PAH2, and one filter in the Q band, Q2, were selected. The table contains measured variations and also indicates the number of data points available.

The results show that selected MIR standard stars display flux variability of no more than 10% in the N band and of less than 20% in the Q band.

Table 1. Variations of the conversion factor

Star Name	PAH1, 8.6 μm SF	PAH2, 11.2 μm SH	Q2, 18.7 μm SF
HD 12524	10% (17)	8% (15)	6% (2)
HD 26967	8% (25)	6% (24)	11% (7)
HD 41047	9% (15)	11% (15)	10% (3)
HD 75691	8% (19)	7% (16)	<20% (5)
HD 99167	7% (20)	4% (15)	15% (4)
HD 145897	6% (19)	8% (17)	<8% (10)
HD 178345	13% (44)	10% (31)	<20% (11)
HD 198048	6% (8)	6% (12)	<20% (8)

This is an ongoing project that includes effort of many members of the VISIR Instrument Operation Team (IOT). We verify stability of the used VISIR standard stars about every 6–12 months.

References

1. M. Cohen, R.G. Walker, B. Carter, et al.: AJ **117**, 1864 (1999)
2. P.J. Quinn, M.A. Albrecht, P. Ballester, et al.: VLT Data Flow System: From Concepts to Operations. In *SPIE Proceedings*, vol 3349, pp 2–10 (1998)
3. R. Hanuschik, D. Silva: Messenger **108**, 4 (2002)
4. A. Smette, L. Vanzi: VLT VISIR User Manual (2006)

Discussion

T. Mueller: Is the calibration factor really that stable ($< 10\%$ in N, $< 20\%$ in Q)? Or did you select only the observations under the best possible conditions?

D. Dobrzycka: Some selection was done (airmass, weather conditions).

Stellar Calibrators for the Medium Resolution Spectrometer of the Mid Infra-Red Instrument on Board the JWST

E. Bauwens[1], L. Decin[1,2], and J. A. D. L. Blommaert[1]

[1] Instituut voor Sterrenkunde K.U. Leuven, Celestijnenlaan 200D, B-3001 Heverlee, Leuven, Belgium; Eva.Bauwens@ster.kuleuven.be
[2] Sterrenkundig Instituut Anton Pannekoek, University of Amsterdam, Kruislaan 403, NL-1098 Amsterdam, The Netherlands

Abstract. For the spectro-photometric calibration of space based instruments, like the Medium Resolution Spectrometer (MRS) of the Mid InfraRed Instrument (MIRI) on board the James Webb Space Telescope (JWST), the accuracy is limited to our knowledge of the spectra of the sources used as calibrators. As the sensitivity of this instrument is much higher than previous ones, there is a need for stars with a more accurately modeled spectrum which can be used as a calibrator. In the selection of such sources we include objects with different spectral types and make sure we cover the whole flux range of the instrument.
Here we will describe a list of candidate sources for this purpose, as well as the preparatory observations needed to further increase the accuracy of the modeled spectra.

1 The James Webb Space Telescope and the MIRI MRS

The James Webb Space Telescope is an infrared space based telescope built by NASA. The primary mirror of the telescope has a diameter of 6.5 m. The four instruments on the JWST are a near-IR camera (NIRCam) and spectrometer (NIRSpec), the Fine Guidance Sensors (FGS) and the Mid InfraRed Instrument (MIRI). The science expected to be done with these instruments is concentrated on planetary systems and the origin of life, birth of stars and proto-planetary systems, assembly of galaxies and first light and reionization. MIRI itself contains an imager, a coronograph, a Low Resolution Spectrometer (LRS) and a Medium Resolution Spectrometer (MRS). The MRS is an IFU spectrometer and has a wavelength range from 5 to 27 µm, with a flux range of 4 to 500 mJy at 10 µm. The field of view ranges from 3.7"x3.7" to 7.7"x7.7" for the longest wavelengths, with a spectroscopic resolution of ∼3000.

2 Calibration Method for the MRS

The calibration will be based on the comparison of high accuracy modeled spectra of stellar sources with the spectra from the MIRI MRS. The spectra will be obtained through theoretical atmosphere modeling codes, like the

MARCS models [7]. These models have a high accuracy in the wavelength range of this instrument and experience with these models for calibration issues is available due to their use for the calibration of ISO-SWS [3] and Spitzer-IRS [4]. In order to be able to model the spectra of the calibration sources with an accuracy better than 5%, the stellar parameters (most important ones being effective temperature T_{eff}, gravity $\log g$ and chemical composition) should be accurately known (for example accuracy of $T_{\text{eff}} < 100\,\text{K}$). For this calibration, sources of different spectral types will be used, as discussed in Decin et al. [5]. This will avoid biases to certain spectral features of one of the types of stars used in the calibration and allow us to use the most accurate calibration for every part of the spectrum. The spectral types commonly used for the calibration of IR spectrometers are (1) A dwarfs, (2) late-type giants (usually G9 - K5 III) and (3) solar analogs. We are well aware of the shortcomings in the theoretical model atmosphere spectra of each spectral type. For an elaborate discussion on this topic, we refer to Decin and Eriksson [6]. We may summarize that for the A dwarfs the possible presence of a debris disk may result in a flux excess in the MIRI wavelength range. Objects with such features should therefore be excluded from the list of calibrators. The main spectral uncertainties for the A dwarfs may arise from the complex hydrogen line predictions. The late type giants have strong molecular absorption bands in the relevant wavelength range. These can be modelled, provided that stellar parameters like effective temperature, $\log g$ and abundances, are accurately known. A mid-IR flux excess might also be introduced due to a circumstellar shell. All objects should be checked for this and excluded as calibrator if an excess is found. For the solar analogs, both atomic and molecular lines are present but have the big advantage of the availability of high resolution solar spectra.

3 The Candidate Calibration Sources

For the selection of the calibration sources we started by composing a first list of more than 50 candidate calibration stars [5]. The objects in this list are well known objects as most of them have already been used as calibrators in previous space missions like Spitzer (IRAC, IRS and MIPS) [8] and ISO (ISOCAM) [1], or are standard stars from IR catalogs (Cohen et al. [2] and references therein). Another selection criterium used for this list of objects is a flux level inside the dynamic range of the MIRI spectrometer (4–500 mJy). Further selection of this long list has been done based on the availability of Spitzer MIPS observations. As these observations provide evidence for the existence or non-existence of a debris disk, they are of major importance for our goal. An extra list of stellar sources was included at this point, namely the A-dwarf and solar analogs from Rieke et al. [9]. All sources in [9] do not have a debris disk.

From these lists a reduced list of 28 sources was constructed (Table 1). All of which have MIPS 24 micron observations and all solar analogs and A-type

Table 1. List of the calibrator candidates for the MRS with position, spectral type and flux at 10 μm in mJy

nr	object	rRA	Dec	Spect Type	Flux 10 μm (mJy)
1	BD+62 1644	18 41 52.4	+62 57 41.2	K3 III	388.6
2	HD 001644	00 20 43.6	03 26 58.0	K0	378.1
3	HD 002811	00 31 18.5	−43 36 23.0	A3 V	56.7
4	HD 017254	02 44 10.6	−52 34 14.0	A2 V	171.8
5	HD 020722	03 18 44.8	−41 09 36.2	K3/K4 III	172.9
6	HD 021981	03 30 37.0	−47 22 30.5	A1 V	203.2
7	HD 040335	05 58 13.5	+01 51 23.0	A0	56.7
8	HD 046819	06 30 22.7	−66 01 23.0	K0 III	273.5
9	HD 101452	11 40 13.6	−39 08 48	A2	63.6
10	HD 128998	14 38 15.2	+54 01 24.0	A1 V	186.5
11	HD 141937	15 52 17.5	−18 26 10	G2/G3 V	440.2
12	HD 15646	02 28 04.4	−64 17 59	A0 V	98.4
13	HD 158485	17 26 04.9	+58 39 06.8	A4 V	134.1
14	HD 159048	17 23 34.5	+76 03 18.2	K0	355.6
15	HD 163466	17 52 25.4	+60 23 46.9	A2 V	136.4
16	HD 73666	08 40 11.5	+19 58 16	A1 V	88.5
17	HD 172728	18 37 33.5	+62.31 35.7	A0 V	187.5
18	HD 196724	20 38 31.3	+21 12 04.2	A0 V	438.7
19	HD 197806	20 47 30.8	−43 10 33.6	K0 III	74
20	HD 20888	03 17 59.1	−66 55 37	A3 V	180.9
21	HD 212291	22 23 09.2	+09 27 40	G5	285.5
22	HD 42525	06 06 09.4	−66 02 23	A0	185
23	HD 57336	07 09 30.5	−79 25 55	A0 IV	46.6
24	HD 64324	07 54 48.5	+34 37 11	G0	315.7
25	HD 73819	08 40 56.3	+19 34 49	A6 Vn	104.7
26	HD 92788	10 42 48.5	−02 11 02	G5	455.3
27	SA 103-526	11 56 54.2	−00 30 13.8	K0 III	18
28	SA 112-275	20 42 35.4	+00 07 20.2	K0 III	63
29	SA 112-595	20 41 18.5	+00 16 28.3	M0 III	53

stars are disk-free. The flux range of the MRS is covered by these sources except for the lowest part (4–40 mJy). Therefore we added one source, SA 103–526, with a flux of 18 mJy at 10 μm. No MIPS observations are available for this source, but it is a well-studied source and has an accurately determined spectral type [2]. Observations of this source with Spitzer IRS are available (5.2–38 μm) and this IRS data will be usefull in further stages of characterization and modelling of the sources.

4 Conclusions and Future Work

This list of candidate stellar calibration sources, including different spectral types and with a full coverage of the instrument flux range, is a candidate list for the calibration sources of the MIRI MRS. The use of high accuracy theoretical stellar atmosphere spectra of each target will yield a calibration with the required accuracy (better than 5%). To obtain these high accuracy theoretical spectra, quite some observational data are needed both to test the reliability of the candidate calibrators and to estimate the stellar parameters being input for the theoretical spectra. Indispensable data for this are (1) UBVRI and (2) NIR photometric data to test the reliability as a fiducial calibrator and constrain $T_{\rm eff}$, (3) optical/near-IR spectroscopy to constrain $\log g$ and chemical abundance. (4) Parallaxes, if possible, can be used to estimate the dereddening and the gravity g.

Based on the results of these observations, the list of candidate calibrations stars presented here, will be reduced to a set of 15–20 fiducial calibrators for the MIRI MRS. For these fiducial stars, the list of observational data should be extended towards (1) high-resolution optical spectroscopy to constrain the stellar parameters even more, (2) narrow-band photometry tracing the metalicity, mid-infrared photometric data to test the predicted theoretical spectra.

References

1. J. Blommaert: ISOCAM Photometry Report, http://www.iso.vilspa.esa.es/users/expl_lib/CAM/photom_rep_fn.ps.gz (1998)
2. M. Cohen, S.T. Megeath and J. Stauffer: AJ **125** 2645 (2003)
3. L. Decin, B. Vandenbussche, C. Waelkens, et al.: A&A **400**, 679 (2003)
4. L. Decin, P.W. Morris, P.N. Appleton, et al.: ApJS **154**, 408 (2004)
5. L., Decin, E., Bauwens, J.A.D.L., Blommaert: InfraRed Standards used for the Spectrophotometric Calibration Application to the Medium Resolution Spectrometer of MIRI. In: *ASP Conf. Ser.: The future of Photometric, Spectrophotometric and Polarimetric Standardization*, vol 364, ed by C. Sterken (2007) p. 443
6. L. Decin, K. Eriksson: A&A **472**, 1041 (2007)
7. B. Gustafsson, R.A. Bell, K. Eriksson, et al.: A&A **42**, 407 (1975)
8. W.T. Reach, S.T. Megeath, M. Cohen, et al.: PASP **117** 978 (2005)
9. G.H. Rieke, M. Blaylock, L. Decin, et al.: ApJ (submitted) (2008)

ISAAC LW Telluric Line Atlas

C. Papadaki[1,2], L. Schmidtobreick[1], W. Hummel[3], and A. Smette[1]

[1] ESO, Alonso de Cordova 3107, Vitacura, Santiago, Chile; cpapadak@eso.org
[2] Vrije Universiteit Brussel (OBSS/WE), Pleinlaan 2, 1050 Brussel, Belgium
[3] ESO, Karl-Schwarzschild-Strasse 2, 85748 Garching, Germany

Abstract. We present a status report on a project with the aim to enhance the wavelength calibration of ISAAC's long wavelength spectroscopic mode. We measure telluric standard stars with high signal/noise and use the comparison with theoretical spectra of the telluric lines for the wavelength calibration. We show that calibrations done with arc-lines only can be wrong by up to 0.2 µm. The new atlas of telluric lines allows a spectral calibration with a precision of 0.003 µm.

1 Project Description

ISAAC's long wavelength spectroscopic (LWS) mode is affected by two problems of the wavelength calibration: (1) the non-reproducibility of the grating produces random offsets up to 20 pixels between the science observations and the corresponding daytime calibration; (2) the atlas of line identification using arc-lamps contains mainly lines in high order whose wavelengths are only estimated, the accuracy is unclear. We hence attempted to measure the wavelengths of telluric emission lines in the LWS mode in order to use them for wavelength calibration purposes.

2 Reduction & Analysis Procedure

The telluric standard star HR 0806 (RA, Dec: 02:39:35, −68:16:01) of spectral type B9V and magnitude $L = 4.3$, was observed with ISAAC on 05-12-2006. Spectra were taken with a 1.0" slit in LW-MR mode centred on 2.65 µm, 2.90 µm, 3.15 µm, 3.40 µm, 3.65 µm, 3.90 µm, 4.15 µm, 4.55 µm, 4.80 µm, and 5.05 µm.

Thus the whole LW range was covered with medium resolution spectra. Arc lamps have been taken directly after each spectrum, with no change in the instrument set-up, to avoid the problem with the non-reproducibility of the grating position. The spectra were reduced using the IRAF package *specred*. In a first step, the extracted spectra were wavelength calibrated through the ArXe line identifications provided by the corresponding atlas available in the ISAAC webpage [1]. However, the effective wavelength λ of the higher order lines are only estimated via $\lambda = k \cdot \lambda_0$, where k is the order, and λ_0 the

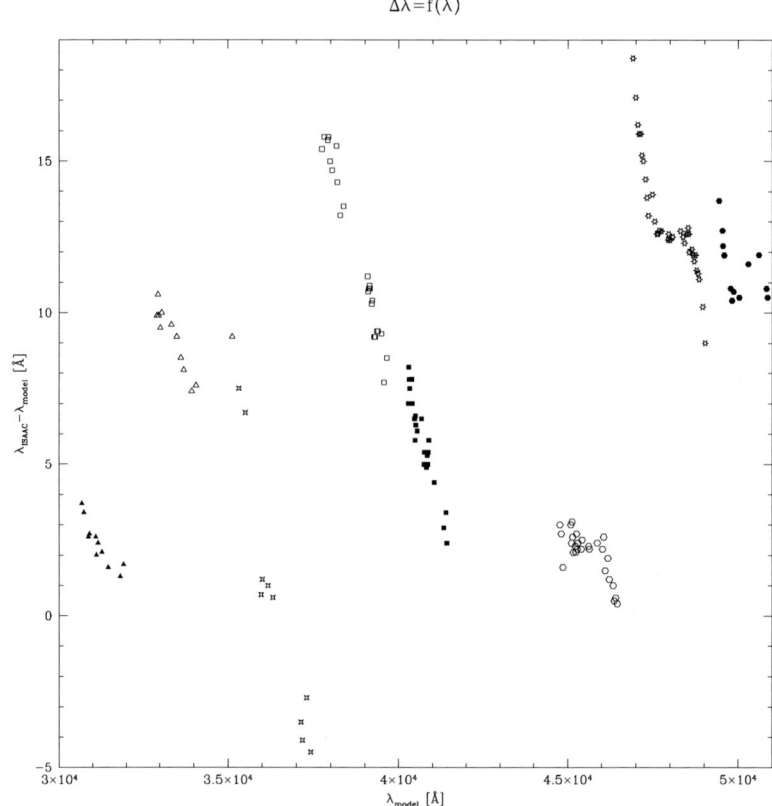

Fig. 1. $\Delta\lambda$ represents the difference between the λ of the telluric lines identified in the arc line calibrated spectra and the corresponding ones identified in the model spectrum. Each symbol represents the λ range of one slit position

original wavelength, leading to an uncertain accuracy. In order to have an estimate for this accuracy, we compared the wavelength calibrated based on the arc-lines with the one based on a theoretical model [2] of the sky emission spectrum. The result is presented in Fig. 1; it is evident that using the arc lines can lead to errors of up to 18 Å in the wavelength calibration, as well as non-linearities.

In order to obtain an atlas based on telluric lines, the model spectrum was used as a reference for finding the effective λ of the telluric lines. By comparing the observed and model spectra, we first identified the telluric lines visible in both, which are thus usable for the wavelength calibration, and created a line list, containing the λ_{model} of those lines. Using the interactive IRAF routines *identify* and *dispcor*, a dispersion function was derived from this line list and applied to our spectra. The order of the fitted functions,

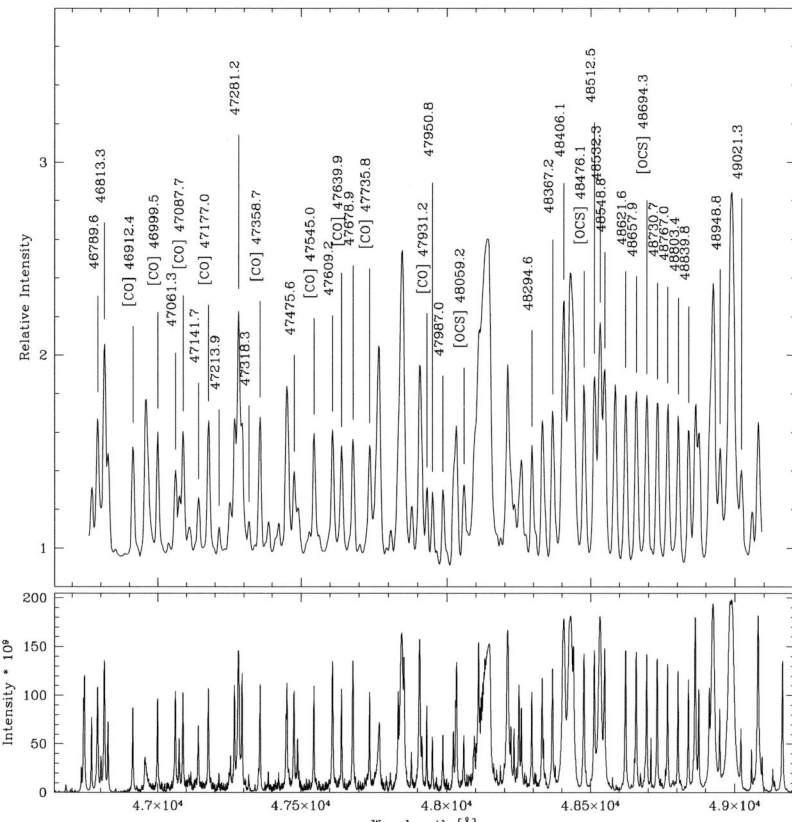

Fig. 2. This plot is an example for the plots in the atlas of telluric lines. The *upper part* contains the sky spectrum (centred on 4.8 μm in this example) with the corresponding lines indicated. The *lower part* shows the model spectrum that was used

in each wavelength range, varied between 1st and 3rd order polynomials and the rms between 0.1 and 0.3 Å therefore greatly increasing the wavelength calibration precision. A sample of the atlas is presented in Fig. 2 for the 4.80 μm setting.

3 Future Work

The next steps related to the work of this project are:

- re-calibrate the higher order arc lines with the telluric ones, so they can be used for decent wavelength calibration;

- include those telluric lines that are observed but not found in the model;
- make an atlas for the low resolution mode;
- extend the line list and atlas to $\lambda = 20\,\mu m$ (for VISIR).

Acknowledgement. We would like to greatly acknowledge A. Seifahrt for providing the synthetic sky spectrum, as well as C. Lidman for his cooperation during this project.

References

1. C. Lidman, J.G. Cuby, ISAAC: Atlas of Xenon and Argon lines, Issue 1.0, 28 July 2000 http://www.eso.org/instruments/isaac/tools/atlas/index.html
2. HITRAN database at http://cfa-www.harvard.edu/HITRAN/

Building-Up a Database of Spectro-Photometric Standard Stars from the Ultraviolet to the Near-Infrared

J. Vernet[1], F. Kerber[1], S. D'Odorico[1], R. Bohlin[2], V. Ivanov[1], C. Lidman[1], E. Mason[1], T. Rauch[3], F. Saitta[1], A. Smette[1], J. Walsh[4], R. Fosbury[4], P. Goldoni[5], P. Groot[6], F. Hammer[7], M. Horrobin[8], L. Kaper[8], P. Kjaergaard-Rasmussen[9], R. Pallavicini[10], and F. Royer[7]

[1] ESO, Karl-Schwarzschild-Strasse 2, 85748 Garching, Germany; jvernet@eso.org
[2] STScI, 3700 San Martin Drive, Baltimore, MD 21218, USA
[3] Institut für Astronomie und Astrophysik, Universität Tübingen, Sand 1, D-72076 Tübingen, Germany
[4] ESA, ST-ECF, Karl-Schwarzschild-Strasse 2, 85748 Garching, Germany
[5] Service d'Astrophysique, Centre d'Etudes de Saclay, F-91190 Gif-sur-Yvette Cedex, France
[6] Radboud Univ. Nijmegen, Postbus 9010, 6500 GL Nijmegen, The Netherlands
[7] GEPI, Observatoire de, Paris-Meudon, 5, place Jules Janssen, F-92195 Meudon, France
[8] University Amsterdam, Postbus 19268, 1000 GG Amsterdam, The Netherlands
[9] Niels Bohr Institute for Astronomy, Blegdamsvej 17, DK-2100 Copenhagen, Danemark
[10] INAF - Osservatorio astronomico di Palermo Giuseppe S. Vaiana, Piazza del Parlamento 1, I-90134 Palermo, Italy

1 Motivation

While there is a well established and widely used set of high quality optical spectro-photometric standard stars [1, 6, 5], we still lack an equivalent set for the near-infrared (NIR). Observations of white dwarfs with HST spectrographs and atmospheric modeling have established 3 stars as primary standards (1% accuracy) for the wavelength range 115–1800 nm [2]. A set of NIR spectro-photometric standards adequate for routine operations of ground based NIR spectrographs simply does not exist at the moment. The currently used methods for estimating the absolute flux calibration of NIR spectra are (see eg. the ISAAC User Manual, Cuby et al.): (i) use the NIR broad band magnitude of the telluric standard that has been observed for a given science exposure (e.g. from 2MASS); (ii) alternatively, when it is known, use the broad band magnitude of the scientific target itself to scale the NIR spectrum. These methods are usually no more precise than 20–30%. Therefore, we decided to start an observing program to remedy this situation by extending the wavelength range of a set of well established optical spectro-photometric standards to the near-IR. While our immediate goal is to support regular science operations of X-shooter, it is also an important step to calibrate future instrumentation associated with Extremely Large Telescopes that are expected to carry out their core science in this wavelength range.

2 Instrumental Setup

The two main difficulties in reliably measuring absolute fluxes in the NIR are the variable telluric absorption and the strongly variable sky OH emission lines. There are actually very few regions completely free of telluric lines. Even the classical J, H and K atmospheric windows are seriously affected by telluric absorption as can clearly be seen in Fig. 1. OH sky lines on the

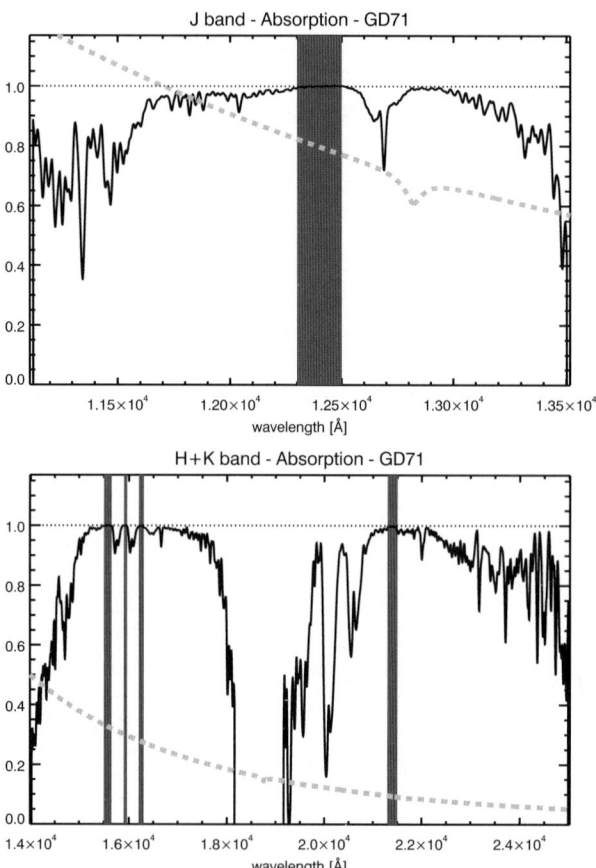

Fig. 1. Telluric absorption in the J (*top*) and the H+K bands (*bottom*) at the resolution of SINFONI. *Shaded areas* show regions almost free of telluric absorption that will be used to perform our absolute flux measurements. The *dashed line* shows the featureless continuum of one of our primary standards GD71

other hand are present throughout the NIR wavelength range up to 2 μm. In order to peer through these forests of lines and perform solid photometric measurements we need to observe at a sufficiently high spectral resolution. By carefully examining the situation in the J, H and K bands using a R=100,000 Gemini atmospheric transmission spectrum together with the list of identified OH sky lines [4], we found that a resolution better than R~1000 is required.

The other constraint to our choice of instrument is the need for a wide aperture to reliably collect the whole flux. In order to avoid slit-losses and to minimize the effect of centering uncertainties, spectro-photometric standards are usually observed using wide 5 arcsec slits or even slitless.

Combining the requirements on resolution with that of a wide aperture led us to select SINFONI IFU as the instrument of choice for our program. It provides a resolution of R~2000 in the J and R~1500 in the H+K bands. It is a rather efficient setup since only 2 shots are needed to cover the whole wavelength range. The observations do not require good seeing conditions and can be performed during bright time and even twilight. Therefore this project does not compete with any science proposals at the VLT/UT4.

3 Strategy

- Use as Primary Standards (PrimS) the two HST white dwarfs (WD) primary standards observable from the southern hemisphere: GD71 and GD153. They provide a uniquely robust reference frame because the measured relative flux between these two stars agrees with the ratio of their model atmospheres SEDs to 1% from 115 to 1800 nm. The models are normalized to the precision Landolt V mag. [2].
- We have selected a set of 13 Secondary Standards (SecS) by choosing WD and very hot stars from Oke [1] and Hamuy et al. [6, 5] optical spectro-photometric standards to provide a homogeneous distribution in RA (see Fig. 2).
- Each SecS will be observed at 5–10 times to build up a solid observational base for their use as standard stars and to allow a proper statistical treatment of the errors on the measured flux whose uncertainty is expected to be dominated by repeatability.
- For each visit, at least one PrimS will be observed to deduce a response function by comparing flux levels in wavelength regions free of atmospheric absorption and sky emission with tabulated values in the HST Calibrations Database. This response will then be used to calibrate the SecS using the same clean wavelength bins.
- We will observe on one occasion a standard at several airmasses to properly derive the effect of atmospheric extinction on our standards.
- All the targets have UVES spectra that will be used to determine stellar parameters to be given as an input to models computed with the Tübingen NLTE Model Atmosphere Package [3]. These models will then be used to

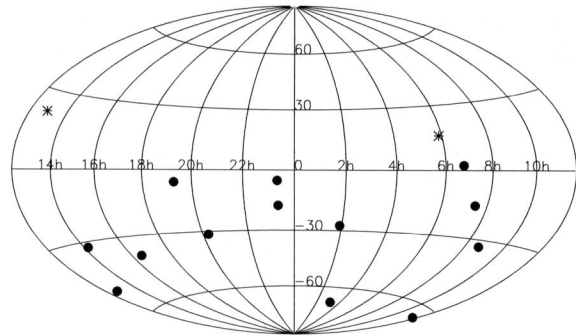

Fig. 2. The distribution of our targets on the sky. The two Primary Standards are marked with a ∗

interpolate in between observed fluxes to derive an absolute flux table for each SecS across the full wavelength range.

4 Current Status

This observing program is being carried out at Paranal as an Observatory Program. The first data package was obtained in May 2007 and is currently being analyzed. Our goal is to provide the full database by the time of the X-shooter commissioning in mid-2008.

References

1. J.B. Oke: AJ **99**, 1621 (1990)
2. R.C. Bohlin: In: The Future of Photometric, Spectrophotometric and Polarimetric Standardization: *ASP Conf. Ser. 364*, ed by C. Sterken, (2007) pp 315–331
3. K. Werner, J.L. Deetjen, S. Dreizler, T. Nagel, T. Rauch, S.L. Schuh: In Stellar Atmosphere Modeling: *ASP Conf. Ser. 288*, ed by I. Hubeny, D. Mihalas, K. Werner, (2007) p 31
4. P. Rousselot, C. Lidman, J.-G. Cuby, G. Moreels, G. Monnet: A&A **354**, 1134 (2000)
5. M. Hamuy, N.B. Suntzeff, S.R. Heathcote, A.R. Walker, P. Gigoux, M.M. Phillips: PASP **106**, 566 (1994)
6. M. Hamuy, A.R. Walker, N.B. Suntzeff, P. Gigoux, S.R. Heathcote, M.M. Phillips: PASP **104**, 533 (1992)

Part V

Session 4: Data Flow and Data Reduction Software

The VLT Data Flow System

M. Peron

ESO, Karl-Schwarzschild-Strasse 2, 85748 Garching, Germany; mperon@eso.org

Abstract. The Data Flow System (DFS) provides the software infrastructure for supporting the end-to-end operation of the VLT. It is a distributed system composed of a collection of components for preparation and scheduling of observations, archiving of data, pipeline data reduction and quality control. The current system, its evolution since the beginning of operations in 1999 and future developments are presented in this chapter.

1 Introduction

The VLT Data Flow System is a distributed system which provides all necessary software components to support the end-to-end operations of the Paranal-La Silla Observatory observing facilities and in particular of the Very Large Telescope. Although a subset of the DFS components are for internal operational use only, some of its components provide scientific services to the scientific community.

The rationale and high-level requirements of the VLT Data Flow system are directly derived from the main objective of the VLT which is to maximize its science output. The original system concepts as well as the underlying software were implemented by the time Cerro Paranal started regular science operations using the first VLT Unit Telescope in April 1999. Important developments have taken place over the last eight years to support the new requirements which came with new instruments and to improve the ease of use and operational reliability. This paper describes some of the existing and future applications, while the operation model itself has been discussed in several other papers [1].

The Data Flow System can be visualized as a closed-loop system, the so called egg diagram (Fig. 1) and consists of a number of subsystems that are being described in the following sections. At its lower layer, the DFS relies heavily on relational database management system technology as databases are used as a persistent state for most of the information flowing through the system. Database replication technology is used to manage the constant two-way information flow across multiple time zones at transcontinental distances required because of the distributed nature of ESO science operations.

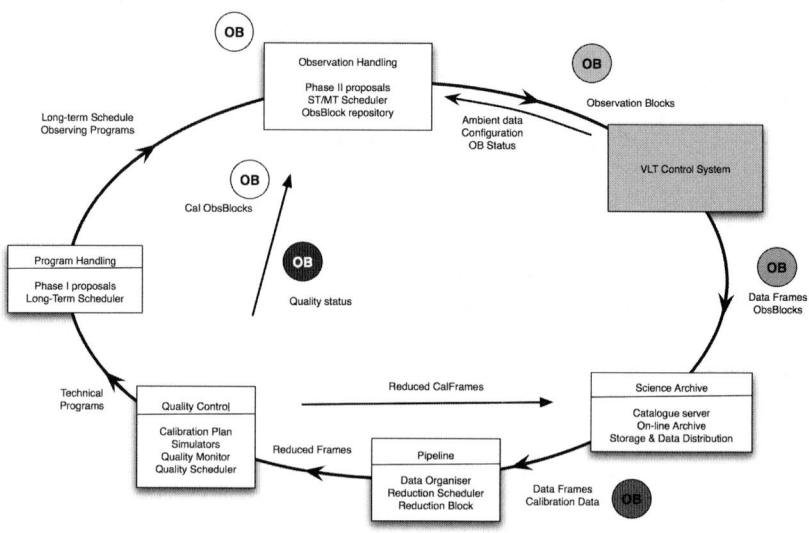

Fig. 1. The VLT Data Flow System

2 Program Handling (Phase I)

Proposals for observations at ESO are invited twice a year. They contain a scientific case, a summary of the proposed observing programme, a list of desired instrument modes as well as a definition of the observing conditions (e.g. seeing). The proposals are reviewed by the Observing Programmes Committee (OPC), which evaluates the scientific merits of the proposals and provides recommendations that serve as a basis for the final decision on the telescope time allocation by the ESO Director General.

Proposals are submitted in the form of LaTex files through a WEB interface to ESO where they are processed and stored in a central database. Exposure Time Calculators (ETC) for each instrument are made available to the users for predicting the signal to noise that will be reached under certain observing conditions. The accuracy of the predictions improves as the instrument is calibrated and tracked under operational conditions. ETCs are implemented as WEB interfaces.

3 Observation Handling (Phase II)

ESO users who have obtained observing time are requested to provide a detailed description of their observations, in the form of **Observation Blocks**.

The preparation differs slightly between those who have been granted Service Mode (SM) observations and the astronomers who will travel to the Observatory for their Visitor Mode (VM) Observations. An Observation Block is the smallest observational unit defining a sequence of high-level operations that needs to be performed sequentially and without interruption in order to ensure the scientific usefulness of the Observation. It contains all the information to perform an observation including the target coordinates, the instrument parameters, scheduling requirements, time constraints and finding charts. Observation Blocks are one of the central concepts of the VLT DFS as they 'flow' through the system from their preparation, execution and delivery of resulting data to the user.

The Phase 2 Proposal Preparation (P2PP) [2] is used by the ESO user community for the preparation of their observations, both in VM and SM mode and is also in use at the telescope to execute VM observations. P2PP is a Java-based client, which runs on the astronomer's desktop and exchanges information with a server located in Garching. The tool is not instrument specific, that is, users are presented with the same look and feel independently of the actual instrument to be used. Instrument specific functionalities such as verification and validation scripts are implemented as plugins to the main application. The Observing Tool, a another Java-based application, is used by the operation teams to create queues of Observation Blocks that could be executed in the medium term. It provides database browsing functionalities and a schedule aid module called ORANG. It ranks a large pool of Observation Blocks, taking in account parameters such as target visibility, priority of the observation, time constraints and all other user provided constraints (moon distance, seeing).

4 Science Archive

The ESO archive is an critical instrument for helping operations to be predictable by providing traceability of instrument performance. Every single raw FITS file generated by the Observatory is stored in the archive while the meta-data describing the information is stored in a database. The bulk storage system for the long-term storage of data has evolved with time in order to handle large data rates, and is now based on clusters of Linux PCs hosting magnetic disks. Those disks are sent to the ESO headquarters where they are stored in a permanent primary archive and duplicated to a secondary one. The ESO archive contains data from ESO telescopes located at both La Silla Paranal and Chajnantor observatory sites. Since June 2005 it also contains data from the UKIRT Infrared Deep Survey taken with the Wide Field Infrared Camera.

Important users of the ESO archive are also astronomers doing science with archive data. Except for a few special cases, all ESO science observations have a proprietary period of one year. After this period the data sets are available worldwide to the community. WEB archive applications provide archive users with browsers to query and retrieve data. In addition, the abstracts of all successful ESO proposals are available publicly. All calibration data are public immediately after the observations.

5 Pipelines and Quality Control

Instrument Pipelines are developed for each VLT/VLTI instrument and its most heavily used modes [3], covering a wide observational domain in imaging, spectroscopy and optical interferometry (Fig. 2). The prime goal of the pipelines as described in the VLT 1st level requirements is to enable the automatic monitoring of instantaneous and long term instrument performance. For that purpose, instrument pipelines run in two operational environments: at the Observatory and at the ESO headquarters in Garching. They focus on processing automatically calibration data and extracting robust parameters (e.g. read-out noise, zero-points) required to monitor the health of the instruments and to ensure that the archive is filled with data that can be calibrated to a known and certified accuracy. On the mountain, the system makes use of certified master calibration data, while in Garching the best available calibration data are used, in most of the cases the closest in time.

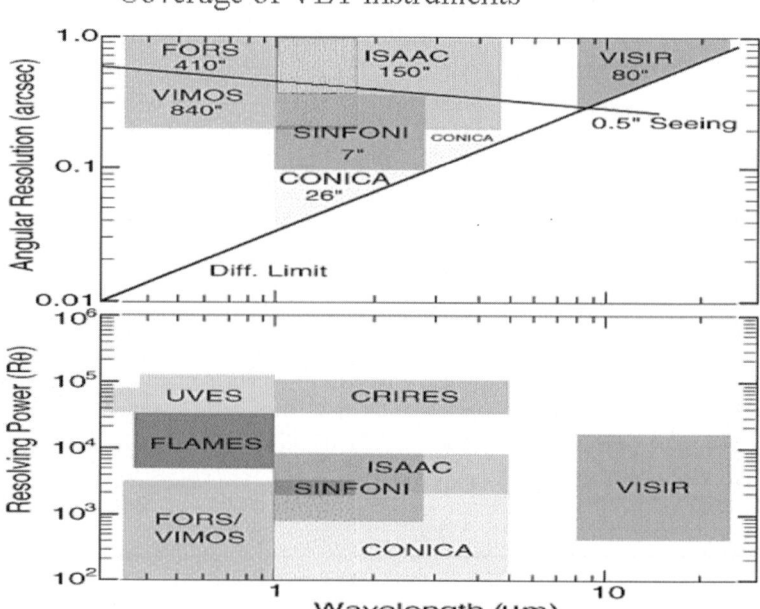

Fig. 2. VLT instrument pipelines

In the ESO jargon, instrument pipelines consists of instrument recipes, each implementing a data reduction step. In addition to the recipes, ESO has developed a software infrastructure that connects the recipes in series and creates a real 'pipeline'. In particular this environment provides means for automatically classifying ("what kind of data am I ?"), grouping ("to which group (or OB) do I belong?"), and associating the data ("which type of calibration are needed to process me?"). The usefulness of the products generated by the instrument pipelines is limited by the quality of the available data, the quality of the algorithms and the choice of parameters. As a consequence, ESO produces better quality science data in Garching than on the mountain where the night calibrations are not used and default parameters taken.

5.1 The ESO Common Pipeline Library

Instrument pipelines are implemented on top of an ANSI-C library, the ESO Common Pipeline Library (CPL, http://www.eso.org/cpl). CPL was built to shorten the development cycle of pipelines and to ease their maintenance. It provides the implementation of data structures and general purpose image and signal-processing functions. It is not designed as a general purpose astronomical data reduction library and does not for instance provide any graphic nor display functionality. While some of the first instrument pipelines were built using various systems such as MIDAS, all of them have been (or are in the process of being) converted to CPL. Pipelines for second generation VLT and VLTI instruments will be built from the beginning on top of CPL. Recipes are built in such a way that they can run efficiently within a pipeline environment. Therefore each of them does, in most cases, perform a large number of operations. As an example, the *FORS spectroscopic science recipe* corrects for bias and flat-field, subtracts the sky, removes the cosmic rays as well as distortions, calibrates in wavelength and extracts the spectrum.

5.2 Interactive Pipeline Environments: EsoRex and Gasgano

The instrument pipeline recipes are made available to the community together with user documentation (http://www.eso.org/pipelines). Two front-end applications, EsoRex and Gasgano, are distributed together with pipelines. **Gasgano** is a Graphical User Interface that provides users with means for organizing, viewing, grouping, classifying, searching and filtering data. Additionally, it can launch any instrument pipeline recipe on selected data (Fig. 3) and allows users to change interactively parameters. EsoRex is a very simple command line application that can be embedded in scripts for the automation of data reduction tasks.

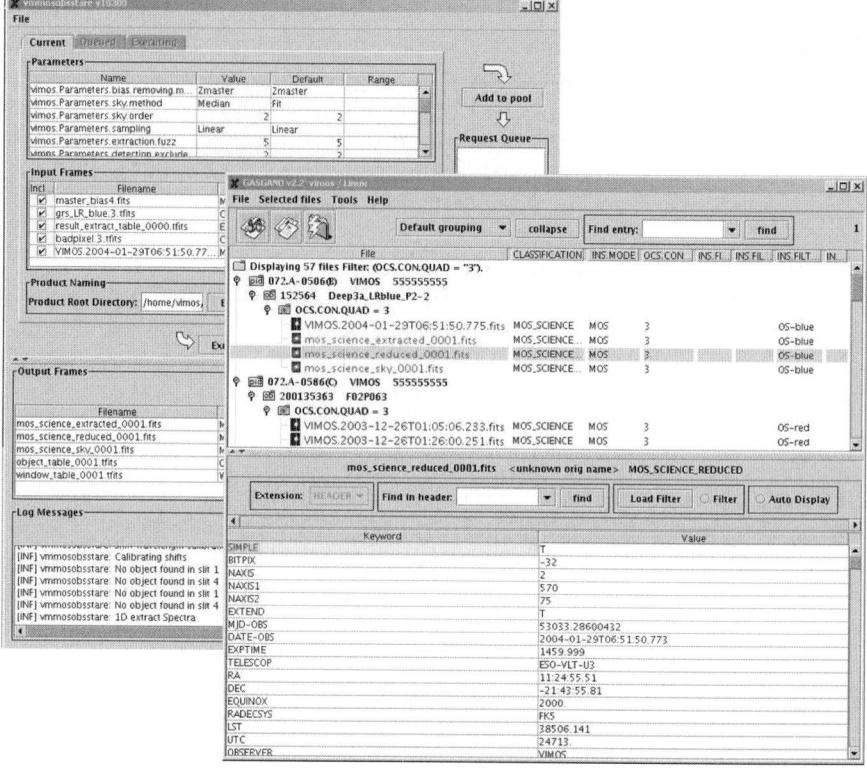

Fig. 3. Gasgano and its interface to pipeline recipes. A set of VIMOS observations are displayed in the main panel while the recipe panel shows a call to the recipe *vmmsobsstare*

5.3 Interactive Pipeline Environments: the SAMPO Project

Even though users' feedback on Gasgano had been positive, it came out clearly in discussions with internal and external users that the level of interactivity provided by Gasgano is not sufficient for many purposes. In addition, each instrument recipes does many things, cannot be broken in smaller modules and is therefore seen as a black box. Users want to check step by step that the data reduction recipes do the right thing. The SAMPO project, an inkind contribution of Finland to joining ESO, has been investigating solutions to this problem and has developed a prototype application called ESO Reflex in which the sequence of data reduction steps is rendered and controlled as a graphical workflow and which allows users to insert any script/application at any stage of the workflow. This work is described in detail in [4].

As already stated, the quality of the science products generated by the pipelines depend of the quality and robustness of the algorithms that have been implemented. While for some instrument modes the resulting products are of adequate quality for immediate scientific analysis, this is not always the case as this would require the pipelines to cover all scientific goals of the hundreds of programs carried out every year with ESO observing facilities. The goal of ESO Reflex is to allow users to make use of CPL-pipelines algorithms and recipes while allowing them to optimize the quality of the results by running dedicated tools without having to implement a new data reduction system.

6 Evolution of the DFS: Support of Survey Facilities

VISTA, the survey telescope that will be installed and commissioned on Paranal in 2007 will offer the community a new way of observing. The scope of the programs that will be executed with this facility will pose new challenges in terms of observation preparation and planning strategies as well as data volume produced. This section addresses briefly the changes that are needed to Phase II applications to support survey programs. These proposals can be characterized as follows:

- They require large numbers of individual observations and involve many OBs per run
- They require observations of large fields (i.e. many pointings)
- They require repeated observations over a time span, with a given frequency
- They require sets of observations to be performed over a short interval
- They require sequencing of the observations so as to maximize the scientific usability of partial datasets.

While the concept of Observation Blocks is still valid, the above requirements impose the introduction of observation concepts on top of the OB, which are also called **containers**. Containers are mainly created for scheduling purposes.

- Concatenation of Observation Blocks: All OBs belonging to a concatenation must be executed without breaks, in no particular sequence.
- Time Link: OBs connected by a time link must be executed in a given sequence, respecting the relative time intervals specified for each link.
- Group of Observation Blocks: Execution of an OB belonging to the group raises the relative priority of all other OBs of the same group. As a result, OBs belonging to the same group tend to get executed close to each other.

It is to be noted that these concepts will be useful not only for the survey facilities but also for the VLT/VLTI. For instance the VLTI pairs target/calibrators could be defined as time links. All existing Phase II applications and in particular P2PP (Fig. 4) and the Observation Tool require upgrades to support these observing concepts.

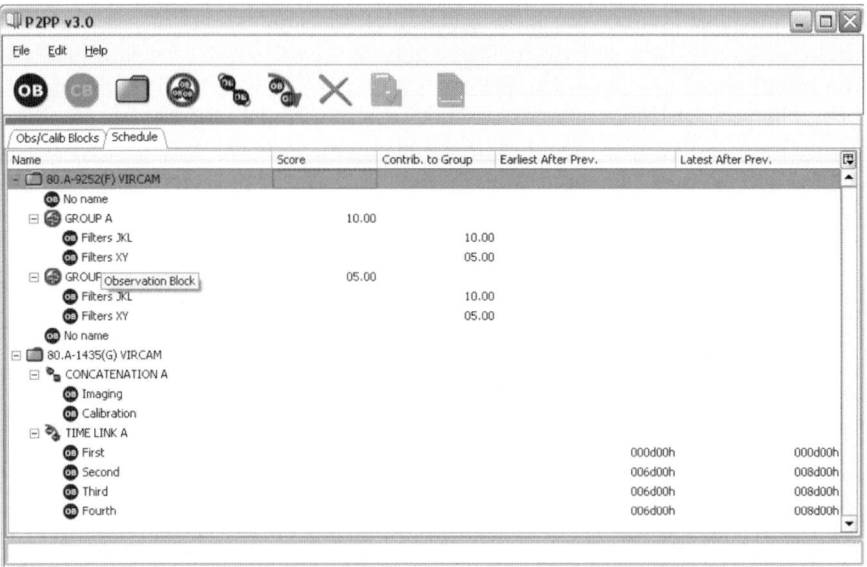

Fig. 4. The P2PP scheduling panel shows two groups, one concatenation and one time link that have been created for a given VIRCAM Observing Run. The user has defined a contribution (score) of the Observation Blocks to their group (5 and 10) as well as the time gaps between the OBs in the time linked sequence (6–8 days)

7 The User Portal

Many of the services provided by ESO to the users of the Observatory are available through the WEB. Different passwords are used for different applications and the correct URL is not always easy to find.

The aim of the User Portal [5], which is currently under development at ESO, is to provide a central access point for all services offered to its scientific community via the WEB. This gateway will provide registered users with a personalized set of service access points, the actual set depending on each user's privileges.

As an example, Principal Investigators (before submitting their first proposal to ESO) will need to register to the system and setup their account and profile. They will then be able to login into the system, upload a proposal, check the progress of the observations and query the ESO archive, all from the same access point. Each account will have one or more profiles that will include affiliation name and address, telephone numbers, and so on. This will allow, for instance, the PIs to choose to which address data products resulting from their observations should be sent. Users having a referee role will be able to review and evaluate the observing proposals assigned to them.

Acknowledgement. The Data Flow System is the result of many years of hard work of the Data Flow System Department of the Software Development Division at ESO. The author of this paper is indebted to all the team members and would also like to acknowledge the support of ESO operation teams without which the developments described in this paper would not have been possible.

References

1. F. Comeron, et al.: VLT service mode operations at seven years. In *Proceedings of the SPIE*, vol 6270, ed by D. Silva, R.E. Doxsey (2006)
2. A.M. Chavan, et al.: A Front-End System for the VLT's Data Flow System. In *Proceedings of SPIE*, vol 4010, ed by P.J. Quinn (2000)
3. P. Ballester, et al.: Data reduction pipelines for the Very Large Telescope. In: *Proceedings of the SPIE*, vol 6270, ed by D. Silva, R.E. Doxsey (2006)
4. R. Hook, et al.: ESO reflex: A Graphical Workflow Engine for running Recipes, these proceedings, (2008)
5. A.M. Chavan, et al.: Unifying access to services: ESO's User Portal Observatory Operations, Strategies and Systems. In *Proceedings of the SPIE,* vol 6270, ed by D. Silva, R.E. Doxsey (2006)

Discussion

Kimeswenger: Did you think about another pillar for the reduction pipelines? Somewhere between ESO-HQ and the users desktop, the use of European Science grid(s)? This is of special interest after changes of rules - recalibration data from the archives.
M. Peron: ESO is currently looking at the future of pipelines and data reduction systems. Options will be presented to the ESO Management in the course of 2007.
P. Bonifacio: Is it necessary to change version of P2PP each semester? I would like the format of the P2PP to be officially released so I can use my own codes to generate them and then simply import them to P2PP.
M. Peron: We have been trying not to release P2PP at each semester and will continue to do so. I am not aware of changes done to the IMPEX format. However, TSF files may change.
M. Horrobin: Are there plans to release Gasgano for more platforms than the current small number?
M. Peron: Gasgano will be released to all Linux platforms and Solaris. There was, in the past, a limitation due to the way CPL and Gasgano were linked.

ESO Reflex: A Graphical Workflow Engine for Running Recipes

R. N. Hook[1], M. Romaniello[1], M. Ullgrén[2], P. Järveläinen[3], S. Maisala[2], T. Oittinen[2], V. Savolainen[3], O. Solin[4], J. Tyynelä[3], M. Peron[1], C. Izzo[1], and T. Licha[1]

[1] ESO, Space Telescope European Coordinating Facility,
    Karl-Schwarzschild-Strasse 2, 85748 Garching, Germany; rhook@eso.org
[2] Observatory, University of Helsinki, Helsinki, Finland
[3] CSC–Scientific Computing Ltd., Espoo, Finland
[4] Space Systems Finland Ltd., Espoo, Finland

Abstract. Sampo is a 3 year project that began in January 2005. It is led by ESO and conducted by a software development team from Finland as an in-kind contribution to joining ESO. The goal of the project is to assess the needs of the ESO community in the area of data reduction and analysis environments and to create pilot software products that illustrate critical steps along the road to a new system.

The Sampo team has been developing new ways in which instrument pipeline recipes can be executed in a more flexible way. This has led to a prototype application called ESO Reflex that provides a new approach to astronomical data analysis. The integration of a modern graphical user interface and field-proven legacy data reduction algorithms aims to give an astronomer the best of both worlds: ease of use combined with optimal scientific results.

Much of the raw data produced by ESO instruments is reduced using compiled recipes based on the Common Pipeline Library (CPL). Each of these recipes performs a single task in the reduction process and they are combined to form reduction pipelines. Currently such recipes are run one at a time using a command line application EsoRex or the GUI-based application Gasgano. Using ESO Reflex, a sequence of recipes can be run as a workflow and the output of a recipe can be used as an input to another recipe. The workflow can be executed either automatically or in interactive mode, where the user can interrupt processing at intermediate stages to visualize the results and also change the processing parameters if needed. The data classification functionality offered by ESO Reflex (based on Gasgano) helps the user to select the right data for input, which also reduces the likelihood of errors. ESO Reflex also detects errors that occur during the execution of the recipes, and appropriate action can then be taken by the user.

ESO Reflex is based on a graphical workflow engine called Taverna that was originally developed for the molecular biology community in the UK. Workflows have been created so far for several instruments on the ESO VLT and VLTI. The easy-to-use GUI allows the user to make changes to these or create their own workflows.

1 The ESO Context

As the volume and complexity of astronomical data increase, astronomers are faced with the challenge of a timely and accurate reduction of their own data. Archival research, data mining and the facilities offered by the emerging

Virtual Observatory allow easy access to vast amounts of diverse data, an indepth knowledge of which may be challenging for the average user.

There appear to be two possible, but not conflicting, approaches. Firstly the user can be provided with fully-reduced data products ready for immediate scientific analysis. Secondly the user could be provided with a powerful, flexible and user-friendly data reduction system so that data may be reduced effectively and easily on a typical desktop machine.

At ESO quality control processing of all pipeline-supported modes for the VLT/VLTI Service Mode stream (virtually 100% of the data volume) involves running the data through standard calibration pipelines [1]. The resulting data are distributed to the principal investigators. The same applies for data from the HARPS and FEROS instruments on La Silla and APEX. In addition advanced data products for selected datasets (e.g., GOODS IR imaging and optical spectroscopy) are produced at ESO and ingested into the archive. Additional, fully-reduced datasets are also returned to ESO from the community (e.g., Public Surveys, Large Programmes, etc.) and also stored in the archive.

For individual users ESO provides pipeline 'recipes' for all VLT instruments. These remove the instrumental signature and, in addition to being used for quality control at ESO, are distributed to the community. All newer recipes are compiled C code using the CPL library and tools to run them offline are available (Gasgano and EsoRex). Many older general purpose reduction and analysis systems remain in wide use (MIDAS/IRAF etc.) and many instrument-specific packages have been developed in the community (e.g., Euro3D tools). Future developments will inevitably lead to enhanced data complexity – ESO instruments cover a wide variety of modes.

So – what is the best way forward for ESO in this area?

2 The Sampo Project

As part of Finland's joining fee a contribution 'in kind', computer scientist staff based at two sites in Helsinki were made available for a period of 3 years. This project was named Sampo [2] and began in January 2005. The overall aim is to assess the requirements for ESO data reduction and analysis software infrastructure over the next decade and beyond and perform a series of pilot projects to assess different options and produce useful tools. The project manager and scientist are at ESO in Garching and the Sampo team are based in Helsinki. The project oversight is provided by a Science Advisory Committee at ESO, representing the ESO science faculty, and the Finnish Astronomical Advisory Group (FAAG) representing the Finnish astronomical community.

The Sampo approach is to regard the requirements of the user to be in three parts. Firstly one needs modular recipes, implementing the best algorithms available and also providing access to intermediate products and allowing greater flexibility. Then one needs interactive tools, defined or customized by the user, to analyze and visualise intermediate and final data products and lastly a user-friendly, intuitive and flexible user interface.

Sampo, and specifically the ESO Reflex tool described in this paper, tries to address the interface issue with a focus on the use case of ESO data. This leads to the requirement of having a common front-end interface to cope with the potentially large number of ESO instruments in a uniform and maintainable way (in a similar manner to the front-end p2pp proposal preparation tool) and the need for a general-purpose invoker for CPL-based recipes complemented by easy-to-use ways to invoke Python scripts (and hence have easy access to IRAF and MIDAS through PyRAF and PyMidas) and other user tools.

3 ESO Reflex

To address these issues we have developed a prototype software tool called ESO Reflex (ESO Recipe Flexible Execution Workbench). ESO Reflex is a graphical workflow engine for executing CPL Recipes and other tools. It is not instrument specific and allows the easy inclusion of external tools and scripts as well as allowing user interaction during execution.

ESO Reflex is based on Taverna [3], an open source Java workflow system developed for the molecular biology community in the UK. It has proved easy to use and adapt and is suitable for science (rather than business applications for which many workflow systems were designed). ESO Reflex is a pilot project to assess in detail the concepts.

After assessing the choices and deciding to use Taverna as the basis for the prototype, it was necessary to add several new features to address ESO-specific requirements. The Sampo project has hence developed Taverna further to allow the execution of CPL Recipes through a general purpose recipe invocation mechanism. To ease the running of astronomical workflows a completely new graphical user interface for launching the workflows has also been written and FITS file handling functionality for data classification and viewing added. In addition, to allow the incorporation of external tools, we have added adaptors between ESO Reflex and commonly used astronomical tools and facilities. These include a Python and IDL invoking mechanisms and access to DS9, FV the Virtual Observatory PLASTIC standard and so forth. We are currently working on providing access also to Web services such as those provided by the Virtual Observatory.

For the initial testers release it was decided to include recipes, workflows and tools for contrasting VLT/VLTI instruments: the spectroscopic modes of FORS [4] and the AMBER VLTI instrument [5]. The workflows, as visualised by ESO Reflex, are shown in Figs. 1 and 2. A sophisticated interactive tool, written in Python and making heavy use of the matplotlib graphics facilities, is available to validate and improve on the wavelength calibration. This is shown in Fig. 3.

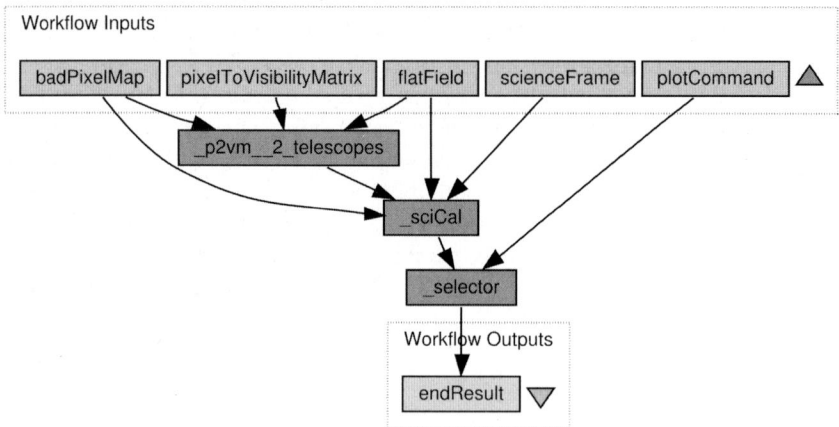

Fig. 1. ESO Reflex for AMBER. This workflow performs a standard reduction of AMBER datasets and has a final MIDAS-based visualisation stage (hidden in the _selector processor) to validate, and if needed repeat, processing with different parameter settings

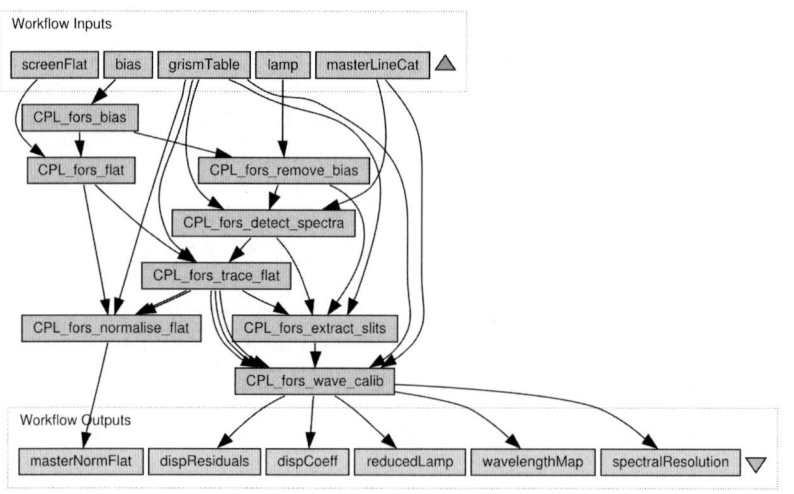

Fig. 2. Workflow for FORS spectroscopic calibration. Each processor 'box' invokes one of the FORS CPL recipes for generalised multi-object spectroscopic reduction. The interactive tool shown in Fig. 3 can also be added to check and improve on the wavelength calibration for each spectrum

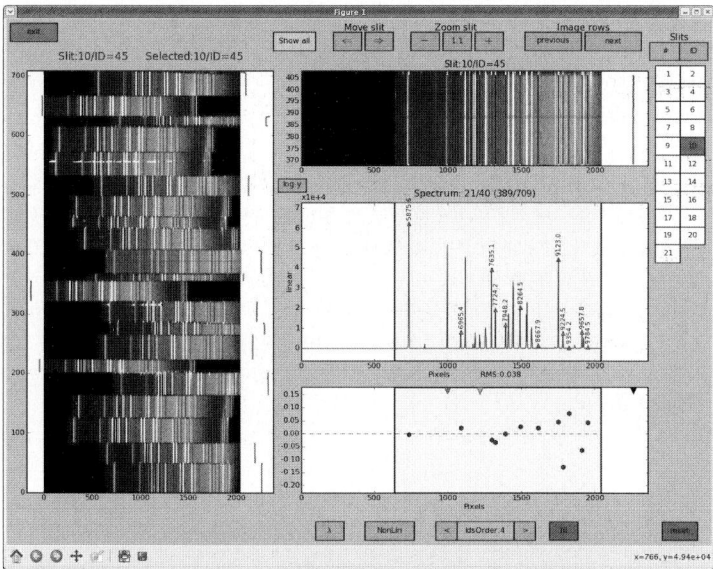

Fig. 3. The FORS wavelength calibration interactive tool. This tool is intended to be used to check and, if needed, improve, the wavelength calibration of a FORS arc exposure. It can be included at the end of the calibration workflow shown in Fig. 2, or run as a stand-alone application. The left-hand panel shows the full detector with the detected spectral lines high-lighted. On the right at the top is the selected single-slit spectrum and below it are a line plot of one line from this spectrum and a plot of the residuals to the dispersion solution. The user can change the fit order, delete lines from the solution and assess the quality of the dispersion solutions in several ways. This application was coded in Python and uses the matplotlib graphics library

4 Advantages of the Workflow Approach

The use of a graphical workflow system, and specifically ESO Reflex, offers many advantages when compared to more traditional ways to invoke recipes. These include the following:

- Modular sub-recipes are run in a sequence. This allows for quality control tools to be easily inserted between the reduction steps and much greater interactivity easily achieved.
- Step-by-step execution. In interactive mode the user can make changes to input data and parameters during execution. This allows full control of each processing step.
- Error detection and recovery. If an error occurs at any processing step the user can choose to re-run the step with different parameters or abort the workflow.

- Parallel Execution. The recipes on a parallel workflow execute in parallel so the full advantage of a multi-processor or multi-core machine can be exploited.
- Customisability. The user can easily modify the workflows. Python, IDL and other scripts can be included in the workflows and any system command can also be invoked from the workflows.
- Incorporation of instrument-specific tools. Sophisticated tools can easily be included in workflows to allow detailed quality assessment. An example is the FORS wavelength calibration tool shown in Fig. 3.
- FITS-file handling. Code from the Gasgano application is used to allow instrument-specific tags to be automatically extracted from the FITS headers. This allows tag checking so that only appropriate FITS files are allowed as inputs.

5 Current and Future Activities

ESO Reflex was released to selected testers in early 2007 and feedback is currently being collected and assessed. The initial release includes FORS spectroscopic reduction recipes, workflows and interactive tools as well as AMBER recipes and workflows. Work is in progress on many enhancements and updates to ESO Reflex and an internal release will be made in May 2007. During the final phase of the Sampo project, based on feedback and a detailed report, ESO will decide on a future course of action.

References

1. P. Ballester, et al.: Data reduction pipelines for the very large telescope. In: *Proceedings of the SPIE*, vol 6270, ed by D. Silva, R.E. Doxsey SPIE (2006)
2. R.N. Hook, et al.: The Sampo project. ESO Messenger **120**, 17 (2005)
3. T. Oinn, et al.: Taverna: a tool for the composition and enactment of bioinformatics workflows. Bioinformatics **20**, 3045 (2004)
4. I. Appenzeller, et al.: Successful commissioning of FORS1 – the first optical instrument on the VLT. ESO Messenger **94**, 1 (1998)
5. R.G. Petrov, F. Malbet, G. Weigelt, et al.: AMBER, the near-infrared spectro-interferometric three-telescope VLTI instrument. A&A **464**, 1 (2007)

Discussion

P. Manzato: According to the VO standard and WP5 VO-DCA (Data Center Alliance) with deals about grid and VO standard (VO store) could easily ESO Reflex be imported in a Grid Worker Node?

M. Romaniello: TAVERNA, the underlying workflow engine of Reflex, has built-in capabilities to handle distributed computing. However, at this stage, there are no firm plans to implement these functionalities within Reflex itself.

P. Bonifacio: I have been using MIDAS for the last 20 years and have developed a number of programs written in FORTRAN which run under MIDAS. Will I be able to import them easily under REFLEX or will I have to re-code everything?

M. Romaniello: REFLEX currently contains a python invocation to incorporate python scripts within workflows. If your scripts are compatible with PyMIDAS, then there should be no problem in incorporating them into REFLEX. If not, you will still be able to call them from within REFLEX, like any other shell command, but input/output handling (parameter passing, etc.) may not be trivial.

S. Kimeswenger: You mentioned IDL interfaces – do you intend to use this expensive commercial product as part of REFLEX? Some institutes can't afford IDL licenses.

M. Romaniello: No, what we're looking into is ways to interface IDL scripts to REFLEX, just because IDL is currently so popular (for those who can afford it . . .). No part of REFLEX itself will sanction software that needs to be paid for.

E. Emsellem: I am a great fan of python (as well as MIDAS for certain purposes) but I admit I was quite disappointed when I tried pyMIDAS because python there is only just (or almost) a script layer on top of MIDAS. The power of python is therefore not really used (dealing with arrays, object oriented, loading data in memory and doing complex manipulations, ...) and could be replaced by any good scripting language (perl ...) In that sense I don't see pyMIDAS's legacy value ? Is the planned development just adding a python layer, or really exploiting its full capabilities?

R. Hook: PyMidas was the first experimental project of Sampo. The aim was to replace the MIDAS procedure language with Python so that scripts could easily combine use of MIDAS tasks with those in other Python-based systems such as PyRAF. Although we appreciated that a deeper integration would be very valuable this would have required much more work, and particularly extensive changes to the underlying MIDAS system itself, and was hence outside the scope of the project. We do not have plans or resources to proceed further in this direction at present.

Astronomical Data Reduction Pipelines at NASA: 30 Years and Counting

D. Lindler

Sigma Space Corporation, 4801 Forbes Boulevard, Lanham, Maryland, USA;
don.lindler@gsfc.nasa.gov

Abstract. NASA has been very active in development of data reduction pipelines for space observatories since the early 1970s. Many of the techniques and architectures used today by the NASA Great Observatories have a pedigree that can be traced back to the pioneering work done by the International Ultraviolet Explorer (IUE) team. These techniques continued to evolve and improve from the manually controlled IUE processing systems to the automated system used today by the Hubble Space Telescope and other NASA projects.

1 Introduction

In the early 1970s, NASA and its international partners began development of a very successful data reduction pipeline for the International Ultraviolet Explorer (IUE). The longevity of IUE resulted in a pipeline that evolved over time giving valuable experience for development of future data reduction pipelines. However, some of the lessons learned were ignored or soon forgotten during the early development of the Hubble Space Telescope (HST) data reduction pipeline. The initial HST pipeline was a step backward in pipeline performance, reliability, and functionality. You might say that the young Space Telescope Science Institute (ST ScI) inherited a "lemon". The ST ScI quickly recognized the short comings of the system and began development of OPUS, a second generation pipeline. OPUS was highly successful and has been in continuous operation at the ST ScI since late 1995. It has proved to be adaptable to the increased requirements of the new HST instruments and to the requirements of many other projects including the Far Ultraviolet Spectroscopic Explorer, The Chandra X-ray Observatory, the Spitzer Space Telescope, and the Gemini Observatory. As such, OPUS provides an effective infrastructure for data reduction pipelines. However, the valuable experience from individual pipeline projects is important to future pipeline development to insure the quality of the archival data in the era of on-the-fly calibration and virtual observatories.

2 The International Ultraviolet Explorer

IUE was a joint project between NASA, UK Science Research Council, and the European Space Agency that was launched on January 26, 1978. IUE exceeded all expectations and continued to operate for almost 19 years until it was turned off on September 30, 1996. The pipeline processing system for IUE was the first astronomical satellite facility to deliver fully reduced data within 48 hours to the worldwide community of users. It was also the first time that a worldwide astronomical reduced-data archive was created that delivered more than 40,000 spectra per year to astronomers all over the world.

The development of the IUE pipeline was a long and difficult road. The pipeline development began in the early 1970s in an era of mainframe computing. The development began on IBM 360 mainframes with computing power much less than today's typical laptop computer. The system was developed under the Video Image Communication and Retrieval (VICAR) system developed in the 1960s at NASA/JPL. VICAR was a batch oriented system developed for analysis of planetary images. It contained the necessary infrastructure for the needed IUE pipeline and allowed re-use of many analysis applications for the geometric and photometric calibration of IUE images.

The IUE project chose an existing Sigma 9 computer available at Goddard Space Flight Center (GSFC) for the pipeline production system. The Sigma computers had a very problematical history and it was a challenge to keep the pipeline system running during the extended life of the IUE mission. The Sigma computers were developed by Scientific Data Systems in the early 1960s. The company was sold to Xerox in 1969 with Honeywell taking over the computer's support in the mid 1970s. No viable replacement became available for the aging Sigma 9 system. In retrospect, it is easy to second guess the choice of Sigma 9 computer considering the cost to convert VICAR from the IBM to Sigma system. However, it is difficult to find fault with the initial decision process if you consider that the longevity of IUE exceeded everyone's dreams and the future hardware support for the Sigma computers would have been pure guesswork.

Despite the hardware problems, the IUE pipeline system must be considered a tremendous success. The reasons for success include:

- The pipeline processing was considered an important part of IUE observatory function and was well funded throughout the lifetime of the spacecraft.
- The code was modular in well defined application units. Application units were easily modified or replaced without impact to the other application modules.
- The pipeline supplied intermediate calibration products which encouraged the astronomical community to become involved in improving the algorithms.

- The evolution of the pipeline algorithms were based on scientific return and controlled by the astronomical community through frequent meetings including the three agencies involved in the project and calibration meetings available to the whole user community.
- Astronomers were well informed on the continuing evolution of the pipeline and the technical issues concerning the calibration process through the IUE newsletter.
- Excellent dialog was maintained between the user community and the observatory staff. Astronomers came to the control center for real-time observing and in many cases returned to use the IUE regional data analysis facilities at GSFC and the University of Colorado.
- Introduction of data quality flags for each output data value to flag anomalous conditions such as detector blemishes.

Despite the success of the IUE pipeline, there were many difficulties. The VICAR system proved very adequate for the pipeline processing but the system itself was not widely used by astronomers. The output VICAR formats were machine independent but astronomers not using VICAR were still required to develop special purpose routines for reading the reduced data files into their analysis systems. This problem was greatly reduced by creation of the IUE regional data analysis facilities. These analysis facilities allowed astronomers to perform scientific analysis of the reduced data using an Interactive Data Language (IDL) based system. At that time, IDL was still in its infancy and not widely used by most astronomers. Astronomers came to the facility, worked with the observatory staff, and hopefully left with results ready for publication. As IDL became more widely used by the astronomical community, the IUE analysis facilities left an important legacy of analysis software adaptable to future astronomical missions including the Hubble Space Telescope.

The longevity of IUE led to other problems. The pipeline system was not completely automated. The reduction of each raw image was initiated and controlled manually. Perhaps, if the initial designers of the system knew that IUE would last so long, an automated system would have been considered. During the lifetime of IUE, the processing algorithms continued to improve. This resulted in inconsistencies between the quality of the early data in the IUE data archive with the newer data. It was not practical to constantly reprocess all of the data for every improvement to the pipeline and on-the-fly calibration was not yet feasible. To insure the legacy of the IUE data archive it was decided that all of the data would be reprocessed with a new version of the processing system. The New Spectral Imaging Processing Pipeline (NEWSIPS) was developed [1]. In addition to improving the consistency between the archived products during the extended mission, the new system changed the data formats from VICAR to the more widely used FITS format and provided improvements in the Signal/Noise, spectral resolution, and absolute flux calibration of the final products.

3 Hubble Space Telescope

The Hubble Space Telescope was launched on April 24, 1990 and has become one of the greatest scientific success stories of all time. It was launched with five initial science instruments all developed by five independent instrument science teams. Each team was led by a group of astronomers which were highly motivated to deliver a first-rate instrument. The motivation was insured by giving each team guaranteed observing time with the telescope. The teams were committed to delivering fully calibrated instruments to allow them to get the maximum scientific return for their guaranteed time. In the late 1970s these teams began work on calibration, reduction and analysis systems for their instruments in anticipation of a 1983 launch. For the most part, each team worked independently on their pipeline and analysis systems with little coordination between the teams. This would later cause problems for the Space Telescope Science Institute (ST ScI) that was not yet in existence.

3.1 The Goddard High Resolution Spectrograph

Each team was free to select their own programming languages and analysis systems. Two of the teams, the Goddard High Resolution Spectrograph (GHRS) and the Faint Object Spectrograph (FOS) contained team members that were very knowledgeable of the experiences of IUE. Both of these teams chose IDL as the principle programming language and analysis tool. In particular, the GHRS team at GSFC began development of a complete data acquisition, archival, calibration, pipeline processing, and analysis system written completely in IDL. Highlights of the system include:

- All data (both pre and post-launch) are quickly and easily searchable by instrument configuration, engineering parameters, and target parameters.
- Data were easily pipe-line processed using "on-the-fly" calibration.
- The pipeline processing included propagation of statistical errors.
- Comprehensive data quality flags were assigned to each output data value.
- Same package can be used for searching, reducing, and analyzing the data.

The basic architecture of the GHRS system was later extended to the FOS, Space Telescope Imaging Spectrograph (STIS), the Advanced Camera for Surveys (ACS), and the Wide Field Camera 3 (WFC3) instrument development team's calibration and reduction systems.

One of the most important legacies of the GHRS experiences is that IDL is a very good language for rapid prototyping of calibration and reduction algorithms and continues to be used for prototyping algorithms by the ST ScI.

3.2 The HST Post Observation Data Processing System

The first version of the HST Post Observation Data Processing System (PODPS) was part of the complete Science Operations Ground System

(SOGS). The initial pipeline requirements were developed by the instrument teams in 1980 and 1981 prior to the completion of the pre-launch instrument calibration and prior to the formation of the ST ScI. In 1981, these requirements were delivered to the SOGS contractor. After which, very little contact occurred between the SOGS contractor and the instrument teams. The interaction of the software developer and the end user was also lacking (the user, the ST ScI, did not yet exist). In 1983 the first software components of the ground system were delivered to the newly formed ST ScI. Problems with the ground system quickly became evident. The ground system was developed using "last-generation" programming technology. The PODPS code was non-modular. Fixing an error in one part of the system would result in un-anticipated problems in another part of the system. The ST ScI basically inherited an unusable system. To make matters worse, the generation of calibration reference files (e.g. flats, darks, wavelength dispersion coefficients) was not included as part of the ground system. The ST ScI scrambled to correct the situation. They negotiated with the instrument teams for a joint effort to re-write the pipeline software and complete a calibration data base system. Early in its existence, ST ScI also decided, with the approval of NASA, that it would be better to change the ST ScI from a strictly service organization to a world class science organization. The ST ScI staffing increased from the proposed 100 people to approximately 400 people at launch. This change in the charter of the ST ScI was instrumental in correcting the problems with the ground system and in making the HST one of the most successful NASA projects to date.

A significant HST launch delay resulting from the Challenger shuttle disaster left plenty of time to correct the ground system problems. The ST ScI was already contracted to develop a Science Data Analysis System (SDAS) for analysis of the reduced science data. The ST ScI chose to create pipe-line modules that would run under the same system. Using prototype software developed by the instrument teams and joint development with the teams, the pipe-line software was rewritten. Significant problems were still encountered. The ground system was a VAX/VMS based system. The ST ScI developed SDAS to run under the Image Reduction and Analysis Facility (IRAF) which they adapted to a VAX/VMS environment. The system had a number of significant problems. The code was machine dependent, the internal data files were a machine dependent FITS-like data format, and the analysis system was slow.

Although the pipe-line system was adequate, the ST ScI was not satisfied. SUN/Solaris systems were becoming popular and the ST ScI was determined to move away from their VAX/VMS dependence. They renamed the IRAF package from SDAS to STSDAS (allowing both systems to run simultaneously during the conversion process) and began migration from VAX specific code to machine independent code while significantly improving the performance of the original SDAS system. Finally in a move to become completely machine independent, the internal data formats were changed to the now popular disk FITS format using IEEE binary data representation.

The Science Institute continues to make advancements in the area of pipeline processing for the HST, other NASA space missions, and in anticipation of the James Webb Space Telescope. These enhancements have included, the development of observation associations which allow multiple files and calibration observations to be processed as a group, automatic generation of observations specific calibration reference files, implementation of generic calibration modules that can be used by multiple instruments, the development of OPUS (sect. 3.3), and the migration to Python/PyRAF.

Throughout the lifetime of the HST, the ST ScI has continued to improve the coordination of effort between the HST instrument development teams and the astronomy community as a whole. Many of the current pipeline algorithms and modules were developed by the Instrument Development Teams, the Space Telescope European Coordinating Facility (ST ECF), and other HST users funded by both science and calibration grants. Improvements are continually made to the pipelines of the archival instruments including significant contributions by the ST ECF. Of particular note is the work done by the ST ECF using model based calibration [2]. This work has shown that significant improvements can be made to calibration of the past and current HST instruments and offers a new approach to calibrating and understanding the performance of future instruments.

3.3 OPUS

Astronomical pipelines have become increasingly more complex. Manual control of the pipeline as done in the early days of IUE is no longer feasible. In the case of the HST pipeline, servicing multiple instruments, significant automation is required. This led the ST ScI to develop the OPUS system [3]. OPUS is a fully distributed system for any series of pipeline activities. It can run in a computing environment with multiple instances of multiple processes running in multiple pipelines on multiple computers. Monitoring tools within the system minimize the manpower required to keep it operational. OPUS is generic and can easily be tailored to other pipeline systems. It is currently operational for many observatories and ground systems including the HST, the Far Ultraviolet Spectroscopic Explorer (FUSE), the International Gamma-Ray Astrophysics Laboratory, the Chandra X-ray Observatory, The Spitzer Space Telescope, and the Gemini Observatory. OPUS has been in operation at the ST ScI since December of 1995. Particularly important is that OPUS does not put any burden on the developers of the pipeline applications. The applications can run under a existing package (IRAF) or run as independent modules. This allows migration of an existing pipeline to OPUS without significant software development. Following the lead of the Canadian Astronomical Data Center and the Space Telescope European Coordinating Facility, OPUS now supports on-the-fly calibration.

4 Summary: Lessons Learned

1. Keep the user community involved in the instrument calibration and pipeline development.

Many enhancements in the IUE and HST pipelines came from the end users of the pipeline processed data as evidenced by calibration workshop proceedings. The IUE three agency meetings, calibration workshops, and the HST calibration workshops were extremely productive. The end-users have more motivation to improve the pipeline than does a service organization with a limited budget. Important services that assist the user community in providing calibration support include making the pipeline code available to the users and supporting intermediate data products.

2. Code should be modular.

Pipelines should use self-contained modules which can be replaced without affecting other modules within the pipeline. It also increases the ability to re-use modules between different instruments.

3. Fast Prototyping is important.

The ability to quickly prototype new pipeline algorithms in languages such as IDL and Matlab decreases the time it takes to implement new algorithms and significantly improves the ability to test the final pipeline code.

4. Code and datasets should be machine independent.

The computing environment used by the astronomer community is highly heterogeneous.

5. Include propagated errors and data quality flags with the output calibrated data products.

This is particularly important in the era of Virtual Observatories where astronomers do not want to become experts on every instrument that they utilize data from. To the extent possible, data products should be self documented. Post pipeline processing analysis tools should provide support for these data quality flags. These flags were very useful for analysis of IUE data largely because the IUE Analysis facilities had software that utilized them. One of the biggest obstacles in creation of analysis software is the inconsistency of the format of these flags. It is time for the astronomical community to develop a consistent set of standards. Currently there are inconsistencies even between the HST instruments.

6. On-the-fly pipeline processing helps to maintain consistency of pipeline products.

This is particularly important in the early stages of an operational pipeline when many improvements to the calibration process occur.

7. The end user should be involved in the early pipeline development effort.

Significant problems with the early HST ground system would have been avoided if the ST ScI was formed prior to the release of the Science Operations Ground System contract.

References

1. J. Nichols, J. Linsky: AJ **111**, 517 (1996)
2. M. Rosa: 1997, In: *ASP Conf. Ser., Vol. 125*, 411, Astronomical Data Analysis Software and Systems VI
3. J. Rose, et al.: 1995, In: *ASP Conf. Ser., Vol. 77*, 429, Astronomical Data Analysis Software and Systems IV.

Discussion

J. Lewis: FITS probably wasn't chosen as the format of choice for disc files at the start of the SDAS project as at that time FITS was strictly a type format. "Disc FITS" only evolved around 1983–84.

The End-to-End Pipeline for HST Slitless Spectra PHLAG

M. Kümmel[1], R. Albrecht[1], R. Fosbury[1], W. Freudling[1], J. Haase[1], R. N. Hook[1], H. Kuntschner[1], A. Micol[1,2], M. R. Rosa[1], and J. R. Walsh[1]

[1] ESO, Space Telescope European Coordinating Facility, Karl-Schwarzschild-Strasse 2, 85748 Garching, Germany; mkuemmel@eso.org

[2] European Space Astronomy Centre, P.O. Box - Apdo. de correos 50727, 28080 Madrid, Spain

Abstract. The Space Telescope-European Coordinating Facility (ST-ECF) is undertaking a joint project with the Canadian Astronomy Data Centre and the Space Telescope Science Institute to build a Hubble Legacy Archive (HLA) that contains science ready high level data products to be used in the Virtual Observatory (VO). The ST-ECF will provide extracted slitless spectra to the HLA, and for this purpose has developed the Pipeline for Hubble Legacy Archive Grism data (PHLAG). PHLAG is an end-to-end pipeline that performs an unsupervised reduction of slitless data taken with the Advanced Camera for Surveys (ACS) or the Near Infrared Camera and Multi Object Spectrometer (NICMOS) and ingests the VO compatible spectra into the HLA. PHLAG is a modular pipeline, and the various modules and their roles are discussed. In a pilot study, PHLAG is applied to NICMOS data taken with the G141 grism, and the first results of a run on all available data are shown.

1 The Hubble Legacy Archive

After a proprietary period of usually one year, all science data from the Hubble Space Telescope (HST) are offered for retrieval at the archives hosted at the Space Telescope-European Coordinating Facility (ST-ECF), the Canadian Astronomy Data Centre (CADC) and the Space Telescope Science Institute (STScI). These archives offer the individual images with calibrated pixel values for download. Any further, time extensive processing such as the coadding of the single images must be done by every archive researcher individually.

The technical expertise on how to generate calibrated images of a certain HST instrument peaks towards the end-of-life (or decommissioning) of the instrument and, as the instrument team personnel move to other projects, that knowledge quickly dissipates.

After seventeen years in orbit, HST is approaching the last phase of its life, and now is the right time to build a Hubble Legacy Archive (HLA) that contains high level science data products which can be immediately be used for research [1] and be exploited by the VO tools currently being developed. In the static HLA the accumulated experience of HST instruments will be

Table 1. Some HST slitless modes under consideration for the ST-ECF HLA activities

Instrument	Disperser	Wav. Range (Å)	Resolution (Å/pixel)	FOV (arcsecond)
NICMOS/NIC3	G141	1100 − 19000	80.0	51 × 51
ACS/WFC	G800L	5500 − 10500	38.5	202 × 202
ACS/HRC	G800L	5500 − 10500	23.5	29 × 26
ACS/HRC	PR200L	1700 − 3900	20(@2500Å)	29 × 26
ACS/SBC	PR130L	1250 − 1800	7(@1500Å)	35 × 31
ACS/HRC	PR110L	1150 − 1800	10(@1500Å)	35 × 31

preserved, and providing science ready products reduces the time archive scientists have to spend on non-science matters.

The ST-ECF has a long record of supporting various spectroscopic modes of HST instruments and therefore has selected slitless spectroscopy as the main ST-ECF contribution to the HLA. With the development of the spectroscopic data extraction package aXe [2], which was designed for slitless data, the ST-ECF has a very good starting point to build an automatic, unsupervised pipeline around. Since slitless data are currently only delivered with the most basic correction applied (bias, dark), the improvement of an HLA with science ready spectra is large.

To generate the high level science data and provide input to the HLA the **P**ipeline for **H**ubble **L**egacy **A**rchive **Gr**ism data (PHLAG) has been developed at the ST-ECF [3]. The properties of slitless HST modes which could be reduced for the HLA with PHLAG are listed in Table 1. In a pilot project, PHLAG is being developed and first applied to NICMOS G141 data.

2 Slitless Spectroscopy

Slitless spectroscopic images have some special features and properties which distinguish this observing technique from the better known spectroscopy with apertures (slits or masks). These features are illustrated in Fig. 1. The right and the left panel show a basic dataset consisting of NICMOS G141 slitless images and their corresponding direct images taken in the H-band (F160W filter). Due to the low spectral resolution, there are multiple spectral orders (three in Fig. 1) of the same object visible in a single Field of View.

The absence of apertures makes contamination, which is the mutual overlap of spectra, a ubiquitous phenomenon. There is contamination in the spatial direction, as can be seen in the dispersed image for the two objects in the lower right circle, contamination in the dispersion direction with the object in the lower left circle, and even contamination across different spectral orders, as it is the case for the two object in the upper circles.

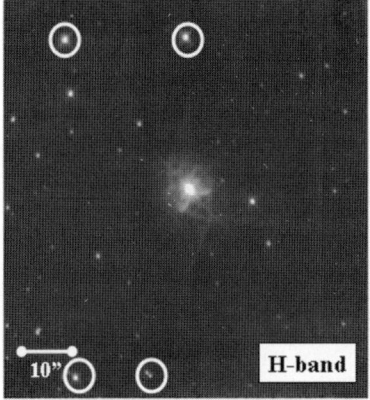 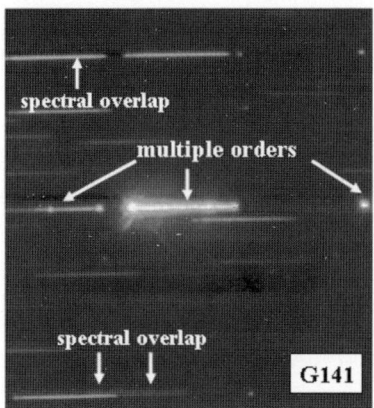

Fig. 1. Basic dataset of slitless spectroscopic images (*right*) and their corresponding direct H-band (F160W) images (*left*). The multiple orders of the same object are marked on the slitless image. The spectra of the objects encircled in the direct image contaminate each other in spatial direction, spectral direction and even across different spectral orders

In slitless spectroscopy the zeropoint of the wavelength calibration must be derived from the object position in the direct image. Therefore a slitless spectroscopic dataset always consists of a set of direct images and dispersed images, which were taken at the same position on the sky (as in Fig. 1), and the processing of the direct images is an integral part of any slitless spectroscopic pipeline such as PHLAG.

Since none of the HST instruments with slitless spectroscopy contains useful on-board calibration lamps, suitable astronomical targets must be used as calibration sources [4, 5]. Depending on the wavelength regime (UV, optical or NIR), planetary nebulae (see Hb12 in the lower right panel of Fig. 2), Wolf-Rayet stars or QSOs are used for the wavelength calibration. The flux calibration is achieved by analysing white dwarf standard stars such as G191B2B (see middle panel in the right column of Fig. 2).

3 The Modules of PHLAG

The inputs to PHLAG consists of calibrated images as provided from the HST archive. The final products of PHLAG are reduced, VO-compatible slitless spectra stored in an HLA archive.

PHLAG consists of a series of modules with each performing a certain reduction step on the data. Whenever possible, PHLAG uses external software (such as PyRAF [6] for basic image processing) to keep the development effort small. The pipeline is implemented in Python [7].

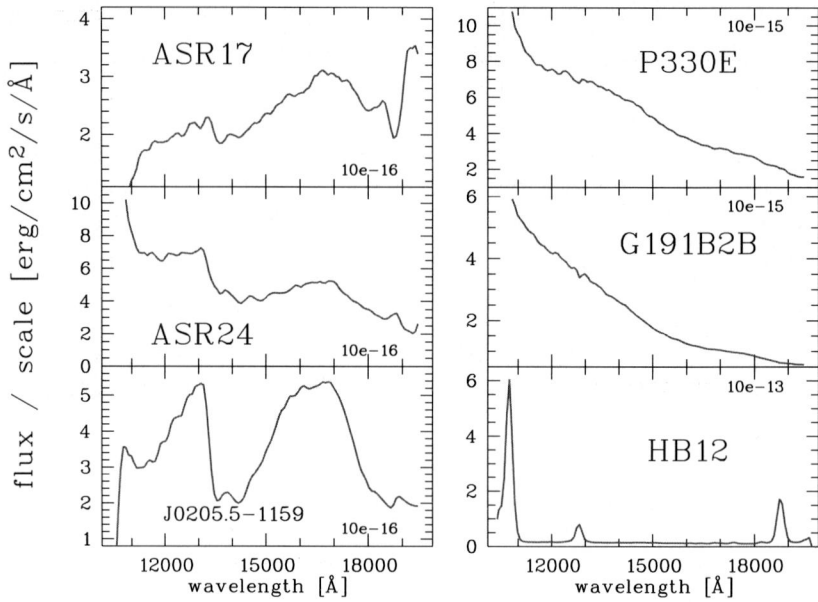

Fig. 2. Spectra of bright objects as reduced by PHLAG. The individual scale factor is given in the plots. The *left column* shows the spectra of two brown dwarfs in NGC 1333 (*upper panels*) and the field dwarf standard J0205.5-1159 (*lower panel*). The full dataset, interactively reduced with a different extraction program, is published in [14]. The spectra of three standard stars are shown in the *right column*. The solar analogue star P330E (*upper panel*) is followed by the white dwarf G191B2B, which is a flux standard (*middle panel*). The *lower panel* displays the spectrum of the planetary nebula Hb12

As in imaging photometry, there usually exist repeated, "dithered" slitless observations of a target field. Coadding the spectra from different images requires high precision in the relative astrometry of the slitless and direct images. For HST data such a precision can only be guaranteed within a so called "visit", a period during which HST is continuously locked on the same guide star. The association of direct images and slitless images taken during the same visit therefore forms the natural dataset on which to run PHLAG.

The pipeline is very flexible, and it is possible to run one dataset through all modules as well as to run several or all existing image associations through a set of selected modules.

PHLAG contains the following modules:

1. **data preparation:** In this step the data are prepared for the pipeline reduction. Every slitless image is paired with the direct image that was obtained with the smallest positional difference. The set of direct image - slitless image relationships created here is an essential input to the spectral extraction.

2. **image combination:** To prepare for the object extraction, the direct images are combined to create a deep direct image. This is done using the MultiDrizzle software [8].
3. **object detection:** The object detection software SExtractor [9] is run on the combined direct image. The spectra of all sources detected here will be extracted from the slitless images. Conservative parameter settings are used, and the prime aim is to detect all objects that also have detectable spectra in the slitless images.
4. **spectral extraction:** The spectrum extraction package aXe [2, 10] is used to extract the object spectra from the slitless images. aXe determines a local background around every object to remove any residual sky background. The flatfielded 2D spectral stamp images extracted from the individual grism images are coadded using the aXedrizzle technique [11]. Finally, deep spectra are extracted from these combined 2D spectral images. For each spectrum, an estimate of the contamination caused by neighbouring objects is derived using the photometric information from the direct images [12].
5. **quality control:** Consistency checks, e.g between the photometry (from the direct image) and the spectroscopy, are performed. Spectra with quality problems are either flagged or discarded.
6. **metadata:** The spectra are post-processed and prepared for ingestion into the HLA archive. Metadata are collected (object positions, extraction geometry) or derived (signal-to-noise estimates for the spectra). To ensure compatibility with the Virtual Observatory, we closely follow the rules and recommendations of the IVOA Spectral Data Model [13] in the selection of the metadata.
7. **data ingestion:** This last module inserts the fully reduced, quality controlled and VO-ready spectra into the HLA archive.

4 Status

As of January 2007, all modules of PHLAG have been implemented, and a solid first set of calibration files has been produced. PHLAG is in the late phase of development and testing. Further improvements to the data is expected from an optimisation of the background estimation, an improvement in the handling of bad pixels and the finetuning of various algorithms and parameters in the aXe extraction.

One complete test run of PHLAG has been done in December 2006 on all available ($\sim$ 1000) NICMOS G141 dataset. From the $\sim$ 5000 grism images $\sim$ 40 000 object spectra were extracted. Out of these we expect $\sim$ 10 000 spectra with a high enough Signal-To-Noise ratio to be included in the data release. The test run needed around ten days of computing time on a single Linux box.

Figure 2 shows the spectra of some bright objects reduced from different datasets as extracted by PHLAG. The left column shows the spectra of some dwarf stars, and the right column displays some standard stars (see caption for details).

A production run on all data and a release of NICMOS G141 spectra is expected in the second half of 2007. This released dataset will become a formal part of the HLA and will be accessible through both, custom interfaces and the Simple Spectral Access interface [15].

References

1. J.R. Walsh, R.N. Hook: ST-ECF Newsletter **40**, 6 (2006)
2. aXe: http://www.stecf.org/instruments/ACSgrism/axe/
3. M. Kümmel, A. Micol, H. Kuntschner, J.Walsh, W. Freudling: ST-ECF Newsletter **41**, 6 (2006)
4. A. Pasquali, N. Pirzkal, S. Larsen, J.R. Walsh, M. Kümmel: PASP **118**, 270 (2006)
5. S.S. Larsen, M. Kümmel, J.R. Walsh: in *The 2005 HST Calibration Workshop*, eds. A. Koekemoer, P. Goudfrooij, & L.L. Dressel, p. 103 (2006)
6. PyRAF: http://www.stsci.edu/resources/software_hardware/pyraf
7. Python: http://www.python.org
8. A.M. Koekemoer, A.S. Fruchter, R.N. Hook, W. Hack, C. Hanley: in *The 2005 HST Calibration Workshop*, eds. A. Koekemoer, P. Goudfrooij, & L.L. Dressel, p. 423 (2006)
9. E. Bertin, S. Arnouts: A&A **117**, 393 (1996)
10. M. Kümmel, S.S. Larsen, & J.R. Walsh: in *The 2005 HST Calibration Workshop*, eds. A. Koekemoer, P. Goudfrooij, & L.L. Dressel, p. 85 (2006)
11. M. Kümmel, J.R. Walsh, S.S. Larsen, R.N. Hook: in *Astronomical Data Analysis Software and Systems XIV*, eds. P. Shopbell, M. Britton, & R. Ebert, p. 138 (2005)
12. J.R. Walsh, M. Kümmel, S.S. Larsen: in *The 2005 HST Calibration Workshop*, eds. A. Koekemoer, P. Goudfrooij, & L.L. Dressel, p. 79 (2006)
13. IVOA Spectral Data Model: http://www.ivoa.net/Documents/latest/SpectrumDM.html
14. J. Greissl, M.R. Meyer, B.A. Wilking, et al.: AJ **133**, 1321 (2007)
15. The SSA interface: http://www.ivoa.net/twiki/bin/view/IVOA/SsaInterface

A Bottom-Up Approach to Spectroscopic Data Reduction

C. Izzo, Y. Jung, and P. Ballester

ESO, Karl-Schwarzschild-Strasse 2, 85748 Garching, Germany; `cizzo@eso.org`

Abstract. Any reference object identification based both on pattern-recognition techniques and on keeping to a minimum any instrument-specific assumption, goes in the direction of fulfilling the essential requisites of robustness and flexibility in automatic data reduction pipelines. This also allows a bottom-up approach to instrument calibration (from data to model, rather than the other way around). Pattern-matching methods extend the palette of tools available to solve calibration problems, including identification of reference objects based on physical models, or correlation methods. In this chapter a qualitative description of the bottom-up strategy is given in the context of FORS1/2, VIMOS, EMMI, GIRAFFE, and CRIRES data reduction.

1 Robustness and Flexibility

Robustness and flexibility, together with reusability, are among the most used words in software presentations. But what is exactly meant by these terms in the context of *automatic* data reduction? Robustness is the capacity to deal with the unexpected, always reporting to what extent the instrument behaviour has changed, rather than just stopping with a generic failure message. As a prime advantage, a more extensive monitoring of the instrument health becomes possible. Flexibility, on the other hand, is obtained by defining algorithms and procedures that are so general to be promptly applicable to different instruments, and in particular to the occasional hardware upgrades on the same instrument. This leads to a significant lowering of the software maintenance costs.

One can immediately see that the qualities of robustness and flexibility described here are not so much related to the software, but they are rather related to a strategy. Some strategies turn out to be inherently fragile, no matter how good their software implementation.

2 Identification of Reference Objects

To calibrate an instrument means to determine a relationship between instrumental and physical quantities. For this purpose reference objects, such as stars and spectral lines, can be used. Once the reference objects are detected and identified, their known physical characteristics (such as position on the sky, intensity, wavelength) can be matched to the observed instrument response, and a transformation can be determined. This ideally completes the instrument calibration task. Failures in instrument calibration are mainly related to the possible misidentification of reference objects that would jeopardise the comparatively simpler task of fitting a transformation between instrumental and physical quantities.

2.1 The Top-Down Approach

How can the reference objects be safely identified? Traditionally a top-down approach is applied, that is based on (and actually requires) a good preliminary knowledge of the instrument behaviour (see Bristow et al. these proceedings). This knowledge should be at least sufficient to provide a first-guess transformation between the physical and instrumental quantities of interest. Alternatively, an empirical first-guess transformation (typically a polynomial function) may be provided.

The first-guess transformation is applied to a catalog of physical quantities relative to the reference objects, converting them in terms of instrument response. The first identifications are performed on the basis of closeness to expectations.

In the specific case of Multi-Object Spectroscopy (MOS) wavelength calibration, for example, the wavelengths of reference emission lines in calibration spectra are converted into positions on the detector, applying a transformation that depends on the position of the slits on the telescope focal plane, and on the dispersion law of the spectrograph. The observed reference lines are identified by their closeness to their expected positions.

This top-down approach is the only possible solution when very few reference objects are available. This is also the safest possible approach, if the instrument is guaranteed to be mechanically and optically stable: the detected objects will never get too far from expectations, hence will always be safely identified.

On the other hand, experience has shown that instrument instability is a rather common issue: a certain amount of mechanical play will always be present (flexures, slit mask positioning). Calibration lamps intensity, filters transmission, detector response, may all change, often in an unpredictable way. The instrument behaviour may be affected by aging, hardware interventions, or even earthquakes. In past years a significant part of the maintenance work on automatic pipelines went into periodic revisions of first-guess transformations. Without this, the reference objects would drift away from their expected positions and cause the identification task to fail. Even worse, the identification may remain apparently successful if alternative objects are found around the expected positions: this potential misidentification may

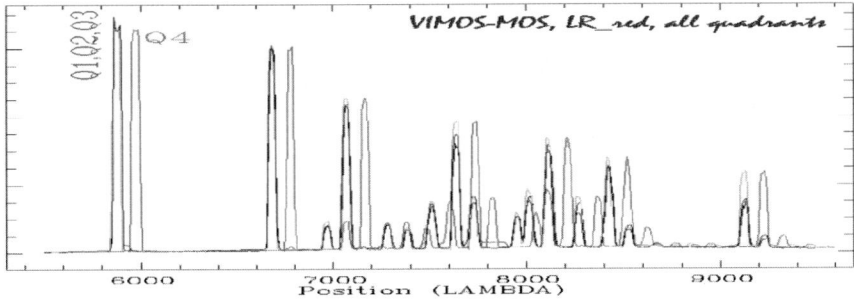

Fig. 1. Calibrated reference spectra from four VIMOS quadrants. In this real example the spectrum from quadrant 4 is not properly calibrated: this was caused by an imperfect insertion of a mask, which led to the misidentification of several lines. Yet no significant variation of the computed transformation was recorded: false matches tend to confirm expectations

silently confirm expectations, impairing a correct instrument health monitoring. This can commonly happen if many reference objects are available (see Fig. 1).

2.2 The Bottom-Up Approach

An alternative approach to reference objects identification may be based on pattern-recognition techniques. A straightforward example of this is point-pattern-recognition applied to the identification of stars, commonly used in target acquisition of telescopes [1]. The task consists in matching the positions of stars in the detector with the celestial coordinates listed in a reference star catalog. If enough stars are available, and if field distortion is not too severe to destroy the pattern, their identification is possible with only a rough knowledge of size and position of the field-of-view: each star can be safely identified on the basis of its relationship with the nearby stars (see Fig. 2). In the case of spectroscopy the same strategy can be applied to matching spectra with mask slitlets, or to identifying arc lamp lines.

With this approach, assumptions on instrument behaviour are minimised. Not only is the instrument model excluded from the object identification phase, its usage being limited to the task of fitting a transformation to the data; but also the calibration procedure does not involve any specific, first-guess model solution. Only a general (i.e., with undefined parameters) model of the instrument is required — and not even that, in case just an empirical transformation is going to be fitted to the data. A physical model, rather than an empirical one, should be preferred for fitting because it offers the advantage of a meaningful and realistic transformation, based on the instrument physical properties. Such transformation is more reliable in

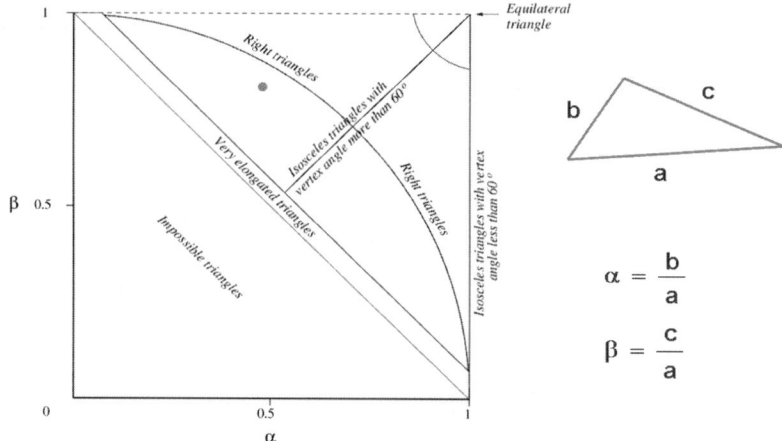

Fig. 2. Stars form patterns (they may be called more appropriately constellations) that are invariant to rotation, translation, rescaling. The simplest of such patterns is the triangle. Similar triangles, found both on the data and on the pattern side, correspond to the same point in the α-β plane. Each triangle identifies three stars, while each star is common to several different triangles: the inherent redundancy of this approach guarantees a robust identification of stars

the intervals where reference data are missing (ideally resolving the problem of interpolation and extrapolation). The use of an empirical transformation may only be acceptable when the (unknown) calibrating transformation is well sampled by several and well distributed reference objects.

One of the main advantages of minimising assumptions on the instrument is that it naturally leads to instrument-independent, and therefore reusable, software. On the other hand, in the same way as the top-down approach depends on the instrument model and fails if the model is inaccurate, the bottom-up approach depends on the data and fails if too few reference objects are available for providing unambiguous patterns. Another important limit is that this approach is a black-box: as for any bottom-up strategy (e.g., based on trained neural networks, or relying on evolutionary algorithms), it is often difficult to find the reason of a failure.

3 Application to the FORS Spectroscopic Pipeline

In October 2006 a first release of a new spectroscopic data reduction pipeline for the FORS instruments, based on a bottom-up calibration approach, was delivered to Paranal. The procedure described below, determining spectral curvature, wavelength calibration, and transformation between telescope focal plane and detector, was successfully applied not just to FORS1/2

MOS/MXU data, but also to VIMOS and EMMI MOS data. The wavelength calibration task alone was also applied to long-slit spectroscopy of the FORS instruments, and to extracted VIMOS and GIRAFFE IFU spectra. The mean accuracy of the determined extraction mask, currently based just on empirical polynomial modeling, is typically better than 0.1 pixel.

We assume here that the dispersion direction is roughly along the x CCD coordinate (rows), while the spatial coordinate is y (columns), as in the case of FORS spectra. For instruments where the dispersion is along other directions, such as VIMOS, the input data is rotated accordingly. The spectra will generally be curved, and may even display a small rotation (up to about 20 degrees, depending on the length and width of the spectra on the detector).

The detector is preliminarily scanned row by row, for recognising in the sequence of detected peaks the pattern of wavelengths listed in a reference catalog. Contiguous detector rows where the pattern is detected at very similar offsets are grouped together, marking the regions corresponding to a single slit spectrum. These regions are used on an associated flat field exposure as a starting guess for determining the real shapes of the spectra, obtained by tracing the flat field spectral edges. The determined spectral curvature model is finally applied to extract the arc lamp spectra, deriving a wavelength calibration that follows the spectral curvature. If enough and well positioned spectra are detected, it would also be possible at this point to apply the point-pattern-matching algorithm for associating each spectrum to its corresponding slit: this is not required to guarantee an appropriate extraction of the scientific spectra, but it provides an identification tag to each spectrum, and it also determines a transformation from telescope focal plane to detector, useful for instrument health monitoring.

At this point the spectral extraction mask is completely determined: for each spectrum a specific coordinate system is defined, where each CCD pixel corresponds to a wavelength and a position on the telescope focal plane. If the instrument were stable, it would be possible to extract the scientific spectra directly applying this extraction mask. In general, however, an extraction mask based on day calibration exposures will be aligned using the available sky lines as reference, before extracting the scientific spectra.

A more detailed description of this procedure is given in the FORS Pipeline User Manual [2].

4 Application to the CRIRES Pipeline

The wavelength calibration method applied in the CRIRES pipeline (Fig. 3) is another successful attempt to minimise assumptions on the instrument. The applied pattern-matching technique is even more flexible than the one described in Fig. 2, because it can overcome extreme distortions of the pattern, finding transformations that can be strongly non-linear. The basic idea

Inputs:
- The Estimate Polynomial
- The Wavelength Error of the estimate (WLerror)
- The degree of the searched polynomial (degree)
- The number of samples (nsamples)
- The lines catalogue (OH, Gas cell, Lamps, Hitran)
- The signal to calibrate (in pixels)

Algorithm:
- Consider degree+1 positions Ai regularly spaced, and nsamples points spread within WLerror around these positions
- For each possible sequence of points (nsamples^(degree+1) possibilities), the interpolation polynomial is created and considered as candidate
- The candidate polynomial is used to convert the signal to calibrate from pixels to wavelengths. This signal is compared to the signal generated from the catalogue. A likelihood coefficient in computed.
- The best likelihood parameter gives the best candidate, i.e. the polynomial that is the closest to the solution
- A second pass (or more) is used to refine the solution with the first pass solution used as estimate, with a smaller WLerror and a higher degree

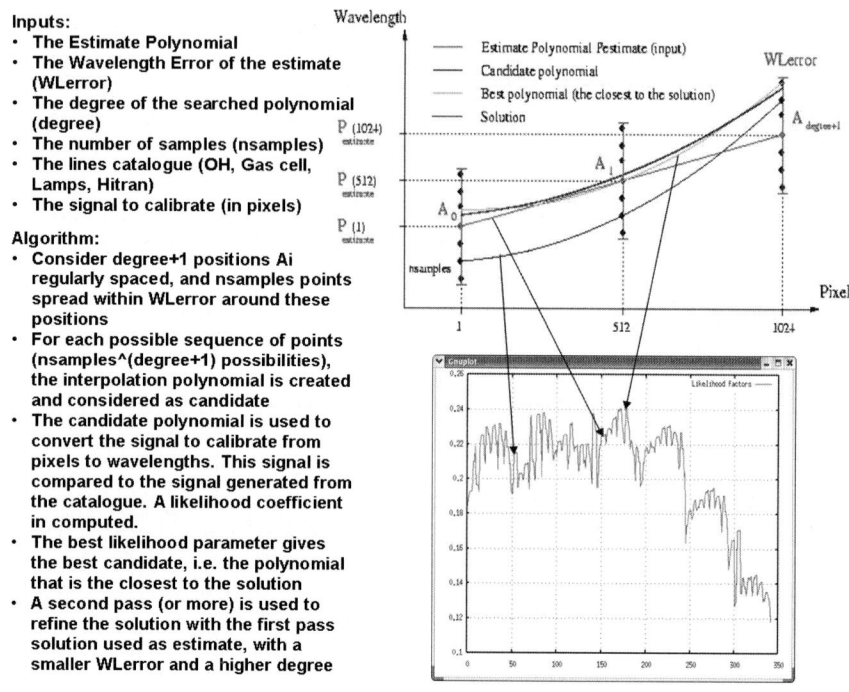

Fig. 3. Wavelength calibration in the CRIRES pipeline

is to begin with a linear estimate of the transformation from pixel to wavelength (a mean dispersion and an expected offset). All candidate polynomials around this first estimate are used to generate from the reference line catalog a model spectrum that is compared to the observed one. Each candidate is associated to a likelihood factor, the highest identifying the best solution. Ideally, it is also possible to do without a reference line catalog, if an accurately calibrated reference spectrum is available. This is in principle a brute force approach, but in practice the search for the best calibration can be made efficient by performing a preliminary search limited to low degree polynomials within a larger region of the solution space, narrowing then the search to smaller regions and higher degree polynomials.

References

1. G.S. Cox, et al.: A New Method of Rotation, Scale and Translation Invariant Point Pattern Matching Applied To the Target Acquisition and Guiding of an Automatic Telescope. In: 2nd South African Workshop on Pattern Recognition, p. 167–172, November 1991
2. FORS Pipeline User Manual, 1.0, 2006-10-10 (VLT-MAN-ESO-19500-4106)

Discussion

W. Hummel: Is the rotation, scaling and moving operation enough to find a match or must non-linearities be taken into account when fitting empirical with model/catalog patterns?

C. Izzo: Non-linear terms are very small if the search is limited to small sub-patterns (e.g. small triangles).

U. Hopp: How many trials in comparing triangles for a given data set do you need to get a stable or secure solution for the pattern recognition? So phrasing differently, is this step time-consuming within your pipeline?

C. Izzo: The pattern-matching method currently applied is extremely fast; because not all possible triangles are built – just a network of small and consecutive triangles is defined. On a typical case, with 20–30 spectra on the CCD, the pattern-matching step takes about 15 s.

M. Rosa: Very nice approach. It seems to be that a thorough combination of our physical model top down concept and your bottom up search would cover all aspects encountered at a large visitor observatory. I am really eager and looking forward to a closer collaboration.

Model Based Instrument Calibration

P. Bristow[1], F. Kerber[1], and M.R. Rosa[2]

[1] ESO, Karl-Schwarzschild-Strasse 2, 85748 Garching, Germany;
bristowp@eso.org
[2] ST-ECF, Karl-Schwarzschild-Strasse 2, 85748 Garching, Germany

1 Introduction

The Calibration Support Group, part of ESO's Instrumentation division, aims to develop and implement advanced calibration techniques for ESO instruments. Our physical models use the engineering and design information employed to construct the instrument as a starting point, and are then optimised to describe the actual configuration of the instrument including the response to environmental effects.

Traditionally the wavelength calibration of spectrographs relies upon an empirical approach. An exposure of a source, usually an emission lamp, with clear, laboratory-calibrated features, is obtained. The location of features on this wavecal exposure are then matched to the catalogued wavelengths of the source, and a low order polynomial is fitted to the data points to provide an empirical relation between positions on the detector and wavelengths. A meaningful polynomial fit will require a sufficient density of useful lines distributed over the wavelength range of interest. Since such an empirical polynomial fit has zero predictive value outside the range defined by data points, a lack of calibration lines at the limits of the wavelength ranges and detector boundaries is particularly critical.

We replace this empirical method of wavelength calibration by using our physical understanding of the instrument. We know from the design process that even sophisticated spectrographs can be accurately modelled [1, 2]. However, with the exceptions of the model driven bootstrap for the still canonical wavelength calibration of the VLT/UVES instrument and the recent application of the method used here to HST's STIS [3], that approach has previously been perceived as to complex and to difficult to become the backbone of the wavelength calibration of a pipeline. Our CRIRES and X-Shooter optical models have model kernels that are fast, simplified ray trace codes. The speed with which these streamlined models can be solved make them suitable for iterative evaluation for many different wavelengths and slit positions. Most importantly, parameters describing the configuration of the optical components can be optimized using the Monte-Carlo type "Adaptive Simulated Annealing" technique [9, 10] so long as appropriate calibration data is available. In order to avoid computationally expensive, but for the present purpose unnecessary level of detail, those surfaces which do not affect the relative geometry on the detector (e.g. plain folding or pick off mirrors) are neglected. However, the ray tracing at all relevant surfaces is performed by the appropriate 3D matrix transformation.

We have recently delivered a wavelength calibration model for CRIRES that enables reliable wavelength calibration even in regions with a scarcity of calibration source features and fits wavelength calibration data from 1.0 to 3.0 μm to an accuracy of better than 0.4 pixels. We discuss the concept, the input required, the optimization algorithm and the practical application of this kind of model. Whilst the investment of effort required to develop such a model for CRIRES has been considerable, this is already paying dividend as we reuse most of the techniques, procedures and algorithms for X-Shooter. Here the predictive power of the physical model is already employed in an early phase of the project supporting the development of both the instrument and data reduction software.

2 CRIRES

CRIRES is a cryogenic echelle spectrograph covering the wavelength range from 950 to 5300 nm at high spectral resolution, R_{max} 100,000 [6]. Pre-dispersion is achieved by means of a massive, reflective ZnSe prism with a wedge angle of 15° and a length of about 150 mm used in double path [4]. An echelle grating with a blaze angle of 63.5° and a groove density of 31.6 lines/mm provides dispersion in the main spectrograph, whilst the detector plane is equipped with four 1024×1024 (512×1024 illuminated) 27 μm pixel InSb detectors.

CRIRES wavelength calibration is particularly difficult regardless of whether we employ an empirical fit or a physical model:

- CRIRES' high resolution, combined with the fact that it is not cross dispersed, results in very few calibration features per exposure.
- CRIRES' high sensitivity results in detection of features that aren't in existing lamp line lists increasing the chance of false matches.
- The gaps of ∼300 pixels between the chips make pattern matching difficult. Often key features that determine a match "disappear" in the gaps.

In addition there are the usual problems of ghosts, hot pixels and cosmics, hardware and software improvements that have ameliorated the former two effects. These problems are partially circumvented by a clever cross-correlation technique [5] that provides an empirical wavelength solution; see also Izzo et al. these proceedings. Even so, it turns out that in only about one third of cases are there enough calibration features to provide sufficient input for an empirical fit.

2.1 Physical Model

Based upon [4], the principle components of the CRIRES model are:
- Orientation and temperature of the prism
- Orientation of the grating and the grating constant
- Orientation of the detector plane
- Position, orientation and distortion coefficients of the detector chips
- Scale factors for the focusing optics

The model utilizes high quality reference data describing the refractive index of the ZnSe prism as a function of temperature [8] and a Th-Ar hollow cathode lamp line list [7] supports calibration.

2.2 Multiple Modes and Slit Positions

Clearly there is a large number of parameters and therefore a large number of data points are required to fix them. In practise we need to obtain calibration data representative of CRIRES's operating range and then optimize the model using data from many of CRIRES operating modes simultaneously.

The modes of CRIRES are essentially defined by the grating and prism angles in the following way. An intermediate slit and adjustable prism allow the pre-selection of a limited wavelength range. This wavelength range is usually selected to match orders of the grating. Rotating the grating then further selects which wavelengths actually make it onto the detector array.

This means that if we use calibration data for multiple modes as input to the calibration process, we must set up the process to produce multiple optimised prism and grating angles each corresponding only to the data from an individual input mode, significantly increasing the number of free parameters.

In practise the interaction with the instrument (i.e. setting the prism and grating for an exposure or reading the prism and grating setting used from the meta data) is done via encoder values rather than physical angles. We have two approaches for dealing with this situation:

- Ignore the meta data encoder values and simply optimise the pair of prism and grating angles for all data from a given exposure.
- Optimise the coefficients of functions relating the encoder values to angles for both prism and grating. This has the advantage of reducing the number of free parameters when many prism/grating combinations are in use and automatically providing the encoder angle relations. However it presupposes that the encoder values accurately reflect the angles.

Further complication arises from the fact that, in order to get a fix on parameters that affect the detector location perpendicular to the main dispersion direction, we need to sample multiple slit positions without moving the prism or grating in between the exposures. Slit positions are treated as free parameters and optimised along with other model parameters. All data from a given slit position will affect the optimisation of that position.

2.3 Annealing

The problem of optimising the large number of parameters involved is clearly much too complex to handle analytically, so we employ the simulated annealing (SA) optimisation algorithm.

As described above we have a list of features that we have matched to wavelengths of known Th-Ar emission lines. We take each wavelength and use the model, with some initial guess parameter configuration, and compute the location at which a feature of that wavelength would arrive on the detector. We then compute the sum of the squares of the x and y offsets between the predicted and observed positions. We repeat this for all of the features in the matched list adding the square sums of all together. This is our metric which describes how well the model, with the current parameter configuration matches the observed calibration data.

The parameter configuration is changed and the process repeated. The SA algorithm determines the changes to the parameter configuration in such a way that it gradually converges upon the optimum configuration.

2.4 Results

We used over 2000 matched features from 350 exposures, at 50 combinations of prism and grating angles corresponding to 1.0–3.5 μm and 7 slit positions. Figure 1a shows x (dispersion) and y (parallel to slit) residuals in the case where all exposures are allowed independent prism and grating angles. The mean residuals are $\overline{|x_{cd} - x_{pm}|} = 0.5\,\text{pix} \simeq 0.007\,\text{nm}$ (at $2\,\mu\text{m}$)$\simeq 1000\,\text{m/s}$ and $\overline{|y_{cd} - y_{pm}|} = 0.5\,\text{pix}$. (subscripts cd and pm indicate the value measured in calibration data and the value calculated by physical model respectively).

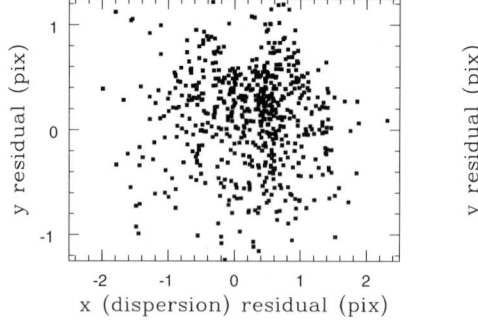 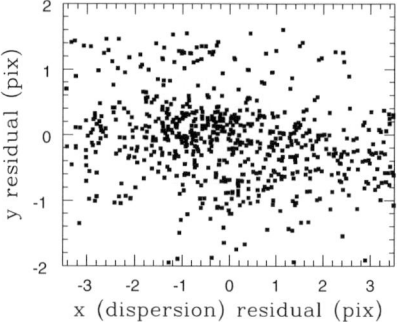

Fig. 1. (a) x and y residuals in the case where all exposures are allowed independent prism and grating angles. (b) x and y residuals in the case where prism and grating angles are restricted to a function of meta data encoder values

However, if the configuration obtained in this way is used in the calibration, this accuracy will only be achieved if a contemporaneous wavecal exposure is available. This is because the prism and grating angles used for the exposure are required and the encoder values recorded in the header are not reliable at the required precision. If we have a calibration lamp exposure that was obtained immediately before or after the science exposure without the prism or grating being moved, then we can run the optimisation again, with all parameters fixed except for the prism and grating angles. Figure 1b shows x and y residuals in the case where prism and grating angles are restricted to a function of meta data encoder values. In this case $\overline{|x_{cd} - x_{pm}|} = 1.5$ pix $\simeq 0.021$ nm (at $2\,\mu$m)$\simeq 3000$ m/s and $\overline{|y_{cd} - y_{pm}|} = 0.6$ pix. The somewhat larger residuals reflect the inaccuracy in the meta data encoder values. The encoder mechanisms have a known repeatability problem resulting in a small random element in the relation between the physical angle of the components and the encoder value recorded. When no wavecal exposure is available the encoder values are our best indication of the actual prism and grating angles and there is no way around this uncertainty. In this case these higher residuals reflect the expected calibration accuracy.

The eventual precision of the fit is inherently limited by the accuracy of the centroids of the detected features. The data used to obtain these results suffered from a number of defects (including distorted pinhole images, non-linear flux readout leading to an apparent odd/even effect and non-uniform illumination of the slit) that degraded this accuracy. We believe that a large component of the remaining residuals is due to these defects and expect that better data will improve the residuals by at least a factor of two.

2.5 Pipeline Implementation

The CRIRES pipeline already includes an implementation that uses meta data encoder values to set the prism and grating angles and returns:

- A 2D wavelength FITS image of the CRIRES detector array in which each pixel contains the value of the wavelength in that pixel.
- The trace of the spectrum on the detector for a given entrance slit position.
- Wavelength calibration of the extracted 1D spectrum.

An implementation that uses contemporaneous wavecals to return the same products is in preparation.

3 X-Shooter

X-shooter is a second-generation VLT instrument. It consists of three medium-resolution (R=4000–10000, depending on wavelength and slit width) cross-dispersed echelle spectrographs that operate in the UV-Blue, the VISible and the Near Infra-Red spectral regions (overall spectral range 300–2400 nm).

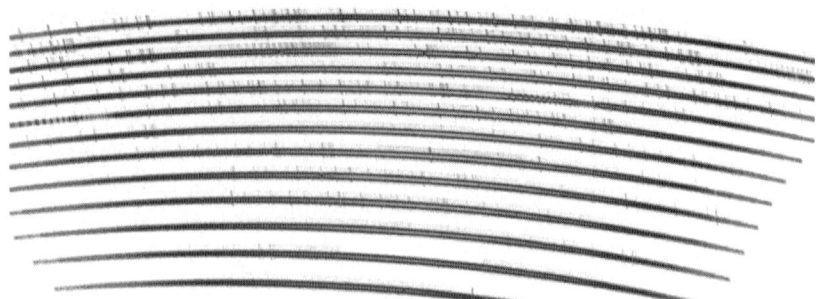

Fig. 2. An example of simulated 2D X-Shooter VIS detector data

3.1 Model

There are a number of aspects of X-Shooter that, in fact, make it simpler to model and optimise than CRIRES.

- The three arms can be described by the same model, they just require independent configuration files.
- It is cross dispersed, meaning that it is much easier to get a fix on component tilts which effect positions perpendicular to the dispersion direction.
- It has only one mode, we do not need complicated optimisation strategies
- Each of the three arms has a relatively large wavelength coverage so that we get many more calibration features in a given exposure

3.2 Simulated Data

There are not yet any X-Shooter data to feed the optimisation process. However a spin-off from this approach allows the production of simulated 2D detector data (Fig. 2) which can be used during the development of the DRS. The simplified model kernel makes the production of this kind of simulated data much quicker than it would be with a full ray trace code like Zemax.

4 Conclusions

The physical model based wavelength calibration technique has a number of advantages. Most importantly its parameters have physical meaning and it has predictive power. This enables us to calibrate in spectral regions where there are few reference features and to anticipate the effects of configuration changes in the instrument.

We are nearing the completion of an implementation for CRIRES that fits wavelength calibration data from 1.0 to 3.0 μm to an accuracy of better than 0.4 pixels. For X-Shooter we have a model that is ready to be optimised.

References

1. P. Ballester, M.R. Rosa: A&AS **563**, 126 (1997)
2. P. Ballester, M.R. Rosa: In *ADASS XIII*, ASP Conf. **Vol. 314**, p 418 (2004)
3. P. Bristow, F. Kerber, M.R. Rosa: ST-ECF STIS Calibration Enhancement Project. In *2005 HST Calibration Workshop*, STScI, p 162 (2005)
4. B. Delabre: VLT Instrumentation Plan: CRIRES Optical Design Report, ESO Internal Document **VLT-TRE-ESO-14500-2096** (2001)
5. Y. Jung, P. Bristow: These proceedings
6. H.-U. Käufl, P. Ballester, P. Biereichel, et al.: In: *Proc. SPIE 5492*, ed by A.F.M. Moorwood, M. Iye, pp 1218–1227 (2004)
7. F. Kerber, G. Nave, C. Sansonetti, P. Bristow, A. Rosa, H-U. Käufl, M.R. Rosa: Ground-based and Airborne Instrumentation for Astronomy. In *Proc. SPIE* **Vol. 6269**, ed by Ian S. McLean, Masanori Iye, 62692O (2006a)
8. F. Kerber, B.J. Frey, D.B. Leviton et al: Ground-based and Airborne Instrumentation for Astronomy. In *Proc. SPIE* **Vol. 6269**, ed by Ian S. McLean, Masanori Iye, 6269 pp 42 (2006b)
9. S. Kirkpatrick, C.D. Gelatt Jr, M.P. Vecchi: Science **220**, No. 4598, 671 (1983)
10. W.H. Press, B.P. Flannery, S.A. Teukolsky, W.T. Vetterling: *Numerical Recipes, The Art of Scientific Computing*, (Cambridge Univ. Press, Cambridge, UK, 1986)

Discussion

D. Baade: Can the directionality of the predictive power of your models be inverted such that it becomes diagnostic power in the case of instrumental failures?

P. Bristow: (including comments from A. Kaufer and F. Kerber): Certainly it can and we hope it will be. This is already being done for UVES. We do not yet have procedures for monitoring the CRIRES and X-shooter parameters, but this is likely to be established as these instruments enter routine operations. Example for CRIRES: the cold structures temperature accidentally changes. We take some calibration data, run the model optimization and note that the new parameter configuration has a significantly changed value for the pre-disperser prism temperature.

U. Hopp: Do you have to measure all the components in the instruments to derive the parameters for your model? Do you include e.g. distortion maps for the lenses and/or mirrors included in the design? Would it be of advantage to measure all optical components in detail for the reduction pipelines and their power?

P. Bristow: The models that we use are simplified versions of the full ray trace design models (Zemax/CodeV). We restrict to principal dispersive and focusing components. These components are not individually measured but the parameters describing them are optimized (simultaneously with all other model components) by comparison to calibration data. Distortion co-efficients for the overall system may be introduced as a final step to help the simplified model match the calibration data.

M. Rosa: A comment on the issue whether you need to measure all components of an instrument. No, no absolute need. But knowing more is always improving your predictive power. So if your scheme gets you through 90%, your likely uncertainties resulting from the 10% residual fudge are cut down a factor 10. The next 90% another leap forward. So - use insight whenever available.

S. Stefl: I suppose that in your model-based calibration it may be convenient to confront on line the values of the instrumental parameters used in the model with those actually measured. Do you assume that your calibration package will communicate with AutRep or a similar database in order to check or correct the used instrument parameters?

P. Bristow: For many parameters it is expected that the values established by the model optimization will be more accurate than the values obtained from physical measurements. I think that the parameter values established by the model should be monitored and used as diagnostics. The procedure for this remains to be established.

From Predictive Calibration to Forward Analysis – Preparing for the ELT Era

M. R. Rosa[1,2]

[1] Space Telescope European Coordinating Facility, ESO, Garching, Germany; mrosa@stecf.org
[2] Affiliated to the Space Telescope Operations Division, RSSD, ESA

1 Introduction

This contribution is about the backgrounds and considerations to a proven concept that takes for real the physical dependencies between observables on our detectors and their true causes in the light reaching us from the astrophysical targets as well their true causes as from the effects of local environmental conditions. It also takes for real the operator nature of our instruments and detectors in the process of mapping the target parameter space into the raw data domain.

In what follows I pretend that there is no room anymore to treat instrumental signatures as something to be wiped off otherwise candid data, but that they are the consequence of the physics of detection in non-ideal instruments that work according to the physics designed into them. As such the whole process (instrument generation, observation, calibration and analysis) can work only as well as the weakest member of the chain permits. Here we now focus on what we can do to make calibration and analysis really strong members.

2 Wavelength Calibration or Approximating Dispersion Data

First, let us take the task "wavelength calibration" as an example and cast some (epistemological) light on what is actually to be achieved and how. Quoting some (almost) randomly selected hits from a Google-search on "dispersion wavelength calibration" (quotations concatenated from longer text in the originals):

- One of the most challenging and time consuming steps in the processing of astronomical spectra is determining the instrumental dispersion function. The dispersion function is determined by identifying spectral lines with known dispersion coordinates in a calibration spectrum (typically an arc-lamp spectrum) and fitting a function to the set of pixel position and dispersion coordinates. We want this capability to be as general as possible so that the correct dispersion function is found even when there is poor or

no knowledge of the dispersion coverage and resolution of the observation [11].
- Note that the wavelength calibration will vary with slit position. Mostly this will be a change in the wavelength zero-point, but there may be subtle effects in higher order terms as the distortions of the optical system become significant. A polynomial fit of pixel to wavelength (or vice versa) will likely require terms as high as sixth order. Such high order polynomials require care when extrapolating the fits beyond the positions of wavelength calibration lines (http://www2.keck.hawaii.edu/inst/esi/lowdmode.html, ESI Modes: Low-Dispersion Mode).
- For very low dispersion spectroscopy one would expect that a linear guess will cause line mismatches at the edge of the detector. One can avoid this, if more then two lines are identified. After the polynomial fit the residual of each line is checked and the line is thrown out, if the residual exceeds the tolerance parameter (MIDAS User Manual on "Fitting the dispersion curve").
- The reduction of raw echelle data is a straightforward, although quite complicated task where the reliability of the final result strongly depends on precise accomplishment of each step. The real challenge is the wavelength calibration. The number of lines required to get an initial fit is quite high. After a number of tests we had to mark about 100 lines to get reliable results. One has to be careful mainly at the edges of orders and in IR orders (where there is a lack of good Th lines and a number of overexposed Ar lines). Optimizing the rejection and matching parameters and increasing the degree of 2D polynomials until the degree 6 (in x) and 5 (in order coordinate) we achieved about the (*demanded, MRR*) RMS. It requires, however, 56 coefficients and it is less reliable at the edges of frames or in orders contaminated by strong Ar lines [9].

We learn that the primary goal seems to be to find a (some suitable, not the true) functional expression of the dispersion relation (for later use to predict the wavelength of a spectral feature from its geometrical location on the detector array). Next we learn that this can be a daunting task because obtaining "the correct dispersion function" by a "polynomial fit" will "likely require terms as high as sixth order". The recipe to accomplish this is "to mark MANY lines to get reliable results", and then, "after the polynomial fit to check the residual of each line" with the goal to "throw the line out, if the residual exceeds the tolerance parameter".

2.1 We Observe

- The "correct dispersion function" is defined through the process by which it is to be found. Since polynomials are imposed, there is no clue whatsoever on the nature of that dispersion function, let alone whether its is correct or not.

- Data points that do not seem to behave properly are discarded on grounds of an imposed "tolerance parameter". Obviously the only constraint on this parameter is an ultimate lower limit to the effect that there will have to remain at least as many points as there are degrees in the chosen polynomial.
- The unspecified reliability of the final result strongly depends on "precise accomplishment of each step" in a given recipe. Recipes differ greatly between observatories (see [9]).

2.2 In Contrast – Our Own Considerations from Opening a Text Book on Optics

The task of "wavelength calibration" is to find a functional representation of the physical process that, in the instrument of choice, throws light of differing wavelengths and spatial origin onto differing locations of the detector – such, that measuring the position of a collected bunch of light on the detector we will be able to assign a wavelength (and spatial position).

3 A Historical Analogon

Kepler, in analyzing Tycho Brahe's superb positional data of Mars, was confronted with the fact that all previous and his own attempts had failed to find suitable geometric arrangements of circular motions for the solar system that would be in accord with ALL data for Mars to within the 1–2 arcmin accuracy of Brahe's data. In the end he resorted to elliptical orbits with the sun in one of the foci.

Kepler had searched for a model that would simultaneously predict positions correctly and provide a physical explanation for the observation that planets moved faster when closer to the sun. However, contrary to popular belief the apparent places of planets can be predicted with off-center circles, equants and epicycles to about the same accuracy (couple of arcmin) that they can with Kepler's ellipses. The subtle, but in the end ground-breaking difference is the fact that these ellipses are the physically correct description, a consequence of the planets' motion in the gravitational field of the sun. Now Kepler in his own words:

It is at least now clear to what extent and in what manner the truth may follow from false principles: whatever is false in these hypotheses is peculiar to them and can be absent, while whatever endows truth with necessity is in general aspect wholly true and nothing else. Further, as these false principles are fitted only to certain positions throughout the whole circle, it follows that they will not be entirely correct outside those positions, except to the extent (as shown in this example) that the difference can no longer be appraised by the acuteness of the senses. This mutual tempering of various influences causes one error to compensate for another, brings the calculation within the limits of observational precision, and makes it impossible to perceive the falsity of this particular hypothesis.

J. Kepler, 1604, in Nova Astronomia, full English title: New Astronomy, Based upon Causes, or Celestial Physics, Treated by means of Commentaries on the Motions of the Star Mars, from the Observations of Tycho Brahe.

4 Predictive Calibration from Physical Instrument Models

On the occasion of the "ESO/ST-ECF Workshop on Calibrating and Understanding HST and VLT Instruments", in 1995 [5], I presented the concept circumscribed in the section heading. At the time it was almost solely an intellectual exercise since I had only very fragmentary small scale tests available that would help to support even my own belief into the practicality. Yet, I was convinced that the arguments in favor, the plausibility for implementation and the anticipated superb results were soundly based.

Siding with Kepler now, 12 years of *applying physical principles to achieve predictive calibration* later, I can say:

– *Whenever the correct physical principles are used in a calibration scheme, that process endows true predictive power and therefore with necessity is in general aspect wholly true and bound to yield superb results.*

The arguments and principles, which all were laid down already at the time of that workshop [5] and supplemented by technical considerations and results from the FOS recalibration soon after [6], are:

1. First of all let there be a radical "change of mind set" – if it is known (textbook), how nature works, why attempt to redo the research of your ancestors every night on mediocre data sets. For example it is close to ridiculous that every night in many observatories the grating equation in combination with predictable camera distortions is painstakingly redetected (not really, only polynomials are fitted) with the help of short duration calibration lamp exposures and inadequate laboratory line lists.
2. In transforming (we call it calibrating) observables (e.g. positions on detector array) into (astro) physical meaningful quantities, use all a-priori information about the observational process (e.g. the optical design of the spectrograph, the air pressure, temperature, gravity vector) as a prescription for functional relationships.
3. Use the minimum amount of physical principle based functionality that produces almost all (say 98%) of the observed effect (e.g. the on-axis optical ray description of STIS, UVES, CRIRES does exactly that). Overdescription of the system with parameters that are outside control sends you back on the "polynomial" road into the realm of epicycles.

4. Then leave the remaining say 2% to a "fudge" polynomial or spline – predictably, with more data analyzed in the new way you will soon see the main physical source for those 2%. In Kepler's case it was good luck that Tycho's data were just a small tick not good enough to see the perturbing effect of Jupiter on Mars's ellipse. How should he have handled that without Newton?
5. In doing so the calibration will have predictive power beyond the range of supporting calibration measurements. In particular, the calibration process – obtaining the projection between the targets physical parameter vector space and the detector event vector space – will yield very valuable insight into the effective engineering parameters of the instrument and current environmental conditions.
6. For that reason – insight into the instrument from even science data analysis – it is beneficial to accompany the instrumentation project from its earliest (design) phase with physical models.
7. If consequently and coherently implemented, a physical model based operation scheme will enable a closed loop process between science observations and operational tasks.
8. It will be straight forward to port software solutions developed using physical principles for a particular instrument to many more representatives of the same class of instrument (eg. spectrographs, from UVES and FOS to STIS and CRIRES and X-shooter).

Following the 1995 workshop the then Director General of ESO, Riccardo Giacconi (Chapter 4 in [2]), together with the Science Directorate and the Astrophysics Department of ESA (Roger Bonnet, Brian Taylor), enabled the funding (from ESA's HST budget) of an "Instrument Physical Modleing Group" attached to the ST-ECF. Primary task was to apply the above principles to HST instruments and archival data to achieve the promised substantial calibration enhancements.

After 6 years, in the wake of the uncertainty of the HST refurbishing, the allocation of funds was terminated and the group disbanded. Ironically, in the year of the shut down, our team, in collaboration with the Atomic Spectroscopy Group at NIST, won the precious Group Achievement Award of NASA, *in recognition of the painstaking efforts to provide maximum scientific value to HST dats using precision laboratory spectral measurements and physical instrument modelling techniques* (ST-ECF Newsletter No. 12 (2006), p. 19).

Florian Kerber and Paul Bristow, former members of the team, are now applying the physical model based calibration concept to ESO/INS supported instrumentation, currently CRIRES and X-shooter. Their papers in the present workshop proceedings provide a lot of detail of the practical aspects and the measurable successes of the concept.

The physical principles implemented in the re-calibration of HST's Faint Object Spectrograph FOS are a physical optics description of the dispersion relations and the electro-magneto-static description of the detector image tubes operating in the variable geomagnetic field conditions in HST's orbit [7]. The so called Post-Operational Archive FOS re-calibrated data are served as the standard product now from all HST archives (STScI, CADC, ST-ECF).

For the STIS echelle modes a fully functional entirely physical principle based model description of the 2D spectral format was developed and tested on demanding data sets, demonstrating the undisputable superiority of the concept [4]. For the STIS CCD modes (imaging and spectral), a physical model for the Charge Transfer Inefficiency CTI was developed and shown to be superior to empirical correction schemes for non-point source data [1].

5 A Striking Lesson on Predictive Power

The most obvious direct impact of applying the predictive power carrying physical principles to the dispersion relation analysis of first the FOS and then later the STIS came from a totally unexpected quarter.

The wavelengths line catalogues in use by the developers and later also by the calibration pipelines for FOS (and STIS, and ...) where found to be non-representative of the actual lamps flown. The predictive power of the FOS dispersion analysis [1] showed that misidentified lines had been the basis for grossly wrong empiricial dispersion solutions in the pipeline. These lines were "misidentified" since the lamps actually flown were of the Pt/Cr–Ne type, while the line catalogue in use was Pt-Ne only. Further background research yielded no useful Pt/Cr–Ne catalogue, and so as a companion to the physical model description we had to start a second project to make a highly accurate Pt/Cr–Ne lamp assessment at NIST [8] comprehensive account is given by Kerber et al. [4]

So, in the end a situation where the physical model makes it plain obvious that not only the empirical approach was amiss (meanwhile you grasp that I firmly believe it almost always is), but also the basis of the empirical approach, the line list for the "fit high order polynomials to many good data points" was inadequate.

Here we close the loop, with the introductory inspection of "polynomial fitting" and Kepler's strong remarks on the capability of false concepts to yield seemingly acceptable, but entirely wrong results.

6 Onwards to the E-ELT Era – Forward Analysis

It is well known that it is always advantageous to start the comparison of hypotheses with reality (measurements) with the noise free models in physical parameter space, and then adjust those models rather than trying to transform noisy and background limited data back into the target domain. Examples from imaging are the Malmquist biases on population diagrams, from spectroscopy the difficulties with curves of growth under conditions of variable spectral resolution. An abundance of theoretical (statistical mathematics, information theory) work exists (keyword Bayesian) that lends strong support to this view.

Now, what hinders us in a direct application of the above principles is our current empirical approach to data calibration and analysis. In the end, since we do NOT imprint the physical "truth" when transforming positions to wavelengths, or microscopic photon detection processes with non-linearities on photometry to observed magnitudes, we have no good handle on inverting our algorithms. Hence, the field of hypothesis testing by mapping astrophysical target descriptions into raw data space (images and spectra) is largely unexplored. Yet it promises to offer tremendous rewards.

It is obvious that an operational environment (calibration pipeline and data analysis sytem) that incorporates the physical principles in the transformation of observables (detector positions and amount of photons registered) back into the parameter space of the target will permit the reverse operation as well. Actually, the software model spectrographs we have developed so far for FOS, STIS, CRIRES, X-shooter, function in exactly this forward direction. They take laboratory line catalogues to produce an expectation value for the positional observation on the detector array. Mismatches with observed data are then used to optimize the engineering parameters (optical configuration and detector setup) of the instrument model configuration.

Already in my 1995 workshop paper I have argued for this strategic move to forward analysis, based on a rigorous application of physical model principles. Seeing that the physical model approach is now becoming a sound part of ESO's instrumentation development I am in good hopes that the way is being paved to be able to do ultimate justice to the precious data from the E-ELT.

7 Concluding Thoughts

If you have followed me up until here, then you might be convinced by the arguments and even by the testimony of successful applications. However, it might seem as if there was only room for physical model based calibration and analysis in "dispersion solutions" for classical single order low resolution slit, or very demanding high resolution echelle spectrographs.

There is plenty of unexplored space in instrumental calibration beyond those examples. When opening this new avenue intellectually, when working on practical implementations and when arguing (like in the present paper) we chose these spectrographic instruments because the concepts are pretty obvious in spectroscopy (dispersion by a grating), and because the results of the new approach can be checked rigorously against high precision empirical work.

Here now is a non-exhaustive list of areas that would immediately enjoy substantial benefits from a rigorous application of predictive power from physical principles:

- *The atmosphere:* Considered a nuisance it is usually treated with undue disrespect. Almost all observations (spectroscopy and filtered 2D photometry) use some sort of empirical scheme to "correct" for an average atmospheric extinction "law". Unless implemented in hardware, no provisions are made for parallactic angle effects. Slit widths in most cases of so-called "spectrophotometry" are insufficient (in order to maintain the spectral resolution). For longer duration exposures and/or at larger zenith distances the simple divison of the observed signal by a single value airmass extinction function is equally inadequate. Obviously, to do real justice to the data, also to those of broad band filter 2D photometry (so-called color terms), forward analysis fed by the proper instrument and atmosphere models is the proper answer.
- *LS/PO extinction tables:* Has anyone ever checked the applicability of the atmospheric in standard use at Paranal. The table now in use is based on the very table that I implemented early in the 1980s in the baby-MIDAS system. That table is based on measurements for a very few epochs on La Silla and the discussion by Tueg [10], and uses the 3 component modeling (ozone, aerosols and Rayleigh-scattering) by Hayes and Latham [3]. A clear cut example for (a) update, and (b) inclusion of the physical formulation as 3 variable components into pipelines steered by an observatory data base of environmental parameter monitoring.
- *Detectors:* The approach taken in the microscopic CTI corrective model for the STIS CCD can be used as a paradigm for understanding the subtle detector effects in CCDs and IR detectors that we have learnt about during this conference. CCD fringe correction has already been approached the "physical" way (interference) be several; e.g. Walsh et al. this volume.
- *And at last:* 3 spectrographs that deserve serious physical treatment. FORS, VIMOS, and KMOS – we have seen during the workshop (papers in this volume) the complexity of the data analysis and calibration tasks. Carlo Izzo (this volume) has presented an impressive tool that enables a completely blind boot-strap to understanding the data of VIMOS by using ingenious methods of pattern recognition. In my opinion such tools are required during the weeding process of getting *any* result before physical insight based rigorous methods are available, but can not be the ultimate goal.

Acknowledgement. This is the proper place to thank an unknown tutor at the Faculty of Physics of the University of Heidelberg. During exercises in a low level course early on in studies I refused to represent the dispersion relation of a monochromator (measured with 3 reference lines) by a linear function as required, but drew instead something curvy that I tentatively labeled "sinish". Taken up on this I was asked to demonstrate why, but since I could not develop the full theory on the spot I got a bad mark.

The support by Ricardo Giacconi, Brian Taylor and Piero Benvenuti to set up a team where the ideas could be turned into practical effect was essential. The Instrument Physical Modeling Group team members during those years 1997–2005, in sequence of arrival, Florian Kerber, Paul Bristow, Anastasia Alexov, and Mauro Fiorentino, all have been superb players, each dominating the field in their area of responsibility, carrying forward the concept beyond even our far reaching initial expectations on expected accuracy gains.

References

1. P. Bristow, A. Alexov, F. Kerber, M.R. Rosa: Modelling Charge Transfer on the STIS CCD, in *The 2002 HST Calibration Workshop*, ed by S. Arribas, A. Koekemoer, and B. Whitmore, STScI, Baltimore, p. 176 (2002)
2. R. Giacconi: ARA&A **43**, 1 (2005)
3. S.D. Hayes, D.W. Latham: ApJ **197**, 593 (1975)
4. F. Kerber, P. Bristow, M.R. Rosa: STIS Calibration Enhancement (STIS-CE): Dispersion Solutions Based on a Physical Instrument Model, in *The 2005 HST Calibration Workshop*, ed by A. M. Koekemoer, P. Goudfrooij, and L.L. Dressel, NASA, GSFC, p. 309 (2006)
5. M.R. Rosa: Predictive calibration strategies: The FOS as a case study, in *ESO/ST-ECF Workshop on Calibrating and Understanding HST and ESO Instruments*, ed by P. Benvenuti, ESO, Garching, p. 43 (1995)
6. M.R. Rosa: Physical Modeling of Scientific Instruments, in *Astronomical Data Analysis Software and Systems VI*, ed by G. Hunt and H. E. Payne, ASP Conference Series, Vol. 125, PASP, p. 411 (1997)
7. M. R. Rosa, A. Alexov, P. Bristow, F. Kerber: FOS Post-Operational Archive and STIS Calibration Enhancement, in *The 2002 HST Calibration Workshop*, ed by S. Arribas, A. Koekemoer, and B. Whitmore, STScI, Baltimore, p. 161 (2002)
8. C.J. Sansonetti, F. Kerber, J. Reader, M.R. Rosa: ApJS, **153**, 555 (2004)
9. P. Skoda, M. Slechta: Comparison of Echelle Spectra Reduction Packages, in *Astronomical Data Analysis Software and Systems XIII*, eds. F. Ochsenbein, M. Allen, D. Egret, ASP Conference Series, Vol. 314, PASP, p. 816 (2004)
10. H. Tueg: Msngr **11**, 7 (1977)
11. F.G. Valdes: Automated Arc Line Identification in IRAF, in *Astronomical Data Analysis Software and Systems V*, ed by G. H. Jacoby and J. Barnes, ASP Conference Series, Vol. 101, PASP, p. 33 (1996)

Discussion

A. Smette: You mentioned the flux determination part in instrument modeling. An aspect of this is scattered light. In the case of STIS, Don Lindler made a nice algorithm to remove scattered light from STIS echelle data. It would be nice to include a physical model for the determination of scattered light into the recipe of the UVES pipeline.

M. Rosa: Absolutely. Yes, the GSFC Scattered Light model is indeed a physical model (Don is one of the enlighted), since it is based on predicting the scatter level using the optical layout of STIS and the targets primary input in an iterative way to obtain a cleaned version. Yes, the pipeline of UVES, which had seen one of the very first implementations of a physical model based boot-strap to the wavelength calibration deserves to be upgraded by all that one has learned in the meantime.

E. Pantin: In forward analysis of data, how can you deal with non-stable components, e.g. the atmosphere (PSF, transparency) or the instrument (fluctuation of temperature)?

M. Rosa: First a return question – how do you deal with those in classical backward analysis? You have no real handle, since your signal is the time integral of photon arrivals under variable conditions. Even if you have monitored those conditions there is no way you can disentangle things.

In forward analysis it is the contrary, you automatically have a handle, since you build up the modelled raw data by a stochastic photon-push-through-the-instrument process. Obviously, you have to have information on those supposed fluctuations of conditions to do that.

P. Nissen: In the forward modelling of raw data one may be "locked" to the model so that one is not discovering new features, e.g. spectral lines not included in the model.

M. Rosa: The opposite is true if we look at the difference/quotient "raw observed data – raw modelled data". All that was in the model already will be taken off, any "unexpected" new find will stick out even more clearly. The probability for detection of weak "unexpected" is greatly improved.

A. Smette: Telluric lines are in practice the only way VISIR spectroscopic data are wavelength calibrated. They are also used in several spectral regions for the reduction of CRIRES data. Another aspect is: can the atmospheric model be used to provide transmission spectra of the atmosphere that are good enough to get rid of the use of telluric standards. We are currently examining this question with CRIRES.

Efits: A New Efficient and Flexible FITS Library

A. Grado[1], M. Pavlov[2,3], and L. Limatola[2]

[1] INAF – Osservatorio Astronomico di Capodimonte, Via Moiariello 16, 80131, Naples, Italy; agrado@na.astro.it
[2] INAF – VSTceN, Via Moiariello 16, 80131 Naples, Italy
[3] Sternberg Astronomical Institute, Moscow, Russia

Abstract. The era of petabyte data-flow regime and precision astronomy calls for an overall efficiency of the software used to process the data and for a careful control on the errors introduced by the processing. We present a new FITS library baptized Efficient fits (Efits) developed to satisfy basically two requirements: one related to efficiency and flexibility, and another related to accuracy. To achieve the first goal, the library contains special mechanisms/layers, which support different I/O methods (disk and shared memory, currently), and consent to easily extend the set of supported methods in the future. The usage of the shared memory allows to efficiently exchange data among independent processes in a modular pipeline. Moreover the library can work on single extension of FITS file extracted directly from Multiple Extensions FITS (MEF) files without previously splitting them. For the second goal a true noise map, mandatory in some science application, is created implementing a noise map propagation mechanism which creates (or modifies) a noise map each time a mathematical operation is performed on an image. Structure and mechanisms adopted are presented.

1 Introduction

Three main elements are to be considered for a fast data processing: the software efficiency, the processor speed, and the throughput of the connection between processor and data mass storage. In a system globally well designed, there must be a balance among these three elements. In practice, however, the speed of processors has increased faster than the capability of feeding data to them. Indeed a highly performing disk such as a 15 krpm FC disk reaches a maximum of 200 I/O operations per seconds in random access, and this figure is basically the same since years. Even striping on several disks does not solve the problem of the bus speed limitation in the disk-processor connection. There are some promising solutions to work around the problem, yet not very well supported and clearly quite expensive. Therefore, the I/O can be presently considered a bottleneck in data handling/processing.

The problem of computational power can be solved using distributed environments. This is particularly true in the wide-field imaging with multi-CCD detectors. Due to the intrinsically parallel nature of the data – most of the processing can be performed independently on each CCD of the mosaic – a distributed computing environment such as a Beowulf Cluster can be straightforwardly used [1, 2]. Clearly this approach does not alleviate the I/O problem because we still need a data storage system, and a connectivity to it, capable of transferring data back and forth to the cluster in a very fast way. The I/O problem can be mitigated by reducing as much as possible the data transfer. A way is to create a monolithic code where the I/O is minimized at the expenses of modularity. If we want to keep modularity, a mechanism is required to exchange data among independent processes. To the best of our knowledge, all the software packages available at present to astronomers for image processing make use of files to exchange data among independent processes, increasing the I/O problem.

In order to attack the I/O problem, we report on a library able to use the shared memory in such a way that independent processes can access data directly in the memory without going through files. The code is also able to work with compressed data, further reducing the I/O. In this way a real software modularity is intrinsically implemented without the need to write monolithic code containing all the functionalities.

Behind the I/O optimization, the library presented here addresses a problem related to the scientific exploitation of an image. In some cases it is necessary to know the true noise associated to each pixel of the analyzed image. Just think to all the scientific cases where morphology is involved, and in particular to future weak lensing surveys where sub-percent precision galaxies ellipticity measurement will be required. In order to increase image analysis precision we have also implemented a mathematical library supporting a true noise map propagation. This makes it possible to perform mathematical operations on images; at the same time, the noise maps associated to the images are updated according to the statistical noise propagation. The next Section provides an overview of the Efits library, followed in Sect. 3 by some details on the shared memory I/O implementation.

2 The Efits Library

The Efits library comes as a low-level software facility to process images in FITS format through the calls to a C library. It is written according to POSIX and ANSI standard with the purpose of ensuring the maximum portability. The main components of Efits library are: FITS_IO, GEN_IO, MEM_IO and ISCIENCE. FITS_IO includes sets of functions to deal with FITS-key representation, FITS-header attributes and history, and the manipulations of FITS header and data. It was not possible to use existing FITS library since we want to work not only with disk files, but with memory also. All the new functions will work with memory buffers, not with I/O streams directly. Such an approach keeps completely separated I/O from FITS manipulations. In other words we first have to put FITS-block(s) in the memory buffer. Then we work with just the memory. In order to write FITS-file, we have to create in some

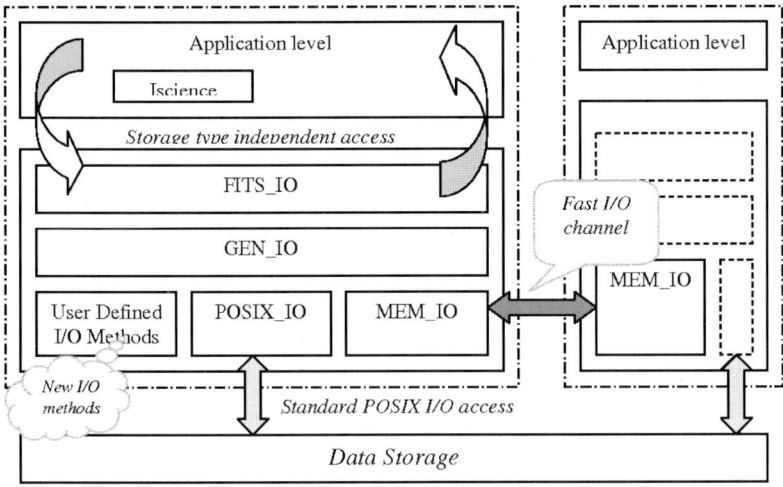

Fig. 1. Efits schematic diagram

way FITS-block(s) in memory. Then this block(s) will be written/flushed to the storage media (disk, memory, etc.). The use of "generic_io" libraries for I/O operations makes our approach independent (algorithmically) from the type of storage. Figure 1 shows the main library components, shortly described in the following.

GEN_IO provides a fixed, pre-defined, storage type independent interface, used by applications in I/O operations. Currently, shared memory I/O and standard file system I/O operations are supported by GEN_IO. The library provides also a program interface,to extend the set of supported low level I/O methods (external "plug-ins") or replace the existing ones. MEM_IO library includes a set of basic POSIX-like functions (open/read/write/seek/...) with the purpose of using the shared memory in the same way as a standard file-system file. The library uses special locking mechanism (semaphore based) supporting multiple/parallel/concurrent access to the same data/memory segment(s) (see below). ISCIENCE contains statistics functions, pre-reduction methods, noise map generation, and propagation functions (still under development).

3 The Shared Memory I/O Implementation

The request to work with the shared memory has made it necessary to design the structure of the shared memory segments and a locking mechanism for concurrent access. One Shared memory Control Block (SCB) contains information about the status of the segment, the attached memory segment ID, and the semaphore ID. The structure was designed to be able to create and

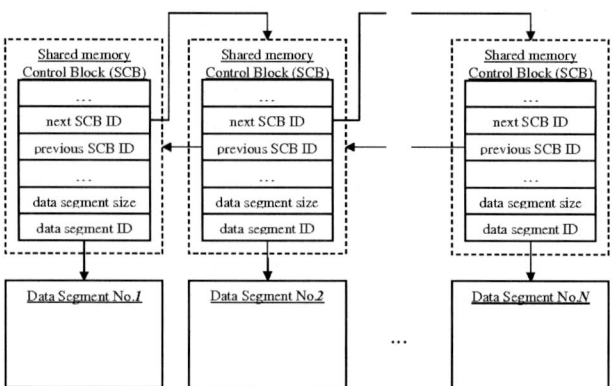

Fig. 2. Organization of the data into the shared memory

manipulate chains of memory segments with the purpose of removing limitation of maximum shared memory segment size (32 M for Linux OS). This is why the SCB structure contains links to next/previous/top/last segments (see Fig. 2). A chain of segments is useful also to separate FITS-headers and FITS pixels information easily. The SCB implementation uses a locking mechanism based on IPC (System V IPC system calls) semaphores to allow concurrent access to the same segment for different processes. The IPC (not POSIX 1003.1b) was chosen, because of different limitation of POSIX semaphores implementation under LINUX.

4 Conclusion

We report the description of the on going work on a completely new set of C-code library to process FITS images. The main peculiarities regard the usage of the shared memory to exchange data among independent processes and the implementation of a true noise map propagation mechanism.

References

1. A. Grado, P. Quinn, A. Wicenec, et al.: "The BEOWFI project: reducing WFI data with a Beowulf cluster", The New Era of Wide Field Astronomy, ASP Conference Series, **232**, 303 (2001)
2. A. Grado, M. Pavlov, E. Cascone, et al.: "New Opteron based Beowulf for VST images processing", High performance parallel image processing on beowulf clusters, G. Sedmak ed., Proc. MIUR COFIN200102544 Project Conference, Università di Trieste, 53 (2004)

Error Propagation in the X-shooter Pipeline

M. Horrobin[1], P. Goldoni[2], F. Royer[3], P. François[3], G. Blanc[2], J. Vernet[4], A. Modigliani[4], and J. Larsen[4]

[1] University of Amsterdam, Postbus 19268, 1000 GG Amsterdam, The Netherlands; horrobin@science.uva.nl
[2] APC/UMR 7164, Paris, France
[3] GEPI, Observatoire de Paris-Meudon, France
[4] ESO, Karl-Schwarzschild-Strasse 2, 85748 Garching, Germany

Abstract. X-shooter will be the first of the second generation VLT instruments. It is a medium resolution spectrograph covering a spectral range of 300–2500 nm in a single exposure. It will be delivered to Paranal in 2008.

The pipeline will perform full error propagation from raw data through to fully reduced science products. This chapter gives a brief overview of the motivations, complications and solutions found during its development.

1 Motivations

The goal of providing full error propagation throughout the pipeline was originally introduced during the definition of the scientific requirements for the instrument. The main motivation for this requirement was to enable the development of a fully accurate optimal extraction routine based on those of Horne [1] and Marsh [2], which require the variance of the data as an input to the algorithm. It is also expected that providing the variance map of the data in the final science products will give scientists an extremely useful tool in the further analysis of the data.

Another use of the error data is in the quality control process. The pipeline will enable observers to calculate a quick-look signal to noise spectrum from a completed observation. Also, QC parameters, used by the observatory to monitor the instrument health, are a natural product of the error propagation process.

2 Calculations and Data Formats

All the error calculations are based on the classical approach of assuming all errors are randomly distributed, giving the standard error formula for $x = f(u, v, ...)$ of:

$$s_x = \sqrt{s_u^2 \left(\frac{\partial x}{\partial u}\right)^2 + s_v^2 \left(\frac{\partial x}{\partial v}\right)^2 + 2 s_{uv} \frac{\partial x}{\partial u} \frac{\partial x}{\partial v} + ...} \qquad (1)$$

Where the first two terms in the square-root are the variance of the variables and the third term is the covariance.

The data format used within the pipeline carries information on both the variance and quality of all pixels. This format is based upon that proposed by the Euro3D group [3], but simplified to flat Fits image planes, as X-shooter does not require the complex spatial information of the full Euro 3D format. Each output file contains 3 extensions:

- The data extension: a 32-bit floating point image containing the result of a pipeline calculation.
- The statistical error extension: a 32-bit floating point image containing the result of the error calculation.
- The quality extension: a 32-bit integer image, which is used to provide 32 boolean flags for defining bad pixels.

3 The Problem of Covariance

As can be seen in Fig. 1, the spectral format of X-shooter is highly distorted, the orders are curved, and the spectral lines have a varying amount of tilt within the orders. This means it is impossible to resample the data using a simple linear kernel, which would preserve the noise statistics. Instead we will have to use a more complex resampling method, which will introduce systematic patterns into the variance map of the data.

The way to formalize this systematic change in variance is to keep track of the covariance matrix of any transformation. Figure 2 shows the covariance matrix for a Lanczos3 resampling with a pixel shift of (0.5,0.5). Using this

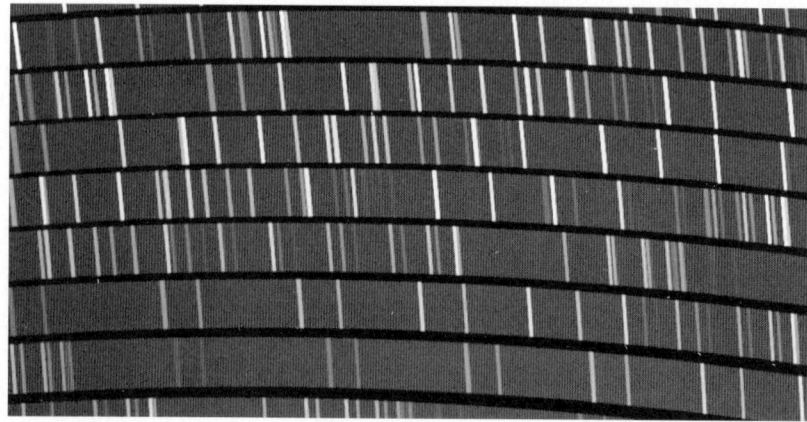

Fig. 1. A section of a simulated sky frame for the NIR arm of X-shooter, see Bristow et al. and Saitta et al. in these proceedings for details of these simulations

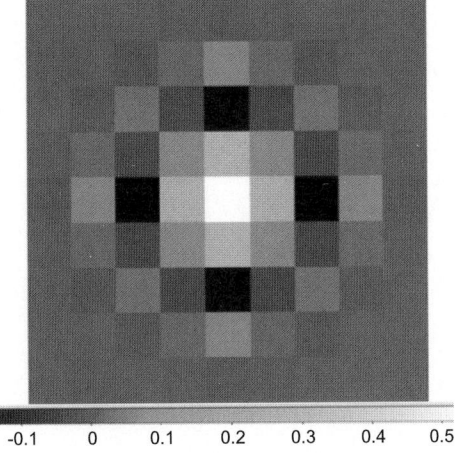

Fig. 2. The covariance matrix for a Lanczos3 resampling with a (0.5,0.5) pixel shift. Data taken from Evans and Irwin [4]

information, any further analysis of the data can correctly calculate resulting variance for any combination of pixels.

For X-shooter, the plan to keep track of the covariance presents some computational difficulties. As each position on the detector will have a different rotation/scaling in the resampling, we need to be able to calculate the covariance matrix for every single pixel in the resampled data. Assuming we need the full 7×7 matrix shown in Fig. 2, this would amount to around 1 GB of data for each reduced frame.

Within the pipeline we plan to avoid this problem by leaving any resampling to the final stage of the data reduction. Flux calibration is the only significant step where we will have to worry about the affects of covariance.

For the end user, the pipeline will be capable of either producing a full covariance matrix for further analysis, or producing a covariance matrix on a pixel by pixel basis for systems that have limited memory available.

The use of the covariance matrix in the further data analysis is left as an exercise to the scientist.

References

1. K. Horne: PASP **98**, 609 (1986)
2. T.R. Marsh: PASP **101**, 1032 (1989)
3. M. Kissler-Patig, Y. Copin, P. Ferruit, A. Pécontal-Rousset, M.M. Roth: AN **325**, 159 (2004)
4. D.W. Evans & M.J. Irwin: Interpolation tests, VDF-TRE-IOA-00016-0002, (2003)

The CRIRES Data Reduction Challenges

Y. Jung and P. Bristow

ESO, Karl-Schwarzschild-Strasse 2, 85748 Garching, Germany; yjung@eso.org

Abstract. CRIRES is a cryogenic high-resolution pre-dispersed infrared echelle spectrograph, developed by ESO. It provides a resolving power of up to 100,000 between 1 and 5 µm. Science operations with CRIRES will start from P80 on, and the data reduction pipeline is in a rather advanced state. It has been used during commissioning and Science Verification runs to evaluate the sensitivities and the stability of the instrument in the different modes. Beside the usual infrared specific calibration difficulties, the high resolution capabilities of the instrument were particularly challenging [3] when it came to the development of a precise, robust and automatic wavelength calibration. This contribution gives an overview of the different tasks achieved by the CRIRES pipeline, and concentrates on the whole wavelength calibration strategy including the physical model development, a new correlation method, and the problems linked to the instrument's high resolution.

1 Data Reduction

The main data reduction steps achieved by the automatic pipeline are frame combination, spectrum extraction and wavelength calibration. Beside these, there are several calibration recipes to correct for flat-fielding, dark current, bad pixels, or photometry procedures that can be derived from standard stars observations: the sensitivity, the throughput, and the conversion factor.

We concentrate here on the spectrum extraction and the wavelength calibration.

2 Spectrum Extraction

The spectrum extraction on CRIRES data is systematically applied with two methods. A simple one that just collapses the signal around the detected spectrum (Rectangular extraction) and a more evolved one (Optimal extraction) which gives better results for faint sources.

2.1 Rectangular Extraction

This extraction is performed by simply collapsing the signal around the automatically detected spectrum. The size of the window can be specified by

```
#
# file             crires_util_extract_extracted.fits
# extensions       4
# ----------------------------------------------------
# XTENSION         2
# Number of columns 7
#
Extracted_RECT|Extracted_OPT|Error_RECT| Error_OPT|Weight_Norm_RECT|Weight_Norm_OPT|Background_noise
            0|            0|         0|         0|             41|         403510|                0
            0|            0|         0|         0|             41|         403510|                0
     0.301236|      0.32826| 0.0907629| 0.0488197|             41|        47.1861|        0.0141587
     0.244521|     0.300889| 0.0963394|  0.051743|             41|        41.8561|        0.0150334
     0.247475|     0.303204| 0.0864468| 0.0465019|             41|         52.003|        0.0134869
     0.185717|     0.271941| 0.0903254| 0.0495425|             41|        47.6034|        0.0149065
```

Fig. 1. Produced extracted table

the user. This method gives satisfactory results for bright spectra, but when it comes to fainter objects, a more evolved method is needed.

2.2 Optimal Extraction

For fainter sources, it is critical to perfectly separate what is the background from what belongs to the spectrum. Even better is it to create a weight profile of the spectrum in order to maximize the signal to noise ratio. This weight profile is one of the products of the Optimal Extraction method.

2.3 Products

The extracted tables produced systematically contain the results of the two methods with dedicated columns for both as shown in Fig. 1.

Figure 2 illustrates the advantage of the optimal extraction compared to the rectangular on a relatively faint object.

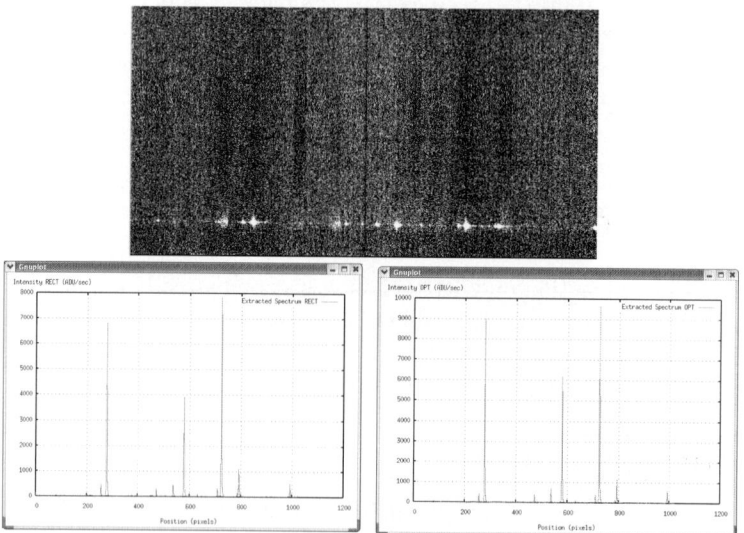

Fig. 2. Rectangular and Optimal extractions on the same source

3 Physical Model

One of the inputs of the wavelength calibration is an estimate of the wavelength range of the observed spectrum. The more this estimate is precise, the faster the calibration. Currently an estimate of the wavelength range is stored in the CRIRES data headers and used for the calibration.

A model of the instrument [1] is being developed in order to produce a precise estimation of the wavelength for every pixel on the detector. Once finalised, this model will eventually be used as input to the wavelength calibration method and allow it to reduce the solution searching space and optimise the computation. Additionally, a 2D wavelength map computed using the model will be systematically produced by the pipeline.

4 Wavelength Calibration

The basic idea behind the wavelength calibration is to apply to the observed (and extracted) spectrum all possible wavelength solutions and evaluate them by comparing to the catalog. The best one is the wavelength solution. This bottom-up approach [2] is giving good results if the solution space is properly defined. An iterative method allows easy refinement of the solution space to finally obtain a precise solution.

4.1 Algorithm

Given a first estimate of the wavelength (A_i points), an error bar around these points, a number of samples around these points, we define a number of candidate polynomials. These $nsamples^{degree+1}$ candidates (see Fig. 3) are all used to generate from the extracted spectrum a spectrum in wavelength. These ones are compared with the catalog. The closer to the catalog, the higher the associated likelihood factor.

At the end of the process, the higher likelihood factor gives the polynomial that is closer to the wavelength solution. This polynomial can be used in a second pass as estimate (with a smaller error bar) to refine the solution. The solution will then converge in an iterative process to the best possible polynomial. An iterative process can also be used to increase the degree of the searched polynomial.

4.2 Products

The wavelength solution produced is a series of polynomial coefficients. They are stored in 4 extension FITS file (1 per chip), each of them containing the coefficients table. This file can be provided as input to the *crires_spec_jitter* recipe, this one would then use this solution.

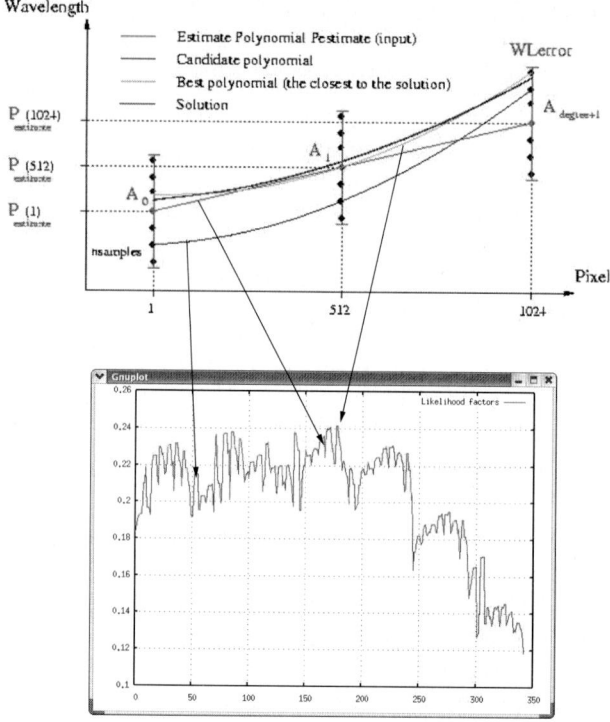

Fig. 3. Wavelength calibration algorithm

The wavelength map is also produced. This one can be used together with the spectrum profile (produced during the extraction) to associate a wavelength to the extracted bins.

References

1. P. Bristow, et al.: Model Based Calibration, these proceedings (2008)
2. C. Izzo, et al.: A Bottom-Up Approach to Spectroscopic Data Reduction, these proceedings (2008)
3. H.U. Käufl, et al.: IR-Spectrograph Calibration Issues: CRIRES as an example, these proceedings (2008)

THELI - A Pipeline for UV- to Mid-IR Imaging Data Reduction

M. Schirmer[1] and T. Erben[2]

[1] Isaac Newton Group of Telescopes, Calle Alvarez Abreu 70, 38700 Santa Cruz de La Palma, Spain; mischa@ing.iac.es
[2] Argelander Institut für Astronomie, Auf dem Hügel 71, 53121 Bonn, Germany

Abstract. *THELI* is a stand-alone package for the automatic reduction of imaging data, mostly based on C/C++ programmes, and can be run on the command line or by means of a GUI. The entire reduction process is fully transparent to the user and essentially instrument-independent. Data from optical multi-chip cameras is handled as well as the intricacies of near-IR and mid-IR imagers.

1 Overview

THELI was initially developed for the reduction of the 20 square degree Garching-Bonn Deep Survey (GaBoDS, see also [2]), conducted with the Wide Field Imager at the 2.2 m MPG/ESO telescope in La Silla. Main purposes of this survey are weak gravitational lensing studies.

To be more efficient, we did not start the development from scratch, but based the work on already available stand-alone software packages such as *SExtractor*, *Swarp*, *WIFIX*, *Eclipse* etc. These are wrapped into shell scripts, and we subsequently interfaced these modules to allow for a continuous data flow. Most of these packages were slightly modified to better suit our purposes, of others (e.g. *Eclipse*) we took only basic functionality or ideas and developed own programmes. The code is mostly based on C/C++, and runs in principle on any up-to-date UNIX/Linux machine. Further advantages of *THELI* are:

- It is entirely instrument-independent. All camera-specific parametres are contained in an instrument configuration file. The reduction steps are very similar, independent of what kind of data one reduces. In particular, this extends from optical to mid-IR wavelengths, and from single- to multi-chip instruments.
- It is fully parallelised. On multi-core PCs the user only has to select the number of CPUs he wants to use, where ideally one CPU is selected per CCD for a multi-chip instrument. Parallelisation for cluster architectures is supported as well.
- A fully documented graphical user interface is available, which allows easy access to all parametres and functions. It makes the reduction process very convenient and preserves full transparency. Yet, *THELI* wants to be learnt as any other reduction software.

– The modular principle of *THELI* allows us to quickly include new routines or to implement improvements suggested by the end user.

2 Instrument Independence

The development of *THELI* was based on WFI@2.2m MPG/ESO data, hence all code is already generalised for the particular structure of multi-chip camera data. Major differences in the reduction of different data types appear only at a higher level, for example in the choice of optimal parametres or strategies for sky background modelling.

The *THELI* GUI (see Fig. 1) takes this into account, having all parameters not more than a mouse-click away, and by offering various methods for a particular task. In this way *THELI* could be extended to the near-IR (Fig. 2), delivering as good or even better results than with other standard reduction packages. Recently, *THELI* was successfully expanded to the mid-IR, based

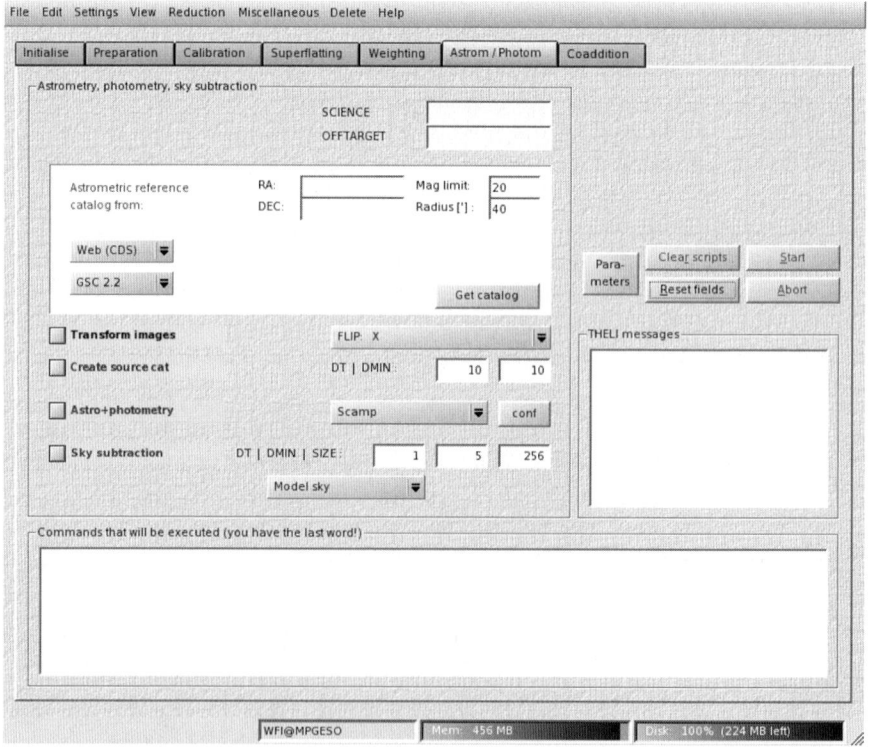

Fig. 1. The graphical user interface for THELI. The user simply marks the tasks he wants to execute, and then clicks the *Start* button

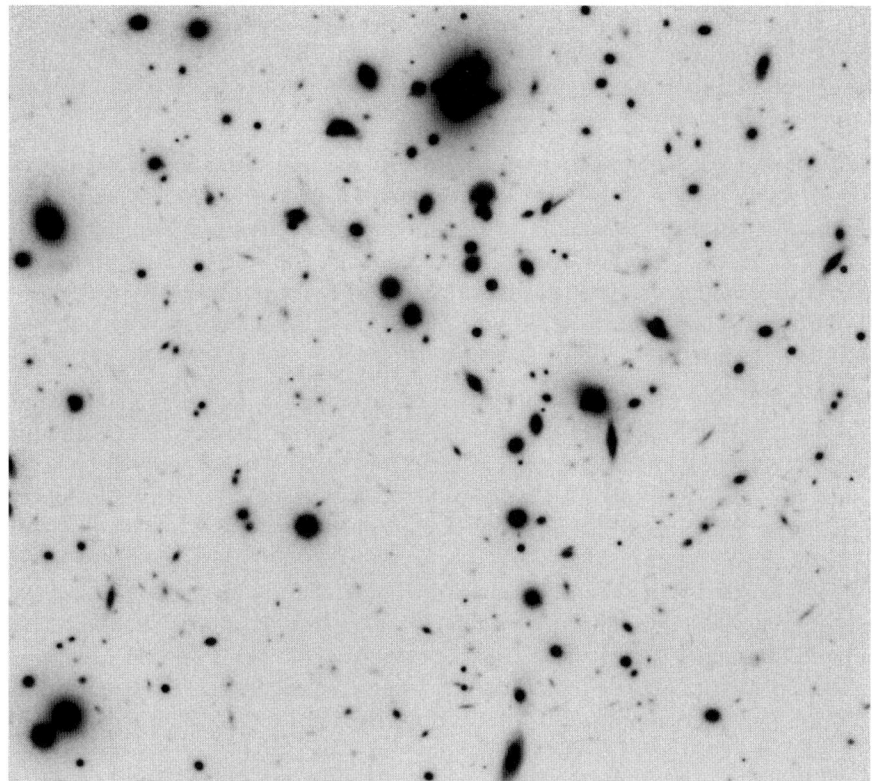

Fig. 2. Example: An 11 ksec ISAAC J-band image of the lensing cluster Abell 1835, reduced with *THELI*. The data was taken from the ESO archive (PI: Pelló)

on different VISIR data sets. However, further testing with different mid-IR imagers is still required to ensure its global usability.

3 Further Capabilities and Availability

Apart from the already mentioned general features, *THELI* offers the user many more tools to obtain best results. Amongst those are for example:

Fully configurable static and dynamic superflats For best pre-processing results, several different kinds of superflats (in the IR: sky backgrounds) can be calculated, taking into account different sky conditions. Furthermore, chop-nod techniques (mid-IR) and the use of blank sky fields (optical and near-IR) are fully supported.

Various sky background subtraction methods A broad range of sky background subtraction methods are available. This ranges from full individual modelling down to the subtraction of a constant estimate, obtained from a portion of one CCD of a multi-chip camera and considered to be representative for all other detectors in that array. In this way virtually all kinds of targets and data can be dealt with.

Full astrometric correction Offsets, de-rotation and distortion corrections are determined, in case of a multi-chip camera for each chip individually. The method is based on automatically created object catalogues and yields an image registration that is usually better than 1/10th of a pixel. It works for very sparse and very crowded fields, using *Astrometrix* or *Scamp*. A relative flux scaling is included, too. Absolute photometry is currently being tested and will be implemented in the future.

Statistically optimised weighting scheme To maximise the S/N of the final image, individual weight images are created for each exposure and taken into account during the coaddition. For the latter, various resampling kernels are available. The resulting coadded mosaics have a global astrometric solution and the same photometric zeropoint for all pixels.

Besides, the GUI comes with full documentation with step-by-step processing examples for multi-chip cameras, near-IR and mid-IR imagers. An integrated online help is available as well. For further help, the user can subscribe to the *THELI* forum[1] which is run by the authors. Technical and performance details of the pipeline itself are given by [1].

THELI can be downloaded from the ftp site ftp://ftp.ing.iac.es/mischa/THELI or is also available directly from the authors who will happily assist you!

References

1. T. Erben, M. Schirmer, J. Dietrich, et al.: AN **326**, 432 (2005)
2. M. Schirmer, T. Erben, M. Hetterscheidt & P. Schneider: A&A **462**, 875 (2007)

---

[1]http://marvin.astro.uni-bonn.de/forums

MUSE: Design and Status of the Data Reduction Pipeline

P. Weilbacher[1], J. Gerssen[1], M. M. Roth[1], P. Böhm[1], and The MUSE Team[2]

[1] Astrophysikalisches Institut Potsdam, An der Sternwarte 16, D-14482 Potsdam, Germany; pweilbacher@aip.de
[2] CRAL Lyon, ESO, IAG Göttingen, Sterrewacht Leiden, AIP Potsdam, LAOMP Toulouse, ETH Zurich

Abstract. We briefly summarize instrument properties of the future second generation VLT instrument MUSE, a giant integral field spectrograph, and describe the layout of the data it will provide. The current design of the data reduction pipeline is presented along with a project timeline.

1 The Instrument

MUSE, the **M**ulti **U**nit **S**pectroscopic **E**xplorer, will be one of the second generation instruments for the ESO VLT [1]. It is scheduled to see first light in 2012. It is designed as the largest integral field spectrograph worldwide with a field of view of $1'\times 1'$ sampled $0.2''\times 0.2''$ spaxels in the standard wide field mode. A narrow field mode for high resolution work will have a sampling of $0.0253''$ per spaxel and will cover $7.5''\times 7.5''$ on the sky. Both modes can be supported by an adaptive optics system to enhance spatial resolution and the wide field mode can also operate in natural seeing.

MUSE is being built by a consortium of 7 European institutes, led by CRAL Lyon, and including ESO, IAG Göttingen, Sterrewacht Leiden, AIP Potsdam, LAOMP Toulouse, ETH Zurich. The main scientific goals of the instrument are presented by McDermid et al. [6].

To meet the design specifications, incoming light is divided into 24 sub-fields. These sections illuminate 24 single Integral Field Units (IFUs) within which the light is again split up into 48 slices. Each of these slices carries the information of a region of $15''\times 0.2''$ on the sky and is projected onto a 75 pixel wide strip on the detector. Each IFU is equipped with a 4k×4k CCD, that allows a wavelength coverage of 4650–9300 Å at a resolution of ∼3000. This extended wavelength coverage results in second order overlap in the red part of the spectrum, the standard wavelength range is therefore restricted to 4800 Å in the blue part.

To visualize what data from MUSE will look like and to aid the first development steps of the data reduction procedure, we have created mock data for a single IFU. This mock data is presented in Fig. 1. It shows the layout of the 48 slices on the CCD. The sky background is modeled using the UVES

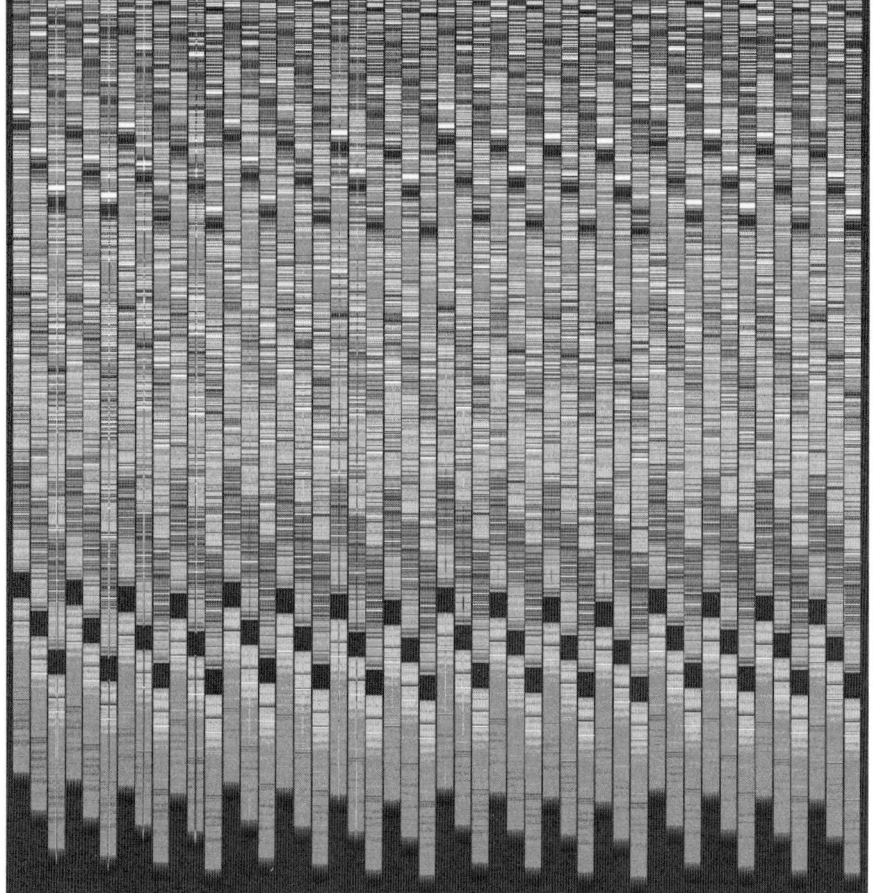

Fig. 1. A greyscale version of raw mock data of one IFU, i.e. 1/24th of one MUSE exposure

sky spectrum from Hanuschik [5], convolved with a preliminary instrument response and a filter blocking the region around the NaD line. Three different artificial spectra are taken from the GALEV models [2] to display three objects in the frame. In later stages of instrument development the mock data will be replaced by raw data from an instrument numerical model.

2 The Data Reduction Pipeline

Goals of the data reduction system for MUSE are to automatically and reliably reduce the large amounts of data the instrument will produce. Two configurations of the pipeline should allow (1) fast and simple reduction to allow a quick look at the reduced data during observations and (2) optimal

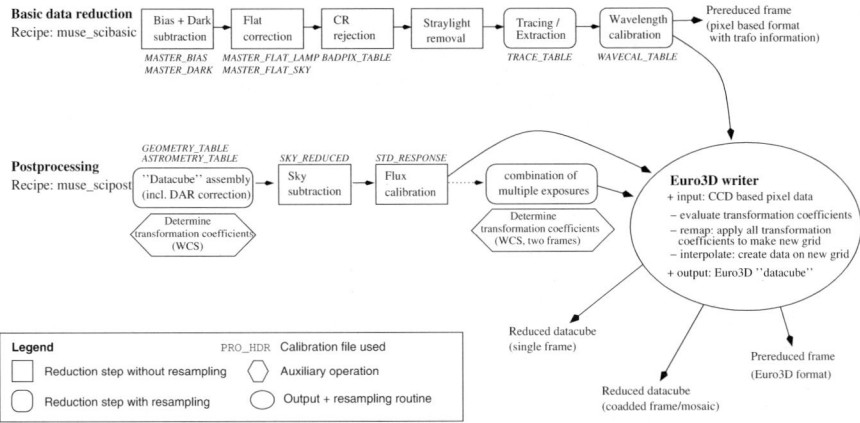

Fig. 2. The outline of the processing steps forseen for the science data reduction for MUSE

reduction of various kinds of data, ready to be used by scientists. The pipeline should also keep track of bad pixels and variance at every step to allow easy error estimates of measurements and to allow discrimination of faint emission line sources from noise fluctuations. Due to the high data rate of the instrument ($\sim$800 MB for a single raw and $\sim$4 GB for a reduced exposure, up to 150 GB of data per night), parallelization of the pipeline seems like a logical way to operate the data reduction.

To meet the first goal, the pipeline design only uses few distinct pipeline recipes. A few are forseen for each type of calibration to create master calibration products and to monitor the instrument properties. Two recipes are needed to handle the science data, one to correct the instrument signature on the level of a single IFU or CCD and a second one to treat overall instrumental effects, distortions due to the atmosphere and to combine multiple exposures. Depending on the input data, paths to integrate flux over a standard star or to do full reduction on the actual observed object are chosen. A simple flowchart of the operation of the two envisioned recipes for science data processing is shown in Fig. 2.

Keeping track of the variance as required by the second goal adds some additional restrictions on a data reduction software. To make this possible and derive a meaningful result, one has to keep the number of resampling steps as low as possible, and ideally only use one step of interpolation. Davies [3] discusses this approach in more detail.

3 Status and Outlook

With the instrument preliminary design review in summer 2007 we have finished the overall design of the pipeline. We have also written several prototype routines for the data reduction, both using the C/CPL framework required by the ESO data flow system and extra test programs using Python and IDL. After the final design review in 2008, programming of the final pipeline recipes will start. This should be finished before the first light of the instrument in 2012.

In addition to the data reduction pipeline as such, we plan to develop a new visualization tool for integral field data, optimized for MUSE but capable to display data from all integral field spectrographs. For details see Weilbacher et al. [7]. We also plan to compile a number of data analysis tools that might work as plugins to the visualization tool or standalone as discussed by Gerssen et al. [4].

Acknowledgement. PMW and JG received financial support through the D3Dnet project from the German Verbundforschung of BMBF (grant 05AV5BAA).

References

1. R. Bacon, S. Bauer, P. Böhm, et al.: Msngr **124**, 5 (2006)
2. J. Bicker, U. Fritze-v. Alvensleben, K.J. Fricke: A&A **413**, 37 (2004)
3. R. Davies: this volume (2008)
4. J. Gerssen, P.M. Weilbacher, P. Boehm, M.M. Roth: in: *ADASS XVI, ASP*, 376, 301 (2007)
5. R.W. Hanuschik: A&A **407**, 1157 (2003)
6. R.M. McDermid, R. Bacon, et al.: these proceedings (2008)
7. P.M. Weilbacher, M.M. Roth, A. Pécontal-Rousset, et al.: in: *ADASS XV* (eds C. Gabriel, et al.), *ASP*, volume 351, PASP p. 347 (2006)

Part VI

Session 5: Adaptive Optics Instruments

IOT Overview: Adaptive Optics

N. Ageorges, C. Lidman, and C. Dumas

ESO, Alonso de Cordóva 3107, Vitacura, Casilla 19001, Santiago 19, Chile;
nageorge@eso.org

1 Introduction

All of ESO's adaptive optics (AO) fed instruments are located on Paranal. This contribution concentrates on them and their respective AO oriented calibrations.

Currently there are seven adaptive optics systems that are in regular use on Paranal: NaCo & SINFONI on UT4-Yepun (Fig. 1), CRIRES on UT1-Antu and the four Coudé multiple applications curvature adaptive optics MACAO units that feed the VLTI instruments AMBER and MIDI.

The AO specific calibrations are not (really) part of the Instrument Operations Team (IOT) agenda, since they are not used to calibrate scientific data. They are, however, essential for the proper functioning of the instrument and are actively discussed in the *parAOgar* meetings. These three-weekly meetings bring together experts from the AO group in Garching and the Paranal instrument scientists and engineers involved in the AO instruments.

On Paranal, there are presently five instruments that use adaptive optics, but only two different kinds of adaptive optics technologies are used. This is summarized in Table 1. Most instruments use a curvature wavefront sensor and a bimorph deformable mirror (DM), while only a single instrument uses a Shack-Hartmann wavefront sensor (WFS) and a piezo-stack DM.

To complete this introduction, we recall briefly the principle of adaptive optics (Sect. 1.1) to stress the differences between these two technologies.

The generalities about the AO calibrations are discussed in Sect. 2. We then discuss separately the MACAO (Sect. 3) and NAOS (Sect. 4) case.

1.1 Reminders of Basic AO Principles

It is not the purpose of this contribution to explain thoroughly the principle of adaptive optics. We nevertheless felt it necessary to recall a number of principles that help to understand the philosophical difference between the two technologies currently used in Paranal.

The principle of Adaptive Optics (e.g. [3]), illustrated in Fig. 2, is to use a point source at infinity to measure the distortion of the incoming wavefront, via the wavefront sensor, due to the atmospheric turbulence and to flatten

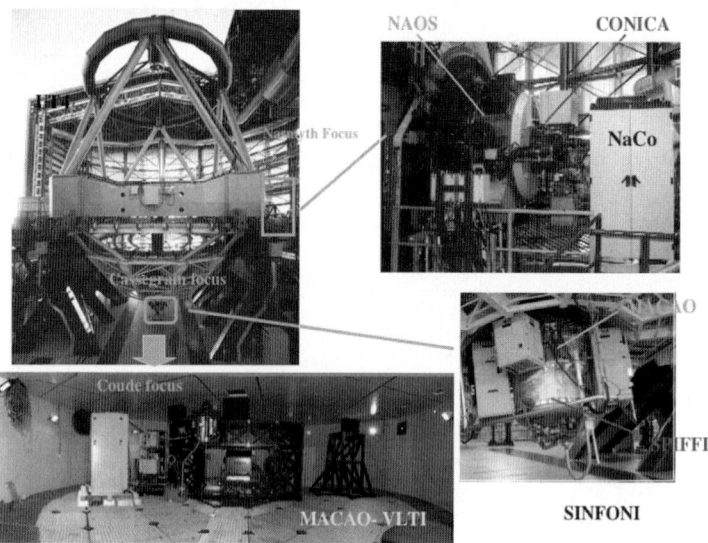

Fig. 1. Image of UT4-Yepun and its AO instruments

it thanks to a deformable mirror. This has the effect to sharpen the image and increase its resolution thus allowing to reach the diffraction limit of the telescope.

The two types of adaptive optics systems in use on Paranal differ in the way the wavefront sensing is done (see e.g. Chapter 5 of [4]). NAOS is based on a Shack-Hartmann system, while the MACAOs use curvature wavefront sensing (Fig. 3). In the former, an image of the AO reference point source passes through an array of micro-lenses. The deviation in the position of the spots compared to the ideal case (perfect alignment tested on reference

Table 1. List of AO fed instruments in Paranal, classified as a function of the AO technology used. With the exception of NaCo, all instruments are also operated in non-AO mode

AO System	Instrument	Tel. Focus
	Shack-Hartmann WFS	
NAOS [1]	NaCo	Nasmyth B (UT4)
	Curvature sensing WFS	
	SINFONI	Cassegrain (UT4)
	CRIRES	Nasmyth A (UT1)
MACAO [2]	VLTI instruments: MIDI & AMBER	Coudé (All UTs)

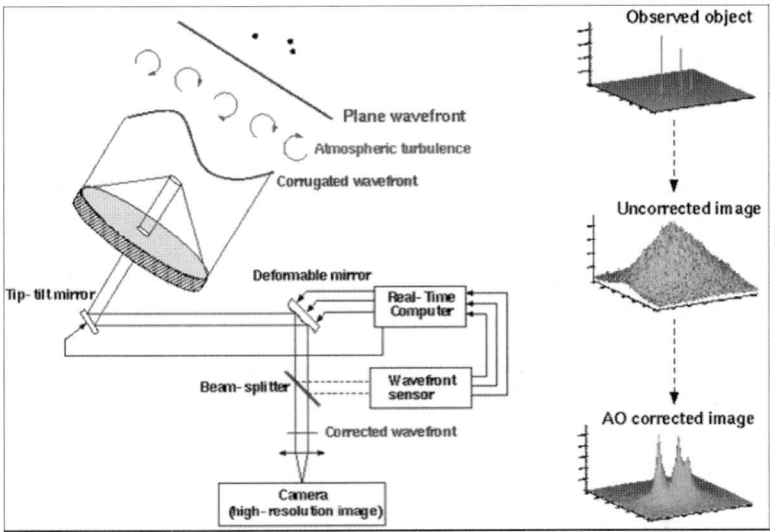

Fig. 2. Illustration of the principle of adaptive optics

fibers) can be used to estimate the distortion the wavefront and thus the deformation that needs to be applied to the mirror to flatten the wavefront. In the case of a curvature system, intrafocal and extrafocal images, taken with the help of a vibrating membrane, are used to compute the second derivative of the wavefront deformation. The technology used by both systems is entirely different, leading to different calibrations plans, as detailed in the coming sections.

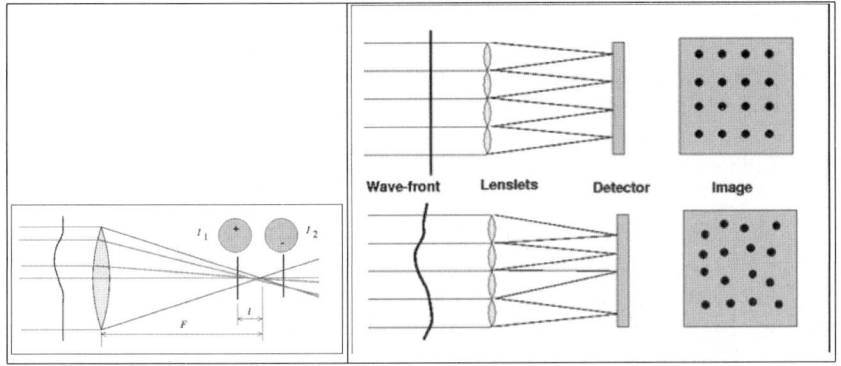

Fig. 3. Illustration of the principle of curvature (*left*) and Shack-Hartmann (*right*) wavefront sensing

2 AO Specific Calibrations

The only science calibration related to AO is related to point spread function (PSF) reconstruction. Indeed, the PSF is never perfect and contains, e.g., residuals due to static aberrations of the system not seen and thus not corrected by the AO system. For the MACAOs nothing is provided so far that would allow to perform PSF reconstruction. In the case of NaCo, two matrices are attached to each science frame. These files are the residual and mirror modal covariance matrices. Please refer to the contribution of Yann Clenet (in these proceedings) for details about this issue.

For NaCo the performance of the AO system is estimated from the Strehl ratio, when it can be measured on a point source. For both NaCo & SINFONI, when working in closed loop, the real time computer estimates the system performance from measurements performed in parallel on the wavefront sensor. This information is then used to classify the observations.

All other AO calibrations performed are instrument calibrations and are fundamental for maintaining the performances of the system.

3 The Case of the MACAO(s)

3.1 Calibration Plan

Currently, the following MACAOs calibrations are performed daily:

- During daytime:
  - The phase lag between the speaker and the membrane mirror. It is optimized daily.

  - The reference voltage vector that is applied to the DM to make the mirror flat. It is optimized daily.

  - The interaction matrix (IM) (Fig. 4) between the WFS and the DM. The results are currently not applied, but, from this measurement, the following quantities are derived and monitored:

    - a0: voltage × Radius of Curvature (RoC)
    - Pupil center/offset
    - Dead electrodes: appears as a column of zeros in the IM
    - Inter-electrode shortcuts: check for interconnection between two electrodes.
    - Slow electrodes: If an electrode is too slow, then an alarm is generated. This is determined as the ratio of diagonals in the fast/slow IM.

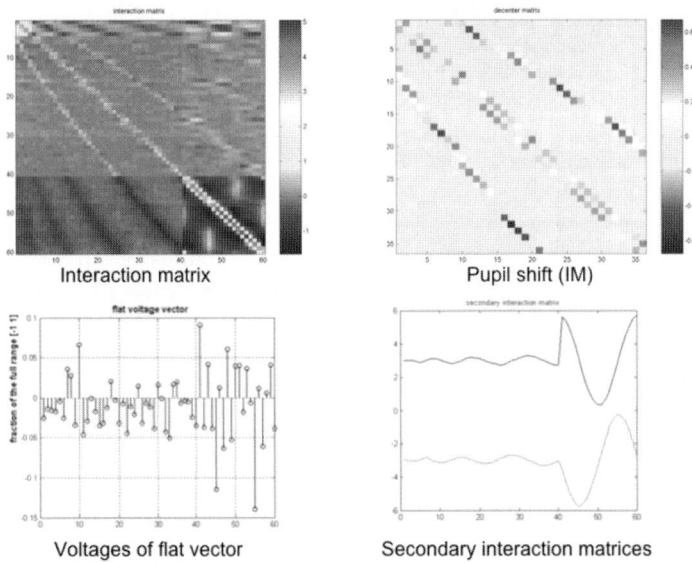

Fig. 4. Illustration of some of the MACAO calibrations: Interaction Matrix (*top left*); the effect of pupil shift on the interaction matrix (*top right*); the 60 voltages of the flat vector in fraction of the full range (−1,1) (*bottom left*) and a plot of the secondary interaction matrices (*bottom right*)

- Dead Avalanche Photo-diodes (APDs): appears as a line of zeros in the IM
- In the evening, typically during the instrument startup: Close loop on fiber and measure strehl (on fiber thus without atmosphere) ==> Quality Control
- At night:
  - For SINFONI /CRIRES: recompute reference vector (flatten the mirror) before non-AO observations. This is particularly critical for SINFONI, which is installed at the Cassegrain focus, and is caused by changes in the telescope inclination, which changes the effect of gravity on the shape of the DM.

All of the above calibrations are done on the MACAO units of CRIRES and SINFONI on a daily basis. On the MACAO VLTI units, these calibrations are done monthly, since the MACAO VLTI units are in the Coudé rooms, which is a much stabler environment.

Figure 5 illustrates a typical result of the automatic reporting system as a result of the daily calibrations run on SINFONI calibrations.

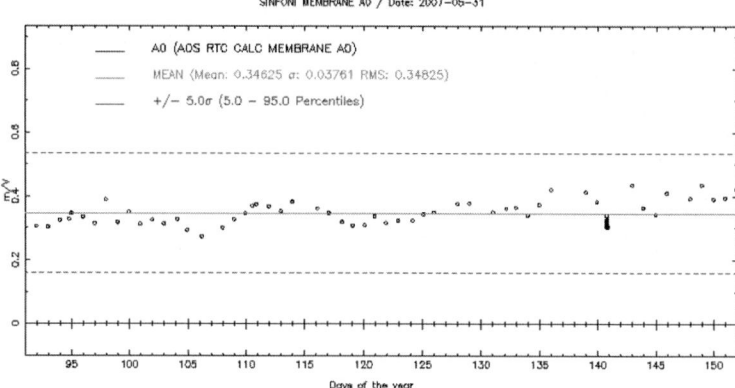

Fig. 5. The result of monitoring a0 of SINFONI over a 60 day period. a0 directly affects the gain of the system and is the product of the RoC of the membrane mirror with the voltage that is applied to the speaker that drives the mirror. The fluctuations are real and are caused by changes in the ambient temperature. If a point were to appear below or above the *blue dashed lines*, an alarm will be triggered

3.2 Monitoring Tools

AutRep (=automatic report), Performance monitor and diagnostic tool:

- Automatic report of daily monitoring (MACAO parameters) over a 60 day period. The mean value (solid line in Fig. 5) and the $\pm 5\,\sigma$ limits (dashed lines in Fig. 5) are used to trigger alarms. The parameters that are currently followed are:

  - Radius of Curvature times voltage (volts × m)
  - Membrane phase lag (rad)
  - Deformable mirror (DM) electrodes response (volts)
  - DM slow electrodes (volts)
- Diagnostic/investigation tool: Dynamic report of key-parameters can be plotted (parameter versus time, parameter versus other parameter) to investigate possible correlations (e.g. a0 versus ambient temperature)

3.3 Planned Improvements

The next steps, which are currently being implemented, are daily revision of a0 and daily monitoring of the interaction matrix (in order to determine when a new set of IMs are needed). In parallel, Laser Guide Star specific technical templates are being developed. The list below is an overview of other topics for which improvements are needed:

- Monitoring of the membrane mirror surface quality (reflectivity)
- Measurement of high order static aberrations. Presently we do not correct for them, only differential tip-tilt & defocus
- Monitoring of the transfer functions.
- Monitoring of the secondary IM (between the Tip-Tilt mount and the DM).

4 The NAOS Case

4.1 The Calibrations

For NAOS, like MACAO, there is a number of technical templates that need to be run regularly. The first steps are to check the instrument alignment (pupil and field or illumination of the WFS). Then the instrument remaining static aberrations and focus are re-measured. Finally the parameters for the adaptive optics loop are re-determined; these are the reference slopes (related to the optical alignment) and the interaction matrices (reaction of the wavefront sensor to a voltage applied to the deformable mirror).

Moreover, on a regular basis, the on-sky alignment of the full instrument (NAOS + CONICA) needs to be checked, especially after an intervention when either CONICA or NAOS or both have been detached from the telescope adaptor-rotator. In this case measurements are performed in open and closed loop. Our aim is to maintain the instrument aligned to better than 0.1 degrees.

4.2 The Monitoring Plan

The monitoring plan of NAOS foresees daily checks of the pupil alignment (detailed hereafter) and of the AO system. Both are done using internal NAOS fibers. The check of the AO consists in making a full instrument set-up, closing the loop and measuring the strehl. Its evolution is plotted into Autrep (Fig. 6). If its value is below a given threshold then clearly the instrument has a problem, possibly an optical element that did not initialise properly, and needs an urgent check before the beginning of the night.

On a weekly basis, the same check is performed, but with longer exposures on CONICA, in order to allow a more detailed check the performance of the system, i.e. check on new static aberrations.

Every month the focus of the dichroics is verified as part of the calibration plan. So far the focus has been very stable and no adjustments have been necessary.

Typically once a year the instrument flexures are measured again.

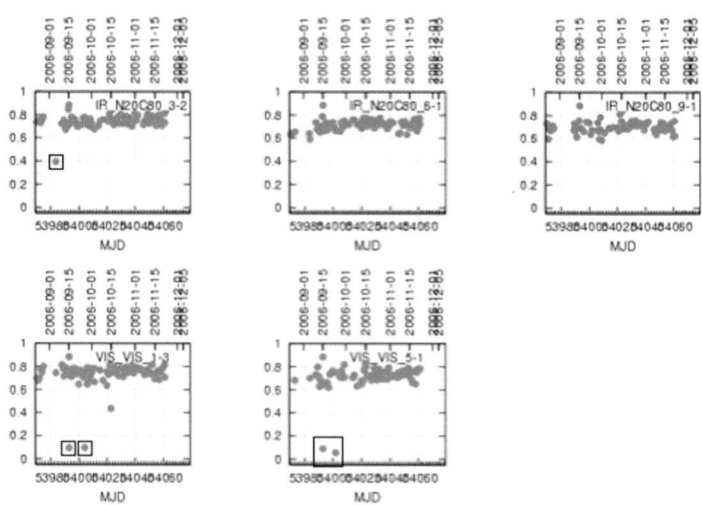

Fig. 6. Example of monitoring of the Strehl ratio resulting from the daily Check-AO template. Outliers indicated by *black squares* are a clear sign that something is wrong with the instrument; most probably a function did not initialise properly

The NAOS Pupil Alignment

The pupil alignment represents the most critical check and is performed on a daily basis. When the mis-alignment increases the system will not only lose performance but also become unstable and, eventually, completely inoperable. It took us some time on NaCo to identify the reason for the poor performance that was sometimes observed on sky as a poor pupil alignment and thus better understand the 'moods' of the system. Regular measurements have helped us realise that the system is temperature sensitive and any major temperature gradient will result in a shift of the optical alignment. Since we cannot afford to always wait for few days (especially if we are in the middle of a visitor run) for the system to 'thermalise', the pupil has to be checked and re-aligned regularly. When we are in a service period, if a measurement is bad for the first time, no special action is taken except informing the night astronomer that this configuration should not be used. Should the measurement remain clearly bad for few days, the pupil will be re-aligned. Our daily monitoring has helped us identify mis-behaving functions before the system became fully unusable.

The pupil check is performed on a calibration fiber and consists in pushing 4 actuators and measuring the corresponding shift of the spots on the WFS. The measurement is repeated for each of the 5 NAOS configuration: visible

WFS with 2 micro-lens arrays and infrared WFS with 3 different micro-lenses. Moreover, as the IRWFS can be used with 5 different dichroics, tests are also performed with these for the 14x14 configuration (the one with the smallest pixel scale thus the most sensitive to pupil shifts). Data are then analysed through a script, which lists the misalignment as a fraction of a sub-aperture, as shown in Fig. 7. These are used by the engineers to decide on the need to re-align the system and by the night astronomer to judge if a configuration can be used. In the given example, if not re-aligned then no observations should be performed with the K dichroic for that night.

For completeness, it should be specified that a fixed mis-alignment is not nice but not critical as long as the system has been calibrated under the same conditions. This is however not the case of NAOS, since we re-align the system before taking new interaction matrices.

4.3 Planned Improvements

We are actively working on a number of known issues missing in the current calibration/monitoring of NaCo:

- We need to provide the on-sky alignment data to the user (or at least the results). This is very important for monitoring programs, for example, when no special astrometric observing block has been provided.
- We are missing regular astrometric measurements. The on-sky alignment is checked but e.g. no distortion map is known.
- Checks of the WFS (noise, gain, ..) are still under development.
- Regular check and update of the static aberrations are missing. Some measurements have been taken but the results have not been compared yet to the current correction applied.

```
-----------------------------------------------------------------
VIS WFS (14×14) - Dichroic VIS : -0.0384784  0.0723252
VIS WFS (7×7)   - Dichroic VIS :  0.0380239 -0.0326174
IR  WFS (14×14) - Dichroic N20C80 :  0.0294762 -0.107975
IR  WFS (7×7V0) - Dichroic N20C80 : -0.0440543  0.0992913
IR  WFS (7×7V1) - Dichroic N20C80 :  0.0249882  0.0111194
IR  WFS (14×14) - Dichroic N90C10 :  0.100903   0.0305231
IR  WFS (14×14) - Dichroic JHK    :  0.0389628  0.0312667
IR  WFS (14×14) - Dichroic K      :  0.192338  -0.0548553

GOOD   :    abs(Delta_x, Delta_Y) < 0.1
Medium :    0.1 < abs(Delta_x, Delta_Y) < 0.15
            To avoid if possible tonight
BAD    :    abs(Delta_x, Delta_Y) > 0.15
            Cannot be used tonight
-----------------------------------------------------------------
```

Fig. 7. Example of result output by the daily CheckPupil monitoring template

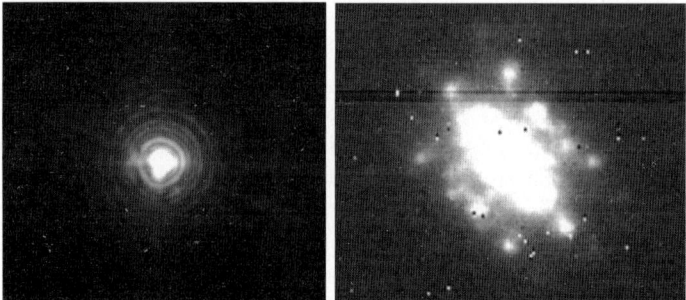

Fig. 8. Illustration of the importance of a good pupil alignment. *Left*: On sky measurement of Feb 02, 2006. 60% Strehl measured on pipeline reduced image in NB_3.74 μm. *Right*: Image of a point source with a 30% mis-aligned pupil

- Once a year measurements of the instrument's flexure are performed but the flexure compensation model has never been updated, whereby it is necessary. In the case of coronography, a long series of exposures will result in the object drifting out of the mask. We have an operational work-around but these are expensive in terms of telescope time.

5 Conclusion

Almost all calibrations mentioned here might not be so transparent to the user since they are not provided and are poorly advertised but they are fundamental for the health of the AO system (Fig. 8) and thus to provide the best performances. Both for the MACAOs and NAOS we are aware of some flaws in our calibration plan and are actively working on filling them in as detailed in Sects. 3 and 4.

References

1. G. Rousset, F. Lacombe, P. Puget, N. Hubin, G. Gendron, et al.: *NAOS, the first AO system of the VLT: on-sky performance*, SPIE 4839, 140, 2003
2. R. Arsenault, J. Alonso, H. Bonnet, et al.: *MACAO-VLTI: An Adaptive Optics system for the ESO VLT interferometer*, SPIE 4839, 174, 2003
3. R.K. Tyson: *Principles of adaptive optics*, 2nd edn (Boston Publisher, Academic Press), 1998
4. F. Roddier: *Adaptive Optics in Astronomy*, F. Roddier Ed. (Cambridge University Press), 2004

AO Assisted Spectroscopy with SINFONI: PSF, Background, and Interpolation

R. Davies

Max Planck Institut für extraterrestrische Physik, 85741, Garching, Germany;
davies@mpe.mpg.de

Abstract. I discuss three widely applicable aspects concerning calibration of the near infrared adaptive optics integral field spectrometer SINFONI: (1) the accuracy with which one needs to quantify the PSF and how this might be achieved in practice; (2) how it is possible to fine tune the background subtraction to minimise the residual OH airglow; and (3) how an altered perspective on calibration data might lead to improvements in interpolation and greater flexibility in reconstructing datacubes.

1 A Short Introduction to SINFONI

SINFONI [11, 3] is a versatile instrument comprising of a 60-element curvature adaptive optics system [2] that feeds a 1–2.5 µm integral field spectrometer [12]. The camera has 3 pixel scales spanning $0.25''$ to $0.025''$, making it adaptable to both seeing and diffraction limited resolutions. The associated fields of view range from $8'' \times 8''$ to $1'' \times 1''$. It can cover the H and K bands together in a single exposure at a spectral resolution of $R \sim 1500$; or a complete single waveband (J, H, or K) at $R \sim 2000$–5000, depending on the pixel scale. Since the highest resolution (associated with the smallest pixel scale) is under-sampled, one has the option of spectrally dithering and interleaving the 2 exposures. Image slicers dissect the field of view and re-arrange the slitlets along a single pseudo-slit. On the detector, the dispersed data from each slitlet appear exactly analogous to standard longslit data, except that there are 32 such 2D spectra next to each other. A dedicated data reduction package *spred* [1] reconstructs the 3D datacube. An excellent tool for viewing them is *QFitsView* (see http://www.mpe.mpg.de/~ott/QFitsView), which displays the spectrum in real time as one moves the pointer across the spatial field. With this tool it is also extremely quick and simple to apply a wide range of processing techniques in real time.

2 The Adaptive Optics Point Spread Function

Misunderstandings about adaptive optics PSF abound: that it must be known in great detail, and that its temporal and spatial variability casts doubt on interpretation of the data. In this section, I attempt to alleviate these concerns by discussing some ideas about the level of accuracy with which one needs to know the PSF, and some ways in which this might be achieved.

2.1 Quantifying the PSF

There will always be some situations where it is necessary to know the PSF in great detail, most obviously in planet searches where one is trying to detect small faint object close around a bright point source. In these cases, it is crucial to distinguish between the object and structure that belongs to the PSF. On the other hand, many – perhaps most – applications do not require such a detailed level of knowledge. Particularly for extragalactic science, where the AO correction is mediocre, a simple combination of two analytical functions will often suffice. For example, the PSF can be generally well matched by the sum of a narrow Gaussian, which represents the core of the PSF, and a Moffat function which can trace the wide wings in the halo. Greater detail in the PSF is unnecessary because the accuracy is limited by the model (kinematic or morphological), which, in contrast to the real intrinsic structure in galaxies, is usually simple, and often symmetric.

2.2 (De)-Convolution

Dealing with a PSF that comprises two contrasting components – a narrow core and broad wings – is an important issue. Clearly one needs to separate the PSF from the instrinsic structure in the observed data. But deconvolution is not always the best solution. It is an inverse problem, and hence mathematically messy, tends to amplify noise, and can easily generate unreal artifacts (e.g. ringing). And at the end, one still has no convenient expression for the intrinsic source shape and one still has to deal with a PSF in the deconvolved data. It may be narrower, but it is probably less well defined and may vary with signal-to-noise across the field. In some cases, such as kinematics (Fig. 1), deconvolution is simply not an option due to the cross-talk between the velocity and dispersion.

An alternative is to convolve a model of the intrinsic structure with the PSF, compare the result to the observations, and adjust the model iteratively. This is the basis of popular galaxy fitting algorithms, such as galfit [17] which has been used in Fig. 2. The method also enables one to make realistic estimates of the uncertainty in the fitted parameters. While it cannot be used if there is no way to parameterise the intrinsic source structure (e.g. the features on the surface of a planet such as Titan), it is still widely applicable.

2.3 Methods to Estimate the PSF

Below are suggested several ways one might try to infer the shape of the PSF. This list is not necessarily exhaustive, but is intended to indicate that many possibilities exist if one can be a little inventive.

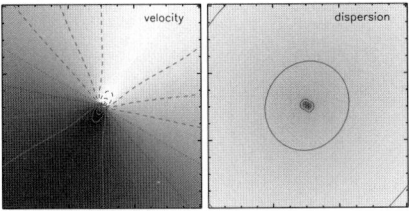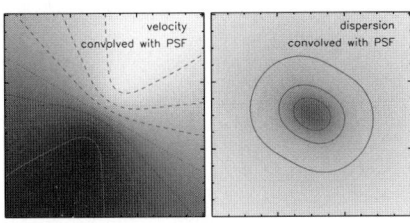

Fig. 1. Velocity and dispersion fields for matter in a self-gravitating disk around a supermassive black hole. In the field shown, the integrated mass of the disk is 5 times that of the black hole. The effect of PSF-induced smearing on the kinematics is very dramatic. Due to cross-talk between these 2 quantities (i.e. velocity gradients on scales comparable to the PSF contribute to the dispersion), the smeared kinematics cannot be deconvolved. The only option is to create a 3D kinematic model (2 spatial and 1 velocity dimension), convolve it with the PSF, and then extract the kinematics. By iterating one can constrain the model parameters

Fig. 2. Example of a faint $K_s = 19.2$ galaxy observed with adaptive optics [5]. The galaxy was 18.3" from the guide star, and so the PSF derived for it had to take into account isoplanatic effects (see point (iii) below). The image of the galaxy (*left*) is too noisy to deconvolve with the PSF. It has therefore been fitted by convolving the PSF with a parameterised galaxy profile (*centre*). The parameters of the profile were adjusted to minimise the residuals (*right*). This can yield not only the best fitting parameters, but also a good estimate of their uncertainties

(i) Reconstruct it from the wavefront sensor data
From the astronomer's perspective, this is the ideal option. PSF reconstruction has been developed for both curvature [20] and Shack-Hartmann AO systems [21, 14], and there are no technical limitations. However, there is no general facility for PSF reconstruction yet available at the VLT. A tool is being developed for NACO (see Clenet's contribution to this proceedings).

(ii) Use an isolated star as a reference
This is generally the path recommended to an observer. However, for practical reasons – specifically the time needed to slew to and observe a separate

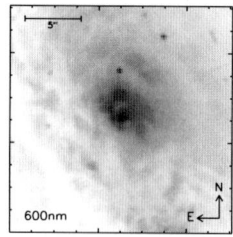 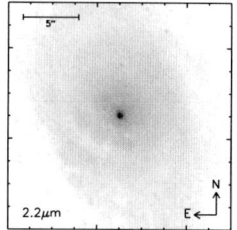

Fig. 3. Images of the Circinus galaxy: *left* optical, *right* near infrared [18]. It is impossible to reproduce the flux distribution seen by a visible WFS using a reference star. On the other hand, what an infrared WFS sees is dominated by a point source and so it may be possible to use a reference star to estimate the PSF

reference star – it is often impractical. Furthermore, a reliable PSF estimate requires that the intensity and distribution of flux on the WFS is the same for the reference star as for the science wavefront reference object. If both are stars, this can work well; but as Fig. 3 shows for AGN and other extended sources, it is simply unreliable.

To complicate the matter further, if the wavefront reference is not the science target, then using it to estimate the PSF is misleading due to anisoplanaticism. Instead one needs to find a pair of stars (e.g. from the Washington Double Star Catalog [15]) separated by the same distance, one of which matches the guide star magnitude and the other of which can be observed by the science camera.

(iii) Extrapolate it from surrounding stars
If one is lucky, it may be possible to measure the PSF from nearby stars [18]. More often, it will be necessary to account for anisoplanaticism. Several methods have been developed to estimate an off-axis PSF; and in principle these could be turned around to derive an on-axis PSF from off-axis stars. Typically, they require knowledge of the C_N^2 distribution through the atmosphere [13, 4] or observations of calibration frames containing many stars [19]. But it is also possible to make a reasonable (and sufficient) approximation to the way the PSF varies across a wider field using the science data alone, as long as at least one or two stars or compact objects are detected [5, 6].

(iv) Extract it from the science data itself
The broad line region in AGN is only a few lightdays across and is therefore always unresolved in 8-m class telescopes. In addition, the near-infrared non-stellar continuum associated with AGN is only 1–2 pc across and hence unresolved in AGN that are at least ~ 20 Mpc away. The spatial distribution of both these quantities can be extracted using the spectral information available in a near infrared datacube, and has been used as an estimate of the PSF in several cases [7, 8, 9].

One might expect that it should also be possible to extract information about the PSF from other science data in an analogous way.

(v) Derive it by comparison to other higher resolution data
If data taken with another instrument at another time exist at the same wavelength and at higher spatial resolution, one might derive the PSF by reference to these [16]. This is because convolution of the PSF P with the intrinsic source S yields the observed source $O = P \otimes S$. One can define a broadening function F which, when convolved with the higher resolution observation O_h, reproduces the lower resolution observation $O_l = O_h \otimes F$. Then by definition the lower resolution PSF is $P_l = P_h \otimes F$.

2.4 Effects of LGS Adaptive Optics

The VLT Laser Guide Star Facility has recently been commissioned, and so the observer will soon have to cope with LGS-AO data. Because the wavefront reference will be the same regardless of the science target, the PSF should in principle be easier to measure using an isolated reference star as in point (ii) above. In addition, because the LGS samples a cone rather than a full column through the atmosphere, the isoplanatic effects will be smaller. The main impact on the PSF will be residual jitter from the tip-tilt star, which depends on how faint and how far away it is. But this is relatively easy to add in afterwards to an initial (better) estimate of the PSF.

3 Improving the Background Subtraction

A method to improve subtraction of the near infrared background, which is dominated by OH emission lines, has recently been described [10]. This is based on the fact that most of the variation between these lines occurs from changes in the vibrational rather than rotational temperature of the OH radical. And grouping the emission lines according to the vibrational part of their transition can be done to a reasonable approximation by wavelength: any particular spectral segment contains all the strong lines for one specific vibrational transition. One can then apply an appropriate scaling to the sky frame for each segment separately before subtracting it. Since the segments span a reasonably wide wavelength range, the method is robust against emission or absorption features being blended with OH lines. Treatment of the rotational part of the transitions is similar, although trickier because they cannot be grouped so easily and also because OH lines from different transitions are blended. The integration of this background subtraction algorithm into the SINFONI pipeline is described by Modigliani in this proceedings.

A quantitative indication of the improvement this method can yield is given in Fig. 4. In this experiment, a long sequence of blank 5 min H-band SINFONI frames were used. For each frame, the successive one was used to

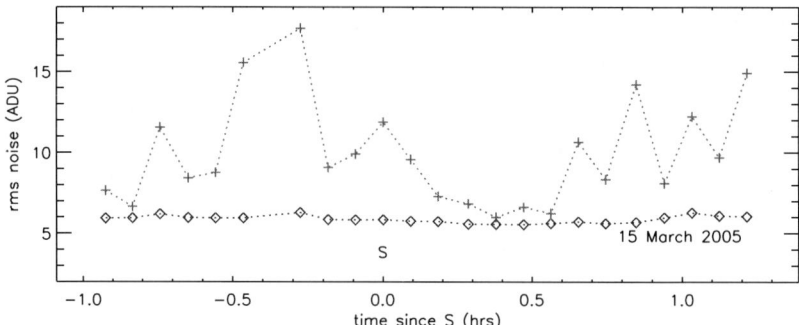

Fig. 4. Noise in sky subtracted H-band SINFONI cubes (arbitrary data units) for consecutive 5 min integrations. Pluses denote cubes which each had a different sky cube, taken immediately afterwards, subtracted. Diamonds show the resulting noise level when this was performed using the scaling algorithm described in the text

subtract the sky background. The noise in the sky-subtracted cube – which is dominated by residual OH emission – was measured as the standard deviation of all data values within the spatial field and spanning 1.55–1.75 µm. The figure shows that the noise is variable and high. Using this new algorithm not only reduced the residual noise but the resulting noise is much more stable across all the frames. In fact, this method allows one to use fewer sky frames, thus significantly increasing the observing efficiency.

One aspect that also needs to be addressed in relation to background subtraction is the accuracy of the wavelength calibration. It is a feature of SINFONI data that there may be a shift of a fraction of a pixel along the spectral axis between frames. If this is not corrected, then subtracting a sky frame will leave P-Cygni residuals for the OH lines. This can be corrected by reconstructing each cube before subtracting the background, measuring the wavelength of each strong OH line, and deriving the mean spectral offset. The whole frame can then be shifted by an appropriate amount to correct the offset. Details of all these aspects are described elsewhere [10].

4 Interpolating in 3 Dimensions

Interpolation is a crucial issue for integral field spectroscopy, since reconstructing 3D datacubes requires a significant amount of interpolation. In the classical approach, one needs to correct for bad pixels, straighten the spectral traces, linearise the dispersion, and finally align the slitlets (or pixels). Tuning the wavelength scale to correct for flexures between frames (as described in Sect. 3) can introduce an additional interpolation step. Poor management of the interpolation strategy or poor choice of the interpolation scheme can degrade the quality of the final data. For this reason, I propose an alternative perspective on the purpose of calibrations which allows one to view the data reconstruction in a different way – enabling one to perform all the interpolation in a single step while at the same time permitting a far greater flexibility. This can be summarised as follows:

Standard View
calibrations allow one to create the mathematical functions necessary (e.g. polynomials) to correct the spectral and spatial curvature on the detector.

Alternative View
calibrations allow one to create look-up tables which associate each measured value on the detector with its spectral and spatial location in the final reconstructed data.

This new perspective, which is incorporated into the design of the KMOS data reduction library, is outlined graphically in Fig. 5. The most important realisation is that in 'detector space' there can be no concept of a wavelength or spatial axis. These concepts apply only to the final reconstructed cube. The detector is nothing more than the medium on which raw data values are

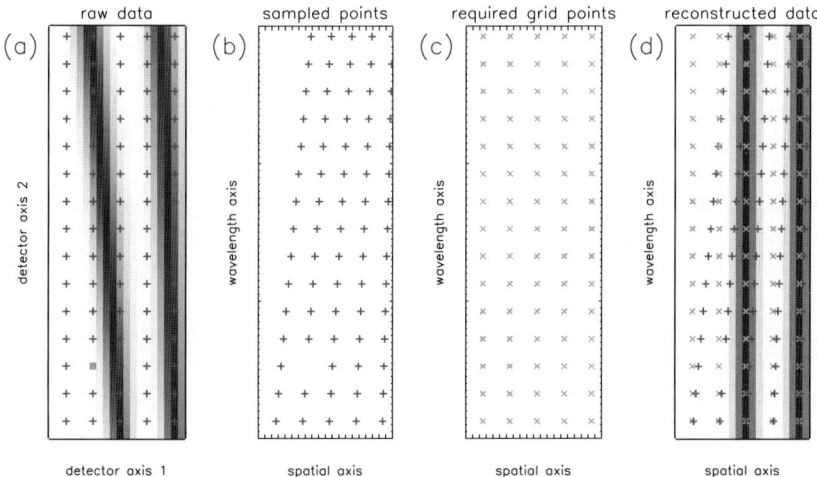

Fig. 5. Illustrative example of an alternative perspective for reconstructing datacubes. (**a**) observed data are sampled regularly in the reference frame of the detector. (**b**) this sampling is irregular in the reference frame of the reconstructed cube; bad pixels can simply be omitted from the set of sampled points. (**c**) one can freely specify the required gridding (i.e. spatial/spectral pixel scale) for the reconstructed data; it is independent of the actual sampling. (**d**) each required grid point is interpolated from sampled points which lie in its local neighbourhood. Any suitable algorithm can be used for the interpolation

recorded. The calibrations allow one to assign each measured value on the detector to a spatial/spectral location in the reconstructed cube. Together, these locations provide an irregularly spaced sampling of that cube. The aim is thus to reduce the raw data and the calibrations to a list of values with their associated locations:

$$\begin{array}{cccc} value_0, & x_0, & y_0, & \lambda_0 \\ value_1, & x_1, & y_1, & \lambda_1 \\ \vdots & \vdots & \vdots & \vdots \\ value_n, & x_n, & y_n, & \lambda_n \end{array}$$

Data associated with bad pixels is simply excluded from the list and so does not contribute to the set of sampled locations. Creation of this list is the first step. The second step is to specify the regular sampling – i.e. the spatial and spectral pixel size – that is required for the reconstructed cube. The third step is to interpolate each of these regularly gridded positions from sampled locations in the local neighbourhood. In a fourth step, one can determine any spectral (or spatial) offsets in the reconstructed cube and feed these back to create a new list with updated locations for each measured value. One can then re-interpolate the regular grid of points, leading to a final cube which has been reconstructed in a single interpolation and which has no offsets. There are a number of advantages of this method:

- Only a single interpolation is required to reconstruct the final cube from the raw data. This leads to improved noise properties in the final cube.
- One can combine separate frames during this interpolation, by concatenating their lists of data values and locations. This is useful because it avoids the need to shift and combine cubes afterwards.
- One has a free choice of spatial and spectral sampling in the final cube. This is useful if one wants to compare the data to that from another instrument: one can reconstruct the cube at the appropriate pixel scale, rather than having to re-interpolate it afterwards.
- One has the option of smoothing the data during the reconstruction. If the data is particularly noisy, one can increase the size of the local neighbourhood around each point to reduce the noise at the expense of resolution.

References

1. R. Abuter, et al.: NewAR **50**, 398 (2005)
2. H. Bonnet, et al.: SPIE **4839**, 329 (2003)
3. H. Bonnet, et al.: ESO Messenger **117**, 17 (2004)
4. C. Britton: PASP **118**, 885 (2006)
5. G. Cresci, R. Davies, A. Baker, M. Lehnert: A&A **438**, 757 (2005)

6. G. Cresci, R. Davies, A. Baker, F. Manucci, M. Lehnert, T. Totani, Y. Minowa: A&A **458**, 385 (2006)
7. R. Davies, L. Tacconi, R. Genzel: ApJ **602**, 148 (2004)
8. R. Davies, L. Tacconi, R. Genzel: ApJ **613**, 781 (2004)
9. R. Davies, et al.: ApJ **646**, 754 (2006)
10. R. Davies: MNRAS **375**, 1099 (2007)
11. F. Eisenhauer, et al.: ESO Messenger **113**, 17 (2003)
12. F. Eisenhauer, et al.: SPIE **4841**, 1548 (2003)
13. T. Fusco, J.-M. Conan, L. Mugnier, V. Michau, G. Rousset: A&AS **142**, 149 (2000)
14. L. Jolissaint, J.-P. Véran, J. Marino: SPIE **4590**, 151 (2004)
15. B. Mason, G. Wycoff, W. Hartkopf, G. Douglass, C. Worley: AJ **122**, 3466 (2001)
16. F. Mueller Sánchez, R. Davies, F. Eisenhauer, L. Tacconi, R. Genzel, A. Sternberg: A&A **454**, 481 (2006)
17. C. Peng, L. Ho, C. Impey, H.-W. Rix: AJ **124**, 266 (2002)
18. A. Prieto, et al.: ApJ **614**, 135 (2004)
19. E. Steinbring, et al.: PASP **114**, 1267 (2002)
20. J.-P. Véran, F. Rigaut, H. Maître, D. Rouan: JOSA **14**, 3057 (1997)
21. R. Weiss: PhD thesis, University of Heidelberg (2003)

Discussion

R. McDermid: How do you propose to propagate the error frame in the scheme of a single transformation from raw to calibrated data?
R. Davies: That depends on the interpolation scheme used. "Nearest neighbor" is simple and the noise is that of the adopted pixel. "Kriging", which is an optional linear interpolation method, produces an estimate of the uncertainty in the interpolated values for other schemes, error propagation will be studied using simulations.
R. McDermid: You give some nice ideas/techniques for PSF determination from observed data. But it is the (numerous) cases when no reference is available that poses a serious question.
R. Davies: This is true, my point is that in many cases, if one is inventive, there is often a way to recover information about the PSF. Ideally, it would, of course, be provided by the AO system.
A. Seifahrt: Concerning the improved OH line compensation: How to deal with the variations in the line strength introduced by spatial-pixel or wavelength shifts between consecutive observations and the fact that you are spectrally under sampled?
R. Davies: In SINFONI, the spectral resolution depends on the pixel scale; and the data is only spectrally undersampled for the smallest scale. If one cannot spectrally dither (which is in principle possible) then one has to cope as well as possible. Fortunately for the smallest scale, the pixels subtend a solid angle on the sky that is 100 times smaller than the seeing limited scale.

So the flux, variations and noise in the OH lines is less important with respect to the read noise; and hence less of a problem.

M. Casali: When the vibrational levels in OH from the sky are re-scaled, are the different scalings consistent with a change in the molecular excitation?

R. Davies: Yes, approximately. In principle one could use this to help constrain the fit; but it's simpler (and certainly not worse) to leave each segment independent.

Imaging and Treatment of the PSF in AO Instruments: Application to NACO

Y. Clénet[1,4], E. Gendron[1,4], G. Rousset[2,4], T. Fusco[3,4], C. Lidman[5], M. Kasper[6], N. Ageorges[5], and O. Marco[5]

[1] LESIA, Observatoire de Paris, 5 place Jules Janssen, 92195 Meudon Cedex, France; yann.clenet@obspm.fr
[2] LESIA, Université Paris 7, 5 place Jules Janssen, 92195 Meudon Cedex, France
[3] DOTA, ONERA BP72, 29 Av. de la Division Leclerc, 92322 Chatillon cedex, France
[4] Groupement d'intérêt scientifique PHASE (Partenariat Haute résolution Angulaire Sol Espace)
[5] ESO, Alonso de Cordóva 3107, Vitacura, Santiago, Chile
[6] ESO, Karl-Schwarzschild-Strasse 2, 85748 Garching, Germany

Abstract. After about 15 years of operation with astronomical instruments, adaptive optics has now reached maturity: it is routinely used at several large telescopes and transparent for the astronomer. Though, adaptive optics assisted images have specificities compared to seeing-limited images, such as their point spread function profile or spatial/temporal variability, that I review in the following. Adaptive optics imaging, unlike seeing-limited imaging, offers the possibility to calibrate the PSF simultaneously to the scientific observations, using real-time measurements of the adaptive optics system itself. I introduce here this PSF reconstruction and review the different PSF reconstruction algorithms developed till now, in particular the one for NAOS.

1 Introduction

Adaptive optics (AO), which provides a real-time correction of the image quality degradation due to the atmospheric turbulence, is a pretty young technique in astronomy. The first AO system available for the astronomical community was ADONIS, offered by ESO to the community in 1993, less than 15 years ago! Despite this relative youth, AO is now sufficiently mature and equips almost all large telescopes and no (extremely) large telescope is designed without AO. Moreover, a great number of new AO concepts, such as Multi-conjugate (MCAO), Laser Tomography (LTAO), Ground Layer (GLAO), Extreme (XAO), Multi-object (MOAO) adaptive optics, have emerged to overcome the limitations of the original "single conjugated" AO concept and have driven several technological breakthroughs, in lasers or deformable mirror (DM) developments for example. AO has also extended over astronomy to be applied in other fields like retinal imaging.

Similarly to classical imaging, AO images are calibrated with the point spread function (PSF), i.e. the image at infinity of a point source. Parameters of this AO PSF, such as the Strehl ratio and the full width at half maximum, are indicators of the image quality or the AO correction. And the AO PSF is used as an input variable for various post-processing software such as astrometry/photometry or deconvolution software.

In the following, we first remind of some AO PSF characteristics, like its shape or its characterising parameters. Then we illustrate how the temporal and spatial variability of the AO PSF affects the image quality. We finally introduce PSF reconstruction and review the different PSF reconstruction algorithms developed till now, in particular the one for NAOS.

2 Adaptive Optics PSF Characteristics

The image of a point source at infinity with a telescope, with no turbulence, is the well-known Airy function: it is made of a bright central peak, surrounded by the Airy rings that get fainter as the distance to the central peak increases. Through a turbulent atmosphere, index fluctuations result in phase lead or delay, and short exposures of a similar source are made of several "speckles". These speckles result from the constructive interference of points in the pupil with equal optical path length. Each speckle has a typical dimension of λ/D, with λ the observation wavelength and D the telescope diameter, and are spread out over a region with a typical dimension of λ/r_0, with r_0 the Fried parameter. The temporal average of these short exposures is the classical seeing-limited long-exposure image: speckles are averaged in a Gaussian with a full width at half maximum of λ/r_0.

2.1 Adaptive Optics PSF Shape

Adaptive optics aims at correcting the degradation of the image quality due to the turbulence atmosphere and a perfect correction would lead to the Airy pattern. Though, the correction is always partial and AO-corrected short exposures are made of a bright central speckle surrounded by remaining speckles. In a long exposure, the residual speckles are averaged out in a halo, around a coherent peak:

$$\mathrm{PSF}(\boldsymbol{r}) = I_{\mathrm{coherent}}(\boldsymbol{r}) + I_{\mathrm{halo}}(\boldsymbol{r}) \tag{1}$$

2.2 Adaptive Optics PSF Parameters

Different parameters exist to characterise the PSF. The three usual ones are:

- the Strehl ratio (SR). It is equal to the central value of the PSF divided by the central value of the Airy pattern:

$$\mathrm{SR} = \frac{\mathrm{PSF}(0)}{\mathrm{Airy}(0)} \tag{2}$$

The SR is a good estimator of the AO correction quality. It is often approximated to the value of the coherent energy E_c, which is always lower than the SR:

$$E_c = \exp(-\sigma_{\text{res}}^2) = \frac{I_{\text{coherent}}(0)}{\text{Airy}(0)} \qquad (3)$$

This approximation is valid only if the energy in the halo is negligible compared to the energy in the coherent peak, i.e. when $E_c \approx 1$ or SR $\gtrsim$ 30%. Note that for NACO, the estimation of the AO performance is given by the computation of the coherent energy, which corresponds in the NACO fits file header to the keyword "HIERARCH ESO AOS RTC DET DST ECMEAN" [4].

- the full width at half maximum (FWHM). It is a good estimator of the achieved spatial resolution and is related to the residual tip-tilt errors of the AO correction. It can pretty easily reach a value close to the diffraction limit of λ/D.
- the 50% energy radius. It is the radius in which are concentrated 50% of the total energy. This parameter is mainly relevant for spectroscopy for which the goal is often to collect as much light as possible in a given aperture, corresponding for example to a fibre aperture.

2.3 Estimation of the Strehl Ratio: a Difficult Task!

In practice, the computation of the SR from a scientific image can be fairly difficult, even for a single isolated star. Indeed, the result of such an estimation highly depends on the background subtraction and on the accuracy of this subtraction. A positive residual background will lead to an underestimation of the SR while a negative residual background will give an overestimated SR. One needs for example to take into account the potential spatial variation of the background.

A solution to limit the influence of the background in the SR estimation could be to window the image but the risk is to loose part of the halo if the windowed image is too small: the SR would then be overestimated.

To illustrate the importance of the background subtraction in the SR estimation, let's consider the two consecutive 1024×1024 NACO images of the same star shown in Fig. 1a and Fig. 1b. We first subtract one from the other in order to subtract the sky emission. After the flat correction, we then extract a 512×512 image around the "positive star": we obtain the image in Fig. 1c. In this image, the background is not equal to zero. Let's consider the median of the image as a first estimation of the background to correct for before computing the SR. We can then estimate the SR for different sub-image sizes. The result is the upper curve in Fig. 2a: the SR have successive quasi-linear rises followed by sharp drops. The SR amplitude in these rise-drop behaviours increases with the sub-image size, reaching about

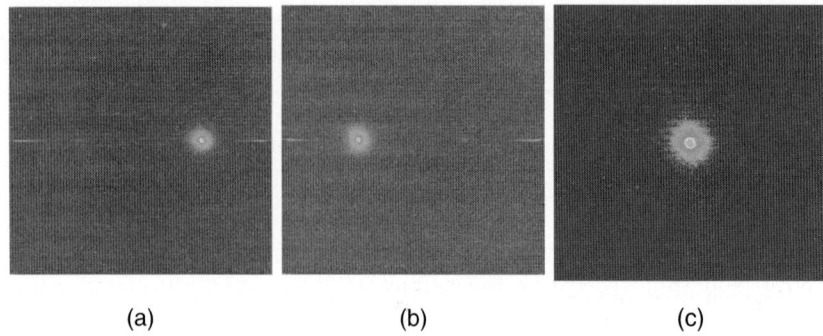

Fig. 1. (**a**): 1024×1024 NACO image of a bright star. (**b**): Image of the same star after an offset. (**c**): 512×512 image of the same star after subtraction of the two precedent images and flat-fielding

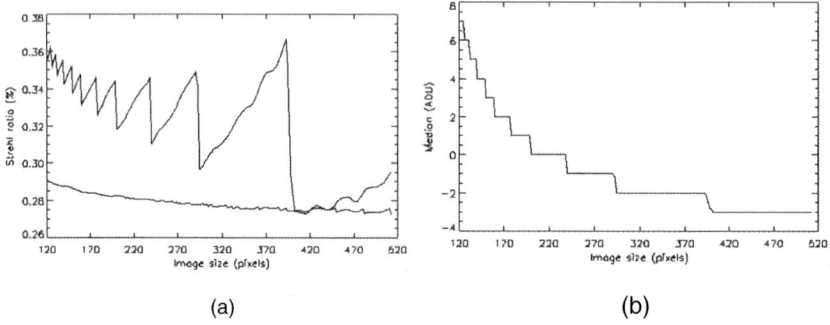

Fig. 2. (**a**): SR vs. the size of the sub-image selected around the star in the 512×512 image of Fig. 1c for different background estimations: median of the sub-image (*upper curve*) or mean value of the pixels at r>60 (*lower curve*). (**b**): Median vs. the size of the sub-image selected around the star in the 512×512 image of Fig. 1c

10%! As shown in Fig. 2b, these SR variations are actually directly related to the variations of the median when the sub-image size increases: the median is not a good estimator of the image background. If we compute instead the background as the mean value of the pixels outside a circle of diameter 60 pixels and centred on the central star, we obtain the lower curve of Fig. 2a, which follows a more reliable behaviour and has a smaller variation amplitude (a few %).

3 Adaptive Optics PSF Variability

3.1 Introduction

The AO PSF is known to be subject to both temporal and spatial variability. A non-exhaustive list of the parameters that affect the AO PSF shape is:

- the reference source brightness,
- the reference source shape,
- the reference source colour,
- the atmospheric conditions (seeing, coherence time, wind speed),
- the airmass,
- the distance between the science target and the reference source,
- the AO system calibrations,
- the AO system direct environment (e.g. vibrations).

These parameters have of course different time-scales and impacts on the AO correction. In the next two sections, we concentrate on the varying atmospheric conditions and then on the science target-reference source distance.

3.2 Temporal Variability

Different campaigns of site testing have shown that the seeing is subject to large variations at different time-scales, in particular at time-scales of seconds or minutes, comparable to the typical image integration times. Hence successive exposures can be acquired under pretty different seeing conditions, leading to different AO corrections and finally to different image qualities (or SR).

In their study of this effect, Rigaut & Sarazin [8] have used the 1994 Paranal seeing data to compute the average seeing change for a given lag between two successive exposures (Fig. 3a). Using a simple AO-correction

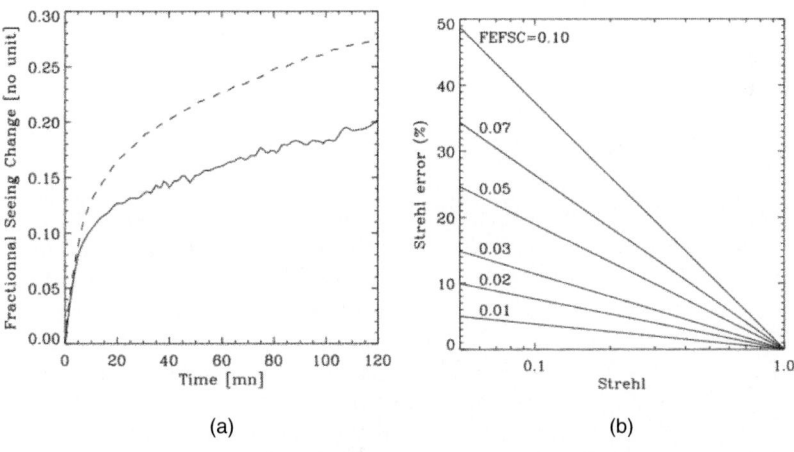

Fig. 3. (a): average seeing change in % for a given lag between two successive exposures. The *dashed curve* is the fractional seeing change (FSC), computed for instantaneous exposures, while the *plain curve* is the finite exposure fractional seeing change (FESC), computed for the finite exposures. (b): for a given FESC, the curves give the SR variations (in %) vs. the SR (in %). These two figures are from Rigaut & Sarazin [8]

model by assuming SR = $\exp(-\sigma_{res}^2)$ and $\sigma_{res}^2 \propto r_0^{5/3}$, they have derived the relative SR variations between two successive exposures for a given SR (Fig. 3b). As an example, for a 10 min lag between two successive finite exposures, the average seeing change is 10% and then the average SR change for a SR of 30% is 20%: this can be seen as a 20% average error on the SR of an exposure acquired 10 min after a 30% SR exposure. A way to decrease the seeing variations for a given exposure time is to increase the alternances of scientific and calibration exposures during this given exposure time but, for a 10 min lag and a goal of seeing variations limited to 2%, 100 alternances of scientific and calibration exposures are required (in 10 min!).

The PSF is often calibrated by alternating scientific exposures with calibration exposures on a calibration star. As demonstrated by Rigaut & Sarazin [8], this method can lead to large SR variations between the scientific and the calibration images and then to a bad calibration of the scientific exposure PSF. It raises the need for calibrating the PSF simultaneously to the scientific images.

3.3 Spatial Variability

The light coming from a source aside the AO reference source does not probe the same part of the atmosphere as the light coming from the AO reference source itself. This anisoplanatism effect results in a degradation of the AO correction as one observes further away from the AO reference source, the stars being elongated in the direction of the AO reference source. The amplitude of the effect also depends on e.g., the atmospheric conditions (seeing, coherence time, ...) or the airmass, as demonstrated in Fig. 4, with two NACO images of the Galactic Centre acquired at different airmasses on the 13 June 2004 [2].

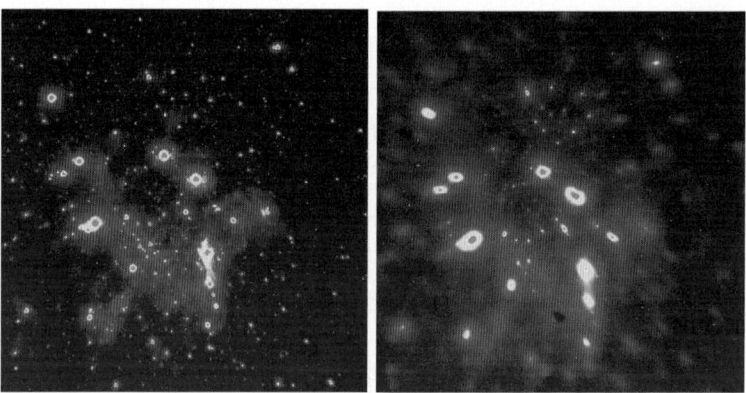

Fig. 4. Galactic Centre NACO L'-band images acquired at different airmasses [2]

4 Adaptive Optics PSF Reconstruction

4.1 Measurement of the PSF

Classically, the calibration of the scientific image PSF is done sequentially: the science object and a PSF calibration star are observed alternatively. This procedure requires that the AO correction is the same during both observations. In practice, this is pretty difficult to achieve or control since, in particular, the level of noise on the wavefront sensor (WFS) depends on several parameters such as the source magnitude, colour, shape, the airmass. Moreover, as shown in Sect. 3.2, the seeing fluctuations can induce fairly large SR variations between the science object and calibration observations.

This sequential calibration procedure is in addition time consuming since the same signal to noise ratio at least must be achieved on the calibration PSF compared to the science object.

In some cases, such as stellar clusters, the science object frames may also include in the field stars that one would like to use as calibration PSFs. Though, the anisoplanatism effect (Sect. 3.3) will lead to different SRs between the science object and these calibration PSFs.

AO provides the possibility to avoid these drawbacks by reconstructing the PSF of the scientific observations using the real time WFS data. In the following sections, we introduce the different "PSF reconstruction" algorithms developed till now. We invite the readers to look at the corresponding references for more details on the algorithms and/or their results.

4.2 Previous PSF Reconstruction Algorithms

PUEO/CFHT

PSF reconstruction has been initiated by Véran et al. [9] who have develop such an algorithm for PUEO, the CFHT curvature AO system. It is made of a 19 actuator DM and 36 modes are controlled.

Few assumptions (high bandpass of the system and not too noisy realtime data) translate into a $V \approx 14$ limit magnitude and few computations, such as those tackling the noise and the aliasing, are dedicated to curvature WFS. Reconstructed PSF are routinely delivered at each AO observations.

ADONIS/La Silla 3.6 m

A first attempt to transpose this algorithm to a Shack-Hartmann (SH) WFS has been undertaken by Harder & Chelli [6] who have written a PSF reconstruction algorithm for ADONIS, the AO system of the ESO La Silla 3.6 m telescope. The AO system is made of a 7×7 SH WFS and a 52 actuator DM.

Table 1. Results of PSF reconstruction for ALFA by Weiss [10]

Guide Star	$m_V=7.14$		$m_V=13$	
	Observed PSF	Estimated PSF	Observed PSF	Estimated PSF
Strehl ratio (%)	45.7±2.0	47.6	13.2±1.9	13.3
FWHM (")	0.14±0.01	0.13	0.24±0.02	0.22

Few tests showed that the reconstruction was performing well under good seeing conditions, with an error in the OTF estimation below 20% at medium and low frequencies, but poorly under bad seeing conditions, with an error in the OTF estimation between 20% and 30% at low and medium frequencies. Though, a variable residual aberration in the first Airy ring affects the reconstruction in almost all cases and limits its accuracy.

ALFA/Calar Alto 3.5 m

Weiss [10] has written a PSF reconstruction software for ALFA, the Calar Alto 3.5 m telescope AO system. ALFA is made of a SH WFS in a keystone geometry with 7 or 28 subpupils and a 97 actuator DM.

This software is adapted from the Véran et al. [9] algorithm to reconstruct on-axis PSFs and can make use of SCIDAR data to reconstruct off-axis PSFs. Results of few tests are reported in Table 1.

Lick Observatory 3 m Shane Telescope

Fitzgerald [3] has written a PSF reconstruction software for the AO system of the Lick Observatory 3 m Shane telescope. This AO system is made of a 40 subpupil SH WFS and a 61 actuator DM. Few tests have been performed and showed a general trend to overestimate the SR.

Altair/Gemini North

Jolissaint et al. [7] have written a PSF reconstruction software for Altair, the Gemini North AO system. Altair is a 4-quadrant 110 subpupil SH WFS with a 177 actuator DM. 128 modes are controlled. The only test we are aware of has lead to a SR of the estimated PSF of 0.13" when the SR of the observed PSF was 0.15".

4.3 NAOS PSF Reconstruction Algorithm

NAOS is a 14×14 or 7×7 subpupil SH WFS with a 185 deformable mirror. It is installed at VLT on the CONICA camera.

Even if no dedicated algorithm had then been written, the development of PSF reconstruction software for NAOS had been considered from the design of the system:

- in the NACO image fits files are attached two covariance matrices that can be used in a PSF reconstruction software following the algorithm of Véran et al. [9],
- few atmospheric parameters are estimated by the AO system and written in the NACO image fits file header, in particular r_0 which is indispensable for any PSF reconstruction software,
- a global noise of the WFS is estimated, written in the NACO image fits file header and can be used by a PSF reconstruction software.

When we have started to write a PSF reconstruction software for NAOS, we first contemplated to adapt the ALFA software. Though, the large differences between the two systems (different modes: KL vs. proper NAOS modes, different wavefront-related measurements available for the reconstruction) lead us to develop a piece of software dedicated to NAOS.

This piece of software is described in Clénet et al. [1]. We won't enter into details. Though we would like to stress that:

- it follows a modified Véran et al. [9] algorithm: the so-called "U_{ij} functions" whose number is proportional to the square number of mirror modes, which are computed once for all, stored before the reconstruction and read during the reconstruction, are replaced by new functions ("V_{ii} functions") which are much less numerous (proportional to the number of mirror modes) and computed on the fly during the reconstruction. This algorithm is then much faster than the Véran et al. one and prevents from the storage of a large amount of data (Gendron et al. [5]),
- the aliasing, which is the reconstruction at low frequencies of the high frequency spectrum of the phase, is not taken into account yet, but will soon be implemented.

The result of a first test, performed on a bright star, is shown in Fig. 5. The SR of the estimated PSF is 39% and the SR of the observed PSF is 32%. This overestimation is certainly due to the aliasing, which is not taken into

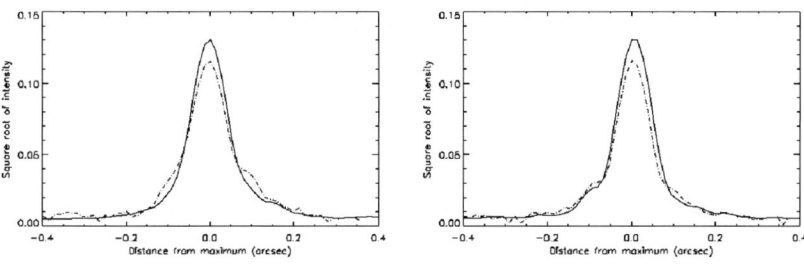

Fig. 5. X- (*left*) and Y-cuts (*right*) of the observed (*dashed line*) and estimated (*solid line*) PSFs, in a square root scale

account in the reconstruction. In addition, part of the difference of the two PSF profiles may be explained by:

- residual static aberrations: the static aberrations are corrected by using a fibre image when an image of a "true" star would probably do better,
- a faint ($\Delta m \approx 3.5$) companion of the observed star: located at PA $\approx 157.4°$ and d ≈ 0.49", this companion may affect the WFS measurement and then the reconstruction.

5 Conclusions

The spatial and temporal variability of the AO PSF, mainly due to the variable atmospheric conditions and to the partial AO correction, is a strong limit to an accurate PSF calibration. PSF reconstruction, which makes use of real-time wavefront-related data, can overcome this limitation and several developments of such software have now been undertaken for various AO systems, including NAOS at VLT. An important next step in this development will be to deliver, in a user-friendly way, these algorithms to the community.

We would like to end by emphasising the need for a specification on the PSF reconstruction accuracy. Such a specification could come for example from photometric accuracy specifications in the final post-processing stage.

References

1. Y. Clénet, M. Kasper, E. Gendron, T. Fusco, G. Rousset, D. Gratadour, C. Lidman, O. Marco, N. Ageorges, S. Egner: SPIE **6272**, 122 (2006)
2. Y. Clénet, D. Rouan, D. Gratadour, O. Marco, P. Léna, N. Ageorges, E. Gendron: A&A **439**, L9 (2005)
3. M. Fitzgerald M.: In "The 1st Victoria Workshop on AO-PSF reconstruction", ed. L. Jollissaint, J.-P. Véran, J. Christou, T. Rimmele, http://cfao.ucolick.org/meetings/psfreconstruction (2004)
4. T. Fusco, G. Rousset, D. Rabaud, et al.: J. Opt. A: Pure & Appl. Opt. **6**, 585 (2004)
5. E. Gendron, Y. Clénet, T. Fusco, G. Rousset: A&A **457**, 359 (2006)
6. S. Harder, A. Chelli: A&AS **142**, 119 (2000)
7. L. Jollissaint, J.-P. Véran, J. Marino: SPIE **5490**, 151 (2004)
8. F. Rigaut, M. Sarazin: In *Astronomy with adaptive optics: present results and future programs*, ESO Conference and Workshop Proceedings, vol. 56, ed by D. Bonaccini, p. 383
9. J.-P. Véran, F. Rigaut, H. Maître, D. Rouan: J. Opt. Soc. Am. A **14**, 3057 (1997)
10. R. Weiss, PhD Thesis, Naturwissenschaftlich-Mathematische Gesamtfakultät der Universität Heidelberg, Germany (2003)

Discussion

R. McDermid: PSF reconstruction offers an important method to provide PSF estimates in the science field. What are the implications of having multiple guide sources?

Y. Clénet: If you are in a "star oriented" configuration you will have a WFS for each guide star. So it will be the same as with one guide star, for the on-axis case. If you are in a "larger oriented" configuration you will have a WFS for each connected turbulence layer. This will have to be taken into account in the algorithm, in a way that I do not know yet.

N. Ageorges: In your test to measure the Strehl ratio, filtering out the noise in your "reduced" image (ima1-ima2) will reduce your noise on median measurement (and if not at least that will be cleaner because your background/sky measurement is presently dominated by noise).

Y. Clénet: Yes, but the effect in the behaviour of the Strehl ratio follows the behaviour of the median. So it comes only from the median computation. The behaviour of the median curve is not linked to the noise. Then neither the behaviour of the Strehl-ratio curve.

D. Osip: For your sample exposure pair, you analysed the Strehl measured for half the frame using different background subtraction and ended up with an unknown result as to which Strehl was correct. What happens if you analyse the other half of the frame? Do you get the same result?

Y. Clénet: I haven't tried but I expect the result to be the same, with possibly a bias in the Strehl value. The effect I have shown on the first Strehl curve comes from the way the background is computed so the effect must be the same on the second half of the frame.

Probing µ-arcsec Astrometry with NACO

A. Seifahrt[1,2], T. Röll[2], and R. Neuhäuser[2]

[1] ESO, Karl-Schwarzschild-Strasse 2, 85748 Garching, Germany;
aseifahr@eso.org
[2] Astrophysikalisches Institut und Universitäts-Sternwarte (AIU), Jena, Germany

Abstract. Relative astrometric measurements with a precision far better than 1 mas (milli-arcsec) are commonly regarded as the domain of interferometry. Pioneering work by Pravdo and Shaklan [8], made in the optical, reached a precision of 150 µarcsec in direct imaging but is ultimately limited by atmospheric turbulence and differential chromatic refraction (DCR) effects. Neuhäuser et al. [6, 7] demonstrated that AO assisted observations with NACO in a near-infrared narrow band filter allow measurements with a precision of ∼50 µas (micro-arsec) on a 0.6 arcsec binary within one hour and are unaffected by DCR effects. This opens new possibilities for astrometric detections of extrasolar planets and the determination of their true masses. We discuss here how to improve the measurements and address the necessary calibrations.

1 Motivation and Introduction

The search for extrasolar planets and their physical characterization has become an important field in today's astronomy and astrophysics. Dedicated instruments, such as HARPS on the ESO 3.6 m telescope on La Silla were designed for precise radial velocity (RV) measurements to find the fingerprints of extrasolar planets, imprinted in the RV signal of their host stars. With this method more than 200 extrasolar planet candidates have been found. *Candidates* – because their masses are determined only to a lower limit, $m \sin i$. The inclination angle of the extrasolar planets are not constrained from RV measurements and the true mass of the planets is unknown.

Only two methods can add enough information to allow the determination of the inclination angle and thus the true mass of an extrasolar planet. In case of a nearly edge-on viewing geometry the planet can be seen in the lightcurve of the host star as a transit. However, such events are rare. Only astrometric measurements of the induced positional *wobble* of the host star can solve for the inclination of the system in all other cases.

Moreover, searches for extrasolar planets by astrometry are sensitive to planets in wide orbits, in contrast to the RV technique. A number of target classes unsuitable for RV measurements, such as early-type stars and ultra-cool dwarfs, as well as fast rotating and otherwise active stars are also unproblematic for astrometry.

Astrometric measurements have the only intrinsic limitation that a large number of suitable reference stars have to be in the field of view (FOV) to define a local restframe. This restricts observation to fields of high stellar density, e.g. to low galactic latitudes. In addition the reference stars themselves have to have known proper motion and parallaxes. Moreover, chromatic refraction effects as well as atmospheric turbulence limit the achievable precision. Thus, the only successful observations of the astrometric wobble of exoplanet host stars have been achieved from space, namely for GJ 876 b, 55 Cancri d and ϵ Eridani b, using the Fine Guiding Sensor (FGS) of the Hubble Space Telescope, see e.g. Benedict et al. [1].

Neuhäuser et al. [6, 7] proposed the use of adaptive optics (AO) in the near infrared to (a) work in a regime where DCR effects are much smaller than in the optical and (b) to suppress the atmospheric turbulence. Since the usable field of view of AO assisted measurements is limited to the isoplanatic angle, measurements have to concentrate on physical binaries or multiple systems where the astrometric wobble of one component is measured relative to the other component(s). With this approach the problem of relative motion of the reference stars is solved since parallax and proper motion are identical for both components, leaving only the orbital motion as an open parameter. First measurements, conducted with NACO on the VLT in December 2004 and October 2006 led to unprecedented precision in ground based astrometry.

2 Results from the Feasibility Study

A first feasibility study, started in 2004, concentrated on the binary systems HD19994 and HD19063. The brighter component in HD19994 has a known exoplanet candidate as RV measurements by Mayor et al. [3] revealed. The expected astrometric signal is at least 131 µas, when assuming the minimum mass $m \sin i$ as true mass and adopting the reported excentricity of the orbit.

Astrometric measurements with NACO were conducted in December 2004. The technique for high precision astrometry is based on a principle used in RV measurements. If the resolution and sampling of a measurement is not sufficient, statistics over many independent measurements have to be used to reach the necessary precision. In the RV technique, measurements over many hundred spectral lines provide the necessary statistical basis. For astrometric measurements a high number of images have to be taken. Separation measurements in each individual frame are checked for a Gaussian distribution and the error of the mean separation can be computed from the standard deviation of the mean divided by the square root of the number of frames. From 120 frames of HD19994 and 60 frames of HD19063 that passed the statistical tests, a precision in the separation of the binary components of $\sim$92 and $\sim$50 µas, respectively, was reached within 1 h. This marks the best relative astrometric precision ever achieved with a single aperture telescope from the ground.

3 Improvements: NACO Cube Mode

Each individual frame was taken through a narrow-band filter and the exposure time was set to the minimum DIT. Hence the measurements in 2004 where fully overhead dominated. Narrow band images with the fine pixel scale of NACO's S13 camera are read-noise limited up to several minutes of exposure time. Hence no jitter observations are necessary for sky subtraction and frames can be taken in staring mode. Thus, we used the cube mode of NACO in our second observation campaign in October 2006 to obtain more frames per given time than in autojitter mode. The chip had to be windowed to handle the high data rate. The readout overheads are strongly suppressed and several thousand frames can be taken per hour. Compared to the results of the previous run, we improved our efficiency by more than a factor 50.

4 Calibration Issues

The precision achieved here marks a breakthrough in ground based relative astrometry. It demonstrates that measurements in the separation of a close binary with a precision to about 4/100,000 are possible with NACO. This raises the question on the calibration of such measurements. In the scheme presented here, we do not need to provide an absolute calibration of the pixel scale. We are aiming to determine relative changes in the separation of a stellar binary of about 1/100 of a pixel, but not its absolute value. Thus, the uncertainties in the determination of the absolute pixel scale, typically of the order of 4/1000 (see [2, 5]) are not the dominating error source, especially since we are not aiming for high accuracy but for high precision. Instead we have to assure the stability of the pixel scale (or reversely the f-ratio of the imager) to better than 4/100,000. The typical calibration sources, HIPPARCOS binaries, have precise coordinates but the uncertainties in their proper motion multiplied by the time since the HIPPARCOS epoch of 1991.25 rule out these sources as calibrators on the needed level of precision. The same holds for the SiO masers in the galactic center. Even though these objects have possibly the best known astrometric properties (from VLBA measurements), they do not fit into the FOV of the S13 camera.

Hence, we have to look for an intrinsically stable reference system that would fit into the FOV of the S13 camera. After a concise literature study, we selected 47 Tuc as a reference system. Forty-Seven Tuc is an old globular cluster with a known and small velocity dispersion. McLaughlin et al. [4] recently determined the two dimensional velocity dispersion in a spherical region around the cluster core within a 20 arcsec radius. For the red giant stars a value of $\sigma_\mu = 0.631^{+0.020}_{-0.025}$ mas/year was reported. To determine the intrinsic astrometric stability, we have measured all separations between any two stars in one sub-field of 47 Tuc for the first time in October 2006 over a large number of frames and we will monitor these quantities in the next observing campaigns.

The relative stability of the pixel scale, obtained with measurements of 47 Tuc depends on the number of stars in the observed sub-field and their mean separation. Due to error statistics, the standard deviation of the separation measurements can be divided by the square root of the number of independent separations, which is $n-1$ with the number n of stars. This relation has been confirmed by a Monte Carlo simulation, tested on a Gaussian distributed field of 20 stars as well as on our true measurements that have the same field density. Adopting a random velocity dispersion of 630 mas/year (hence, no symmetry), the stability of the pixel scale can be assured with a precision of $\sim 60\,\mu$as for one year. Hence, the intrinsic relative stability of the reference system is mainly limited by the velocity dispersion and the number and geometry of the stars within the sub-field.

We conclude that high resolution astrometry with NACO is neither limited by the atmosphere nor by the instrument itself but by suitable reference objects with a high intrinsic stability. This is true as long as the stability of the pixel scale on a level of a few in hundred thousand can not be implied but has to be confirmed and monitored with on-sky measurements.

References

1. G.F. Benedict, et al.: AJ **132**, 2206 (2006)
2. G. Chauvin, A.-M. Lagrange, C. Dumas, B. Zuckerman, D. Mouillet, I. Song, J.-L. Beuzit, P. Lowrance: A&A **425**, L29 (2004)
3. M. Mayor, S. Udry, D. Naef, F. Pepe, D. Queloz, N.C. Santos, M. Burnet: A&A **415**, 391 (2004)
4. D.E. McLaughlin, J. Anderson, G. Meylan, K. Gebhardt, C. Pryor, D. Minniti, S. Phinney: ApJS **166**, 249 (2006)
5. R. Neuhäuser, E.W. Guenther, G. Wuchterl, M. Mugrauer, A. Bedalov, P.H. Hauschildt: A&A **435**, L13 (2005)
6. R. Neuhäuser, A. Seifahrt, T. Roell, A. Bedalov, M. Mugrauer: *IAU Symposium*, Vol. **240**, p. 261 (2006)
7. R. Neuhäuser, A. Seifahrt, T. Roell, A. Bedalov, M. Mugrauer: MNRAS, submitted (2008)
8. S.H. Pravdo, S.B. Shaklan: ApJ, **465**, 264 (1996)

Discussion

N. Ageorges: The pixel scale of on-sky orientation of NACO are checked regularly by Paranal (for all cameras).

A. Seifahrt: Thanks for noticing that. I surely appreciate these efforts. However, I'd like to encourage the Observatory to better advertise these calibrations and make the raw data and results available and/or better "visible" to the user.

R. McDermid: What is the priority of ongoing efforts to deliver PSF estimates in the science field, coming from PSF reconstruction, and models of atmospheric turbulence?

N. Ageorges: Work is ongoing with NACO to characterize the PSF reconstruction quality, atmospheric monitoring (MASS, DIMM) and so on. Hopefully, this will yield results in a few months, and presumably feedback to the ESO community.

Part VII

Session 6: Integral Field Spectroscopy

IOT Overview: Integral Field Spectroscopy

P. Amico[1], S. Bagnulo[1], C. Dumas[1], C. Izzo[2], G. Marconi[1], and C. Melo[1]

[1] ESO, Alonso de Córdova 3107, Vitacura, Casilla 19001, Santiago 19, Chile;
   pamico@eso.org
[2] ESO, Karl-Schwarzschild-Strasse 2, 85748 Garching, Germany

Abstract. Integral Field Spectroscopy at the La Silla Paranal Observatory (LPO) is offered in three flavours:

1. VIMOS on UT3 is equipped with an integral field unit made of 6400 fibers operating in the 390–1000 nm wavelength range. Two spatial scales per fiber are available: 0.67" and 0.33"; they correspond to fields of view from 13"×13" up to 54"×54" depending on the chosen spectral resolution (R=200–2500).
2. FLAMES-GIRAFFE on UT2 allows the observation of up to 15 targets simultaneously in the 370–940 nm wavelength range: it uses 30 small fiber bundles, each covering 2"×3" on the sky and positioned in the 25' diameter circular field of view. Alternatively, FLAMES-ARGUS uses a fiber bundle located in the centre of the FLAMES field of view, covering 6.6"×4.2" on the sky. In both cases the spectral resolution $R = 11{,}000\text{--}39{,}000$.
3. SINFONI is a near-infrared (1.1–2.45 µm) integral field spectrograph fed by an adaptive optics module, currently installed at the Cassegrain focus of UT4. The spectrograph operates with four gratings in J, H, K and H+K ($R = 2000\text{--}4000$) and three different scales (0.25", 0.1" and 0.025"/pixel), corresponding to a field-of-view of 8"×8", 3"×3", or 0.8"×0.8" respectively. SINFONI can also be used in seeing limited mode (no AO) and, starting April 2007 is offered for use with an artificial laser guide star, which will increase the sky coverage dramatically.

In this paper we review the use and science applications of these instruments, their standard calibration and maintenance operations plans and the most important calibration issues affecting science.

1 Introduction

Traditional spectroscopy is based on dispersing the astronomical light that enters a slit or a fiber (or several slits or fibers, in the case of multi object spectroscopy) of a spectrograph. Both slit-spectrographs and (traditional) fiber-fed spectrographs are commonly used for observations of point-like sources. Extended objects may be more conveniently observed with spectrographs equipped with an extended slit, which allows obtaining the spectra from a target slice. However, if one is interested in the spectrum of the entire target, it will be necessary to step in position across the object until the whole field of interest is covered, which is a very time consuming procedure.

This problem is solved by Integral Field Spectroscopy (IFS), which produces spatially-resolved spectra – simultaneously – over a two-dimensional field, and generates data cubes of flux density as a function of RA and DEC coordinates in the field and the wavelength. Integral Field Spectrographs (IFSs or IFUs, IF Units) allow a more robust wavelength calibration with skylines than slit models. If the target is a point-like source, one obtains the spectra of the object and the adjacent sky, which are then subtracted. With extended objects, one can obtain spectra from the slice of the target cut by the slit and measure the sky on an offset position. In both cases, the object might not be perfectly centred in the slit, thus introducing a small wavelength shift. Additional advantages of IFS are: slit losses are eliminated,[1] the actual target position is available from the data by reconstructing an image and, in the case of relatively bright objects, it makes the reconstruction of mosaic sequences easier. Finally, from an operational point of view, target acquisition is easier with IFS techniques, since there is no more need of accurate centering of the object in a narrow slit.

IFS techniques are particularly suited for observations of single *small* individual objects. The typical science applications include: solar system and extra-solar planets, star formation and resolved stellar populations, emission line regions (galactic nebulae and extended emission in galaxies), centres of galaxies (active galactic nuclei and Milky Way galactic centre), nearby galaxies (kinematics and chemistry), (lensed) Distant Galaxies (kinematics and chemistry), galaxy clusters, compact HII regions and high-z galaxies.

Given the high efficiency and the wide variety of scientific applications, Integral Field Spectrographs are becoming more popular and most observatories are equipped with at least one of them. The La Silla Paranal Observatory has three instruments of this kind, two optical spectrographs with integral field capabilities, VIMOS and FLAMES-GIRAFFE, and a *pure* infrared integral field spectrograph, SINFONI.

IFSs are based on a classical spectrograph fed by either one of the following equipments:

1. **Lenslet arrays.** The field is imaged on a microlens array, which sends images of the telescope pupil onto the dispersing element. This design is not currently used at the LPO.
2. **Fiber bundles.** The image is formed on fibers bundles (using arrays of lenslets), which bring the light to the slit of the spectrograph. The fibers allow flexible reformatting of any field shape (e.g. round) into a slit, so that the spectra are aligned in wavelength. Both VIMOS and FLAMES-GIRAFFE use this technique in their IFUs.

---

[1]This is true for image-slicers models. Fiber based IFUs may suffer from spatial degradation when compared to longslit ones, due to the smearing of light across adjacent fibres. Light losses are usually limited, depending how the fibres are assembled together. In the case of VIMOS the light losses are around 10%. Image slicer IFUs do not suffer from any loss or degradation.

3. **Image slicer.** The input image is formed on a mirror made of thin horizontal sections, askew from each other, which send the portion of the image in different directions. A second segmented mirror reformats the slices so that they are now laid out end to end to form the slit of the spectrograph. SINFONI uses an image slicer. The advantage of the slicer over the fibers bundle model is that it covers the field of view with no gaps. It is also particularly suited to infrared observations since it uses mirrors.

2 Integral Field Spectrographs at the VLT

The Resolution (R) versus wavelength chart (Fig. 1) shows the options available to the ESO users who want to take advantage of IFS. In the near infrared (NIR), SINFONI is the only choice: it offers relatively high resolution, up to 4000, higher than what is possible with the other NIR instruments (ISAAC and NACO in long slit spectroscopy mode) for the same wavelength range. Although the field of view (FoV) is small, it is well matched to the science applications in non AO mode (FoV 8"×8") and in AO mode (3"×3" and 0.8"×0.8").

In the optical, the choice between VIMOS and GIRAFFE depends primarily on the nature of the objects to be observed and the desired spectral resolution. For any given wavelength, the ARGUS and IFU components of FLAMES-GIRAFFE deliver a much higher resolution than the VIMOS IFU. The latter produces spectra of a contiguous area of up to 54"×54" on sky, which is almost a factor of 5 larger than that of ARGUS and therefore suitable for relatively large objects (e.g. galaxies). The 15 deployable mini-IFUs of FLAMES cover each a small 2"×3" area on the sky, and are therefore suitable for small objects, but can be independently positioned anywhere in a 25 arcmin FoV.

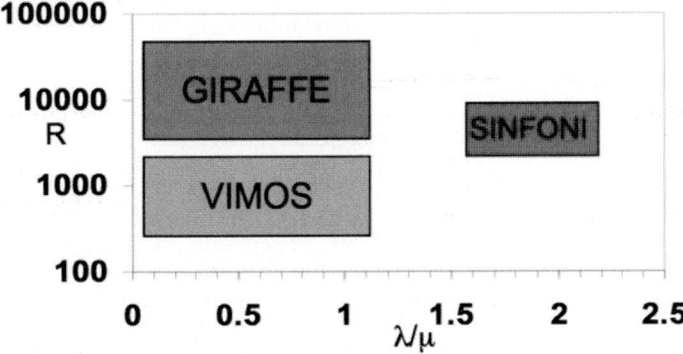

Fig. 1. Resolution versus wavelength chart for GIRAFFE, VIMOS and SINFONI

3 VIMOS

VIMOS [1] is the Visible Multi-Object Spectrograph for the Very Large Telescope (VLT) of the European Southern Observatory (ESO). This instrument has been built to provide the ESO community with a wide field spectrograph with high throughput and multiplex, dedicated to deep surveys. VIMOS operates in the 0.37–1.0 µm domain in three main observing modes: direct imaging (IMG), multi-slit spectroscopy (MOS), and integral field spectroscopy (IFU). VIMOS is the outcome of collaboration between ESO and several research institutes in France and Italy, under the responsibility of the Laboratoire d'Astrophysique de Marseille (CNRS, France). Its spectroscopic modes achieved first light in March 2002 and was offered to the community shortly thereafter.

The VIMOS Integral Field Unit is made of 6400 (80×80) fibers, coupled to microlenses. Six grisms are available in IFU mode. The grisms are the same as used in MOS mode, but spectral resolution and wavelength coverage are slightly different. The fiber FWHM is about 3 pixels. The spectral resolution is 1.25 times the spectral resolution corresponding to a 1" slit in MOS mode.

The field of view is a square, and the spatial sampling is continuous. The dead space between fibers/microlenses is below 10% of the fiber-to-fiber distance. Two different spatial samplings (magnifications) are available: 0.67" and 0.33" per fiber, respectively. With the low-resolution grisms (LR blue and LR red), the entire IFU FoV can be used (i.e. 80 fibers each side). The FoV is divided into four quadrants, each of 1600 fibers (40×40) that are bundled up and re-assembled at the other end into four parallel slits. These so called *pseudo-slits* include 400 fibers each and provide 4 stacked, horizontal spectra along the dispersion direction. The field of view corresponds to 54"×54" (when the magnification is set to 0.67") or to 27"×27" (when the magnification is set to 0.33"). A shutter must be used to partially mask the IFU head in case of HR and MR spectroscopy to prevent overlapping of spectra from different pseudo-slits, leaving a 40×40 fibers square. In this case, the FOV corresponds to 27"×27" or to 13"×13", and only one (central) pseudo-slit is illuminated. For more information see the VIMOS User's manual [2].

The calibration plan for the VIMOS IFU unit includes standard bias (to measure mean bias level and read-out noise), dark (dark level), arc (to calibrate in wavelength) and flat (to identify and trace the fibers and compute their relative transmission) frames. In particular, arc and flat frames are taken at night immediately following a science observation. All the calibrations, with the exception of standard stars, are taken with internal lamps.

Observations of spectro-photometric standard stars with the same settings used for science are normally performed within a few nights from the observations. Additionally routine calibrations are taken by the operation staff for the purpose of maintaining and monitoring the instrument configuration and performance. Calibration monitoring data can be found on the pages of the Quality Control group [3]. For the IFU unit, quality control monitors the following parameters:

- IFU Stability: the X displacement of the IFU spectra at a reference row is measured. The displacement may be due to IFU mask play, grism alignment variations and instrument flexure.
- IFU Dispersion: monitors the RMS of dispersion solution as measured by the pipeline.
- IFU Flux: monitors the efficiency of lamps and optical components.
- IFU grism alignment.

Nighttime calibrations are executed with the rotator in exactly the same position as at the end of the last science exposure. The telescope is moved to the Zenith to safely close the Nasmyth shutter without moving the rotator. This procedure ensures that the instrument flexures are as close as possible to those affecting the science exposure(s) but it is significantly time consuming: it takes from 5 to 8 min. Moreover, even when identical OBs are executed consecutively, presets must be repeated.

Fringing in the frames significantly affects IFS science with VIMOS. It is usually detected at 10% level, does not strongly depend on instrument rotation, it is independent of the instrument setup (filter/grism combination) and it seems to be related to the IFU mask unit movement. At difference with *classical* detector fringing, it is seen also at blue wavelengths, red-wards of 4500 Å. Analysis of the data has shown encouraging results for fringes elimination by means of sinusoidal fitting techniques [4].

A major impact on IFS science comes from instrument flexures. In Fig. 2 the amount of flexures along the x-direction is shown for quadrant 2. The dependency on the rotator positions is particularly significant in all those cases when the scientific target is close to zenith (e.g. declination $\delta \simeq -30°$): as shown in Fig. 3; in this case, the rotator moves fast, from PA=30° to PA=130° in 1 h. For example, the amount of flexures measured in the second quadrant (Q2) in Fig. 3 between 30 and 130° corresponds to a total of 4 pixels, enough to markedly smear the spectra on the detector.

It is extremely important that astronomers who plan an observation with the IFU unit, design their observational strategy preferring targets with *favourable* declinations (i.e. avoid objects too close to zenith) and keeping the total exposure times short enough to avoid extreme flexures. Note that, in *service* mode observations at ESO, it is not possible to foresee the time at which an observation is executed. However, it is common practice with VIMOS to keep the hour angle small.

Additional discussion on calibration effects and their impact on science can be read in two papers presented at the conference [5, 6].

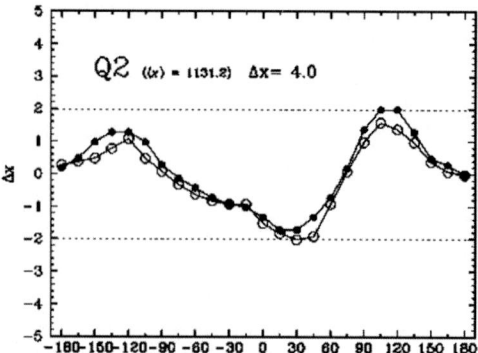

Fig. 2. Amount of flexure measured in quadrant 2 (Q2) along the detector x-direction as a function of the rotator angle. Flexures are measured taking images of a pinhole mask at different position angles of the rotator. Flexures are rotator angle dependent, and suffer from hysteresis. Fiber tracing would be impossible were flats not taken immediately after the science frames

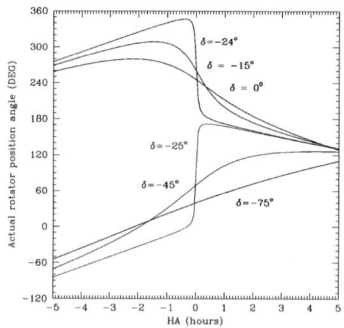

Fig. 3. Variation of the instrument's rotator position angle at Nasmyth B as a function of the hour angle

4 FLAMES-GIRAFFE

FLAMES [7] is the multi-object, intermediate and high-resolution spectrograph of the VLT. Mounted at the Nasmyth A platform of UT2, FLAMES was built and assembled in about four years through an international collaboration between ten institutes in six countries and three continents. First light for the various components was achieved between April and July 2002.

FLAMES can access targets over a large corrected field of view (25 arcmin diameter). It consists of three main components: (1) a fiber positioner (OzPoz) (2) a medium-high resolution optical spectrograph, GIRAFFE, with three types of feeding fiber systems: MEDUSA, IFU, ARGUS. The latter two provide the IFS capabilities. (3) A link to the UVES spectrograph (Red Arm) via 8 single fibers of 1" entrance aperture.

The IFU mode of the instrument consists of 15 deployable mini-IFUs, each one made of a rectangular array of 20 square microlenses with an aperture of 3"×2". The mini-IFUs are complemented by 15 single-fiber IFUs for measurements of the sky background. The ARGUS–IFU is a stationary unit positioned in the centre of the FoV, with a 22×14 microlens array feeding 301 fibers, which are re-arranged into a pseudo-slit fed to the GIRAFFE spectrograph. In addition, five sub-slits contain an additional calibration fiber fed by the GIRAFFE calibration box. The centre-to-centre distance between the fibers in the sub-slits is only 1.47 times the fiber diameter core, which implies that the contamination between adjacent fibers is rather high (about 10%); in any case, an even higher level of contamination is always present at the fiber entrance level in normal observing conditions.

Only a single ARGUS–IFU is available and is fixed to the centre of Plate 2 of the Fiber Positioner. Fifteen deployable single-fiber IFUs are available for sky measurements in the ARGUS mode. The pre-optics of the ARGUS–IFU offer two different magnifications of the focal plane. The scale 1:1 projects 0.52"×0.52" on each microlens resulting in a total aperture of the IFU of 11.5"×7.3"; with the scale 1:1.67 each microlens corresponds to 0.3"×0.3" and the total aperture is 6.6"×4.2".

For both the IFU Unit and ARGUS the calibration plan includes the following: bias and dark frames for detector performance monitoring, arc frames for wavelength calibration, dispersion relation and monitoring of the resolution, robotic flats and flux standards with attached screen flats. The robotic flats are taken with the calibration lamp on the robot arm and are used to determine the fiber location (which is needed for the subsequent data reduction), to correct for fiber-to-fiber efficiency, to flat field the 1-D extracted spectrum, etc. Nighttime calibrations include screen flats taken with the Nasmyth screen illuminated, which has the advantage of providing uniform illumination and therefore allows better determination of the fiber-to-fiber transmission factors. The flats are usually attached to standard stars observation, used for flux calibration. The standard star is directly measured for a few central fibers only and then extrapolated to the other fibers using the fiber-to-fiber efficiency function. In the particular case of ARGUS the screen flats are fundamental, since they provide a better, more uniform, illumination than the robotic flats. However, the screen flats cannot provide much flux at blue wavelengths. Sky flats are a better choice, but are not yet totally supported at the LPO: sky flats are taken with ARGUS only when any of the 3 bluest setups are used. More information on the instrument's calibration plan and the quality control parameters can be found in [8].

5 SINFONI

SINFONI [9, 10] is an Adaptive Optics assisted cryogenic near infrared spectrograph developed by ESO and MPE (Max Planck Institut für Extraterrestrische Physik) in collaboration with NOVA (Nederlandse Onderzoekschool Voor Astronomie). It was commissioned between June and August 2004 at the Cassegrain focus of VLT UT4 (Yepun). The instrument was offered to the community from April 2005 onwards as a unique facility in the field of high spatial and spectral resolution studies of compact objects (star-forming regions, nuclei of galaxies, cosmologically distant galaxies, galactic centre etc.). The instrument was conceived as the combination of an Adaptive Optics facility (the MACAO AO-Module), developed by ESO, and SPIFFI (SPectrometer for Infrared Faint Field Imaging), a Near Infrared Integral Field Spectrograph developed by the MPE.

The SPIFFI IFU is an image slicing design, which uses a series of mirrors to slice the field of view into 32 sections, which make a single long pseudo-slit. Each slice is imaged on 64 detector pixels. Thus, this design provides a square grid of 32 by 64 rectangular *spaxels* (spatial elements, each 1/2 by 1 the slice width).

SINFONI can be operated in both adaptive optics (AO) assisted mode or in no AO mode, the latter using only the largest scale. As of April 2007, the instrument will also be offered together with the Laser Guide Star Facility (LGSF), which will allow to perform AO corrected observations over a much larger sky area than that possible in natural guide star mode. For more information on the instrument refer to the SINFONI User's Manual [11].

The instrument's pipeline takes the 2-D spectra generated by the instrument, performs sky subtraction[2] and reconstructs the 3-D cube, using dedicated algorithms.

From a calibration point of view, SINFONI is a particularly stable instrument, thus not requiring complex data and making it a particular good candidate for an automatic data reduction pipeline. For examples, flexures due to the varying gravity load of a Cassegrain instrument, are compensated in real time by means of control loops. Filter offset compensation is also performed in a transparent way for the users.

The calibration plan of SINFONI defines the default calibrations obtained and archived by the Paranal Science Operations without being explicitly requested by the astronomers. In a nutshell, the observatory provides daily standard calibration frames: dark frames (to derive bad pixel maps), arc frames (for wavelength calibration), internal lamp flat fields (to determine pixel to pixel variation and large scale illumination) and telluric star calibrators (to allow the correction for the Earth's atmosphere telluric features). In addition, the observatory provides monthly characterization of the

---

[2] Often residual sky lines are still present in the pipeline reduced spectra. A discussion on how to improve the OH emission lines can be found in [12].

detector characteristics (gain, read-out noise, linearity, etc). The quality control groups analyses the calibration data and constantly monitors the health of the instrument. In particular, it derives RON, dark current, linearity, distortion maps, resolving power, and fixed pattern noise. Values and trend plots for these parameters can be found in the QC web pages for SINFONI [13].

In order to perform an accurate image reconstruction, the pipeline uses special NORTH/SOUTH frames (series of images of the calibration fiber moved along the y-direction over the image slicer), which are used to measure the orientation of the slitlets and arc frames for precise mapping in the (x,y,λ) cube. The mapping is stable and these calibrations are only needed after hardware interventions, when the instrument is warmed up and opened.

In the particular case of extremely faint objects, such as high-redshift galaxies, where several hours of integration must be accumulated on the targets in different nights, registration of the jittered sequences of images is particularly challenging. The coordinate system provided by the guide stars catalogue is not good enough to ensure an accurate stacking of the images. The observatory recommends the observers to make use of reference stars: it is possible to save images of a relative bright star nearby the science target before and after the jitter sequence (science object + sky frames). Those images allow very precise image stacking during data reduction and eliminate the errors introduced by the telescope offsets in the jitter sequence. An additional aid to the users will come from the GUIDECAM software, soon to be distributed as part of the SINFONI observation preparation kit. This software is routinely used at the telescope and allows the operators to select guide stars. Different guide stars, possibly taken from different catalogues, introduce small offsets in the pointing, reducing the overall accuracy of the image stacking. Through GUIDECAM, the users will be able to indicate the best guide star for their project and, at the same time, will ensure that the same star will consistently be used during the observations of their targets, with significant improvement on the accuracy of pointing.

Another important calibration issue concerns the SINFONI science detector, an Hawaii-2RG produced by Rockwell (now Teledyne). This detector is known to have problems of image persistence, which is the signal that remains on a detector array after the illumination source has been removed. Persistence is affecting science taken with SINFONI, since the observation of relatively bright objects (such a PSF star or a telluric calibrator) leaves ghost images on the array. Subsequent observations of faint targets can be completely compromised by this effect. In the worst cases, when a star saturates the detector, the ghost image may remain for several hours, thus preventing any type of scientific observations or calibrations. Thorough testing by the IR detector group in Garching [14] proved that the latent charge is a function of fluence (i.e. Flux × Exptime) and that the residual charge decays with time, which is roughly proportional to the signal. It has also been confirmed that it does not depend on the instrument configuration (for example, the filter) and a slight persistence (disappearing in a short time) can even be seen at relatively low flux levels with respect to the saturation limit. At present, no known cure for the problem is available. Therefore, particular care is taken at the observatory to keep the signal level of calibrators below a threshold of 2000 counts, and to observe bright science targets only at the end of the night.

Acknowledgement. The authors would like to thank Marina Rejkuba, Dominic Naef, Reinhard Hanuschik, Ric Davies and Martin Roth for the fruitful discussions and comments.

References

1. O. Le Fevre, et al.: Commissioning and performances of the VLT-VIMOS instrument, in *SPIE*, Vol. 4841, 2003, p. 1670.
2. VIMOS User's Manual, http://www.eso.org/instruments/vimos/doc/
3. QC web pages for VIMOS, http://www.eso.org/observing/dfo/quality/index_vimos.html
4. E. Jullo, et al.: Origin and Processing of Fringing in the VIMOS IFU, these proceedings (2008)
5. M.M. Roth: Integral Field Spectroscopy with VIMOS, these proceedings (2008)
6. H. Kuntscher: A User's View of VIMOS-IFU Calibrations, these proceedings (2008)
7. L. Pasquini, et al.: Installation and commissioning of FLAMES, the VLT Multifiber Facility, ESO Messenger **110**, 1 (2002)
8. QC web pages for GIRAFFE, http://www.eso.org/observing/dfo/quality/index_giraffe.html
9. F. Eisenhauer, et al.: SINFONI – Integral field spectroscopy at 50 milliarcsecond resolution with the ESO VLT, in *SPIE*, Vol. 4841, 2003, p. 1548
10. H. Bonnet: First light of SINFONI at the VLT, ESO Messenger **117**, 17 (2004)
11. SINFONI User's Manual, http://www.eso.org/instruments/sinfoni/doc/
12. A. Modigliani, et al.: Improvements in the Residual OH Emission Removal in SINFONI Pipeline Spectra, these Proceedings (2008).
13. QC web pages for SINFONI, http://www.eso.org/observing/dfo/quality/index_sinfoni.html
14. G. Finger, et al.: Test results with 2K × 2K MCT arrays, in "Scientific Detectors for Astronomy – the beginning of a new era", Waimea (Hawaii/USA), June 16–22, 2002, eds. P. Amico, J.W. Beletic, and J.E. Beletic, Kluwer Academic Publisher, 2004, p. 497–500.

Discussion

D. Osip: Is the measured flexure for the IFU mode in VIMOS intrinsic and distinct from the "normal" VIMOS flexure or is it another example of the same flexure?

P. Amico: It is the same flexure. In this respect, the IFU mode behaves the same as the other modes.

Integral Field Spectrographs: A User's View

E. Emsellem

Université de Lyon 1, Centre de Recherche Astrophysique de Lyon, Observatoire de Lyon, 9 avenue Charles André, 69230 Saint-Genis Laval, France; CNRS, UMR 5574; ENS de Lyon, Lyon, France; `emsellem@obs.univ-lyon1.fr`

1 Introduction

We easily tend to think of Integral-Field Spectrographs (IFS) along two opposing trends: as either the beautiful combination between photometry and spectroscopy, or as our worst nightmare including the dark side of both worlds. I favour a view where each IFS is considered individually, as one instrument with specific performances which can be used optimally for a certain range of scientific programs. It is indeed true that data-wise, IFS do sometime merge the characteristics of classic (e.g., long-slit) spectrographs with annoying issues associated with Imagers. This is in fact the price to pay to access a drastically different perspective of our favourite targets. The challenge is then to provide the necessary tools to properly handle the corresponding data. However, this should certainly not be thought as something specific to IFS: such a challenge should be accepted for any instrument, and most importantly solved prior to its delivery at the telescope.

2 Specifics

Aperture and long-slit spectroscopy obviously provide a limited access to the spatial information. The spectral and spatial characteristics being mixed in the datasets (via e.g. the effect of seeing), this can sometimes severely restrict our interpretations. By acceding to the third dimension, we open the door to a wide variety of powerful treatments such as: a posteriori detailed evaluation and correction of seeing effects, atmospheric diffraction, or testing the presence of artifacts (e.g. cosmic ray impacts).

With IFS such as VIMOS (http://www.eso.org/instruments/vimos), we reached a rather impressive multiplex level, with the possibility of acquiring more than 6000 spectra covering more than 500 wavelength pixels. We are therefore nowadays reaching the limit of what we can do in a non fully automated way: defects or singularities can still be detected via a manual search or visualisation of individual (or group of) spectra, we can repeat complex analysis processes many times without worrying (too much) about the CPU cost, and processing algorithms are allowed to crash once in a while without endangering the project. When instruments such as MUSE will start gathering photons at the VLT (http://www.eso.org/instruments/muse), we will collect the equivalent of more than 100 VIMOS exposures in one single shot. For most scientific programs, there is therefore no way we will be able to systematically look at individual spectra. We will then need robust algorithms to analyse and process our data. More importantly, this implies that we should acknowledge the presence of errors in our results, which should be readily evaluated.

3 Pushing the Limits

As mentioned above, there is a strong need for analysis tools in the context of IFS data: basic software including slicing and visualisation techniques, data mining are still not felt as standard, easy tasks, even though there exists a number of successful (but not universal) pieces of software. The data format is one obstacle, only partly answered by noticeable efforts such as the one from the Euro3D consortium (see e.g., http://www.aip.de/Euro3D). We must now realise this situation may soon become untractable in a rather near future if we do not prepare for it. We then need to invent new or refine existing analysis tools to perform mosaicing, binning (Cappelari and Copin [1]), optimal summation, normalisation, smoothing (spatially and spectrally), deconvolution, in 3D. We could for example think of a drizzling (Hook and Fruchter [4]) technique generalised for 3D data.

With the full 3D information in hand, we can proceed with efficient mosaicing procedures, both spatially or spectrally (see Fig. 1), the optimisation of deep fields by taking into account the specific characteristics (resolution varying over the field, signal-to-noise, etc.) of each individual exposure, the use of super-resolution see e.g. Garcia et al. [3], or at last obtaining reasonable spectrophotometry. However, pushing the limits is also demanding. If the spatial two-dimensional ingredient of such data implies that models may then become much more constrained (Cappellari and McDermid [2]), it also means that the uncertainties in the output scientific results will even more critically depend on the errors in the dataset.

This is often, and wrongly in my opinion, seen by many as opening Pandora's box. On the contrary, it should represent an additional motivation for a robust estimate and *control* of such errors. One dangerous asset of IFS data is that they can be used to build two-dimensional maps of various quantities (e.g., emission line flux, kinematics, stellar population indicators), and in general these maps look GOOD. These nice reconstructed images are impressive but may hide severe defects in the datasets impairing the subsequent scientific interpretation. It is therefore of the utmost importance that we systematically evaluate errors associated with the signal we wish to analyse.

Noise must be propagated so that the user can keep track of it. An accurate assessment of the noise is required for any optimal stacking, binning, etc. Published measurement should always include errors bars, which in turn

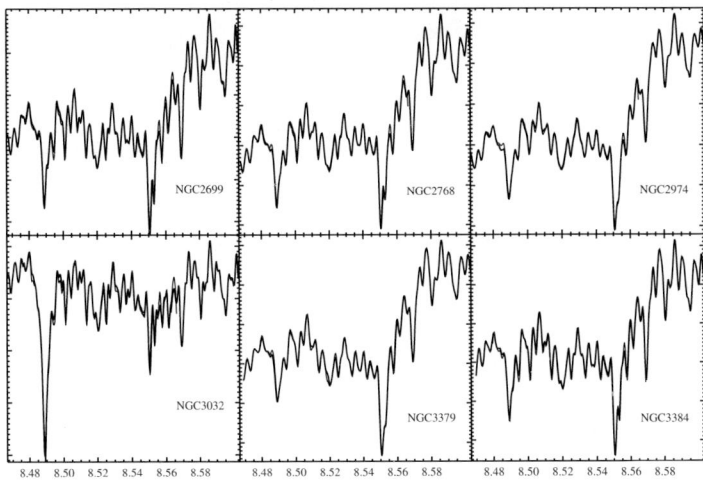

Fig. 1. Comparison of OASIS (*thin solid line*) and SAURON (*solid line*) optical spectra of the central regions of galaxies, showing an impressive agreement when specifics of each instrument are properly taken into account. Extracted from McDermid et al. [5]

depend on our ability to estimate the noise pattern in the data. Formats such as the Euro3D files do include a description of the noise (per pixel). However, it can actually only keep track of the variance (the trace of the noise matrix). Ignoring the noise covariance may sometimes be unavoidable, but this comes at a cost which should be, again, evaluated and taken into account.

4 Implications and the Fear of Resampling

We therefore need a better characterisation of our data. This calls for a good, validated calibration system, with a detailed assessment of the stability of the instrument and telescope. It also means extra caution when e.g., taking into account the contamination by stray light, and more importantly the thorough characterisation of the detectors. Subtle effects such as very low uncorrected fringes may significantly impair the science we wish to conduct (see Fig. 2).

The Data Reduction and Analysis pipelines represent another source of concern: they for example usually contain a number of steps which require a resampling of the dataset (e.g. wavelength calibration). Resampling is often seen as evil, because it may spread artifacts over several spaxels, and makes it difficult to follow the noise pattern. Indeed, resampling means that pixels, spaxels, spectra, may not be independent from each other anymore

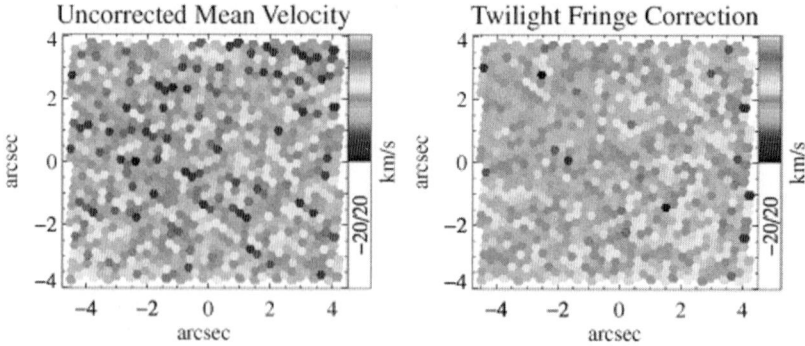

Fig. 2. Effect of fringe pattern as seen in the stellar velocity field of a twilight exposure (*left*): the "zebra" pattern disappears after proper correction of that effect. Extracted from McDermid et al. [5]

(as illustrated e.g., in the noise covariance). The spreading of artifacts also dilutes the unwanted signal, making it less likely to be detected.

Spatial resampling can usually and should be avoided. We should therefore develop an approach when we minimise the number of steps including some resampling. This is naturally done for example when we apply the correction for the trace distortions and the wavelength calibration at the same time. One way out is to always keep working with the detector pixels, and design the reduction and analysis steps via a global model approach (deriving transformation matrices for each step and combining them before applying them). This is certainly a nightmare for the software development people (and one in 3D!), but maybe not an impossible task specially for densely-packed (fiber) systems and image slicers. The other implication of such an approach is, again, to consider the data analysis tools as closely *associated* with the data reduction software, something we are for example seriously pushing in the context of the MUSE project.

We must finally redefine carefully what we do wish to call a science data product. Is flux calibration a requirement for all such datasets, and should flux calibration be a procedure applied systematically? This obviously comes with an additional operational cost, and should again be discussed long before the first mounting of the instrument behind the telescope, and not a posteriori when we realise we lack both the understanding of our instrument, and the data for that specific calibration.

5 Conclusions and Perspectives

From this very brief overview, we can already draw a number of conclusive statements which may be important in the context of calibration procedures for IFS data. Critical needs include:

- The development of the (automated?) tools to analyse the huge data sets of the coming generation of instruments.
- Keeping track of noise and systematic errors (a tricky task, but good for our scientific health).
- A proper characterisation of the instrument (including the reduction software which must be seen as *part* of the instrument).
- An adapted calibration plan, and the motivation for researchers to request telescope time for *Calibration Proposals*.
- The development of a good (parametric?) instrument model.

In fact, most of these statements are absolutely not specific to IFS and should be applied more generally to most instrument. Two final notes regarding software. Firstly, it is important to realise that the software should be tested on (and developed with) realistic data. This obviously requires the development of a good Instrumental Numerical Model, an approach which already exists for space instruments, and seems to become the rule for ground-based ones. In that context, the Data Reduction Software (DRS) and the Data Analysis Tools (DAT) should be considered/designed in a consistent way so that our understanding of the instrument and the DRS is also reflected in the optimisation of the DAT. Secondly, there is clearly a need for more coordination in the development and follow-up upgrading of data reduction softwares. As a user, I do not want to face the fact that several complete (and significantly different) versions of a pipeline exist in different places, specially when the best routines are privately owned. This requires that a strategic plan exists for the maintenance of the software, and that the time-scales for the (official) implementation of new recipes are short enough to avoid recurrent frustration on the user's side.

I would like to warmly thank Pierre Ferruit for fruitful discussions, and the SOC for inviting me to this constructive Workshop.

References

1. M. Cappellari, Y. Copin: MNRAS **342**, 345 (2003)
2. M. Cappellari, R.M. McDermid: Classical and Quantum Gravity **22**, 347 (2005)
3. P.J.V. Garcia, E. Thiébaut, R. Bacon: A&A **346**, 892 (1999)
4. R.N. Hook, A.S. Fruchter: In N. Manset, C. Veillet, and D. Crabtree (eds.), *Astronomical Data Analysis Software and Systems IX*, Vol. 216 of *Astronomical Society of the Pacific Conference Series*, pp. 521 (2000)
5. R.M. McDermid, E. Emsellem, K.L. Shapiro, R. Bacon, M. Bureau, M. Cappellari, R.L. Davies, T. de Zeeuw, J. Falcón-Barroso, D. Krajnović, H. Kuntschner, R.F. Peletier, M. Sarzi: MNRAS **373**, 906 (2006)

Discussion

R. Hanuschik: I can confirm that ARGUS standard stars are important for the GIRAFFE flux calibrations, but I am puzzled by your statement that it was impossible to find such data in your case. Usually a flux standard is taken in the same night or a few nights later. If not, the science OB would need to be repeated.

E. Emsellem: Although this could be a specific problems of a specific mode, my point was more general. What is really a science data product? Shouldn't flux calibration be one of the required steps for data reduction pipelines? Also many standard stars are just not okay: their reference flux tables are often low spectral resolution, badly calibrated in wavelength (with offsets up to 10 Å). Since the flux calibration was such a critical step for me, I wanted to insist on the fact that "standard methods/approaches" are sometimes just not adequate and we should not assume that doing what others do is just okay for our science without prior checking.

G. Marconi: I found interesting your question about closing the loop consortia/ESO-pipeline to update recipes on 2 to 3 weeks scale. Of course, we have to figure out a way to exchange at least knowledge also after the commissioning is over.

Specsim: A Software Simulator for Integral Field Unit Spectrometers

N. P. F. Lorente, A. C. H. Glasse, G. S. Wright, S. K. Ramsay, and C. J. Evans

UK Astronomy Technology Centre, Royal Observatory, Blackford Hill, Edinburgh, 3H9 3HJ, UK; npfl@roe.ac.uk

Abstract. As the scale and complexity of each generation of telescopes and their instruments increases, the requirement for a means of furthering our understanding of their properties and limitations, from the initial design to the point of commissioning also grows. An effective way of learning about the behaviour of a new system is to employ a software simulator to generate synthetic astronomical data, based on a given set of telescope and instrument characteristics. The Specsim tool has been developed to model, in software, the operation of Integral Field Unit (IFU) spectrometers, so as to give the science, engineering and operations teams responsible for designing, building and running such instruments a preview of the data products before the system is operational. Specsim generates synthetic data frames approximating those which will be taken by the instrument. The program models astronomical sources and generates detector frames using the predicted and measured properties of the telescope and instrument. These frames can then be used to illustrate and inform a range of activities, including refining the design, developing calibration strategies and the development and testing of data reduction pipelines. Specsim is currently used to model the Medium Resolution Spectrograph on JWST-MIRI, and KMOS on the ESO VLT. The software has been designed in a modular fashion, thus allowing the tool to expand easily to model future instruments, by incorporating new models into the existing infrastructure.

1 Introduction

Specsim, a software application developed in IDL[1], models the operation of Integral Field Unit (IFU) spectrometers, generating FITS frames which approximate the images which will be produced by the selected instrument. This provides the astronomical community the opportunity to study the behaviour and performance of an instrument throughout its pre-commissioning phases, and therefore be an aid to the design, build and commissioning processes.

Activities such as the design and development of data reduction and analysis software, which require an understanding of the data properties to inform the design process as well as access to datasets with which to test the software tools subsequently developed, can greatly benefit from the availability of simulated data. Similarly, the development of calibration strategies for an

---

[1]IDL is a registered trademark of Research Systems Inc. for their Interactive Data Language software.

instrument relies on the understanding of the properties of the instrument itself and their effect on the astronomical data images. Therefore the quality of calibration strategies can be improved if good data models are available.

The process of modelling IFU Spectrometer data is carried out by Specsim in two main stages: First, a model of the instrument's field-of-view is created in three dimensions (two spatial and one spectral). This model is then used to generate a detector image, using Specsim's internal model of the instrument, telescope and observing environment.

Originally, Specsim was created to model the operation of the MIRI Medium Resolution Spectrometer (MIRI-MRS) on the JWST. Following successful use in this project and interest generated in the VLT-KMOS and E-ELT design teams, Specsim has been redesigned and further developed into a general IFU spectrometer modelling tool.

1.1 VLT-KMOS

KMOS is a near-infrared multi-object IFU spectrometer being built for the ESO VLT [1]. The instrument has three identical channels. Eight IFUs, each dividing the field of view into 14 slices of 14 pixels each, form a long slit at the input of one of three spectrographs. Specsim is being used to generate synthetic astronomical sources and from these produce 2048 × 2048 pixel detector images (see Fig. 1).

1.2 E-ELT Design Studies

With a critical dependence on adaptive optics correction in this project, Specsim has applications both in refining the science case and in the integrated

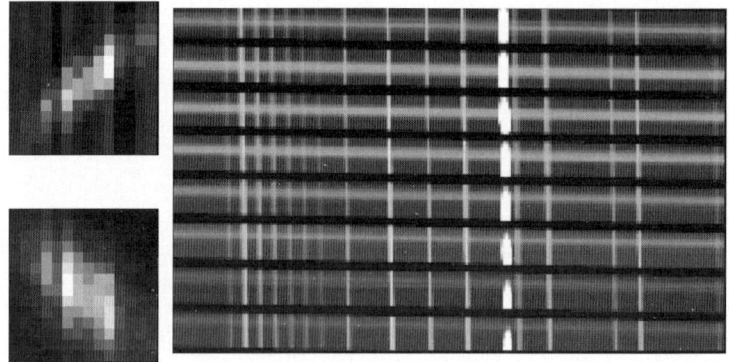

Fig. 1. *Left*, *upper* and *lower*: Two views of the Sky Model (i.e. the synthetic field-of-view) containing a galaxy as observed by the VLT KMOS instrument. The images show a galaxy in *white-light* (*top*) and at the wavelength of the emission line (*bottom*). *Right*: A section of the modelled detector frame. A bright emission line can be seen, as can the OH spectrum and faint galaxy continuum

Fig. 2. Synthetic *white-light* E-ELT images created using Specsim, each showing the modelled 2 × 2 arcsec field of view for three adaptic optics modes: no adaptic optics - i.e. the seeing-limited field (*left*), ground layer correction only (*center*) and laser-tomography (*right*). The model PSFs (provided by Miska Le Louarn - ESO) correspond to K-band observations taken by a 42 m E-ELT, with 0.8″ seeing

modelling of the instrumentation. In the present design study, Specsim is used to investigate the effect of various levels of adaptive optics on image quality. As shown in Fig. 2, a single field of view is created and several alternative AO methods are modelled, and the resulting detector images are then used to study the instrument design concepts.

1.3 JWST MIRI-MRS

The MIRI Medium-Resolution Spectrometer [2] operates over the spectral range 5–8 µm, with a resolution of R∼3000. The band is divided into 4 spectral channels, each equipped with an IFU image slicer designed to match the width of each slice to the diffraction-limited point-spread function of the telescope, at the wavelength of each channel. Figure 3 (bottom-right) shows the Specsim simulated MIRI-MRS spectra for 2 of the instrument's 4 channels and one of 3 exposures needed to observe the instrument's spectral band. The spectra for each pair of channels are mapped onto each of two 1024 × 1024 pixel detectors. The Specsim-generated field of view from which the spectra were generated is shown in the top-left of the figure.

2 Modelling the Field of View

The main contribution to the Specsim model of an instrument's field-of-view (the Sky Model) is the definition of the field. This is created by the user using a set of primitives with which the morphological and spectral characteristics of each target in the field may be defined. Currently available target primitives include those for defining the geometrical characteristics of the target, the continuum spectral profile, broad and narrow spectral lines and other

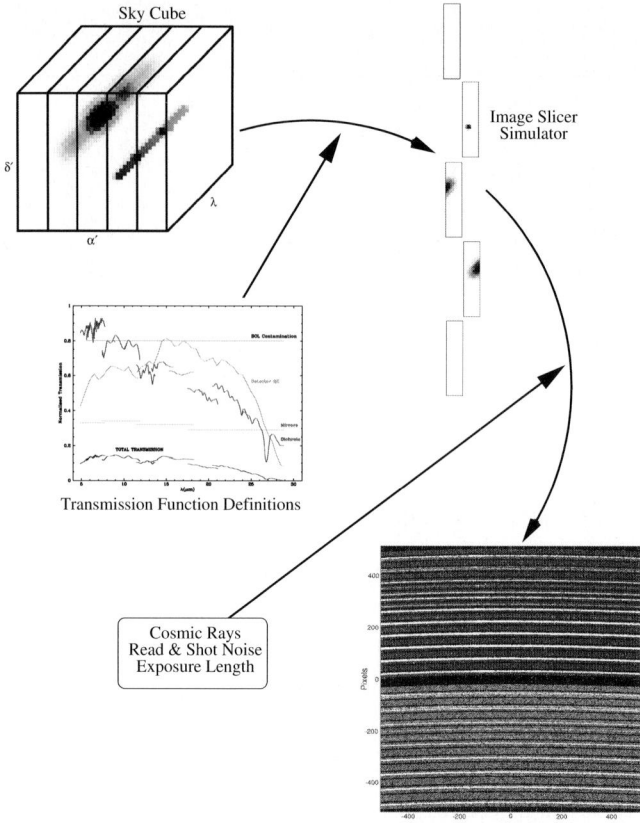

Fig. 3. Schematic of the process of generating a simulated detector frame from the Sky Model. The standard or user-provided transmission functions describing the instrument components are loaded and applied to the Sky Model (*centre-left* and *top-left*). The Sky Model is sliced in the spatial dimensions, and the slices are aligned as for input into the diffraction grating (*top-right*). Dispersion by the spectrometer is then carried out on the vertically aligned slices, to produce a detector image. Finally, the effects of cosmic rays, integration and noise are implemented, and the simulated spectra are mapped onto the detector (*bottom-right*)

common spectral elements. These may be used individually or in combination, to construct target fields of arbitrary complexity. The definitions are stored in a text file and imported into Specsim at run-time. These, together with Specsim's internal models of the telescope and spectrometer, are then used to generate the Sky Model, representing the instrument's field of view over the required spectral range. This process is illustrated in Fig. 4.

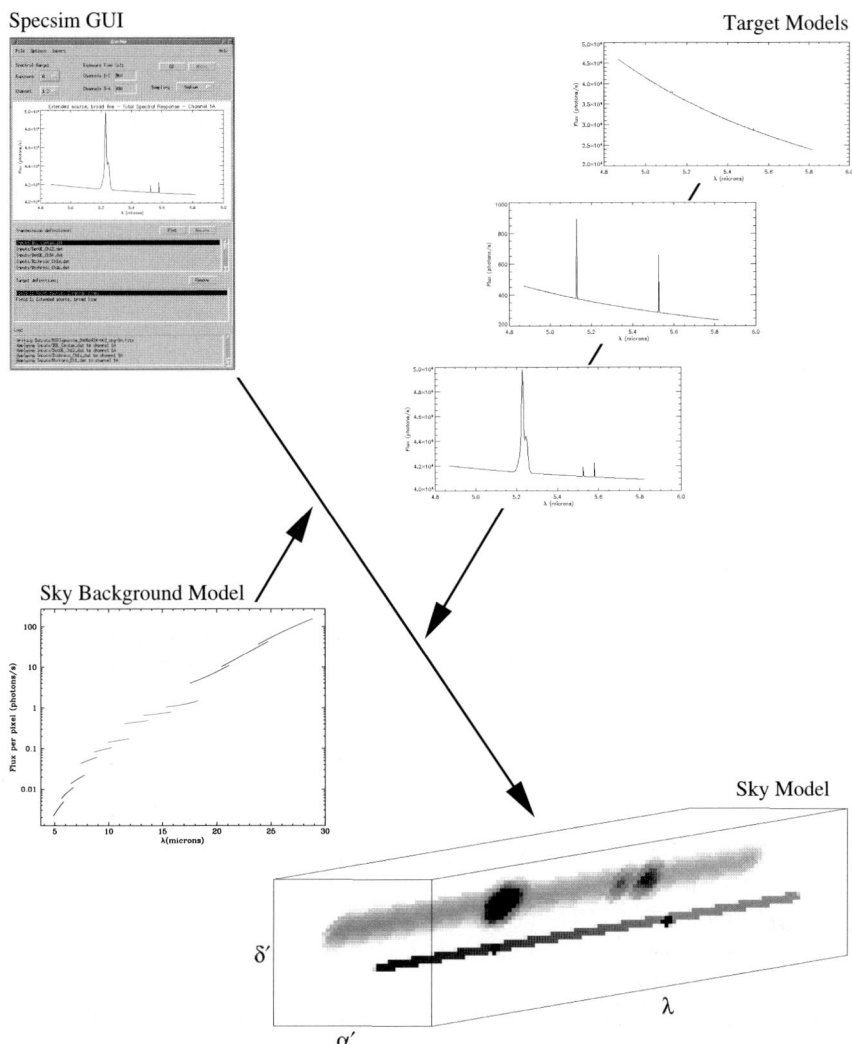

Fig. 4. Schematic showing the process of generating the Sky Model. The user specifies the required observing parameters (channel, exposure, exposure time, data quality) and loads previously created astronomical target definition files via the user interface (*top-left*). When the simulation is started, Specsim creates a model of the field-of-view (the Sky Model, *bottom-right*) using its internal models (*centre-left*) and the spatial and spectral target definitions imported by the user (*top-right*). As can be seen in the Sky Model, here we are simulating a point-source in continuum with two narrow spectral lines, and an extended source with a broad line and two narrow spectral lines

3 Modelling the IFU Spectrometer

Once a model of the target field of view has been produced, Specsim simulates the function of the spectrograph, producing a simulated spectroscopic observation of the field. Figure 3 shows a schematic of this process.

The spectrometer model is generated in several stages. First, the contribution to the observed flux by the instrument's optics and electronics is applied to the sky model. The behaviour of the instrument's components is described by means of transmission functions, consisting of ASCII files listing the normalised transmission of the subsystem (mirror efficiency, detector behaviour, etc.) at each wavelength across the band. The values for the transmission functions may be derived from models of the behaviour of the components (i.e. design values) or be actual measurements taken during laboratory testing of the individual components. Standard transmission functions created for each instrument may be used by program or these may be provided by the user, affording extra flexibility in modelling specific aspects of the instrument, simulating faults, and allowing the effect of specific instrument characteristics on the final spectrometer image to be closely studied.

The next step in the process is to simulate the effect of the image slicer, done by assigning each pixel in the sky model to an IFU slice, depending on its location in the field of view. Geometric deformations are then applied, simulating the optical path of the telescope, the spectrometer's optics and the image slicer. Dispersion of the image slices is then carried out, by mapping each pixel in the Sky Model onto the corresponding pixel in the detector image, based on wavelength.

Once the detector image is constructed, on-sky integration is implemented and cosmic rays, etc. are added as appropriate. Finally, noise (photon, read, etc.) is added to the detector image, and both the Sky Model and the detector image are provided as FITS files for the user. This allows the detector frames to be processed, analysed and compared with the input targets, thus providing an useful test for the refining of instrument design, development of data reduction software, testing calibration strategies and observation planning.

References

1. R. Sharples et al: Design of the KMOS multi-object integral field spectrograph. In: *Ground-based and Airborne Instrumentation for Astronomy*, Proc. SPIE, vol 6269, ed by Ian S. McLean and Masanori Iye, pp 62691C (2006)
2. G. S. Wright et al: The JWST MIRI instrument concept. In: *Optical, Infrared, and Millimeter Space-Telescopes*, Proc. SPIE, vol 5487, ed by John C. Mather, pp 653–663 (2004)

Integral Field Spectroscopy with VIMOS

M. M. Roth, A. Monreal-Ibero, L. Christensen, T. Becker, J. Gerssen, and P. Weilbacher

Astrophysikalisches Institut Potsdam, An der Sternwarte 16, D-14482 Potsdam, Germany; mmroth@aip.de

Abstract. In addition to its standard direct imaging and multi-object spectroscopy modes, VIMOS offers an option for integral field (3D) spectroscopy using an unprecedentedly large integral field unit (IFU). With a total of 6400 spectra covering a maximum field-of-view (FoV) of 54"×54", this mode presents some challenges in terms of data reduction and calibration. Following an introduction and a selected number of illustrative science results obtained from VIMOS-IFU observations, several aspects of data processing and performances are discussed.

1 Introduction

VIMOS, the "Visible Multiobject Spectrograph" at VLT-UT3, is a 4-channel imager and spectrograph, covering the wavelength range 0.36–1 µm [8]. The direct imaging mode provides 4 fields, each 7×8 arcmin^2 wide. The same four fields are available in MOS mode, where 20 to 200 slits per field can be accommodated, depending on the distribution of sources. From a variety of grisms, resolving powers of R = 200–2500 can be selected. 3D spectroscopy is possible through an add-on IFU, consisting of a lensarray of 80×80 lenses which samples the focal plane in two available magnifications with a pitch of 0.67" and 0.33", respectively. The square input FoV is reformatted through optical fibers whose output ends are aligned as "pseudo-slits" to feed the four VIMOS spectrographs. There are 20 bundles of 80 fibers per quadrant, located on 4 deployable masks (Fig. 1). The IFU head creates thus a total of 6400 spectra, which makes VIMOS the largest 3D spectrograph presently available. The layout of the fully multiplexed 6400 spectra on the detectors is illustrated in Fig. 5 and further discussed below (Sect. 4). Depending on the choice of spectral resolution and spatial magnification, either the full set of 6400 spectra can be utilised, or a subset of 1600 spectra. For the more demanding applications, it may be advisable to use the smaller number of spectra in order to achieve high quality results.

2 Science with the VIMOS IFU

In the following a selection of sketches from scientific results obtained with VIMOS-IFU observations are presented in order to highlight the versatility of

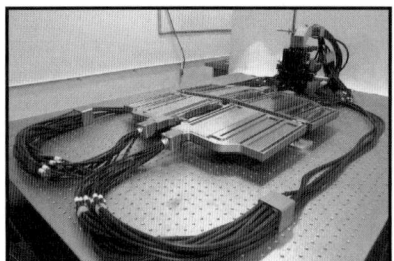

Fig. 1. The VIMOS integral field unit is an optional module of four swing-masks which are deployed in front of the four-quadrant spectrograph system to switch from imaging/MOS to 3D operation

this instrument, which can be useful virtually anywhere from nearby Milky Way objects to high redshift galaxies (the selection here is purely random and not intended to be representative for the use of the VIMOS IFU mode).

2.1 Haloes of Galactic Planetary Nebulae

The use of IFUs to obtain high signal-to-noise (S/N) spectra from low surface brightness objects through binning was introduced and refined for the SAURON survey by Capellari & Copin [2]. The idea to combine many spatial elements (spaxels) in low surface brightness regions where no significant structure is to be expected can yield a factor of $\sqrt{n}$ S/N improvement with respect to a single spectrum, where n is the number of binned spaxels. In this regard, VIMOS is the ideal instrument, given the fact that there is indeed a large FoV with a large number of spaxels available. This property was used for a study of the faint haloes of planetary nebulae, where plasma diagnostic techniques fail with normal slit spectrographs, as diagnostic line intensities are faint (often of order $10^{-17}\,\mathrm{erg/cm^2/s/arcsec^2}$ or less).
From the results of Monreal-Ibero et al. [9], Fig. 2 shows maps of NGC3242 in [O III] and H_β at different contrast levels (left), and various halo spectra (right). The most noisy spectrum (second from top) corresponds to a single

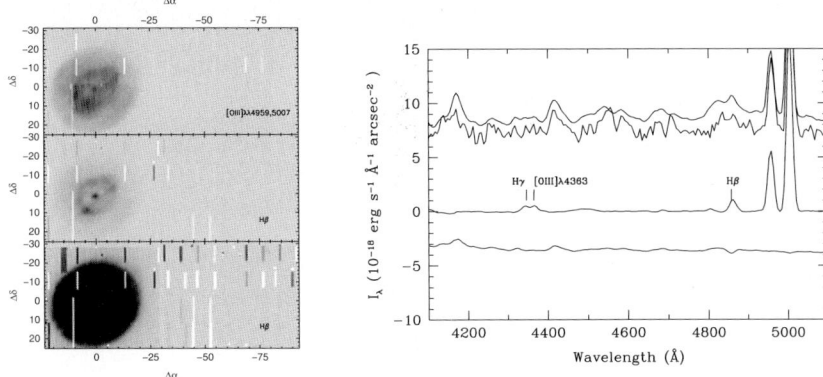

Fig. 2. Monochromatic maps of NGC3242, created from VMOS IFU datacube

spaxel at a radial distance from the central star of ≈ 1 arcmin. The spectrum above is from the same region, however averaged over 4547 spaxels, and shifted by 1 intensity unit for clarity. The third plot is the same spectrum, however sky-subtracted, revealing the emission lines of [O III] $\lambda\lambda 4959,5007$, of H$_\beta$, H$_\gamma$, and of [O III] $\lambda 4363$ at high S/N. Note that the intensity scale is on units of 10^{-18} erg/cm^2/s/Å/arcsec^2. An interesting result from the [O III] line ratio is the detection of a gradient in the electron temperature over the halo from 15,700 K in the inner part to 20,300 K near the edge. Another result of this work is the confirmation of the rings as reported by Corradi et al. [5].

2.2 Luminous Infrared Galaxies

Ultraluminous Infrared Galaxies (ULIRGs), discovered by IRAS, are defined as objects with infrared luminosities similar to the bolometric luminosity of QSOs ($L_{ir} \geq 10^{12} L_{sun}$). They are locally twice as numerous as QSOs. They are rich in gas and dust with emission lines both in the optical and infrared. Most of them show signs of mergers and interaction. They could be the progenitors of intermediate-mass ellipticals as well as of quasars. They also seem to be local counterparts of high-z galaxies. 3D spectrocopy is a perfect tool to observe their complex structure. As a follow-up to a previous study with INTEGRAL at WHT [3, 10], the VIMOS-IFU was used to observe a total of 20 galaxies, constituting a significantly larger sample than the previous one. Besides ULIRGs, also Luminous Infrared Galaxies ($10^{11}L_{sun} \leq L_{ir} \leq 10^{12}L_{sun}$) were observed because of (1) the fundamental properties of (U)LIRGs (frequency of interaction, type of ionization etc.) seem to correlate with luminosity, (2) LIRGs offer the possibility to extend this analysis over a wider luminosity range, and (3) the local density of LIRGs

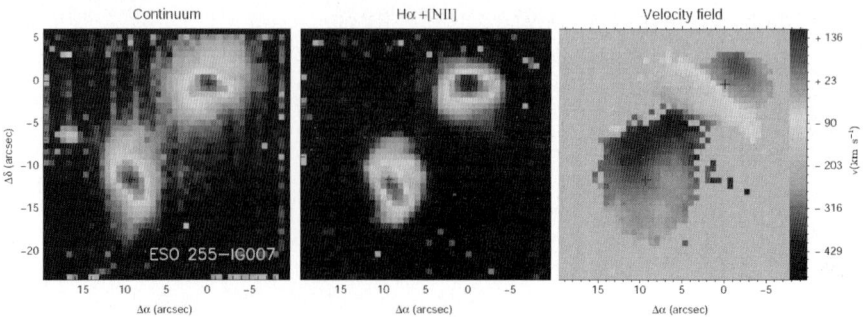

Fig. 3. LIRG ESO255-IG007

is two orders of magnitude higher than that of ULIRGs. As an example for the quality of data obtained in this study, Fig. 3 shows for the interacting system ESO 255-IG007 a continuum map (left), H_α+[N II] (middle), and a velocity map as derived from H_α (right). From a preliminary analysis of a subsample of 10 objects, the following conclusions can be drawn (Monreal-Ibero et al., in prep.): (1) rotation seems to play a more important role for the LIRGs sample: 8 out of 10 systems show signs of rotation; (2) like in ULIRGs, peaks of H_α emission (probably associated with very young stars) are typical. Several systems show line emission peaks which are not associated with the nucleus.

2.3 SDSS Galaxies

Initial results from a VIMOS IFU study of galaxies selected from the Sloan Digital Sky Survey were presented by Gerssen et al. [7]. The motivation for this study stems from the consideration that large fiber-based surveys like SDSS have made a major contribution to our understanding of processes that shape galaxies – that the results, however, are derived from integrated properties over the finite area of the fiber. As the angular extent of galaxies is usually considerably larger than the fiber diameter, the SDSS results are biased towards the nuclear properties of galaxies, and are potentially affected by color and metallicity gradients. With 3D observations, however, the bright emission lines are spatially resolved and can be traced over the whole galaxy. These data are therefore free of aperture effects. Figure 4 shows four galaxies in the continuum (top) and H_α (bottom), as well as a velocity map derived from a three-component gaussian fit to H_α+[N II].

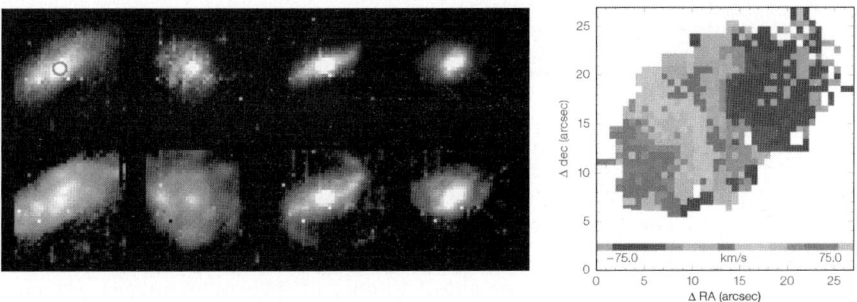

Fig. 4. VIMOS-IFU maps for SDSS galaxies (*top row:* continuum, *bottom:* H_α). The *circle* in the *upper left panel* indicates the SDSS fiber diameter as projected on the sky. *Right panel:* example for a velocity map.

3 Data Reduction

Typically, the following steps must be accomplished in the process of data reduction:

- Pre-reduction (bias, cosmics)
- CCD Response Calibration
- Tracing
- Image Shift
- Straylight Correction
- Extraction
- Wavelength Calibration
- Fiber Response Calibration

While a full account on the details of data reduction is outside the scope of this presentation, it may be useful to list a number of existing pipelines, which have been used successfully for published results:

VIMOS Pipeline (ESO) http://www.eso.org/projects/dfs/dfs-shared/web/vlt/vlt-instrument-pipelines.html
P3d   Becker [1], Roth et al. [13]
VIPGI   Scodeggio et al. [15], Foucaud et al. [6]
R3D   Sánchez [14]

4 Instrument Performance

4.1 Data Reduction Robustness

Experience from data reduction of VIMOS-IFU data has shown that the complexity of multiplexed spectra in the presence of instrumental effects presents a peculiar problem. Figure 5 illustrates the arrangement of the fiber bundles in the IFU mask, feeding into four parallel pseudo-slits (top), and the corresponding pattern of 4 banks of spectra on the detector (bottom, indicated as "Line A",...). The parallel placement of 4 banks of spectra generates zero and second order spectra which overlap with the first order spectra proper. This contamination is scene-dependent and difficult to remove (Fig. 5). Moreover, a noticeable instrumental instability (see Amico, these proceedings) between science exposures and calibrations makes it difficult to trace the spectra reliably and discriminate artifacts. The low resolution fully multiplexed modes (6400 spectra) are most sensitive to such problems, requiring special attention in the process of data reduction, where visual inspection at intermediate steps is often recommended (or even required). Those modes which present just 1 bank of spectra instead of 4 per CCD are by far less critical and not affected by contamination.

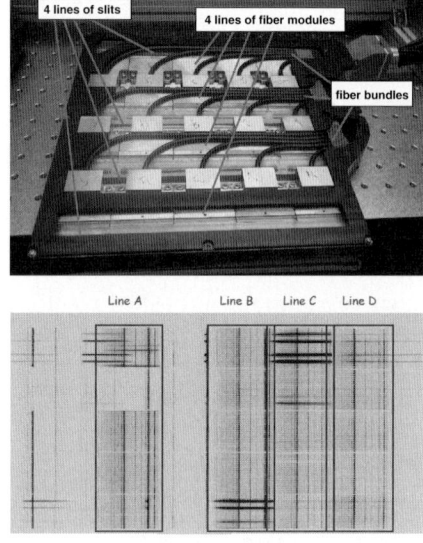

Fig. 5. IFU data

4.2 Response Variation

Similar to pixel-to-pixel response variations in CCDs, which are conventionally corrected through the flatfield exposures, IFUs are also affected by response variations from spaxel to spaxel. Because of the more complex optical train including foreoptics, lensarray, fibers, spectrograph optics, detector, the circumstances are more difficult than with CCDs. In particular cross-talk between adjacent spectra and scattered light behave in non-linear ways, making it difficult to correct by means of the conventional division by a flatfield exposure. Figure 6 shows typical flatfield residuals which appear as diagonal patterns in reconstructed images (left, middle). For comparison, the intrinsic reponse distribution derived for the lensarray with fiber bundle is shown in the right panel (from Prieto et al. [11]). Whenever robustness with regard to flatfield residuals is an issue, it is advisable to not only rely on internal flatfield exposures, but to also to take twilight skyflats.

4.3 Wavelength Calibration

Figure 7 demonstrates the influence of short term mechanical instability (flexure) on the accuracy of wavelength calibrations. The lower left panel shows a small part of an arc lamp exposure, obtained on Nov 3, 2005, at 07:30 UT. The corresponding right panel is an overlay with this latter exposure with another arc lamp exposure, obtained 1h:12m later, on Nov 3, 2005, at 08:42 UT. While the elevation of the telescope was the same between these two exposures (zenith), the azimuth had changed from 253° to 7°. As a result from mechanical instability between these two exposures, the spectra have shifted on the detector by 3 and 5 pixels in the spatial and dispersion directions, respectively. The image is not showing another set of emission lines, but rather indicates a significant amount of image motion. The direction of image motion is indicated by the arrow. The magnified detail shows another peculiar feature of one fiber, which exhibits a distorted image. Such effects are commonly caused by scratched or otherwise affected fiber end faces. As a result, such fibers exhibit lower throughput, but also a skewed profile is likely to occur, with possible effects on the wavelength calibration accuracy in the corresponding spectrum. Nonwithstanding the effects of image motion,

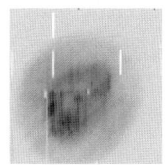 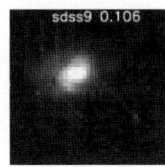 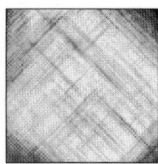

Fig. 6. Reconstructed maps, affected by spaxel-to-spaxel response variations

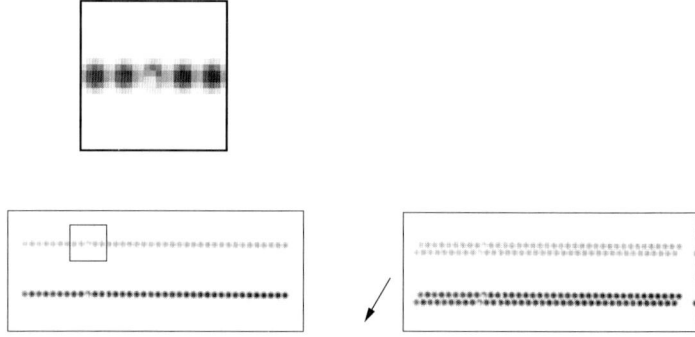

Fig. 7. Flexure and image motion between exposures

one can ask the question whether the shifts introduced by flexure are consistent over the whole face of the IFU, i.e. whether one can calibrate the effect through some kind of normal. To this end, the centroids of the bright night sky lines at 6300 Å and 5577 Å, respectively, were determined from gaussian fits and plotted as histograms (Fig. 8). Results for the examples shown here:
FWHM=0.79±0.03Å, λ=5577.3±0.2 Å, Flux=3431±658 ADU;
FWHM=0.76±0.06Å, λ=6300.25±0.07 Å, Flux=762±176 ADU.

4.4 Flux Calibration

The accuracy of flux calibrations by means of standard stars was tested during our observing run in May 2004 (Sect. 2.1), where subsequently several flux standard exposures were taken at different locations over the face of the IFU. The insert in Fig. 9 indicates the 3 times 4 pointings of the flux standard EG274 on different locations of the lensarray, corresponding also to

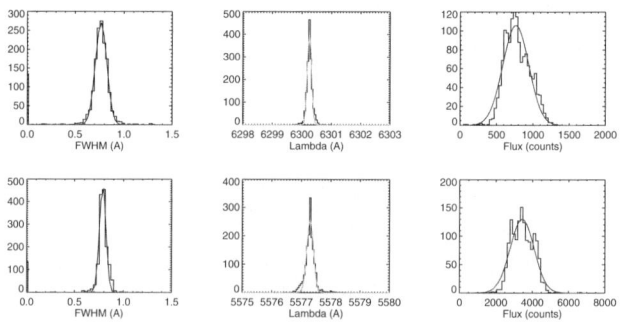

Fig. 8. Intrinsic wavelength calibration accuracy as measured from gaussian fits to night sky emission lines (*left*: FWHM, *middle*: wavelength, *right*: flux)

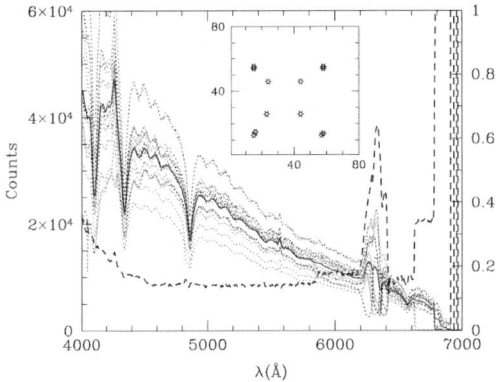

Fig. 9. Flux calibration

different quadrants of the spectrograph system. The 12 exposures were taken immediately one after another under photometric conditions. The analysis of this data yields a 15% r.m.s variation, which imposes some constraints on the accuracy of fluxes. Note, however, that this finding does not necessarily reflect the instrumental behaviour in settings other than the one used here (LR blue).

4.5 Image Quality / Scattered Light

The occurrence of scattered light, appearing as faint extended wings of the spatial PSF – e.g. of interest for the PN halo study (Sect. 2.1, Corradi et al. [4]) – was investigated through a relatively simple method, which is capable to increase the S/N, in particular in wings, where for reasons of limited dynamic range in CCD exposures, normally photon counts are low. However, by co-adding many slices of a datacube, thus essentially creating

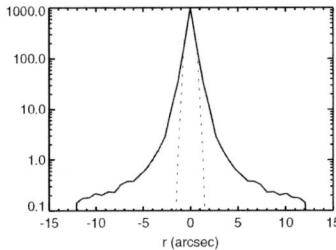

Fig. 10. High S/N stellar point-spread-function, obtained from coadded images through datacube

a broadband image, one can improve S/N considerably. Figure 10 shows the result from such a procedure as a radial plot, averaged over an interval of 4100 – 5800 Å. The extended wings are clearly visible. The dashed curve shows a gaussian of identical FWHM for comparison. Note however, that the scattered light contribution is not severe: 0.02% peak intensity at a radius of 10 arcsec from the PSF centroid.

Acknowledgement. We gratefully acknowledge excellent support and insight into instrumental issues provided to the authors by ESO staff at Paranal. MMR acknowledges financial support from the German Verbundforschung through grant 05AE2BAA/4, and from the EC under contract number HPRN-CT-2002-00305.

References

1. T. Becker: Thesis, University of Potsdam, Potsdam, Germany (2002)
2. M. Cappellari, Y. Copin: MNRAS **342**, 345 (2003)
3. L. Colina, et al.: ApJ **621**, 725 (2005)
4. R.L.M. Corradi, et al.: MNRAS **340**, 417 (2003)
5. R.L.M.Corradi, et al.: A&A **417**, 637 (2004)
6. S. Foucaud, et al.: New Astr. Reviews, Vol. **50**, 401 (2006)
7. J. Gerssen, L. Christensen, D. Wilman, R. Bower: The Messenger **126**, 2 (2006)
8. O. LeFevre, et al.: SPIE 4841, 1670 (2003)
9. A. Monreal-Ibero, M.M. Roth, D. Schönberner, M. Steffen, P. Böhm: ApJ **628**, L139 (2005)
10. A. Monreal-Ibero, et al.: ApJ **637**, 138 (2006)
11. E. Prieto, et al.: SPIE 4008, p. 510 (2000)
12. M.M. Roth, T. Becker, A. Kelz, J. Schmoll: ApJ **603**, 531 (2004)
13. M.M. Roxth, et al.: PASP **117**, 620 (2005)
14. S.F. Sánchez: AN **327**, 850 (2006)
15. M. Scodeggio, et al.: PASP **117**, 1284 (2005)

Discussion

D. Osip Recognizing the consequences of flat fielding using internal flat fields as opposed to flats that follow the same optical path through the telescope as science observations - needs to either: (a) consider an improved empirical model to account for the sky-to-internal differences, and/or (b) document fully for the user community the limitations of the pipeline flat fielding.

KMOS: Design Overview and Calibration Requirements

R. M. Sharples[1], S. K. Ramsay[2], R. Davies[3], and M. Lehnert[4]

[1] Department of Physics, University of Durham, Durham DH1 3LE, UK;
   r.m.sharples@durham.ac.uk
[2] Astronomical Technology Centre, Royal Observatory, Blackford Hill, Edinburgh EH9 3HJ, UK
[3] MPE, Giessenbachstraße 1, 85748 Garching, Germany
[4] GEPI, Observatoire de Paris-Meudon, 92195 Meudon, France

Abstract. We present an overview of the KMOS instrument, currently under construction by a consortium of UK and German institutes, which will provide a unique multi-object near-infrared integral field spectroscopic capability on the VLT. We discuss the instrument architecture and the demanding requirements for calibration imposed by multi-object near-infrared integral field data. Details of the specific calibration procedures being developed for KMOS are presented in a companion paper.

1 Introduction

KMOS is a near-infrared multi-object integral-field spectrometer which has been selected by the European Southern Observatory (ESO) as one of a suite of second-generation instruments to be constructed for the Very Large Telescope (VLT) at the Paranal Observatory in Chile. The instrument will be built by a consortium of UK and German institutes working in partnership with ESO and is currently in the preliminary design phase. In this paper we describe the baseline instrument concept derived from the KMOS science case and the particular calibration requirements of this complex multiple-channel IFU instrument.

2 Science Case and Functional Specification

The focus of cosmological studies at the start of the 21st century is rapidly shifting from accurate determinations of the parameters of the world model into investigations of the physical processes which drive galaxy formation and evolution. To achieve this goal requires a capability to map the variations in star formation histories, spatially resolved star-formation properties, merger rates and dynamical masses of well-defined samples of galaxies across a wide range of redshifts and environments. A few of the brightest examples e.g. [1] are now being observed using single integral field unit (IFU) spectrographs on 8-metre telescopes but statistical surveys of these galaxy properties will

Table 1. Baseline capabilities for the KMOS instrument

Requirement	Baseline Design
Instrument Throughput	J=20%, H=30%, K=30%
Sensitivity (5σ, 8 h)	J=21.2, H=21.0, K=19.2
Wavelength coverage	1.0 to 2.5 µm
Spectral Resolution	R=3400, 3800, 3800 (J,H,K)
Number of IFUs	24
Extent of each IFU	2.8×2.8 arcseconds
Spatial Sampling	0.2×0.2 arcseconds
Patrol field	7.2 arcmin diameter circle
Close packing of IFUs	> 3 within 1 sq. arcmin
Closest approach of IFUs	edge-to-edge separation of 6 arcsec

require a multi-object approach. This is the capability which will be delivered to the VLT with KMOS.

For any instrument to address these fundamental questions about how galaxies evolve it should: (1) have a substantial multiplex capability, commensurate with the surface density of accessible targets; (2) have the ability to obtain more than just integrated or one-dimensional information; (3) be able to resolve the relatively small velocity differences observed in rotation curves, velocity dispersions, and in merging galaxy pairs; (4) have the ability to observe several targets concentrated in a small area of sky; (5) have the capability to observe high-redshift galaxies using the well-studied rest-frame optical diagnostic features used at low redshift. These general characteristics imply a near-infrared multi-object spectrograph using deployable integral field units (dIFUs). The specific choices in delivering these capabilities involves a complex trade of cost and scope which is reflected in the baseline capabilities listed in Table 1.

3 Instrument Description

KMOS will mount on the VLT Nasmyth rotator (Fig. 1) and will use the Nasmyth A&G facilities. The top-level requirements are: (i) to support spatially-resolved (3-D) spectroscopy; (ii) to allow multiplexed spectroscopic observations; (iii) to allow observations across the J, H, and K infrared atmospheric windows (extension to shorter wavelengths down to 0.85 µm has been incorporated at lower priority). The baseline design employs 24 configurable arms that position fold mirrors at user-specified locations in the Nasmyth focal plane. The sub-fields thus selected are then anamorphically magnified onto 24 advanced image slicer IFUs that partition each sub-field into 14 identical slices, with 14 spatial pixels along each slice. The anamorphic magnification preserves a rectangular 0.2×0.2 arcsec spatial sampling, whilst having Nyquist sampling of a spectral resolution element. Light from the IFUs is

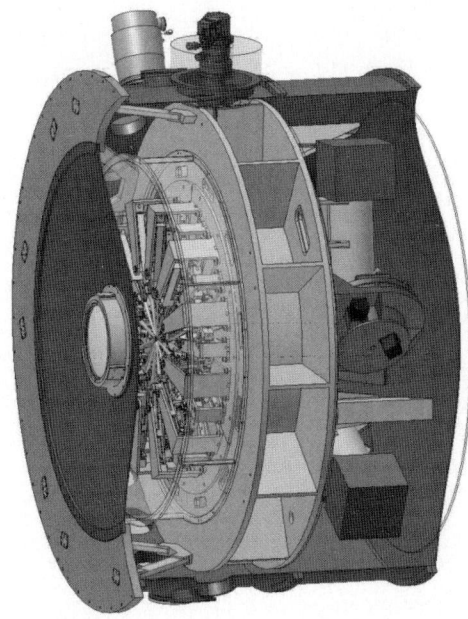

Fig. 1. Cutaway view of the main KMOS cryostat showing the entrance window and the pickoff arm module at the *front*, the IFU module in the *middle* and the spectrograph module at the *rear*. The cryostat will be an aluminium/stainless steel hybrid to reduce weight

dispersed by three identical cryogenic grating spectrometers which generate 14×14 spectra, each with 1000 spectral resolution elements, for all of the 24 independent sub-fields. The spectrometers each employ a single 2k×2k substrate-removed HgCdTe detector. The goal is to employ careful design choices and advances in technology to ensure that KMOS achieves a comparable sensitivity to the current generation of single-IFU infrared spectrometers. Figure 1 shows the overall assembly of the instrument which is partitioned into three layers (Pickoff module, IFU module and Spectrograph module). There is also a natural 3-fold symmetry to many of the opto-mechanical assemblies which has implications for the calibration requirements.

3.1 Pickoff Module

One of the key KMOS elements is the pickoff module which relays the light from 24 selected regions distributed within the patrol field to an intermediate focus position at the entrance to the integral field unit module (Fig. 2). The method adopted for selecting these subfields uses robotic pickoff arms whose pivot points are distributed in a circle around the periphery of the

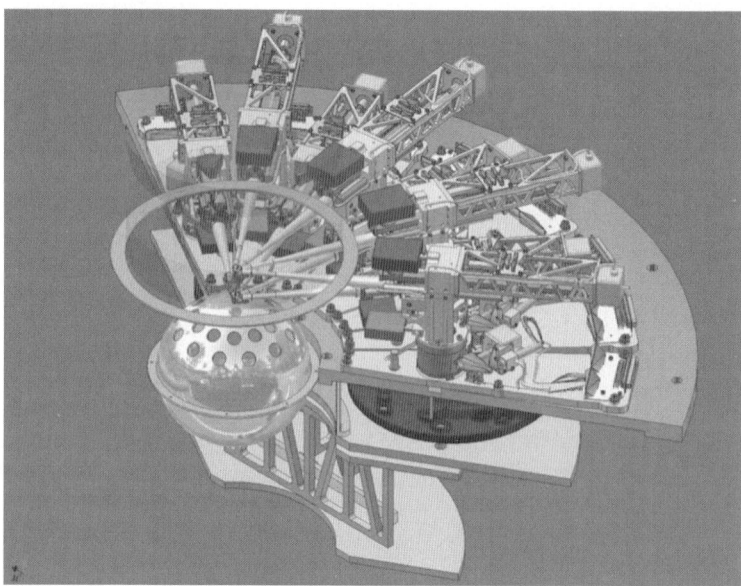

Fig. 2. One of the three integrated pickoff and IFU sub-modules showing the three mounting plates for the pickoff arms, the filter wheels and the IFU optics. Each sub-module is attached to the main cryogenic optical bench within the cryostat. At the centre of the unit is shown the integrating sphere of the calibration unit and the ring mirror which reflects light from the calibration sources in to the pickoff arms

patrol field and which can be driven in radial and angular motions by two stepper motors which position the pickoff mirrors with a repeatable accuracy of < 0.2 arcsec. The arms patrol in one of two layers positioned either side of the Nasmyth focal plane. The changing path length within the arm is compensated via an optical trombone which uses the same lead screw, but with a different pitch, as for the main radial motion. The pickoff module also contains a central integrating sphere which relays the light from the external flatfield and wavelength calibration lamps into the pickoff arms, and a filter wheel which acts as a focus compensation device between the different bands. The cold stop for the instrument is at the base of the arm, after which the intermediate image is formed by a K-mirror assembly which also acts to orientate the pickoff fields so that their edges are parallel on the sky.

3.2 Integral Field Unit Module

The IFU subsystem contains optics that collect the output beams from each of the 24 pickoffs and reimages them with appropriate anamorphic magnification onto the image slicers. The slices from groups of 8 sub-fields are aligned and reformatted into a single slit for each of the three spectrographs.

The optical design of the IFU sub-systems is based on the Advanced Image Slicer concept [2] and draws heavily on experience developed in building the GNIRS integral-field unit for Gemini South [3]. Three off-axis aspheres are used in the fore-optics to facilitate a production method based on diamond-turning, rather than raster fly-cutting, in order to improve the surface roughness. Important considerations in developing the design for 24 optical trains, have been the need to incorporate manufacturability into the optimisation process, and a desire to use monolithic optical components wherever possible. In the current design the slicer mirrors are all spherical with the same radius of curvature, and so are the pupil mirrors. The slit mirrors are toroidal with the same radius of curvature in the spectral direction, but different radii of curvature in the spatial direction. This configuration was chosen because it is well adapted to the available methods of machining. Each IFU sub-module produces a 256 mm long slit containing 112 separate slices from 8 subfields.

The mechanical design of the whole pickoff-IFU module is shown in Fig. 2 which emphasises the three-fold symmetry of the KMOS system and the advantages from a mechanical perspective of positioning common components in a single plane.

3.3 Spectrograph Module

The three identical spectrographs use a single off-axis toroidal mirror to collimate the incoming light, which is then dispersed via a reflection grating and refocussed using a 6-element transmissive achromatic camera. The gratings are mounted on a 5-position turret which allows optimized gratings to be used for the individual J,H,K bands together with two lower resolution gratings and the option of a z-band grating to enhance versatility [4]. Each spectrograph contains a 2048×2048 HgCdTe array which is mounted on a three-axis translation stage in order that focus can be adjusted and, if required, some components of flexure can be compensated. All three spectrographs are mounted in a plane perpendicular to the Nasmyth rotation axis for maximum stability.

4 Calibration Requirements

KMOS will require all of the standard calibration templates used by single integral field unit instruments such as SINFONI. These include bias frames, dark current/thermal background frames, bad pixel masks, wavelength calibration frames, flat-field frames, atmospheric telluric calibration and flux calibration. In addition there will be specialised engineering calibration routines needed to check the metrology of the positioning mechanisms and to map out the geometrical distortions of the spatial/spectral dimensions of the data cubes. These are discussed in more detail in a companion paper [5]. Because of the complex optical layout (Fig. 3) each of the $24 \times 14 \times 14$ spatial

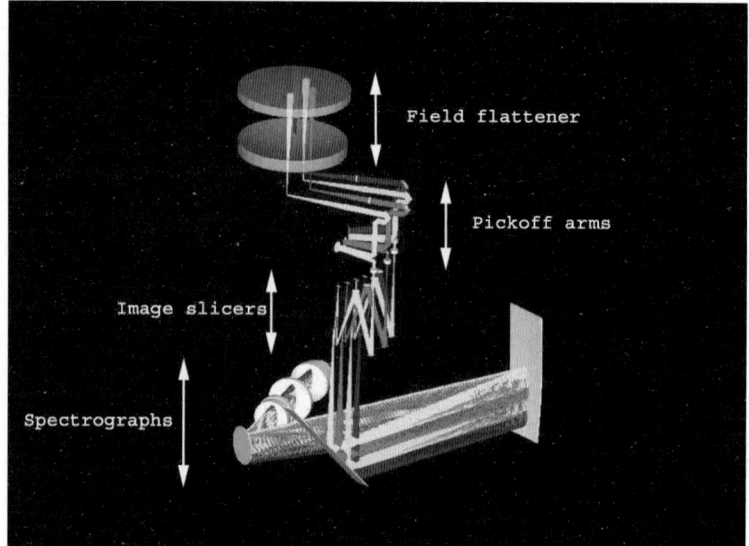

Fig. 3. Optical raytrace through four pickoff arms, their associated IFUs and one of the spectrometers. Light exiting the pickoff arms is brought to an intermediate focus using a 3-element K-mirror, which aligns the edges of all 24 IFU fields on the sky so that they can be put into a compact sparse array configuration for blind surveys of contiguous areas on the sky

channels of KMOS will have subtly different response characteristics. Calibrating these differences using a combination of internal lamps and blank sky exposures will be crucial in removing systematic effects in the wavelength and flatfield response functions, which will be essential if KMOS is to achieve its full scientific potential.

References

1. R. Genzel, R. Hofmann, D. Tomono, N. Thatte, F. Eisenhauer, M. Lehnert, M., Tecza, R. Bender: In *Proceedings of ESO Workshop on Scientific Drivers for ESO Future VLT/VLTI Instrumentation (2002)* (astro-ph/0108318)
2. R. Content: Proc. SPIE **2871**, 1295 (1997)
3. M. Dubbeldam, R. Content, J.R. Allington-Smith, S. Pokrovsky, D.J. Robertson: Proc. SPIE **4008**, 1181 (2000)
4. I.J. Lewis, J. Lynn, W. Lau, S. Yang, M. Wells: Proc. SPIE **5492**, 1395 (2004)
5. S.K. Ramsay Howat, S. Rolt, R. Sharples, R. Davies: these proceedings (2008)

Discussion

D. Osip: The KMOS arm positioners are commanded via pulses to stepper motors and expected steps that are subsequently counted. Do you make use of any absolute encoders to 'warn' of missed steps ?

R. Sharples: Yes, the arm positions are monitored after movement using a pair of LVDT sensors. These are not used to actively position the arm, but in a passive error monitoring mode as you suggest.

Calibration of the KMOS Multi-Field Imaging Spectrometer

S. K. Ramsay[1], S. Rolt[2], R. M. Sharples[2], and R. Davies[3]

[1] UK Astronomy Technology Centre, Blackford Hill, EH9 3HJ Edinburgh, UK; skr@roe.ac.uk
[2] Centre for Advanced Instrumentation, Durham University, NetPark, TS21 3FB Sedgefield, UK
[3] Max Plank Institute for Extraterrestrial Physics, Postfach 1312, 85741 Garching, Germany

Abstract. When it is delivered to the VLT KMOS, will be the first near infrared (0.8–2.5 µm) spectrograph to use multiple integral fields units. Pick-off arms provide the optical relay to image objects selected from the patrol field at the Nasmyth focus onto 24 image slicing IFUs. The output from 8 IFUs forms a single slit which feeds one of three identical spectrometers. The calibration of such an instrument presents a number of challenges. Flat field and wavelength calibration for the instrument will be provided by an internal calibration system of unique design. External light sources are fed via integrating spheres and high transmission light pipes to a calibration sphere with 24 output ports addressed by the 24 arms. Obtaining a high degree of field flatness is key to the sensitivity of NIR instruments. The KMOS calunit is predicted to deliver flatnesses of ∼0.1% over the KMOS IFU fields.

1 An Introduction to KMOS

The K-band multi-field spectrograph (KMOS[1]) is a near-infrared (0.8–2.5 µm) spectrometer for 3D observations of 24 fields of 2.8"×2.8" distributed over a 7' patrol field. The instrument will be located on the Nasmyth platform of the VLT and is due for delivery to the telescope in 2010. Details of the instrument are presented by Sharples (these proceedings, [1]). Objects are selected using pick-off mirrors mounted on arms that patrol the field. The arms are located on two planes, one above and one below the focal plane of the telescope and are positioned using two stepper motors. Each arm relays one pick-off field into an integral field unit (IFU). The output of eight IFUs is formed into a long slit that is the input to a grating spectrometer. There are three identical spectrometer modules. A single channel of KMOS (pickoff+IFU+spectrometer) is a very similar system to the single IFU spectrographs already used in the near-infrared (e.g. SINFONI, UIST, GNIRS). The calibration issues are the same to first order and are increasingly well understood as use of these instruments is becoming standard on large telescopes

---

[1]The KMOS consortium members are: Max Planck Institut für Extraterrestrische Physik, Universitäts-Sternwarte München, ESO, Oxford University, Durham University and UKATC.

[2, 3]. In this paper, we discuss the unique calibration issues encountered in observing with KMOS, the first facility class multi-field NIR spectrometer. We present this in the context of the users experience of KMOS, showing how the principal KMOS observing modes set the calibration requirements and define the pipeline reduction requirements. Finally, we present the unique design of the KMOS calibration system and its expected performance.

2 Planning KMOS Observations

KMOS observers will employ a custom-built add on to the ESO Phase 2 Proposal Preparation Tool (called KARMA) to allocate arms to objects in a given field (Fig. 1). Users provide an input catalogue of coordinates for their science target and reference stars which must all be on the same astrometric reference frame. The arms are then allocated to those positions using an automatic algorithm which handles target priorities. The automatic allocation may be overidden by the user. The translation of input target coordinates (RA, Dec on the sky) to arm motor steps for two motors per arm for 24 arms requires a number of calibration stages, the last of which is to be carried out during the on-sky commissioning of the instrument. These are described in the sections that follow.

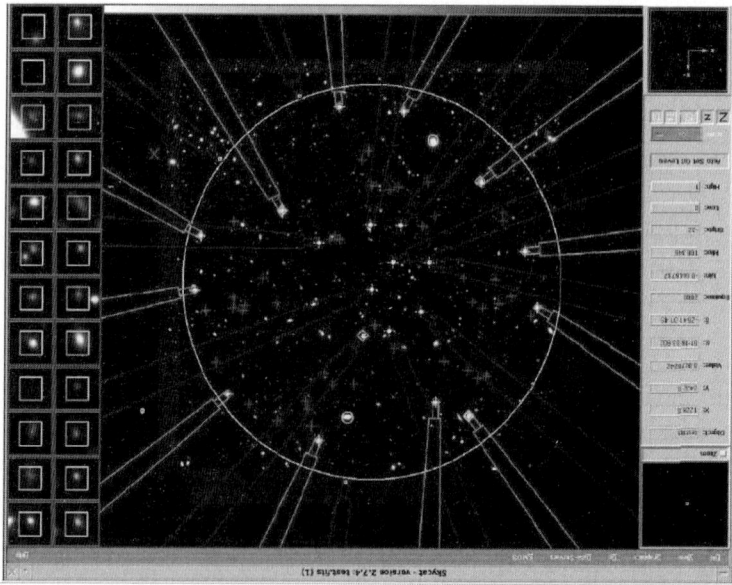

Fig. 1. The KARMA tool for allocation of arms to targets under development at the Universitäts-Sternwarte München

2.1 Calibration of Motor Steps to Arm Positions

Individual arms move in radius (R) and angle (θ) driven by one stepper motor for each axis. Once the range of travel and initial calibration of steps to (R, θ) is obtained for each arm, the arms are assembled into the cryostat. The laboratory calibration of the arm position is then carried out using a reference source that projects an f-15 beam to the position of the focal plane, forming a point source. The arm is driven to acquire the source, the offset from the optical axis at the output of the arm is determined. The f-15 source is then translated on an accurate (x,y) stage and the arm steps to (x,y) on the focal plane is mapped out for the entire patrol field of each arm. The VLT focal plane is curved and therefore the KMOS foreoptics (a powered window and field lens) flatten the field and render the input beam telecentric. Rather than attempt to simulate the telescope field curvature, these calibrations will be carried out using a flat window and the correction for the residual field distortion calibrated at a later stage. This procedure calibrates any repeatable non-linearities of the arm motion, such as would be introduced by e.g. eccentricity of the motor shaft.

2.2 Transformation of (RA, Dec) to Focal Plane (x,y) Position

Once the procedure above is complete, the arms can address any (x,y) on the focal plane accurately and repeatably. A transformation from sky coordinates to focal plane (x,y) is required to link this to the user's input catalogue. This transformation may be estimated for the perfect telescope+KMOS system, as follows.

- Mean (RA, Dec) for the observations transformed to apparent (RA, Dec) using an atmospheric model of the differential refraction.
- Using the field centre, apparent (RA, Dec) are transformed to positions on the tangent plane (ξ, η)
- The coordinates in the tangent plane are corrected to perfect (x,y) in the telescope focal plane using the plate scale and the rotation of the field relative to the instrument focal plane.
- The perfect (x,y) position above must be corrected to the real (x,y) by correcting for predictable residuals in the field distortion from the telescope after correction by the field corrector. These values are obtained from Zemax and are less than 3 pixels at the detector focal plane (0.3").

These steps are calculated in the instrument control software and may be estimated before use of KMOS at the telescope. During the on-sky commissioning, a final confirmation that these calibrations are correct will be made using observations of globular clusters. Stars with H magnitudes between 13 and 11 are ideal for obtaining good centroids in a reasonable time. Globular clusters containing source densities of stars meeting this criterion and that are visible from Paranal throughout the year have been identified and include

47 Tuc, M 22 and M 5. On the telescope, stars are acquired on to pick-offs using the nominal calibration. The arms will be allocated to stars and the offset of the star relative to the centre of the IFU field calculated and used to correct a look-up table of arm positions. This final calibration step confirms that the optical model of the KMOS field corrector does indeed provide the small residual field distortions predicted by the Zemax model. The procedure will be carried out whenever KMOS is remounted on the telescope.

3 Calibration of Spatial and Spectral Distortions

To reconstruct a datacube of an astronomical target, every detector pixel must be mapped onto the correct spatial and wavelength position. The spatial and spectral distortions introduced in the instrument must therefore be calibrated.

These distortions arise in the IFU (the build up of manufacturing tolerances in the optics of the integral field unit introducing small offsets of the field) and in the spectrometer offsets (which introduce slit curvature and curvature of the spectrum). For KMOS, these offsets are tightly controlled, particularly within the IFU as the optics (slicing mirror, pupil mirror and slit mirror) are all manufactured as monolithic pieces [4]. To measure this mapping, the f-15 source used for arm calibration will also project an image of three slits that run parallel to the dispersion direction. These slits will be sliced by the IFU and dispersed by the spectrograph. The offsets required to reconstruct the slitmask image will be determined and applied in a single transformation in the data pipeline, maximising the data quality from KMOS by minimising the number of times the data are resampled.

4 KMOS Sky Subtraction Modes

The KMOS operational concept includes two distinct sky subtraction methods. In the first, the telescope is offset and the sky frame measured in through the same optical path as the source frame. For the second, a sky measurement is obtained from an arm which is not deployed on an object, from an area on the field of the IFU that does not have sky signal or from a combination of both. The first of these methods does not place strong requirements on flat-fielding, as the sky signal is measured along the same optical path and with the same pixels as the source+sky. The flatfield should be accurate to 1%. However, this method is inefficient in observing time. The second method requires the measured sky signal to be corrected for any changes in vignetting or pixel gain between the pixels on which the sky is measured and that on which the source+sky is measured. Thus the calibration and reduction requirements depend on the mode chosen. In this case, the sky subtraction accuracy must be 0.1%. This will be achieved through a combination of flat-fielding accuracy and application of data reduction techniques developed for SINFONI [3]. This second mode is expected to be the standard mode of observing for KMOS.

5 Calibration Unit Design and Performance

To measure accurate flatfields and arc spectra efficiently, KMOS has a built-in calibration system. Figure 2 shows a top view of the internal modules of the instrument with calibration unit running up the centre of the structure. At the top of the calunit, an integrating sphere with 24 ports provides the illumination from either a flat field source (Halogen lamp) or from one of two arc lamps (argon or krypton). Light emitted from the ports is reflected from small mirrors mounted around the field lens. To obtain calibration frames, the arms are moved to a calibration position outside the field where they intercept the reflected beam. A schematic of this is shown in Fig. 3. The uniformity of illumination from this system has been calculated for an integrating sphere of this geometry and using the characteristics of the Spectralon diffuse reflectance coating for the NIR. We find better than 0.01% variability across the individual ports and better than 0.1% from port to port. A high throughput for the system is also a key requirement. As for other ESO instruments, the KMOS calibration plan is to take calibration frames during daylight hours. Nevertheless, maintaining the option for efficient nighttime

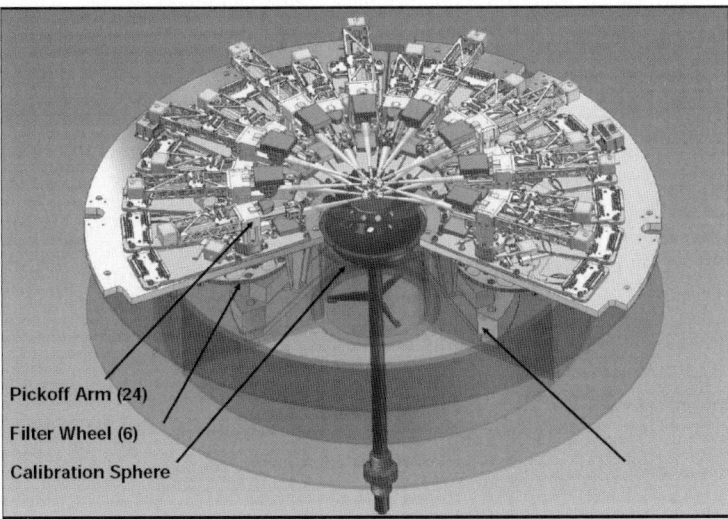

Fig. 2. An internal view of the KMOS cryostat showing the pickoff arms and the calibration unit fed through from the rear of the cryostat

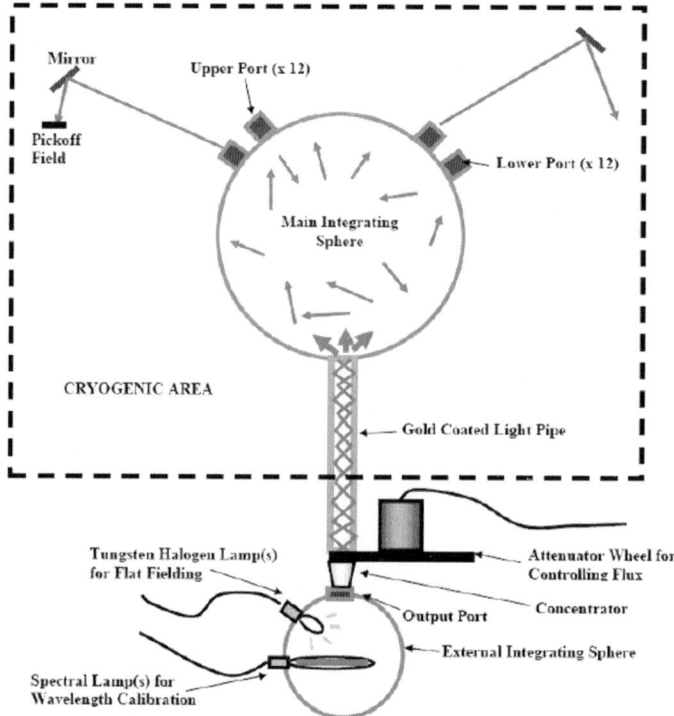

Fig. 3. This schematic of the calibration unit shows key features of the operation and design. Lamps are fed into the cryostat from an external integrating sphere. The light is reflected from mirrors mounted around the field lens into arms located at the calibration position

calibrations is desirable. The light pipes running the length of the KMOS cryostat are gold coated light pipes from Epner. A flux concentrator at the input to the cryostat also boosts the throughput. The predicted throughput of this system is better than 12% for IJH and 12% reducing to 6% over the K band (1.9–2.5 µm). The photon flux from the flatfield lamp in this case is ample. For the pencil beam arc lamps (Ar, Kr), exposure times of 150 s will provide e.g. 15 lines with more than 500 photons in the K band.

References

1. R. Sharples, et al.: *SPIE, Vol. 6269*, SPIE p. 44 (2006)
2. R.I. Davies: MNRAS **375**, 1099 (2007)
3. S.K. Ramsay Howat, et al.: *SPIE, Vol. 5492*, SPIE p. 1160 (2004)
4. C.M. Dubbeldam, et al.: *SPIE, Vol. 6273*, SPIE p. 105 (2006)

MUSE: A Second-Generation Integral-Field Spectrograph for the VLT

R. M. McDermid[1], R. Bacon[2], S. Bauer[3], P. Boehm[3], D. Boudon[2],
S. Brau-Nogué[4], P. Caillier[2], L. Capoani[2], C. M. Carollo[5], N. Champavert[2],
T. Contini[4], E. Daguisé[2], B. Delabre[6], J. Devriendt[2], S. Dreizler[7],
J. Dubois[2], M. Dupieux[4], J. P. Dupin[4], E. Emsellem[2], P. Ferruit[2],
M. Franx[1], G. Gallou[4], J. Gerssen[3], B. Guiderdoni[2], T. Hahn[3],
D. Hofmann[7], A. Jarno[2], A. Kelz[3], C. Koehler[7], W. Kollatschny[7],
J. Kosmalski[2], F. Laurent[2], S. J. Lilly[5], J. L. Lizon[6], M. Loupias[2],
A. Manescau[6], C. Monstein[5], H. Nicklas[7], L. Parès[4], L. Pasquini[6],
A. Pécontal-Rousset[2], E. Pécontal[2], R. Pello[4], C. Petit[2], J.-P. Picat[4],
E. Popow[3], A. Quirrenbach[1], R. Reiss[6], E. Renault[2], M. Roth[3], J. Schaye[1],
G. Soucail[4], M. Steinmetz[3], S. Stroebele[6], R. Stuik[1], P. Weilbacher[3],
L. Wisotzki[3], H. Wozniak[2], and P. T. de Zeeuw[1]

[1] Leiden Observatory, Leiden, The Netherlands; mcdermid@strw.leidenuniv.nl
[2] Centre de Recherche Astrophysique de Lyon, Lyon, France
[3] IAP, Potsdam, Germany
[4] LAOMP, Toulouse, France
[5] ETH, Zurch, Switzerland
[6] ESO, Karl-Schwarzschild-Strasse 2, 85748 Garching, Germany
[7] IAG, Goettingen, Germany

Abstract. The Multi Unit Spectroscopic Explorer (MUSE) is a second-generation instrument in development for the Very Large Telescope (VLT) of the European Southern Observatory (ESO), due to begin operation in 2011/12. MUSE will be an extremely powerful integral-field spectrograph fed by a new multiple-laser adaptive optics system on the VLT. In its usual operating mode, MUSE will, in a single observation, produce a 3-dimensional data cube consisting of 90,000 R 3000 spectra, each covering a full spectral octave (480–930 nm), and fully sampling a contiguous 1×1 arcmin^2 field with 0.2×0.2 arcsec^2 apertures. A high-resolution mode will increase the spatial sampling to 0.025 arcsec per pixel. MUSE is built around a novel arrangement of 24 identical spectrographs (each comparable to a 1st generation VLT instrument), which are fed by a set of 24 precision image slicers. MUSE is designed for stability, with only 2 modes, and virtually no moving parts, allowing very long exposures to be accumulated. Together with high throughput, this ensures that MUSE will have extreme sensitivity for observing faint objects. We overview the technical and scientific aspects of MUSE, highlighting the key challenges for dealing with the unprecedented quantity and complexity of the data, and the integration with the VLT adaptive optics facility (AOF) – a key development on the path to extremely large telescopes (ELTs).

1 Imaging and Spectroscopic Surveys

Imagers and spectrographs are the most common tools of optical astronomers. In most cases, astronomical observations start with imaging surveys in order to find the interesting targets and then switch to spectrographic observations in order to study the physical and/or dynamical properties of the selected object. Thanks to the excellent throughput and large format of today's detectors, large fractions of the sky can be surveyed in depth with imagers. The most limiting factor is the spectroscopic observations, which are time consuming and tend to have small multiplex capabilities. Development of large multi-object spectrographs such as VIMOS at VLT [3] has somewhat improved the situation. However the total number of sources in a typical imaging survey is much larger than what is possible to observe with spectroscopy. The selection of sources is then mandatory. Usually the selection criterion is based on a series of multi-color images and is intended to select the appropriate spectral characteristics of the population of the searched objects. This incurs a direct cost in telescope time since more than one exposure must be made at each sky location. As another disadvantage, the selection process is never 100% efficient, and thus a fraction of time of the follow-up spectroscopy is lost due to misidentifications. The major weakness of this approach, however, is the a priori selection of targets. This pre-selection severely biases the spectrographic observations and limits considerably the opportunity for new discoveries.

An alternative to the classical approach is to perform simultaneous imaging and spectroscopy. The idea is to merge into one instrument the best of the two capabilities: from the imaging world its field of view and high spatial resolution; and from the spectrograph's world its high resolving power and large spectral range. Such an instrument would overcome the difficulty inherent to the classical method. And because there is no longer the need to pre-select the sources, one can even detect objects that would not have been found or pre-selected in the pre-imaging observations. In the most extreme case, such as objects with very faint continuum but relatively bright emission lines, the objects may *only* be detectable with such an instrument.

2 MUSE

MUSE (Multi Unit Spectroscopic Explorer) is a 2nd generation VLT instrument that aims to fill this role. It is an integral field spectrograph (or IFU) which combines large field of view, high spatial resolution, medium resolving power and large simultaneous spectral range. Nowadays, integral field spectroscopy is part of the panoply of modern telescopes. However, most of the currently operating integral field spectrographs have only a small field of view and are thus devoted to the detailed physical study of single objects. Some multi-IFUs, like Giraffe at VLT [4], have multiplex capabilities of a dozen objects, which increase their efficiency. This however does not break the operational three steps (imaging, selection and spectrography) paradigm. MUSE has three operating modes: a wide-field mode (WFM) with and without adaptive optics correction and a narrow-field mode (NFM) with adaptive optics. The observational parameters of these modes are given in Table 1.

The total number of information elements is given by the product of the number of spaxels (90,000) with the number of spectral pixels (4,000), resulting in 360 million elements in the final data-cubes. Such a large number of pixels is not feasible with a single piece of optics and a single detector. MUSE is thus composed of 24 identical modules, each one consisting of an advanced slicer, a spectrograph and a $(4k)^2$ detector. A series of fore-optics and splitting and relay optics is in charge of derotating and splitting the square field of view into 24 sub-fields. These are placed on the Nasmyth platform between the VLT Nasmyth focal plane and the 24 IFU modules. AO correction will be performed by the VLT deformable secondary mirror,

Table 1. Overview of MUSE operating modes and capabilities

All Modes	
Simultaneous spectral range	0.48–0.93 µm
Resolving power	2000 @ 0.46 µm
	4000 @ 0.93 µm

Wide Field Mode	
Field of view	1×1 arcmin^2
Spatial sampling	0.2×0.2 arcsec^2
Spatial resolution @ 0.75 µm (median seeing)	0.46 arcsec (AO)
	0.65 arcsec (non-AO)
AO condition of operation	70%-ile
Sky coverage with AO	70% at galactic pole
	99% at galactic equator
Limiting magnitude in 80 h	$I_{AB} = 25.0$ (full Resolution)
	$I_{AB} = 26.7$ (R=180)
Limiting Flux in 80h	3.9×10^{-19} erg s^{-1} cm^{-2}

Narrow Field Mode	
Field of view	7.5×7.5 arcsec^2
Spatial sampling	0.025×0.025 arcsec^2
Spatial resolution @ 0.75 µm (median seeing)	0.042 arcsec
Strehl ratio @ 0.75 µm	5% (10% goal)
Limiting magnitude in 1h	R_{AB}=22.3
Limiting flux in 1h	2.3×10^{-18} erg s^{-1} cm^{-2}
Limiting surface brightness	R_{AB}=17.3 arcsec^{-2}

part of the VLT Adaptive Optics Facility [1], via the GALACSI ground-layer AO system. Four sodium laser guide stars are used, plus a natural star for tip/tilt correction. All guide stars are taken outside the scientific field of view in order to minimize the amount of scattered light, while the only additional optic located within the scientific field of view is a Na notch filter, dramatically reducing transmission losses with respect to traditional AO systems.

The MUSE narrow field mode uses an additional optical system inserted into the fore-optics to change the spatial sampling from 0.2 arcsec to 0.025 arcsec. The field of view is proportionally reduced to 7.5×7.5 arcsec^2. The most significant change is in the AO optimization and configuration (laser guide star are moved closer) and the tip/tilt, which is performed at IR wavelengths on either a natural guide star within the field of view or the object itself. With such a configuration, the AO facility is expected to deliver a diffraction-limited image with a Strehl ratio of 5–10% at 0.75 µm.

3 MUSE Science Case

MUSE has a broad range of astrophysical applications, ranging from the spectroscopic monitoring of solar system's outer planets to very high redshift galaxies. We give in the following sections a few examples of scientific applications that are considered to be important instrument drivers.

3.1 Wide-Field Mode

The most challenging scientific and technical application, and the most important driver for the instrument design, is the study of the progenitors of normal nearby galaxies out to redshifts z>6. These systems are extremely faint and can only be found by their Lyα emission. MUSE will be able to detect these in large numbers ($\sim 15,000$) through a set of nested surveys of different area and depth (Fig. 1). The deepest survey will require very long integration (80 h each field, Fig. 2) to reach a limiting flux of 3.9×10^{-19} erg s^{-1} cm^{-2}, a factor of 100 times better than what is currently achieved with narrow band imaging. These surveys will simultaneously address the following science goals:

- Study of intrinsically faint galaxies at high redshift, including determination of their luminosity function and clustering properties,
- Detection of Lyα emission out to the epoch of reionization, study of the cosmic web, and determination of the nature of reionization,
- Study of the physics of Lyman break galaxies, including their winds and feedback to the intergalactic medium,
- Spatially resolved spectroscopy of luminous distant galaxies, including lensed objects

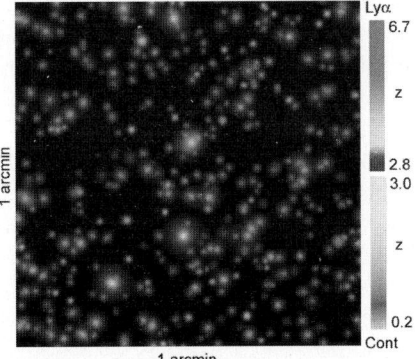

Fig. 1. Simulated MUSE 80 h deep field exposure, showing both the continuum and Lyα emission-line sources detected

- Search of late-forming population III objects,
- Study of active nuclei at intermediate and high redshifts,
- Mapping of the growth of dark matter haloes,
- Identification of very faint sources detected in other bands, and
- Serendipitous discovery of new classes of objects.

Multi-wavelength coverage of the same fields by MUSE, ALMA, and JWST will provide nearly all the measurements needed to answer the key questions of galaxy formation. At lower redshifts, MUSE will provide exquisite two-dimensional maps of the kinematics and stellar populations of

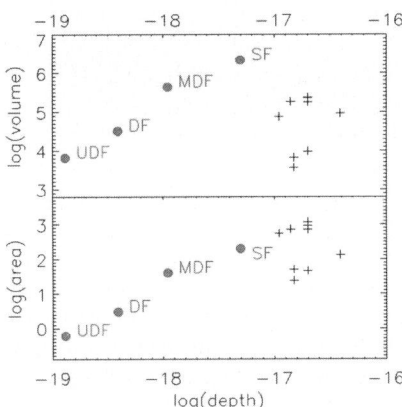

Fig. 2. Big symbols indicate planned MUSE deep surveys. Several wide-shallow and narrow-deep surveys will probe a range of volumes and limiting magnitudes. *Crosses* represent similar surveys currently completed. The MUSE surveys will be conducted as a major part of the guaranteed time allocation

normal, starburst, interacting and active galaxies in all environments, probing sub-kiloparsec scales out to well beyond the Coma cluster. These will reveal the internal substructure, uncovering the fossil record of their formation, and probe the relationship between super massive black holes and their host galaxy. MUSE will enable massive spectroscopy of the resolved stellar populations in the nearest galaxies, outperforming current capabilities by factors of over 100. This will revolutionize our understanding of stellar populations, provide a key complement to GAIA studies of the Galaxy, and a preview of what will be possible with an ELT (Fig. 3).

3.2 Narrow-Field Mode

Contrary to the Wide Field Mode, the Narrow Field mode science is dedicated to detailed study of single objects at very high spatial resolution. We give in the following a few examples.

The study of super-massive black holes: During galaxy mergers, super massive black holes sink to the bottom of the potential well, forming

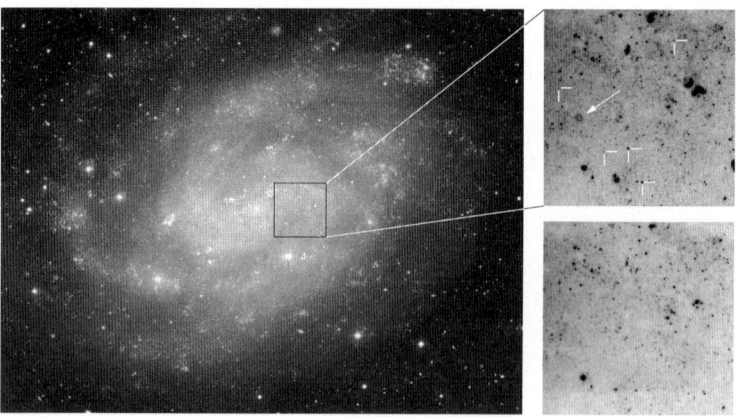

Fig. 3. *Left*: composite image of the southern spiral galaxy NGC 300, illustrating the power of massive spectroscopy with MUSE. The frames to the *right* are a narrowband [O III] 5007 Å exposure (*top*), and a corresponding nearby continuum exposure (*bottom*), obtained with the NTT over a FOV of 2.2 arcmin^2 [5]. MUSE will cover the same field in a total of 4 exposures. Unlike the narrowband imaging example (intended for the purpose of discovering planetary nebulae), the MUSE datacube will provide full spectral information for each spatial element, with a huge discovery potential for massive stars, super bubbles, H II regions, PNe, SNRs, novae – virtually the full inventory of the stellar and gaseous constituents of the galaxy. As a complement to GAIA, application to LMC/SMC and the bulge of the MW will provide kinematics and abundance information for the detailed study of stellar populations and the formation history of the host galaxy, similar to e.g. the RAVE survey [6], albeit orders of magnitude more efficiently

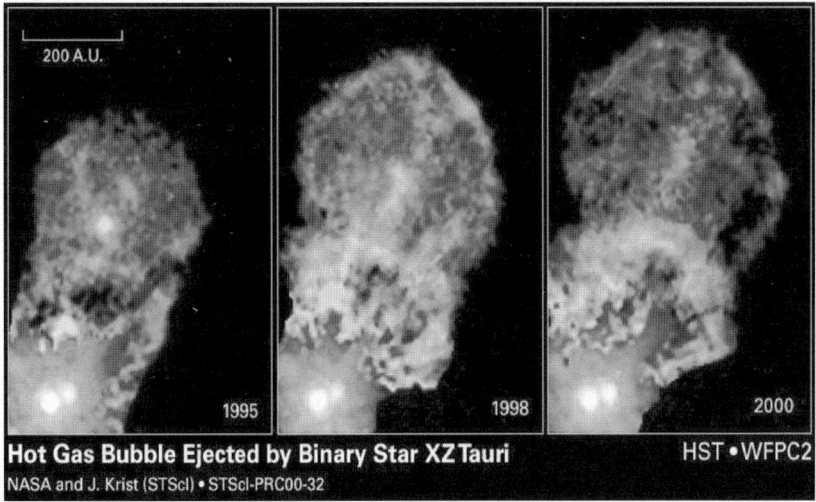

Fig. 4. Time evolution of a Herbig-Haro bow shock in the pre-main-sequence binary system, XZ Tauri, using HST to image emission from ionized gas [2]

binary systems which 'scour out' lower-density cores in the central regions of the remnant. Such processes should leave detectable signatures in the environment of the super-massive black hole. Likewise, accretion of mass onto super massive black holes should trigger activity and feedback to the local regions and beyond. However, observationally, very little is known about this environment, either in terms of stellar orbital structure or chemical enrichment history.

Young stellar objects: The key contribution from MUSE will be both in spectral grasp (covering key diagnostics of density, temperature and ionization) and the ability to provide very high spatial resolution over a relatively large field of view. This will allow the physical processes involved in the formation and structure of the jets to be investigated in detail, as well as tracking the temporal changes in these objects (Fig. 4).

Solar system: MUSE NFM would allow observation of various bodies within our solar system at a spatial resolution approaching that of more costly space missions. Applications are: monitoring volcanic activity on the Galilean satellites, spectral monitoring of Titan's atmosphere, global monitoring of the atmospheres of Uranus and Neptune, internal structure and composition of comets and mineralogical surface heterogeneities of asteroids.

4 Opto-Mechanical Concept

The opto-mechanical concept has to fulfill the following challenging requirements:

- Replication of modules at low cost in order to achieve the required number of spatial and spectral elements;
- High throughput despite the required number of optical surfaces;
- High image quality in order to optimally use the image quality delivered by the AO facility;
- High stability and reliability over long exposures;
- Maintain cost, mass and volume.

The 24 IFUs are central to MUSE. They have been designed to achieve an excellent image quality (85% enclosed energy within a $15\times30\,\mu m^2$ in the detector plane), and make use of innovative slicer and spectrograph concepts. The slicer is based on a 2-mirror compact design, suitable for diamond machining. Recent progress of the manufacturing process has enabled high precision metal surfacing with good surface roughness (3 nm RMS). Such mirrors are now compatible with optical wavelength requirements and are much more cost effective than other approaches for the large-scale production foreseen for MUSE. The compact spectrograph design achieves an excellent image quality over the large spectral bandwidth of MUSE. In this design, the tilt of the detector compensates for the axial chromatism, which then does not need to be corrected optically. This is a cost effective solution, avoiding the use of expensive optical materials, e.g. CaF_2.

To maintain a high throughput (40% for the whole instrument) despite the relatively large number of required surfaces, attention is paid to use state-of-the art transmission and reflection coatings. Detectors are $4k\times4k$ 15 μm deep depletion devices with improved quantum efficiency in the red. Furthermore we will use new volume phase holographic gratings with a high efficiency over the large (one octave) spectral range. To simplify the interfaces between GALACSI and MUSE, all AO components, including the tip/tilt sensor, are mounted in the Nasmyth derotator. There is however a risk of misalignment of the AO reference system with respect to MUSE, which is located on the platform. To mitigate this risk and to maintain the optical axis within the tight tolerances required by the spatial performances and stability, a metrology system has been designed. It is an on-sky closed-loop system that utilizes the regions between the square MUSE science field and the edges of the full transmitted circular beam.

The cryogenic system is based on pulse tubes, which are compact and which avoid to refill 24 dewars with liquid nitrogen. The accompanying compressors are located outside the Nasmyth platform on the telescope floor which avoids any possible transmittance of vibrations onto the instrument. The instrument weight is approaching 8 metric tonnes in total and its size will fill basically the entire volume of the Nasmyth platform of roughly $50\,m^3$. This makes MUSE an impressive instrument, being larger and more massive than any other instrument built so far for the VLT (Fig. 5). With these dimensions, assembling and providing the necessary access to all the components is a challenge. The main instrument structure is designed as a single unit to

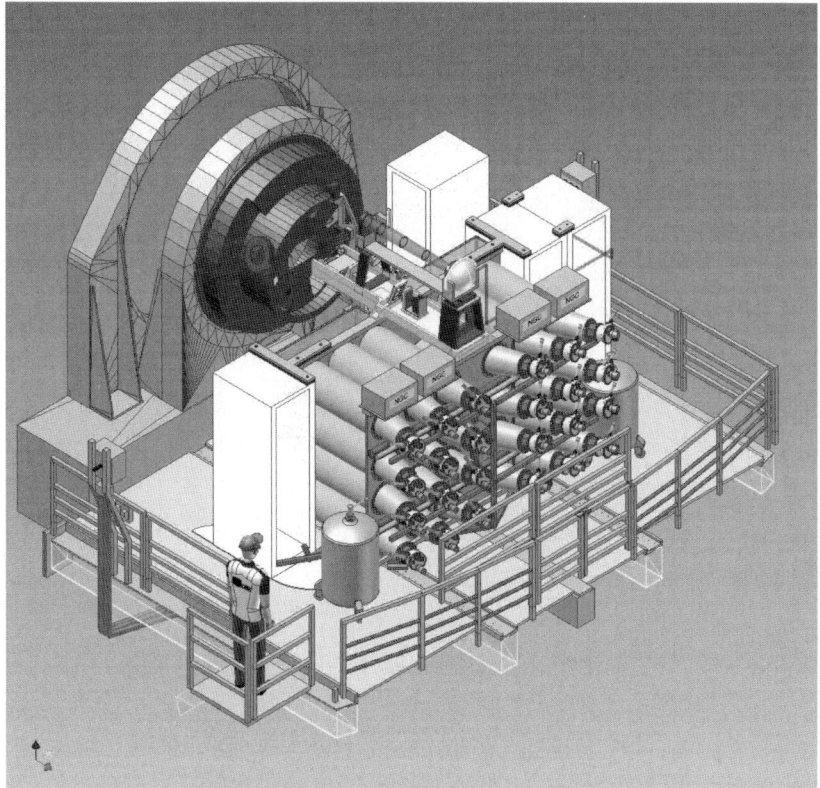

Fig. 5. Current visualisation of MUSE mounted on the VLT Nasmyth platform

fulfil the highly demanded stability of all optical components regarding each other in order to maintain the superb image quality given by GALACSI on long exposures.

The latter is done with a complex optical system that has to derotate and to split the observing field and to distribute and feed the spectrographic units with these sub-fields. Despite its 24 spectrographs mounted into a monolithic structure, MUSE will act as a single instrument with respect to the telescope and the AO system. Nevertheless, the instrument is set up with a highly modular character for the assembly, maintenance and any operational exchange.

5 Operation and Data Reduction

Despite its impressive number of optomechanical elements, MUSE shall be an instrument easy to operate. There are no moving parts in the 24 modules and the switch between wide to narrow field mode implies only the addition of some optics within the fore-optics train. MUSE has only three operating

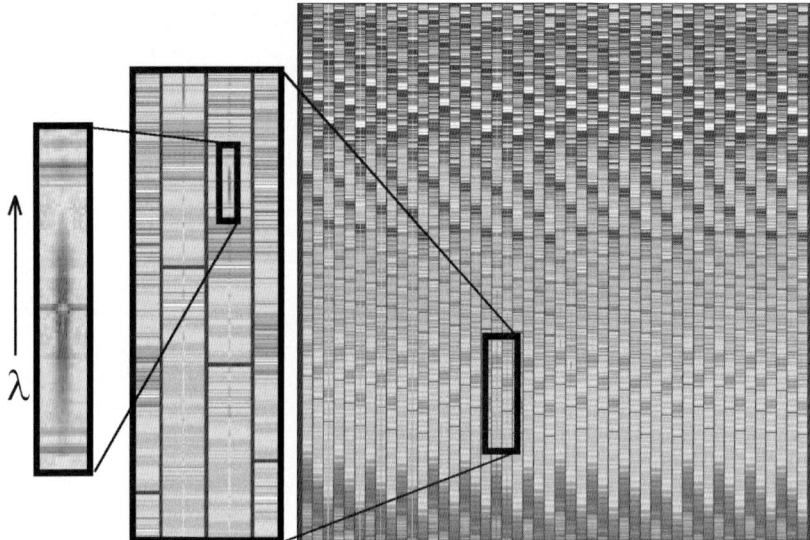

Fig. 6. Example of simulated MUSE data. The large frame shows one exposure from one of the 24 IFU modules, with inserts to show an emission feature. Dealing with the data volume and complexity from MUSE is a major challenge for the data reduction software

modes: non-AO and AO wide-field mode, and AO narrow-field mode. The three modes differ only by the presence of AO and the spatial sampling. In the wide field non-AO mode, operation shall be limited to the simple point-and-shoot scheme. In the other modes, the complexity is related to the operation of AO including the lasers. All modes share the same spectroscopic configurations (wavelength range and resolution).

On the other hand, with 1.6 GB per single exposure, the data reduction (Fig. 6) is a challenge, not only because of this data volume, but also because of its 3D characteristics. The handling of such large data cubes is not straightforward. As an example, one can mention the optimal summation of a series of data cubes obtained with AO and different atmospheric conditions. This is intrinsically a 4-dimensional problem because the AO-delivered PSF changes with time, location within the field of view, and wavelength.

6 Project Status

The MUSE Consortium consists of groups at Lyon (PI institute, CRAL), Göttingen (IAG), Potsdam (AIP), Leiden (NOVA), Toulouse (LAOMP), Zurich (ETH) and ESO. The project is currently in its preliminary design phase. The full preliminary design review is scheduled for July 2007. The first stage will be a prototype IFU module, which will be analyzed for the final design review in 2008. Manufacturing, assembly and integration will then take place up to mid 2011. First light is scheduled on Paranal early 2012.

7 Conclusions

Astronomy is to a significant degree still driven by unexpected discovery (e.g. dark matter and dark energy). These discoveries are often made by pushing the limit of observations with the most powerful telescopes and/or opening a new area of instrumental parameter space. MUSE is designed to push the VLT to its limit and to open a new parameter space area in sensitivity, spatial resolution, field of view and simultaneous spectral coverage. We are convinced that it fulfils all the required conditions to have a large potential of discoveries:

- It will be the first spectrograph that could blindly observe a large volume of space, without any imaging pre-selection.
- It will be the first optical AO-assisted IFU working at improved spatial resolution in most atmospheric conditions with large sky coverage.
- It will be the first spectrograph optimized to work with very long integration times and to reach extremely faint emission line detection.

MUSE will thus be able to discover objects that have measurable emission lines, but with a continuum that is too faint to be detected in broad-band imaging. For example, the deepest broad-band imaging available today is the HST Ultra Deep Field (UDF) with $I_{AB} < 29$. According to CDM simulations, however, only 15% of MUSE high-z Lyα emitters (z $>$ 5.5) will have a continuum bright enough to be detected in the UDF. MUSE is also the only instrument capable of detecting faint diffuse ionized gas, like extended halos or filaments. Finally, objects with unusual spectral features should also be detected by MUSE, whatever their broad band magnitude and colors are. The unprecedented capabilities of MUSE should also lead to discoveries far away from our present expectations.

By many aspects, MUSE is a precursor of future ELT instrumentations. For example manufacturing, integration and maintenance of a large number of identical, high performance optical systems at low cost and on reasonable time scale will be a critical aspect for most of ELT instruments. MUSE will also be the first AO-assisted IFU to address a key science case of future ELTs: massive spectroscopy of resolved stellar populations in nearby galaxies, employing crowded field 3D spectroscopy over a large field-of-view.

References

1. R. Arsenault, et al.: SPIE, **6272**, 62720V (2006)
2. J.E. Krist, et al.: Astrophysical Journal **515**, L35 (1999)

3. O. LeFevre, et al.: SPIE, **4841**, 1670 (2003)
4. L. Pasquini: The ESO Messenger **110**, 1 (2002)
5. T. Soffner, R.H. Mendez, G.H. Jacoby, et al.: A&A **306**, 9 (1996)
6. M. Steinmetz: Proceedings of the Monte Rosa Conference "GAIA spectroscopy, science and technology", astro-ph/0211417 (2002)

SPHERE-IFS: A Tool for Direct Detection of Giant Planets

R. U. Claudi[1], J. Antichi[1], R. G. Gratton[1], S. Desidera[1], A. Berton[2], D. Mesa[1], M. Turatto[1], M. Feldt[2], K. Dohlen[3], J.-L. Beuzit[4], D. Mouillet[5], and P. Puget[4]

[1] INAF – Osservatorio Astronomico di Padova, Padova, Italy;
 riccardo.claudi@oapd.inaf.it
[2] Max Plank Institut für Astronomie, Heidelberg, Germany
[3] Laboratoire Astrophysique de Marseille, France
[4] Laboratoire Astrophysique de l'Observatoire de Grenoble, Grenoble Cédex 9, France
[5] Laboratoire Astrophysique de Toulouse, Toulouse, France

Abstract. The 2nd generation VLT instrument for Spectro-Polarimetric High-contrast Exo-planet Research (SPHERE) will include an integral field spectrograph to enhance the capabilities of detection of planetary companions close to bright stars. SPHERE-IFS is foreseen to work in near IR (0.95–1.35 µm) at low spectral resolution (R = 50). We present the concept of the instrument and the plans to achieve extreme luminosity contrast using it.

1 Science Goal

The search for other planetary systems and the fundamental understanding of how planets form are among the key questions of modern astrophysics. The detection of extra-solar planets via the Doppler wobble of their host stars has been one of the great astronomical achievements of the past years. The detection of direct light from extra-solar giant planets (EGPs) and the characterization of their atmospheres is the next great challenge in this field.

The VLT Planet Finder SPHERE [3] is designed with the primary science goal to image and characterize extra-solar planets around nearby stars ranging in age from 10 Myr to 10 Gyr. In order to overcome the brightness contrast and the speckle noise, the VLT-PF IFS concept consists in a high-order AO+Coronograph system feeding the three instruments optimized for Simultaneous Differential Imaging (SDI).

2 SPHERE

The design of SPHERE, the VLT Planet Finder, includes extreme Adaptive Optics, phase-mask or apodized Lyot Coronagraphy, and three instruments providing complementary differential imaging capabilities: ZIMPOL (differential polarimetry), IRDIS (differential imaging) and IFS (spectroscopic differential imaging). All sub-systems will be mounted on a common optical

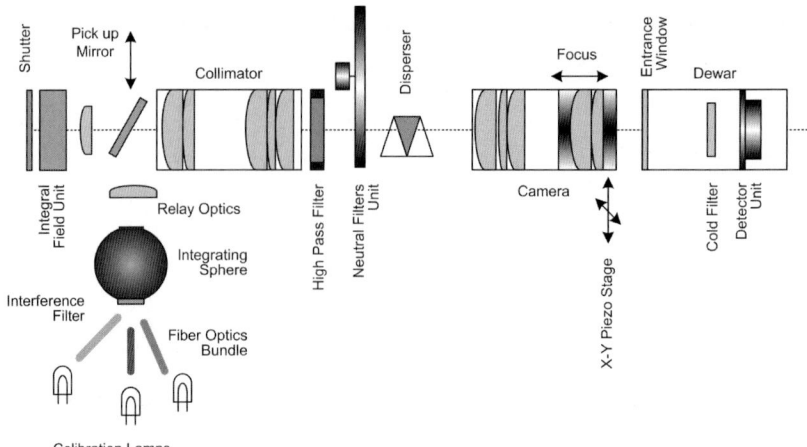

Fig. 1. Preliminary optical layout of SPHERE-IFS

bench fixed to one of the Nasmyth platforms at the VLT-UT4. PDR of the instrument took place in Sep 2007. Commissioning is scheduled in 2010.

3 SPHERE IFS: Instrument Concept

The scheme of SPHERE IFS is shown in Fig. 1. The optical design includes the integral field unit (IFU), the collimator, the disperser, the camera and the 2 k×2 k detector. Furthermore, an internal calibration arm is foreseen to provide the detector flat fielding. Additional calibrations (IFS flat fielding, wavelength calibration) will be provided by sources located in the common path of SPHERE. The instrument covers the spectral range between 0.95 to 1.35 µm with a spectral resolution of about 50. Adequate spatial sampling of the VLT diffraction limited PSF is provided. The field of view is about 1.8×1.8 arcsec. More detailed descriptions of preliminary versions of the design are in [5, 1].

SPHERE IFS is foreseen to be used in parallel with the other NIR instrument of SPHERE, the differential imager IRDIS. The two instruments are complementary in terms of wavelength (Y and J bands for IFS, H for IRDIS), field of view (larger for IRDIS), contrast (better for IFS), sensitivity to non-methane planet spectral features. The simultaneous acquisition using the two channels will also allow us a better rejection of false alarms.

4 Calibrations

Calibration of SPHERE IFS data will include detector calibrations, wavelength calibration, IFS normalization. Flat field calibration is expected to be crucial for the achievement of instrument performances in the central regions of the field of view. To improve flat field accuracy down to a few 10^{-4} dithering will be performed by moving part of the camera. Furthermore, nighttime flat field calibration during pointing and adaptive optics set-up is foreseen, to take into account the time dependence of the flat field calibration. Cross talk is expected to be negligible thanks to the optimized IFU optical design.

5 Spectroscopic SDI Technique

Speckle noise overcomes other source of errors in the ground based search for faint companions around bright stars. The speckle noise can be significantly reduced using the SDI technique. This consists in the simultaneous image acquisition at adjacent wavelengths where the spectra of the central star and the companion differ significantly. The subtraction of the narrow band images, after suitable scaling, should remove the speckle noise. Giant planets and brown dwarfs have suitable spectral features in the NIR spectral region (J and H bands) and SDI devices have been realized in the past years. With NACO-SDI it is now possible to achieve a contrast of > 10 mag at 0.5 arcsec from the central star [4]. SDI was originally proposed using two wavelengths, but extension to three wavelengths has been proved useful in improving performances [6].

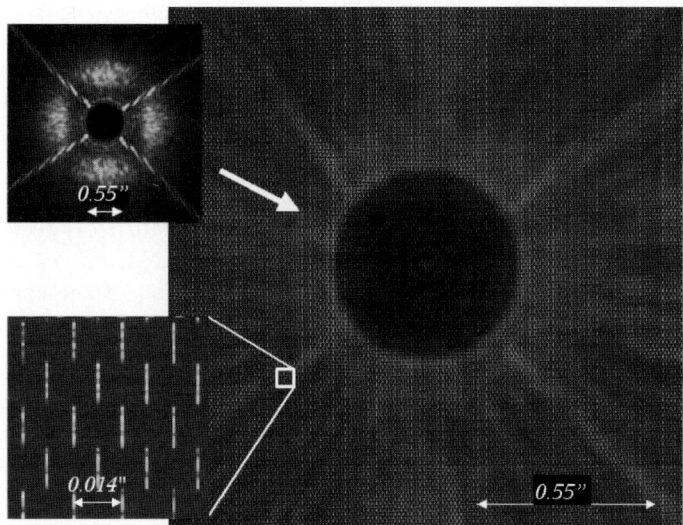

Fig. 2. Simulated images of SPHERE IFS. From Berton et al. [2]

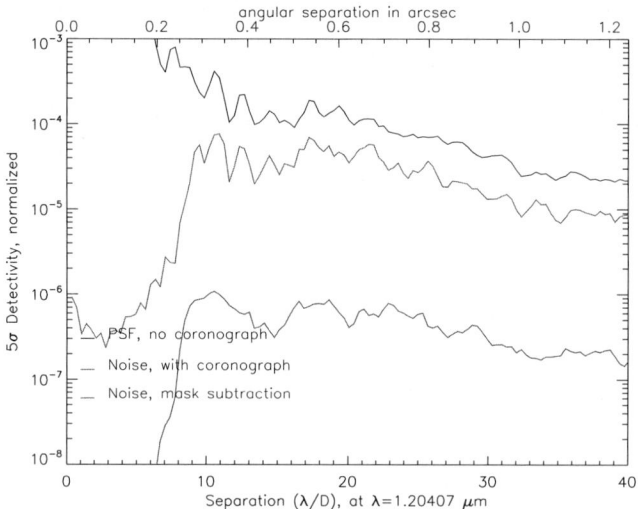

Fig. 3. Preliminary detectability limits for SPHERE-IFS resulting from instrument simulations. *Upper line*: PSF without coronagraph; *central line*: detectability limits with a Lyot coronagraph but without image subtraction (no correction for speckle noise); *lower line*: detectability limits adopting a suitable combination of monochromatic images

The use of an integral field spectrograph allows to work on a larger number of wavelengths and to reduce non-common path aberrations that somewhat limit the performances of standard differential imagers. The spectroscopic SDI has the additional advantage of working on several spectral features and in being sensitive to a variety of companions spectra, beside that characterized by deep methane bands typical of cool substellar objects. First tests on sky using SINFONI confirm the usefulness of IFS to achieve high contrast at small separations from the central star [7]. Detailed simulations of SPHERE-IFS (Figs. 2 and 3) show that speckle noise can be effectively reduced and that contrast fully adequate to detect Jupiter-mass planets around young stars can be achieved. We are currently exploring different algorithms to perform the subtraction of the speckle pattern, minimizing the number of interpolations required on the original data cube.

References

1. J. Antichi: PhD Thesis, Università di Padova (2007)
2. A. Berton, R. Gratton, M. Feldt, et al.: PASP **118**, 1144 (2006)
3. J.-L. Beuzit, M. Feldt, K. Dohlen, et al.: ESO Messenger **125**, 29 (2006)

4. B. Biller, L. Close, E. Masciadri, et al.: SPIE **6272**, 74 (2006)
5. R. Claudi, M. Turatto, R. Gratton, et al.: SPIE **6269**, 93 (2006)
6. C. Marois, R. Doyon, R. Racine, D. Nadeau: PASP **112**, 91 (2000)
7. N. Thatte, R. Abuter, M. Tecza, et al.: MNRAS **378**, 1229 (2007)

The Origin of Fringing in the VIMOS IFU

E. Jullo[1], L. Christensen[1], A. Smette[1], S. Bagnulo[1], C. Izzo[2], and G. Marconi[1]

[1] ESO, Alonso de Cordóva 3107, Vitacura, Casilla 19001 Santiago 19, Chile; ejullo@eso.org
[2] ESO, Karl-Schwarzschild-Strasse 2, 85748, Garching, Germany

Abstract. In addition to detector fringing, the VIMOS IFU suffers from additional, not reproducible fringing affecting some pseudo-slits. It can severely decrease the sensitivity of the instrument. Its origin has not been clearly determined. Here, we present the analysis of calibration data obtained in October 2006. We show that fringing originates either from inside the mask at the pseudo-slits level or from an imperfect positioning of the masks themselves. The lack of reproducibility of the fringes may be caused by the imperfect fixation of the prisms in the pseudo-slits or a loose positioning of the masks.

1 Introduction

In addition to detector fringing, the VIMOS Integral Field Unit (IFU) suffers from additional fringing whose origin has not been clearly determined. It is most clearly visible in the blue part of the spectra where the detector fringing contamination is low. In order to precisely determine the origin of the fringes, we adopt an empirical strategy consisting in changing one parameter of the instrument at a time. Thus, we are able to identify the influence of each parameter on the fringing patterns.

2 Grism and Filter Effect

First, we consider the effect of the grisms and filters in producing the fringes. Figure 1 clearly shows that the HR-Orange spectrum shows continuity with the HR-Blue one taken immediately before: the fringes show the same period and phase. Since the fringe periods and phases are identical in the two settings, the fringes do not originate in the spectrograph.

3 Shift due to Instrument Movements

Then, we look for changes of the phase of the fringing pattern produced by the rotation of the instrument and the masks' movements in the removal

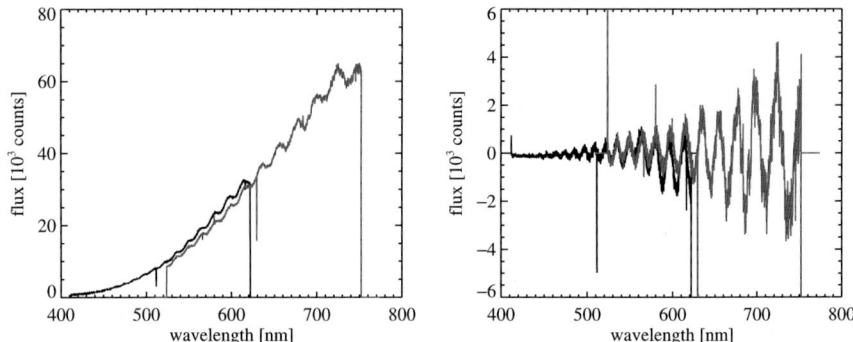

Fig. 1. *Left panel:* Comparison of the screen FF spectrum of fibre 90 in Q2. *In black (in grey)*, the spectrum obtained with the HR-Blue (HR-Orange) grism. The same lamps have been used for the screen FF. *Right panel:* The fringes pattern have been emphasised by subtracting the extracted median spectrum to the spectra of the *left panel*

and insertion of the IFU head. On the left panel of Fig. 2, we note that the maximum shift between fringes appears between the spectra taken at 26.6° and 83°. The last acquisition back to 26.6° is also shifted to the red relatively to the first spectrum at 26.6°. This clearly indicates that the fringes shift in wavelength with the instrument rotation angle. However, some hysteresis

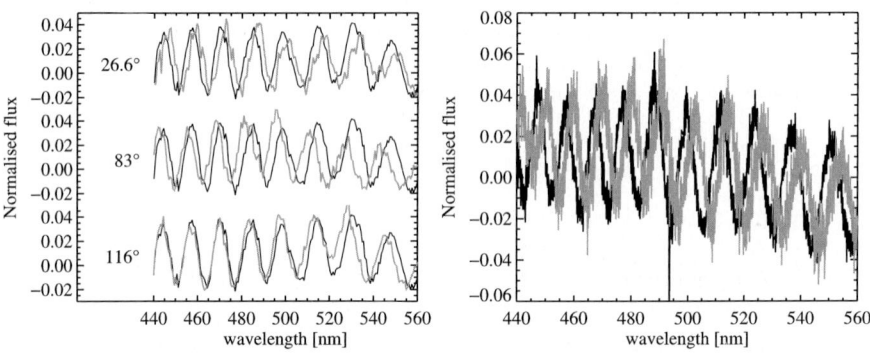

Fig. 2. Four sky spectra taken at three rotation angles divided by the median spectra of all the fibres in Q2 taken during a same sequence with the LR-Blue–OS-Blue (grism–filter) combination and a pixel scale of 0.67". *Left panel:* The solid line spectrum was obtained at 26.6° (usual calibration position). The bottom, middle and upper spectra (*in grey*) were taken at 116°, 83° and finally back at 26.6° respectively. *Right panel:* The black and the grey spectrum were taken before and after the IFU mask movement respectively

exits, as data obtained at the same rotation angle after an instrument rotation do not show the same phase in the fringe pattern.

On the right panel of Fig. 2, a similar shift is observed when comparing spectra taken before and after the removal and insertion of the IFU masks.

4 Difference Between the Sky FF and the Screen FF

Third, we consider the 2 types of illumination of the fibres: from the sky or by reflection of a halogen lamp on the Nasmyth shutter used as a screen. Figure 3 shows that the fringing patterns not only does not show the same phase but does not even show the same period between screen and sky FF.

5 Variations Inside a Single Mask

After looking at the extracted spectra, we consider the raw images of screen FF. In Fig. 4, the fringing patterns shift slowly in phase inside a given module but strong breaks appear from one module to the next. This implies that the fringing pattern appear at the module level in the pseudo-slits likely at the interface between the output lenses of the fibres and the prism used to fold the beams towards the spectrograph.

Another hypothesis is the following. The VIMOS fibres are multimode [1]. They could generate a wavelength dependent pattern of modes at the output of the IFU masks. A misalignment between the spectrograph and the IFU masks could produce a vignetting of the spectrograph entrance pupil to the IFU masks output pupil and thus create the fringing patterns.

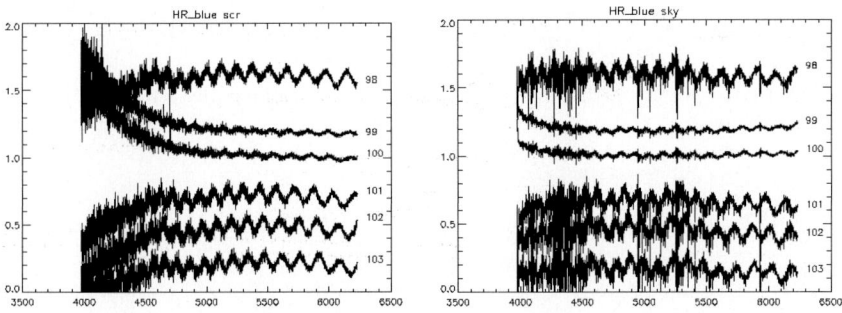

Fig. 3. Comparison of fringing patterns in consecutive sky (*left panel*) and screen FF (*right panel*) in Q2 obtained with the HR-Blue grism with no change in the rotation angle of the instrument. Each spectrum has been normalised

Fig. 4. Variations of the fringing pattern in screen FF raw image between the 5 modules of pseudo-slit 2 of mask Q2 observed with the HR-Blue grism. Each column of the raw image has been divided by the median column spectrum of all the columns of the image

6 Conclusion

The fringes origin remains uncertain and additional tests are necessary. Their lack of repeatability may be caused by the imperfect fixation of the prisms or a problem in positioning the masks themselves.

Reference

1. E. Prieto, O. Le Fevre, M. Saisse, C. Voet, C. Bonneville: Proc. SPIE **4008**, 510 (2000)

A User's View of VIMOS-IFU Calibrations

H. Kuntschner

Space Telescope – European Coordinating Facility, Karl-Schwarzschild Strasse 2, 85748 Garching, Germany; hkuntsch@eso.org

Abstract. The ESO VIMOS-IFU pipeline produces wavelength calibrated spectra for all working fibres and each quadrant separately. Using a recently carried out science program with VIMOS-IFU (HR-blue, 13" × 13" field-of-view, hereafter FoV) I analyse the quality of the basic calibrations.

Furthermore, I study the instrumental resolving power as function of wavelength and position within the FoV. A rather large variation between 1.8–3.0, Å (FWHM) is found. There are clear trends with wavelength, but most importantly, significant differences between Fibre Modules are visible. The differences between modules translate into sharp steps within the FoV and can jeopardize the scientific exploitation if not corrected for.

1 VIMOS IFU Mode

VIMOS is the Visible Multi-Object Spectrograph for the Very Large Telescope (VLT) of the European Southern Observatory (ESO). This instrument has been built to provide the ESO community with a wide field spectrograph with high throughput and multiplex, dedicated to deep surveys. VIMOS operates in the 0.37–1.0 µm domain in three main observing modes: direct imaging (IMG), multi-slit spectroscopy (MOS), and integral field spectroscopy (IFU). Because of the large field of view, VIMOS optical path is split into four channels. The four quadrants of the instrument are operated in parallel, and each of the four channels has a 2048×4096 pixels EEV CCD, with the 4k pixels being used along the dispersion to maximize the spectral coverage.

The Integral Field Unit (IFU) is a dedicated opto-mechanical system aimed at producing spectra of a contiguous area of up to 54"×54". The sky image is projected onto a micro-lens array with a choice of two spatial samplings (magnifications): 0.33" or 0.67" per resolution element (fibre). Each 2D input micro-lens is coupled to an optical fiber. The output of the fibers is re-arranged on a linear set of micro-lenses to produce an entrance slit to the spectrograph at the mask focal plane location. Each IFU mask (one for each channel of VIMOS) has a set of 4 fibre-slits (or pseudo-slits), for a total of 6400 fibers. A masking-shutter (IFU shutter) is provided in front of the input micro-lens array; this allows the exclusive use of the central 1/4th of the field when the IFU is used with intermediate and high spectral resolution grisms

Fig. 1. Cutout of wavelength calibrated VIMOS-IFU science spectra in quadrant 4. The wavelength increases along the x-axis with the bright 5577 Å skyline visible at the *left side*; the total wavelength range shown extends from approximately 5500 to 6000 Å. Each row in y-direction corresponds to one fibre. Several traces of the science object (an E+A galaxy) are visible in the *lower part* of the image

(HR grisms). In this configuration, only 1 fiber-slit per quadrant is used, for a total of 1600 fibers.

In this contribution we study recent IFU observations in the HR-blue mode providing a spectral coverage of 4150–6200 Å and a dispersion of 0.5 Å/pixel (P076.B-0637, PI: Kuntschner). With a spatial sampling of 0.33″ the total field coverage is about 13″×13″. The nominal spectral resolution of this mode is given as R = 2550 in the VIMOS user manual [1]. This corresponds to 1.8 Å and 2.4 Å (FWHM) at 4700 and 6000 Å, respectively.

In Fig. 1 we show an example of pipeline reduced (version vimos/2.0.7) VIMOS-IFU science data for quadrant 4 (only a cutout is shown). The spectrum from each fibre is extracted and wavelength calibrated.

2 Accuracy of the Wavelength Calibration and Resolving Power

The pipeline reduced spectra of VIMOS-IFU data are well aligned in wavelength direction. The mean wavelength calibration error of the 5577 Å sky line as measured in this study is ±0.1 Å corresponding to 1/5th of a pixel (see also Fig. 1). However, it was noted that the width of the emission lines changes as function of position on the detector. We therefore decided to map the spectral resolving power of this VIMOS mode by using the arc lamp calibrations observed during the night together with the science frames.

The arc lamp observations show about 15 strong emission lines for which we determined the width by fitting a Gaussian to the data. In Fig. 2 we show the results for two arc lines at wavelengths of 4713.2 Å and 6030.0 Å,

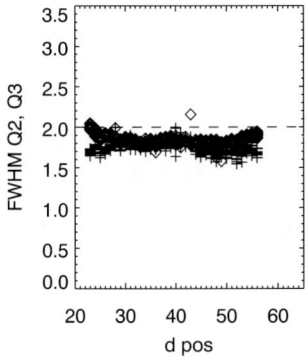 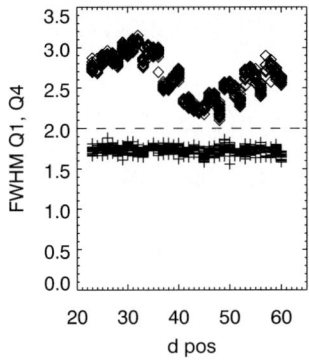

Fig. 2. The plot shows vertical cuts through Fig. 1 for each quadrant at the arc lines at 4713.2 Å *(plus signs)* and 6030.0 Å *(open squares)*. The x-axis shows the fibre number in spatial direction. Note the degradation of spectral resolving power for quadrant 1 and 4 at the wavelength of 6030 Å. Significant steps connected to fibre modules and a large scale sinusoidal wave pattern are visible

and as function of spatial position given in fibre number (this corresponds to declination for our data-set). While for quadrants 2 and 3 we find a mean line width of 1.8 Å (FWHM) with little scatter, there are much larger linewidths measured for the 6030 Å arc line in quadrants 1 and 4. There are also significant steps as function of position which appear to be connected to Fibre Modules.

The same measurements of the emission line widths can also be shown as function of wavelength which is demonstrated in Fig. 3. For quadrants 1 and 4 there is a clear degradation of resolving power towards larger wavelength, whereas this is not seen for quadrants 2 and 3.

3 Conclusions

- The basic wavelength calibration of the VIMOS-IFU mode is satisfactory after pipeline data processing. The errors are on the order of 0.1 Å rms as measured from the 5577 Å sky emission line.
- The spectral resolving power varies in the HR-blue mode with a spatial sampling of 0.33" as a function of (1) field-of-view, (2) fibre module and (3) wavelength.
- The absolute spectral resolving power varies from about 1.8 to 3.0 Å (FWHM). A few individual spatial positions can deliver even worse resolutions.
- Although it is the responsibility of the user to ensure scientific integrity of the data, it would be of help to provide more detailed information on

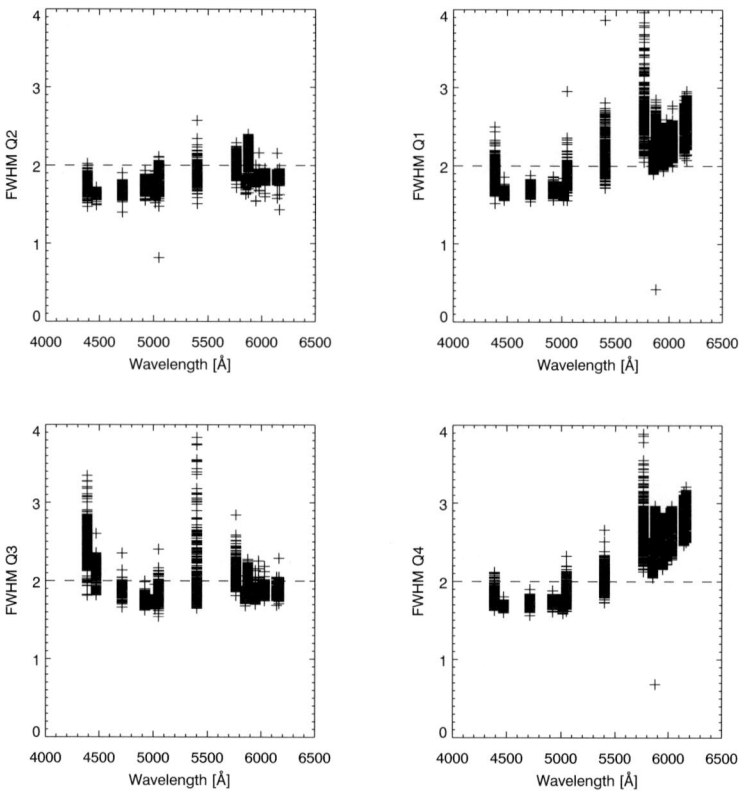

Fig. 3. Measurements of the spectral resolving power (FWHM) of arc lines shown as function of wavelength for all four quadrants. While quadrants 2 and 3 show a median resolution of about 1.8 Å (FWHM), quadrant 1 and 4 show significant trends with wavelength and line widths in excess of 3 Å (FWHM) for wavelengths around 6000 Å

the spectral resolving power in the manual. Furthermore, measurements of the resolving power for each night calibration should be included in the pipeline products as a guideline for the user.

Reference

1. VIMOS User Manual, Doc. No. VLT-MAN-ESO-14610-3509 Issue 79, Date 13/12/2006.

Part VIII

Session 7: High Resolution Spectroscopy

IOT Overview: High Resolution Spectrographs

G. Lo Curto[1], P. Francois[1,2], C. Ledoux[1], C. Melo[1], E. Pompei[1], A. Smette[1], and L. Vanzi[1]

[1] ESO, Alonso de Cordóva 3107, Vitacura, Santiago, Chile; glocurto@eso.org
[2] Observatoire de Paris-Meudon, GEPI, 92195 Meudon Cedex, France

1 Introduction

ESO operates a suite of nine High Resolution (HR) spectrographs in Chile, covering the wavelength range from the optical to the medium infrared. They explore a wide spectra of scientific goals, which ultimately define their specific characteristics. All the spectrographs are operated both in service and in visitor mode, and obey a similar operation scheme in common with all the ESO instruments, which is driven by the concept of Observation Blocks (OB), that allows users to entirely pre-define the instrument configuration for their observations.

Calibrations, which are also used in part to monitor the instruments performance, are executed following dedicated calibration plans, presented in the instrument user manuals and web pages.

This work is organized as follows: Section 2 presents a summary of the instrumentation, Section 3 discusses the main common calibration issues, Section 4 presents the issue of stability for instruments that are used in Radial Velocity (RV) measurements, and the last section presents some future prospects and the conclusions.

2 Instrumentation

The ESO instrumentation for high resolution spectroscopy covers a wide range of formats and wavelengths. The suite of instruments includes both fibre and slit spectrographs, in fixed configuration or with tunable central wavelength. All the ESO high resolution spectrographs use echelle reflective diffraction gratings, either in pre-dispersed or in cross-dispersed format, for high dispersion and optimum efficiency.

As always, the accent has been put on providing high data quality and efficiency: technical downtime due to the spectrographs is generally well below 2%. This is achieved through the execution of well aimed maintenance plans.

The science coverage of the ESO high resolution spectrographs is extremely large, ranging from solar system to extra-galactic astrophysics. In the years 2005 and 2006 the ESO high resolution spectrographs alone have contributed to more than 300 refereed publications.

Apart from CES and EMMI-echelle, which only have "quick-look" scripts, all ESO high resolution spectrographs are equipped with online pipelines, which, apart from performing the reduction of the data, supply the basic quality control parameters. The HARPS pipeline is developed and maintained by the consortium who build the instrument, and besides delivering science products and precise radial velocities also performs online quality control.

2.1 Optical Spectrographs

Fibre-Fed, Extended Range, Echelle Spectrograph (FEROS)

The FEROS spectrograph [10] is a fibre fed, cross dispersed echelle spectrograph. It uses two fibres whose diameter covers an angle of 2" in the sky, one fibre is normally placed on the science target while the other one can be exposed either to the empty sky for background subtraction or to a reference light for drift measurement and optimum radial velocity precision. The configuration of the optical bench is fixed and the echelle grating is operated in quasi Littrow conditions. FEROS was commissioned in 1998 at the ESO 1.52 m telescope targeting time dependent phenomena in stellar atmospheres as well as extra-solar planet science and the fast growing field of asteroseismology. The key figures for its main science goal are the high efficiency and the large spectral coverage between 350 and 920 nm. In 2002 the instrument was moved to the 2.2 m telescope; this upgrade increased even more its already high efficiency, reaching the level of 20% at 500 nm (total system efficiency: instrument + telescope). With its resolving power of 48,000 and thanks to the thermal stability of its enclosure, FEROS can achieve a radial velocity precision better than $25\,\mathrm{ms}^{-1}$ when using the ThArNe reference light for drift measurements and once the appropriate set of day calibrations is performed.

High Accuracy Radial Velocity Planet Searcher (HARPS)

The HARPS spectrograph [16], installed at the 3.6 m telescope, is the ESO instrument dedicated to very high precision Radial Velocity (RV) measurements. Its main scientific goal is the search for extra-solar planets, and its high RV precision makes it well suited for asteroseismology as well. The large spectral coverage (390–690 nm) and the high resolving power (115,000) make it also a desired general purpose spectrograph. The configuration of the optical bench of HARPS, likewise to FEROS, is fixed and the echelle grating is operated in quasi Littrow conditions. The HARPS spectrograph is an evolution of the concept already adopted for ELODIE (OHP) and CORALIE (La Silla, Swiss national telescope) and, similarly to FEROS, its design follows along the lines traced by Baranne [1] for a compact fibre spectrograph with high resolution, large spectral coverage, high stability and above all high radial velocity precision. Its two 1" fibres are used to collect the object light and the reference light from the ThAr lamp, or alternatively the sky emission for background subtraction. Its extreme mechanical and thermal stability (temperature variation in one night at the grating <1 mK, optical bench under vacuum), joined with the image scrambler at the spectrograph's entrance and the optimized data reduction software, allow to achieve an unsurpassed radial velocity precision of $1\,\mathrm{ms}^{-1}$ or better. Such precision has been demonstrated, up to now, on a time scale of 3 years [15], since the commissioning of the instrument. The measured drift of the spectral lines of a reference source on the detector is less than $1\,\mathrm{ms}^{-1}$ in one night.

Coude Echelle Spectrometer (CES)

CES is ESO's highest resolution spectrograph providing a resolving power of 220000 in the 346–1028 nm region [7]. It is available with use with the 3.6 m telescope in its present configuration (Very Long Camera) since 1998. Its main scientific goals are the study of chemical abundances in the Galaxy and the study of line profiles in stellar atmospheres and in the interstellar medium. It is fed through an optical fiber mounted at the f/8 Cassegrain focus of the 3.6 m telescope. The diameter of the fiber is 2" on the sky. The CES echelle grating is in a Czerny-Turner mount, allowing adjustment of the central wavelength; only a part of one spectral order, between 2.5 and 6.5 nm wide, can be recorded at one time on the CCD. In order to optimize efficiency over the whole range, the instrument is composed of two paths, optimized for red and blue wavelengths. The entrance slit is constituted by an image slicer, that allows diffraction while minimizing light losses. A pre-dispersing prism in conjunction with a slit selects the narrow wavelength range which reaches the echelle grating. CES can be used in parallel with HARPS, and the two instruments share the same calibration unit.

ESO Multi Mode Instrument (EMMI)-Echelle

EMMI is a versatile instrument mounted at the Nasmyth B focus of the NTT, allowing a wide range of observations, including high resolution spectroscopy thanks to its echelle gratings [4]. Long slit spectroscopy with EMMI-echelle is possible with the use of an order sorting filter to isolate the wavelength range of interest, while allowing to resolve spatially the source in one direction. Another possibility is to use a low resolution grism as a cross disperser: in this case, due to the limited amount of space available on the CCD, a long slit cannot be used to its full extension, to avoid overlapping of the orders, hence the spatial coverage is limited to a few arcseconds. Grating #14 provides the topmost resolving power of $\approx$ 90,000 and, in conjunction with different cross dispersing grisms, yields a wavelength coverage from 385 to 1080 nm. Chemical abundances, stellar atmospheres, 2D high resolution spectroscopy of Planetary Nebulae, study of the QSO absorption lines as tracers of the

Intergalactic Medium and study of stellar winds cover most of the science goals of this instrument.

Ultraviolet and Visual Echelle Spectrograph (UVES)

The UVES spectrograph [5] installed on the Nasmyth B platform of UT2 at the VLT, is a cross dispersed, folded, two arms slit spectrograph, whose optical design was developed by Delabre [6]. This design optimizes the efficiency of the instrument at red and blue wavelengths separately, using optimized coatings for the optics, and CCDs with high quantum efficiencies in the required wavelength range. The spectrograph covers the wavelength range between 300 and 1100 nm and reaches a resolution of 115,000 with a 0.3" slit in the red arm, or 80,000 with a 0.4" slit in the blue arm. The highest spectral resolutions can be achieved under typical seeing conditions by means of image slicers which distribute the star light along the corresponding spectrograph entrance slit. The UVES optical bench is fixed on the Nasmyth platform. This increases the stability of the instrument. Field rotation is compensated through the use of a de-rotator. The UVES spectrograph at the VLT couples the light collecting power of an 8 m telescope with its high dispersion and high efficiency capabilities to pursue scientific goals such as the study of the absorption spectra of high redshift QSOs, kinematics and chemistry of star clusters, chemical composition and atmospheric models of stars, extra-solar planet searches via the RV method, asteroseismology. Precise radial velocities (below $5\,\mathrm{ms^{-1}}$) are obtained via the use of an iodine absorption cell [13].

UVES can be coupled to the fibre link of FLAMES, the instrument located on the opposite Nasmyth platform, for multi objects spectroscopy. Up to 8 fibres can be used in this mode. Another fibre can be used for simultaneous wavelength calibration.

GIRAFFE

GIRAFFE is a medium-high resolution spectrograph (R = 6000–33,000) for the entire visible range (370–950 nm) [19]. It is equipped with two different echelles for low and high resolution spectroscopy and uses interference order sorting filters to select the required spectral range within an order. In total, GIRAFFE is operated with 30 fixed set-ups (22 high-resolution + 8 low-resolution modes). The typical spectral coverage in one exposure is 60–100 nm in low resolution and 20–40 nm in high resolution.

The fibre system feeding GIRAFFE consists of the following components: (i) MEDUSA fibre slits, one per positioner plate. Up to 135 separate objects (including sky fibres) are accessible in MEDUSA single fibre mode, each with an aperture of 1.2 arcsec on the sky; (ii) 15 deployable Integral Field Units (IFU) per plate, each consisting of an array of 20 square microlenses, for a total (almost rectangular) aperture of 3×2 arcsec. For each plate there are also

15 IFU dedicated to sky measurements (see P. Amico's contribution to this proceedings); (iii) A large integral unit (ARGUS) consisting of a rectangular array of 22 by 14 micro-lenses, fixed at the centre of one positioner plate. Two scales are available: one with a sampling of 0.52 arcsec/micro-lens and a total aperture of 12 by 7 arcsec, and one with a sampling of 0.3 arcsec/micro-lens and a total coverage of 6.6 by 4.2 arcsec. In addition, 15 deployable ARGUS single sky fibres are also available.

GIRAFFE is the natural instrument to address scientific issues regarding galactic (star clusters, Galaxy's center) and extra-galactic (galaxy clusters) structures with high density.

2.2 Infrared Spectrographs

Observations in the infrared are generally background dominated, requiring telescope nodding at the NIR and secondary mirror chopping and telescope nodding at MIR wavelengths in order to subtract the local environment and the sky background respectively.

Cryogenic High-Resolution IR Echelle Spectrograph (CRIRES)

The CRIRES spectrograph is already described in detail by H.U. Käufl in his contribution to these proceedings and in [11]. Here it will suffice to say that it is a pre-dispersed, scanning 2D spectrograph. It can be used at wavelengths between 1 and 5 µm, and only one order at a time, spanning a wavelength range of $\approx \lambda/70$, will be recorded in the 4 Aladdin III detectors. The MACAO adaptive optics located before the spectrograph entrance and the 46" long slit allow for high spatial resolution, high efficiency and "large" field of view along one direction. A resolving power of $\approx 100,000$ can be achieved when using the 0.2" slit. The central wavelength can be set during observations. Currently the reproducibility of the setup is of the order of ≈ 3 pixels. Studies requiring high accuracy wavelength calibration should execute a wavelength calibration template for each setting. This unique instrument addresses science topics such as solar system planets atmospheres, star formation, ISM chemistry, fine structure constant, AGN velocity structure and extra-solar planets searches and atmosphere's studies.

VLT Imager and Spectrometer for Mid-Infrared (VISIR)

The high resolution spectrograph arm of VISIR [14] provides spectra with a resolving power of up to $\approx 30,000$. The instrument, thanks to the use of order sorting filters, allows 2D spectroscopy at ≈ 8, 12.8 and 17 µm with a slit 32" long. It also provides cross dispersed spectroscopy over a wider set of wavelengths via the 4" long slit. Two echelle gratings with different rulings allow to cover most wavelengths well away from the blaze minima.

The gratings can be rotated to set the desired wavelength. The sensitivity of the instrument at this resolution is of the order of 1 to several $Jy/10\,\sigma/h$. VISIR is the right tool to measure thermal emission by warm dust, broad band features from silicates as well as Poly-aromatic Hydrocarbon bands and gas vibrational transitions, for example H_2. This applies to star formation, circumstellar disks, ultra-compact H_2 regions and solar system physics.

3 Calibrations

Calibrations, which are also used for performance monitoring, are executed following dedicated calibration plans, which are presented in the instrument user manuals, published through the La Silla and Paranal web pages.

The typical calibration scheme for echelle optical spectrographs involves, apart from bias and darks, order definition frames and flat field frames (to correct for the pixel to pixel gain variations and to "fit out" the blaze function), both acquired via tungsten lamps providing a continuum spectrum. Next is the wavelength solution, which might assume slightly different flavors for different instruments. The high resolution spectrographs, mainly CRIRES and HARPS, due to their stability and high resolving power, are stressing the need for more accurate calibrators with larger wavelength coverage.

The frequency with which the calibrations are taken depends both on the stability of the instrument and on the needed accuracy, and it is defined in the calibration plan of each instrument.

Calibrations and monitoring of the detectors are presented in the contribution of P. Amico to these proceedings and are not treated here.

3.1 Flat Fielding and Fringing

The standard calibration plans allow flat fields to reach a signal to noise (S/N) of the order of few hundreds. Programs requesting higher S/N for their data (e.g. profile variation of diffuse interstellar bands in the interstellar medium, precise quantitative spectroscopy) should explicitly require high S/N flats (to be taken in day-time). A known issue in flat fielding echelle spectra is the lower signal at the edges of the orders, away from blaze maximum, which under-constrains the fit to the blaze function and introduces errors, which are especially noticeable in the order-merged spectra. The obvious solution to this inconvenience is to acquire a large number of flat fields.

Flat fielding at medium infrared wavelengths does not improve the data quality. Generally the noise introduced by the varying sky is larger than the short scale pixel to pixel gain variation of the detector. Therefore no flat fielding is performed on VISIR data.

Echelle spectrographs which are not strictly stabilized, thermally or mechanically (mechanical vibrations, stretching due to temperature variations, air pressure changes which in turn changes the index of refraction) suffer from drifts during the night (up to 0.7 pixels peak to peak for CES). In this case not only the wavelength calibration, but possibly even the flat field should be repeated.

Telluric absorption features are seen in flat fields red-ward of ≈ 800 nm in UVES, CES and EMMI. These are due to the long path in air between the calibration unit and the detector. The quality of the flat field calibration is therefore compromised in this region of the spectrum.

Fringes appear in the optical spectrographs red-ward of ≈ 650 nm and in the MIR. No fringes are reported in CRIRES. While HARPS is generally immune to fringing due to the smaller wavelength coverage with respect to the other optical spectrographs, fringing is generally well corrected via flat fielding. The $\approx 25\%$ amplitude of the fringes in UVES can be reduced to below 5% using flat fields taken just before or after the science exposure. In the MIR the main component of the fringes is effectively removed by co-adding the chopped frames, although some modulation remains in the spectra in the case of varying sky background. This can be removed by dividing the object spectrum by a standard star spectra; this operation will remove simultaneously the fringes and the telluric features in the spectrum.

3.2 Wavelength Calibration

The definition of a wavelength solution is not an issue generally at optical wavelengths, even at the very high resolution of CES. Hollow cathode lamps with a metal (generally Thorium) deposited on the cathode in a bulb filled with a noble gas (generally Argon) are adequate to provide a large number of narrow lines across the whole spectrum; although the density of Thorium lines decreases towards the red part of the spectrum, their number is large enough for a wavelength solution of good quality in the optical.

However, when seeking the ultimate radial velocity precision (1 ms^{-1} with HARPS), a long term stability of 10^{-9} is required. Although this is achieved with today Th–Ar lamps (at least on the time scale of ≈ 3 years), it is a major challenge. The quality of the lamps, i.e. the absence of contaminants, becomes critical. The choice of the lines to use to compute the wavelength solution is crucial as well: narrow (Th), unblended lines should be preferred, and an adequate de-blending algorithm should be used to increase the number of usable lines (see F. Pepe's contribution to these proceedings).

HARPS, FEROS and FLAMES (coupled both with UVES or GIRAFFE) can use the method of simultaneous Th calibration to correct for variations of the instrument response function due to temperature or pressure changes within the spectrograph. HARPS reaches radial velocity precision better than 1 ms^{-1} with this method.

At larger wavelengths, in the near and mid infrared, calibration is more problematic due to the absence of an adequate source. The number of Th lines drops towards the red, and red-ward of 2 μm the number of lines is not sufficient to achieve a wavelength calibration with CRIRES. For this reason CRIRES does not use a single source for wavelength calibration. Depending on the wavelength and the density of lines, the most appropriate source is chosen (see H.U. Käufl's contribution to these proceedings). The alternative "sources" which are being used in the NIR for the calibration are the lines from a N_2O absorption cell in the wavelength range between 2 and 5.4 µm, and the sky lines red-ward of 4 µm. ESO and the National Institute of Standard and Technology (NIST) in the USA are collaborating to establish near-IR wavelength standards in the spectrum of Th–Ar lamps [12].

VISIR also suffers from the absence of a wavelength calibrator. The instrument model alone gives a wavelength solution good within 15 pixels. A solution with a sub-pixel precision can be achieved using the sky lines as measured in the half-cycle chopped frames (which are part of the data package) and comparing them with an atmospheric model.

Common with all instruments which offer movable gratings to fine tune the wavelength range to be used, is the need to repeat the wavelength calibration after the wavelength setting, as this involves movement of several optical elements.

3.3 Standard Stars (Flux, Telluric, RV)

Absolute flux calibration is performed by observing spectrophotometric standard stars. At optical wavelengths the targets are usually extracted from the catalog compiled from the lists of UV HST spectrophotometric standards, the Oke and the Hamuy works [2, 8, 9, 18, 20].

In the infrared there are no standard stars tabulated spectra as in the optical. There are however works aiming to define a set of spectrophotometric standards in the NIR (contribution by J. Vernet et al., these proceedings). For calibrations up to 2 µm, using the Hipparcos catalogue and by means of the spectral type of the star, an IR magnitude can be inferred. This leads finally to an absolute flux uncertainty of 5–20%. For larger wavelengths, up to 5 µm, observers should refer, e.g., to the works of McGregor and van der Bliek et al. [17, 22]. The same considerations apply in the MIR. The reference work in this case is from Cohen [3]. In all these cases, the spectra should be convolved with the filter band pass and then normalized to the published, or inferred, photometric fluxes.

Telluric features are corrected by dividing the spectrum of the science target by a featureless spectrum, typically a hot, fast rotating star (early B stars). As the atmosphere is changing in space and time, both the telluric and the spectrophotometric standards should be observed as close as possible, in space and time, to the science target.

Radial velocity standards can be used to determine the absolute radial velocity calibration. Users are referred to the work of S. Udry on CORAVEL and ELODIE, whose precision is of the order of $50\,\mathrm{ms}^{-1}$ [21].

3.4 Absorption Cells

The usage of absorption cells is made possible in UVES and CRIRES. The UVES cell contains Iodine crystals that sublimate when the cell is put in operation. The cell supplies a dense forest of deep and narrow absorption lines between 500 and 600 nm. Variations of the positions of these lines can be used to correct for variations of the instrument profile. When using the iodine cell for precise radial velocity measurements, the cell is placed into the beam at the spectrograph entrance, just before the slit. In this way the iodine absorption lines, when the cell is illuminated by the star light, will be distorted, by changes in the instrument profile, in the same way as the star lines. Detailed knowledge of the iodine spectrum on the instrument allows the reconstruction of the instrument profile variation and the recovery of the precise position of the spectral lines on the detector and, therefore, the precise measurement of the radial velocity. Using UVES with the iodine cell a radial velocity precision of $\approx 2\,\mathrm{ms}^{-1}$ on the time scale of 2 years was achieved [13]. Apart from precise radial velocity measurements, the absorption cells can be used to model the instrument profile or to extract a wavelength solution if the wavelength information is available for most of the lines. This is indeed the case of the N_2O cell used in CRIRES. To insure stability, gas cells must be leak free and, in the case of iodine cells, thermal stability has to be achieved at a temperature where all the crystal is sublimated.

4 Instrument Monitoring and Stability

The instruments are continuously monitored to insure stability and optimum performance at any time. The pipelines, or dedicated monitoring tasks, provide the quality control parameters. Examples of the monitored quantities include the system efficiency, the amount of parasitic light and "ghosts", the resolving power, the reproducibility of the set-ups for the tunable spectrographs, the stability of the wavelength solution for the fixed format spectrographs, the flux from the calibration lamps, the relative fibre efficiency in the multi-fibre instruments. These data are published on the instrument web pages.

The requirements towards instrument stability are extremely tight when it comes to precise radial velocity measurements. To give an idea, in the case of HARPS, a Doppler shift of $1\,\mathrm{ms}^{-1}$ corresponds to a shift of the spectral lines of 17 nm on the CCD. It was also noticed that a 7 mK temperature change of the optical bench produces a radial velocity effect of $\approx 1\,\mathrm{ms}^{-1}$. Strict temperature control is therefore mandatory for HARPS. Thanks to the control system and to the fact that the optical bench is located in vacuum (P< 10^{-2} mbar), the temperature variation in the optical bench is < 2 mK/month. Several factors limit the achievable radial velocity precision: the absolute wavelength calibration, the photon noise, the object de-centering on the fibre or on the slit, and finally the instrument thermal and mechanical stability, which can be corrected for by the use of the simultaneous Th technique [16] or by the absorption cell method [13].

5 Conclusions

We presented the main calibration issues of the ESO high resolution spectrographs. While most of the scientific goals can be achieved following the standard calibration plans, more demanding programs, for example high precision quantitative spectroscopy or very high S/N spectroscopy, might need additional and/or special calibrations.

The performance of the spectrographs, which are operating very efficiently and reliably, is constantly monitored in detail through appropriate tasks, both from the observatory and the headquarters.

The high stability of the present instruments and the ambitious science goals put forward for future spectrographs, such as the search for terrestrial extra-solar planets, or studies of the cosmic dynamics and of the constants of physics, drive the need for new wavelength calibration sources, with stabilities over several years at the level of 10^{-11} or better, at optical as well as infrared wavelengths.

References

1. A. Baranne, et al.: AAS **119**, 373 (1996)
2. R.C. Bohlin, et al.: AJ **110**, 1316 (1995)
3. M. Cohen, et al.: AJ **117**, 1864 (1999)
4. H. Dekker, et al.: SPIE **627**, 339 (1986)
5. H. Dekker, et al.: SPIE **4008**, 534 (2000)
6. B. Delabre, et al.: SPIE **2198**, 414 (1994)
7. D. Enard: SPIE **331**, 232 (1982)
8. M. Hamuy, et al.: PASP **104**, 533 (1992)
9. M. Hamuy, et al.: PASP **106**, 566 (1994)
10. A. Kaufer, et al.: Messenger **95**, 8 (1999)
11. H.U. Käufl, et al.: SPIE **5492**, 1218 (2004)
12. F. Kerber, et al.: ASPC **364**, 461 (2007)
13. M. Kuerster, et al.: A&A **403**, 1077 (2003)
14. P.O. Lagage, et al.: Messenger **117**, 12 (2004)
15. Lovis, et al.: A&A **437**, 1121 (2005)
16. M. Mayor, et al.: Messenger **114**, 20 (2003)
17. P.J. McGregor: PASP **106**, 508 (1994)
18. J.B. Oke: AJ **99**, 1621 (1990)
19. L. Pasquini, et al.: Messenger **110**, 1 (2002)
20. D.A. Turnshek, et al.: AJ **99**, 1243 (1990)
21. S. Udry, et al.: ASPC **185**, 367 (1999)
22. N.S. van der Bliek, et al.: A&AS **119**, 547 (1996)

Discussion

D. Baade: Quite a few high-resolution spectroscopists would probably argue that the largest limiting circumstance is the imperfect continuum rectification. What is the corresponding experience with ESO instruments?

G. Lo Curto: The quality of the continuum rectification is affected by the qualify of the extraction and the flat fielding, by possible fringing and finally by the star: the continuum is difficult to define for a late M star. To quote an example, HARPS pipeline copes very well with continuum rectification, but due to the restricted spectral range (390−690 nm) it does not suffer from fringing. Care has to be taken to achieve a high SNR in the flats, especially at the edges of the blaze function.

Quantitative Spectroscopy with UVES

P. E. Nissen

Department of Physics and Astronomy, University of Aarhus, DK-8000 Aarhus C, Denmark; pen@phys.au.dk

Abstract. UVES has proven to be a very competitive high-resolution spectrograph and many important scientific results based on UVES spectra have been obtained. Still, there are limitations in the accuracy of flat-fielding, removal of telluric lines and continuum rectification, which affect programs requiring very high signal-to-noise spectra. Examples of such programs are briefly reviewed and ways to obtain better calibrated spectra are discussed.

1 Introduction

Since its completion in 1999, UVES has proven to be a reliable and highly competitive high-resolution spectrograph. According to the ESO Telescope Bibliography about 500 refereed publications based on UVES data have appeared in the period 2000–2006 and important scientific contributions have been made in such different fields as: stellar atmospheres and abundances; stellar oscillations and pulsations; exoplanets; stellar and gas dynamics; elemental abundances in dwarf galaxies and damped Lyman-alpha systems; tests of possible variations of the fine structure constant and the electron-to-proton mass ratio as a function of redshift. Highlights include: The abundance distribution of elements in extremely metal-poor stars [4]; detection of uranium in two metal-poor giant stars and determination of their ages from the U/Th abundance ratio [6, 7]; detection of solar-like oscillations of α Centauri A and B [3, 10]; study of abundances of K giant stars in nearby dwarf spheroidal galaxies [19]; high resolution spectroscopy of intervening galaxy absorbers towards a gamma-ray burst source [5].

In most cases, standard UVES observations and pipe-line reduction of the data are adequate for obtaining high-resolution spectra of sufficient quality. For programs requiring very high S/N spectra and accurate continuum rectification, one should, however, give special consideration to flat-fielding and removal of CCD fringing and telluric lines. In this review, I will focus on a few examples of such programs and discuss ways to obtain better calibrated spectra.

2 Removal of CCD Fringing

In the red part of the spectrum, interference of light in the thin layers of the UVES CCD detectors creates a highly irregular modulation of the signal as a function of wavelength. The amplitude increases from a few percent at 6000 Å to about 25% around 9000 Å. The wavelength scale of the variations is in the range of 1–10 Å. Hence, the strength and profile of both narrow and broad lines are affected. Traditional flat-fielding, carried out by pixel-to-pixel division of the stellar spectrum with the spectrum of a flat-field lamp, does not remove the fringes completely. A residual fringe structure remains as seen in the example of Fig. 1. The reason is probably that the light from the flat-field lamp does not have exactly the same optical path through the spectrograph as the light from the star.

The residual fringing pattern may be removed by dividing the spectrum of the program star with the spectrum of a bright B-type star observed on the same night and under the same conditions, e.g. at exactly the same position in the spectrograph slit. As shown in Fig. 1, this method leads to a well corrected spectrum of the program star (HD 84937) with a S/N $\sim$ 500.

Residual fringing can be a major problem when deriving stellar abundances from weak lines in the red spectral region. A good example is the current debate on the trend of [S/Fe] vs. [Fe/H] for halo stars. Some studies

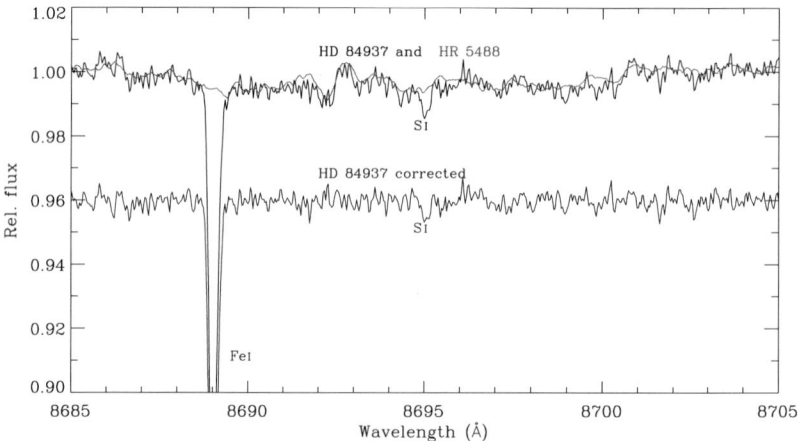

Fig. 1. Residual fringing problem in a UVES spectrum of the metal-poor turnoff star HD 84937 around the S I line at 8695 Å [15]. After flat-fielding, the extracted spectrum of HD 84937 has residual fringes with an amplitude of about $\pm 0.5\,\%$. A 5-pixel smoothed spectrum of the B2III star HR 5488 observed in the same way as HD 84937 is overplotted. Below is the spectrum of HD 84937 (shifted to a continuum level of 0.96) after division with the spectrum of HR 5488

[14, 17] indicate that [S/Fe] is approximately constant at a level of +0.3 dex in the metallicity range $-3 <$ [Fe/H] < -1. Other investigations [8, 18] point, however, to an increasing trend of [S/Fe] towards lower metallicities with [S/Fe] reportedly reaching as high as +0.8 dex at [Fe/H] ~ -2. The sulphur abundances in these studies are, however, based on the weak $\lambda 8694.6$ S I line, and as discussed by Korn & Ryde [12] and Nissen et al. [15], it is likely that the strength of this line has been overestimated due to fringing residuals in the spectra.

Determination of lithium isotopic abundances is another example of the importance of residual fringing correction. The ^{6}Li/^{7}Li ratio can be determined from the profile of the Li resonance line at 6708 Å because the components of the ^{6}Li doublet is shifted 0.16 Å relative to the ^{7}Li doublet. Hence, the width and asymmetry of the Li line depend on the isotope ratio, and a model-atmosphere fit to the observed profile leads to the determination of ^{6}Li/^{7}Li. A recent analysis of UVES spectra of 24 metal-poor halo stars by Asplund et al. [1] suggests the existence of a ^{6}Li plateau at a level of 5% of the well known ^{7}Li plateau. If confirmed, this points to a revision of the standard Big Bang nucleosynthesis model [9] or to pre-Galactic production of ^{6}Li by e.g. cosmological cosmic rays created by Population III stars [16].

In order to measure ^{6}Li/^{7}Li with a precision of a few percent, one should observe the Li resonance line with a resolution of $\sim 10^5$ and S/N $>$ 500. In addition, the continuum around the Li line should be set to an accuracy better than 0.1%. In this connection it is of utmost importance to correct for residual fringing in the Li line region. Figure 2 shows an example, where the continuum of the flat-fielded spectrum varies with an amplitude of about 0.3% across the Li line. If this pattern is not removed by dividing with the spectrum of a bright early-type star, which has no lines in the Li region, the derived isotope ratio will be wrong.

3 Removal of Telluric Lines

The presence of numerous telluric lines in some wavelength regions is another problem in the red and near-IR part of UVES spectra, e.g. O_2 lines near the forbidden oxygen line at 6300 Å and water vapor lines in the region from 9000 to 9300 Å, which contains several C I and S I lines used to measure carbon and sulphur abundances in metal-poor stars.

As shown in Fig. 3, telluric lines can be removed by dividing the program star (HD 84937) with a scaled version of the spectrum of a B-type star (HR 5488) observed on the same night as the program star. The IRAF task `telluric` is used to optimize the fit of the telluric lines in the two spectra by allowing a scaling in airmass and a small shift in wavelength. As seen the intrinsic lines of HD 84937 emerge very clearly after removal of the telluric lines also in the case of the $\lambda 9238$ S I line, which is completely overlapped by a telluric line in the original spectrum of HD 84937. A complication in this

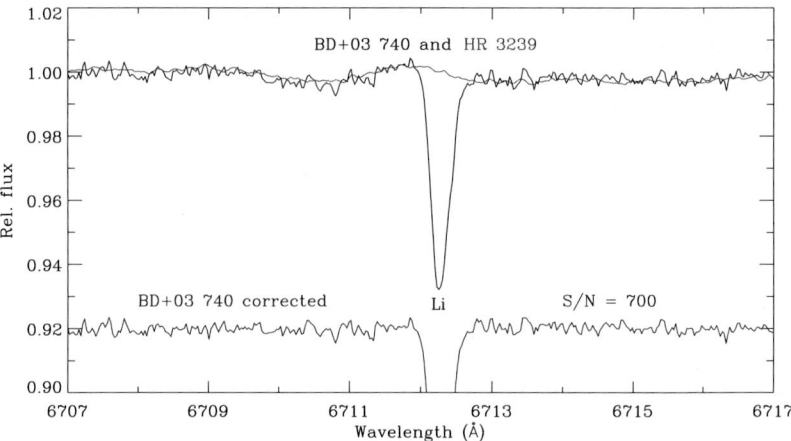

Fig. 2. Residual fringing problem in a UVES spectrum of the metal-poor turnoff star BD +03 740. After division with a 5-pixel smoothed spectrum of the B1.5V star HR 3239 observed on the same night as BD +03 740, a corrected spectrum (shifted to a continuum level of 0.92) is obtained. Note that the Li line in BD +03 740 is radial-velocity shifted by $+200\,\mathrm{km\,s^{-1}}$

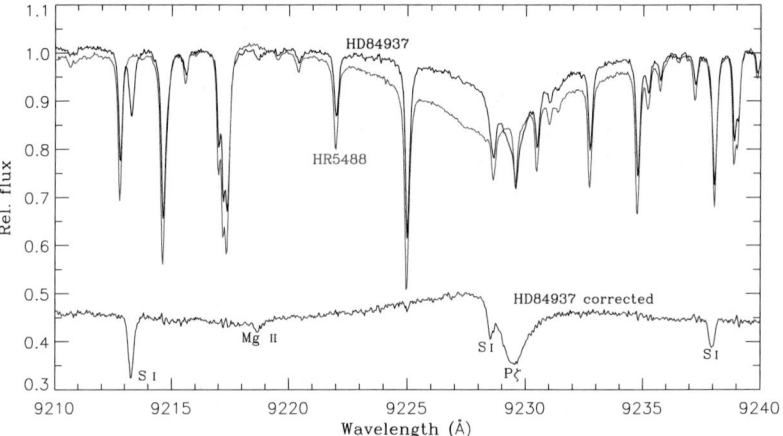

Fig. 3. UVES spectrum of HD 84937 in the spectral region 9210–9240 Å overlaid with the B2III spectrum of HR 5488. Below is the spectrum of HD 84937 (shifted ~ 0.5 units in relative flux) after removal of the telluric lines by dividing with a scaled version of the spectrum of HR 5488

spectral region is the presence of the broad Paschen ζ hydrogen line centered at slightly different wavelengths in the two spectra due to the radial velocity difference of the stars. In the corrected spectrum of HD 84937, the Pζ line is therefore seen as a broad feature in 'emission' from HR 5488 overlaid with a narrower absorption line from HD 84937. This means that the equivalent widths of the S I lines can only be measured relative to the local continuum. Hence, when using their equivalent widths to derive S abundances, the opacity contribution from the Pζ line should be taken into account.

In order to remove telluric lines successfully by this method, the telluric standard star should be observed under the same atmospheric conditions as the program star. In particular, it is important that the seeing is about the same for the two exposures, because the spectral instrument profile is determined by a convolution of the slit width and the seeing profile. Hence, if the seeing changes between the exposure of the program star and the telluric standard, it may be impossible to get a good match between the two sets of telluric lines.

An alternative to using a B-type star to remove telluric lines, is to apply a synthetic transmission spectrum calculated for a model of the Earth's atmosphere. This method has been successfully applied by Lallement et al. [13] for the H_2O lines in the region of the NaD doublet, and has also been tried for the 9200–9240 Å region of the sulphur lines by A. Seifahrt (private communication) with promising results. The advantage of this method is that one does not loose observing time on telluric standards. In addition, there are no disturbing non-telluric features in the synthetic spectrum. It may, however, be difficult to model the spectral instrument profile with sufficient accuracy, because when the star is not filling the slit of the spectrograph, the profile becomes seeing dependent as mentioned above. Furthermore, one has to optimize several free parameters characterizing the distribution of H_2O with height in the atmosphere in order to get a good fit between the telluric lines in the synthetic spectrum and in the program star.

Instead of deriving the sulphur abundance of metal-poor stars from either the very weak λ8694.6 S I line or the S I lines affected by telluric absorption lines, it has recently become possible to use the S I triplet at 1.046 μm observed with CRIRES. Figure 4 shows this triplet in a spectrum of the turnoff star G 29-23 ($V = 10.19$, [Fe/H] $= -1.7$) obtained during the commissioning of CRIRES. The entrance slit width of CRIRES was set at 0.4 arcsec, which corresponds to a resolution of $\lambda/\Delta\lambda \simeq 50\,000$ with four detector pixels per spectral resolution bin $\Delta\lambda$. The exposure time was 2400 sec. The seeing was rather poor (about 1.3 arcsec) but adaptive optics was applied to improve the stellar image, and the combined spectrum has a very satisfactory S/N of 330 per spectral dispersion pixel. For comparison, we note that a UVES spectrum of G 29-23 (exposure time 1800 sec) has S/N ~ 200 around the S I lines at 9212–9238 Å. The UVES spectrum also covers the 1.046 μm S I triplet but due to the low efficiency of the MIT CCD at this wavelength, the

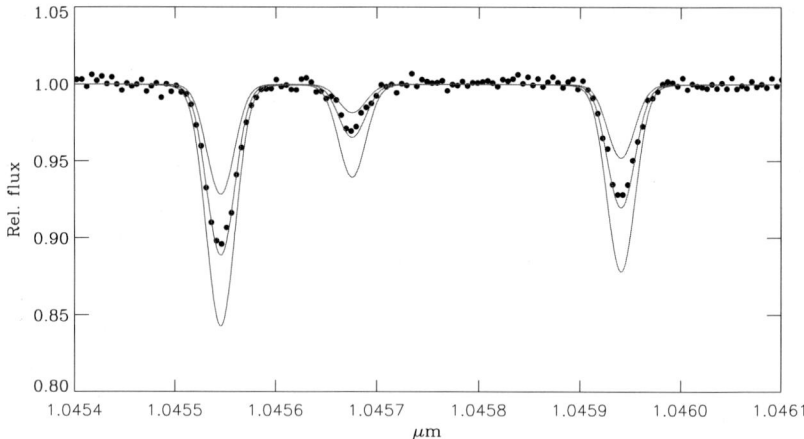

Fig. 4. The CRIRES spectrum of G 29-23 around the 1.046 μm S I triplet (*dots*) compared with synthetic profiles for three sulphur abundances corresponding to [S/Fe] = 0.0, 0.3 and 0.6, respectively. As seen, the [S/Fe] = 0.3 case provides an excellent fit to the observations in agreement with [S/Fe] derived from S I lines in UVES spectra of the same star

quality is far too low (S/N < 25) for a precise determination of the sulphur abundance. Furthermore, unlike the UVES near-IR spectrum, the CRIRES spectrum at 1.046 μm is not plagued by fringing residuals and telluric lines. For other wavelength regions removal of telluric lines is, however, a major challenge when reducing CRIRES spectra.

4 Continuum Rectification

Due to the strong variation of the blaze function of echelle spectrographs, continuum rectification is difficult especially when merging overlapping orders. As an example, Fig. 5 shows the Hβ line in the UVES Paranal Observatory Project (POP) [2] spectrum of HD 84937. As seen the continuum has variations with an amplitude of several percent. In addition there are small discontinuities in the flux at wavelengths where the echelle orders have been merged, e.g. at 4833 and 4892 Å. Similar problems are seen in UVES POP spectra of the Hα line, although the amplitude of the continuum variations are smaller than in the case of Hβ.

Hydrogen lines are sensitive indicators of effective temperatures of stars and unlike colour indices their profiles are not significantly affected by interstellar reddening. Hence, precise values of $T_{\rm eff}$ may be determined from a comparison with synthetic profiles calculated for a set of model atmospheres. In many studies, e.g. of the dispersion of Li abundances among metal-poor

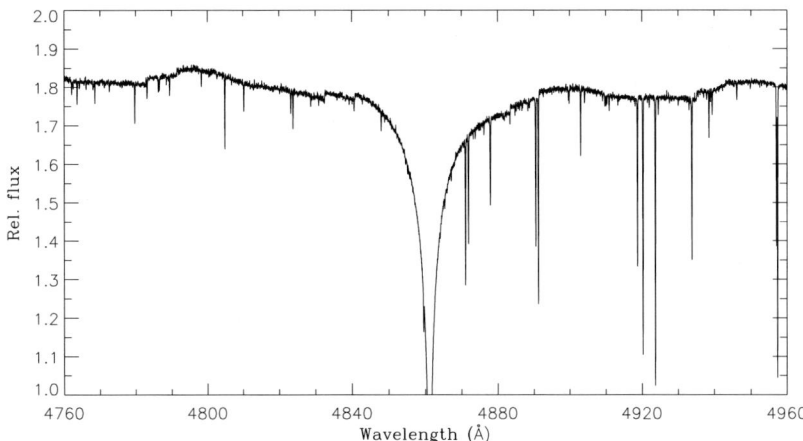

Fig. 5. The UVES Paranal Observatory Project (POP) [2] spectrum of the metal-poor turnoff star HD 84937 in the region of the Hβ line

stars, such precise values of $T_{\rm eff}$ are of critical importance as discussed in the recent work of Asplund et al. [1].

The Hβ line profile in Fig. 5, which was obtained by using the UVES pipeline reduction program, cannot be used for deriving an accurate value of $T_{\rm eff}$. A better reduction procedure has been proposed by Korn [11], who shows that the continua of the flat-fielded echelle orders in spectra obtained with FOCES at the 2.2 m Calar Alto telescope can be well fitted by low order polynomials, which change smoothly from one order to the next. Hence, the continuum of an order containing a Balmer line may be determined by interpolation between the continua of adjacent echelle orders.

Figure 6 illustrates Korn's method for the case of the Hβ line in order 29 of a UVES spectrum of HD 84937 observed by Nissen et al. [15]. As seen, the continua of orders 28 and 30 show only small deviations from a straight line between pixel 300 and 2500. Note that the slope of this line is mainly due to the difference in color temperature of the star and the flat-field lamp. For the first and last 300 pixels, the continua have, however, irregular variations. This must be due to a mismatch between the paths of the light from the star and the flat-field lamp, and the difficulty of merging overlapping echelle orders in UVES spectra is probably due to this problem.

Despite the irregular variations of the continua at the ends of the UVES echelle orders, the central part (pixels 300–2500) may be used to determine the continuum for part of the Hβ line by interpolating between the continua of the adjacent orders. Figure 7 shows the resulting Hβ line profile for two stars from [15] in comparison with synthetic model atmosphere profiles for three values of $T_{\rm eff}$. The symmetric appearance of the profiles shows that the procedure of rectifying the Hβ echelle order has worked well, and values of

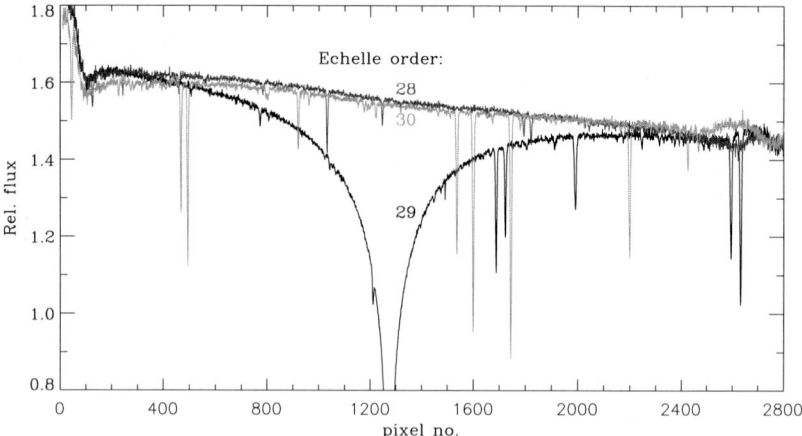

Fig. 6. Flat-fielded echelle orders in a UVES spectrum of HD 84937. Order 29 containing the Hβ line and adjacent orders 28 and 30 are shown

the effective temperature of a given star derived from spectra obtained on different nights agree within ± 20 K. On the other hand, a better accuracy of the absolute value of $T_{\rm eff}$ would have been obtained if a wider part of the Hβ line could have been observed.

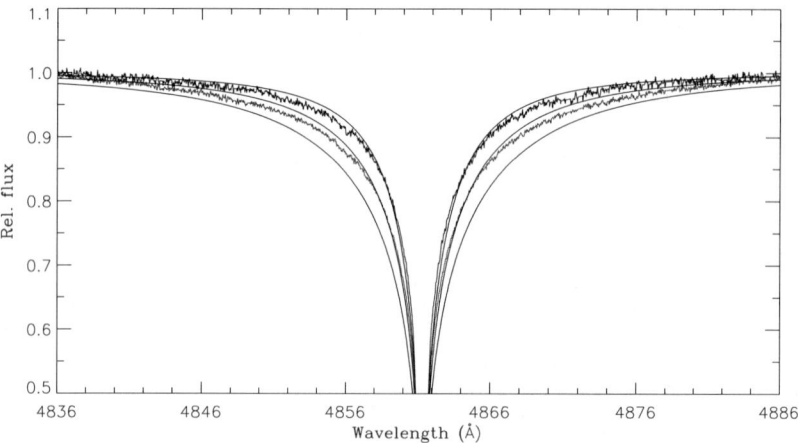

Fig. 7. The Hβ line profile in UVES spectra of CD $-42\,14278$ ($T_{\rm eff} = 6085$ K, [Fe/H] $= -2.0$, *upper jagged line*) and HD 84937 ($T_{\rm eff} = 6360$ K, [Fe/H] $= -2.1$, *lower jagged line*). Note that narrow metallic absorption lines have been removed from the observed profiles by interpolation across these lines. Synthetic profiles calculated for model atmospheres with $\log g = 4.3$ and [Fe/H] $= -2.0$ are shown for $T_{\rm eff} = 6000$, 6300 and 6600 K

5 Conclusions

Although UVES has proven to be a very reliable instrument producing a lot of interesting science, there is still room for improvements of the calibrations especially in connection with programs requiring very high S/N spectra. A better matching of the optical path and illumination of the optics by light from the star and the flat-field lamp may help in solving the residual fringing problem and improve the continuum rectification at the edges of the echelle orders. A more radical way of getting rid of the fringing problem in the red and near-IR would be to replace the MIT detector with a thick deeply or fully depleted CCD, which would also increase the near-IR sensitivity of UVES. The presence of numerous telluric lines is another problem in the red and near-IR. Considerable amount of observing time is presently spent on observing telluric standards for removal of these lines. Forward modelling of the transmission spectrum of the atmosphere of the Earth would be an interesting alternative that should be further explored.

References

1. M. Asplund, D.L. Lambert, P.E. Nissen, et al.: ApJ **644**, 229 (2006)
2. S. Bagnulo, E. Jehin, C. Ledoux, et al.: ESO Messenger **114**, 10 (2003)
3. R.P. Butler, T.R. Bedding, H. Kjeldsen, et al.: ApJ **600**, L75 (2004)
4. R. Cayrel, E. Depagne, M. Spite, et al.: A&A **416**, 1117 (2004)
5. S.L. Ellison, P. Vreeswijk, C. Ledoux, et al.: MNRAS **372**, L38 (2006)
6. A. Frebel, N. Christlieb, J.E. Norris, et al.: ApJ **660**, L117 (2007)
7. V. Hill, B. Plez, R. Cayrel, et al.: A&A **387**, 560 (2002)
8. G. Israelian, R. Rebolo: ApJ **557**, L43 (2001)
9. K. Jedamzik: Phys. Rev. D **70**, 063524 (2004)
10. H. Kjeldsen, T.R. Bedding, R.P. Butler, et al.: ApJ **635**, 1281 (2005)
11. A.J. Korn: In Scientific Drivers for ESO Future VLT/VLTI Instrumentation, eds J. Bergeron & G. Monnet, ESO Astrophys. Symp. Ser., p.199 (2002)
12. A.J. Korn, N. Ryde: A&,A **443**, 1029 (2005)
13. R. Lallement, P. Bertin, E. Chassefière, N. Scott: A&A **271**, 734 (1993)
14. P.E. Nissen, Y.Q. Chen, M. Asplund, M. Pettini: A&A **415**, 993 (2004)
15. P.E. Nissen, C. Akerman, M. Asplund, et al.: A&A **469**, 319 (2007)
16. E. Rollinde, E. Vangioni, K. Olive: ApJ **627**, 666 (2005)
17. N. Ryde, D.L. Lambert: A&A **415**, 559 (2004)
18. M. Takada-Hidai, Y. Takeda, S. Sato: ApJ **573**, 614 (2002)
19. E. Tolstoy, K.A. Venn, M. Shetrone, et al.: AJ **125**, 707 (2003)

Discussion

J. Pritchard: Do you use the ESO MIDAS pipeline for reductions?
P.E. Nissen: No, IRAF.

S. Moehler: If you ascribe the temperature difference between Hβ and H (inf) to problems with the UVES data, does that mean that you do not see such differences for data from other spectrographs?

P.E. Nissen: In fact, the T_{eff} difference between Hβ and H (inf) could also be caused by errors in the model atmosphere calculations of the Balmer line profiles; the difference depends e.g. on the mixing/length parameters.

P. Bonifacio: I think the only perspective for improving the flat-fielding is to do the flat-fielding using a light source which follows the same optical train as the star light.

P.E. Nissen: Yes, I agree.

M. Rosa: Your requirement (desideratum) on wavelength calibration stability - is it on absolute or on the shape, i.e., the dispersion relation?

P.E. Nissen: Both, but in particular on the absolute stability.

High Resolution and High Precision-Spectroscopy with HARPS

F. Pepe and C. Lovis

Observatoire Astronomique de l'Université de Genève, 51. ch. des Maillettes, CH-1290 Sauverny, Switzerland; `Francesco.Pepe@obs.unige.ch`

Abstract. Extra-solar planet search at a level of precision below $1\,\mathrm{ms}^{-1}$ sets strong requirements to the quality and stability of the wavelength solution. It also forces us to understand the effects of instrumental stability, on the one hand, and the quality of the wavelength reference, on the other hand, since both will have an impact, although in a different way, on the short- and long-term precision of the instrument. This chapter presents the calibration principles of HARPS, which lead to its extraordinary wavelength solution and, as a direct consequence, to its unique radial-velocity precision. In particular it will focus on the improvements of the thorium-lamp calibrations we made during the past three years, but it willl also discuss the present limitations. Finally, we give an outlook on further possible improvements which can be made in view of the extreme precision required by instruments like CODEX@ELT.

1 Introduction

"High precision" and "high resolution" are terms in spectroscopy which are quite relative to the field of application and the astronomer referring to it. We shall therefore immediately clarify that this article will focus on the use of high-resolution spectroscopy for the measurement of stellar radial velocities, and in particular in the field of extra-solar planet search, where the most demanding requirements come from. In this specific case, "high precision" means measuring the stellar radial velocity with a resolution and repeatability in time of about $1\,\mathrm{ms}^{-1}$. This value translates to a relative precision of about 10^{-9} on the wavelength measurement of the stellar spectral lines. High resolution is mandatory to achieving this performance. Experience has shown that spectral resolution above $R = 80,000$ provides us with the best results. "High resolution" shall therefore refer, in this article, to values of typically $R = 100,000$ and above.

Finally, we have to point out that our chapter treats only "classical" spectrographs, consisting basically of optically dispersive elements producing a spectrum eventually imaged on a 1 or 2-D detector such as a CCD. Most of the instruments used in the search for extra-solar planets are cross-dispersed spectrographs. Our experience is actually mainly based on this type of instrumentation and in particular the HARPS spectrograph [4]. We are fully aware that other high-resolution spectrographs or spectrometers exist, such as Fourier-Transform Spectrometers, Fabry-Perots, instruments based on heterodyne detection, and many others. Given their diversity and complexity, as well as our lack of experience, we will *not* include them in the following discussion.

2 The Doppler Measurement

2.1 Principles

The Doppler measurement consists in determining the wavelength of an identified spectral line and to compare it to the theoretical value it would have when transferred into the solar-system rest frame. The Doppler equation links the measured wavelength with the theoretical wavelength via the relative velocity vector, finally delivering the projection of this vector in the direction of the line of sight (radial velocity). In order to increase the precision, an average over several thousands spectral lines is built up. There are different ways of determining the precise wavelength of a given spectral line and of building up an average over them in order to compute a single radial-velocity value, but we shall not discuss them here, being it out of the scope of this chapter. There is however one aspect which is common and critical to all spectroscopic methods: the wavelength reference. Indeed, the observed spectrum, which is recorded in some arbitrary scale (e.g. CCD pixels), must be placed into a calibrated wavelength frame. This frame is typically provided by a spectral reference source. The link between the arbitrary scale and the wavelength reference is called the "wavelength calibration".

2.2 State of the Art

For the search of extra-solar planets, the Doppler measurements are carried out in the visible wavelength range. The best precisions obtained nowadays are of the order $0.3\,\mathrm{ms}^{-1}$ [2]. It must be remarked that this value expresses a relative precision on the final radial-velocity computed and averaged over several thousands spectral lines, arising from different depths in the stellar atmosphere and frequency-shifted due to the stellar gravitational potential. The only thing which counts for the Doppler search of exoplanets, however, is the repeatability of the radial-velocity measurements, the absolute accuracy of the actual wavelength of the spectral line being therefore of minor importance.

The mentioned $0.3\,\mathrm{ms}^{-1}$ Doppler precision may be converted into wavelength precisions. We deduce that the wavelength of a line must be determined in average with a precision of 10^{-9}. In order to achieve this value, the wavelength solution (the function which related the wavelength λ to the CCD pixel x) needs a local precision of $10^{-9} \times \sqrt{N}$, where N is the number of spectral lines. For a typical observed star (solar type) we estimate that about 3000 to 4000 lines are used, implying thus a local wavelength accuracy of better than 6×10^{-8}.

Again we have to point out that the absolute value of the wavelength is not very critical, since one is interested in the *variations* of the velocity, and thus the wavelength, as a function of time. However, we must consider that, due to the orbital motion of the Earth, the stellar lines "move" by the equivalent of $v_{Earth} = \pm 30 \, \mathrm{km s}^{-1}$ or about 10^{-4} of their wavelength. An absolute error in wavelength would thus induce a "stretch" of the velocity scale, producing an annual effect of about $\epsilon \Delta v = \lambda_{effective}/\lambda_{theoretical} * \Delta v_{Earth}$. If we want to avoid errors larger than $0.3 \, \mathrm{ms}^{-1}$ we have thus to ensure an absolute accuracy in wavelength of about $\pm 10^{-5}$ or ± 0.05 Angstroem. As mentioned above, this precision and accuracy has been actually achieved and proven by recent observations and exoplanet discoveries.

2.3 Error Sources

We can distinguish three main types of error sources:

Stellar noise groups all error sources which are produced at the *emission* of the light, i.e. by the observed source itself. It includes, for example, the stellar pulsations, surface granulation and stellar jitter due e.g. to spots or other "instabilities" in the stellar atmosphere.

Errors due to the intermediate medium are introduced on the way of the light from the source to the observer. The main error source is the Earth's atmosphere, but other effects may be introduced by the interstellar medium (e.g. absorption, polarization) or by general relativity (frequency shifts and lensing).

Instrumental errors are finally those related to the detection process. They include the whole light path starting at the telescope and ending on the detector. Among these instrumental effects we shall distinguish two categories:
   1. errors arising on the measurement, actually due to the limited stability or repeatability of the measurement when supposing a perfect wavelength reference, and
   2. calibration errors, i.e. errors due to the wavelength reference.

In the following section, we shall focus on the main subject of this chapter, i.e. the calibration errors.

3 The Wavelength Calibration

3.1 Building the Wavelength Solution

Before entering into technical aspects of the wavelength calibration, we would like to give here one possible definition for the calibration process, and use that as a guideline for the following discussion: *"The calibration is a process that establishes, under specified conditions, the relationship between the values indicated by the measuring system, and the corresponding values of a quantity realised by a reference standard or working standard."* In the concrete case of todays high-resolution spectrographs, the output of the wavelength calibration process is the wavelength solution, i.e. a function $\lambda(x)$ which assigns, in a unique way, a wavelength λ to each (extracted) detector pixel x.

In order to build up a wavelength solution, we need to "observe" a well-known spectral reference source with the same instrument we want to use for the actual observations. The spectral reference will deliver a number of lines with known wavelength λ_i, which will be imaged by the spectrograph on the detector around a given pixel x_i. This measurement will thus deliver a certain number of pairs (λ_i, x_i). A model function $\lambda(x)$ will then be searched for, which fits best all the observed data pairs simultaneously while keeping to a minimum the number of free parameters.

As for the measurement itself, we may have "instrumental errors" on the calibration process, too. As for the measurement, the errors may reside on both axis, i. e. on the "independent" variable λ_i, which is supposed to be known, but of course only to a certain accuracy, or on the "dependent" variable x_i, which results from the line-position measurements on the detector.

3.2 Uncertainties on the Spectral Reference

At visible wavelengths, the majority of spectral features arise from electronic transitions in atoms or ions. A good spectral source has to provide many narrow spectral lines along a broad wavelength range. Only few well-adapted sources are known for the visible wavelength range, and actually only two of them are commonly used as precision wavelength reference:

- The first is the thorium(-argon) hollow-cathode lamp (ThAr). The ThAr lamp presents several thousands thorium emission lines from the UV to the Infrared. The spectrum is very rich and dense. The drawback is that many lines are blended and several unresolved lines superimpose. In addition, the dynamical range of the line intensity is very large (10^4 and more) especially if we take into account the very strong and unstable "parasitic" argon lines. Given these characteristics, the ThAr spectrum cannot be superimposed on the stellar spectrum. It is therefore shifted either spatially or temporally, or both, with regard to the stellar spectrum. This opens the door to possible instrumental error sources given the fact that the instrumental profile (IP) might change and thus be different when recording the stellar and the reference spectrum. Another important disadvantage of the thorium lamp is that it is an active element and as such possesses only a limited lifetime.
- The second "calibrator" is iodine. Being packed in a "absorption cell", the iodine is very suitable to use it in combination wit the self-referenced

technique, in which the absorption cell in inserted in the stellar light beam "branding" on it many absorption features of known wavelength. As in the case of thorium, the spectrum is dense and almost all the lines blended. Since used in absorption, the dynamic range of the spectral lines is better controlled and, in particular, the line intensity will by definition not exceed the stellar continuum. On the other hand, the iodine absorption lines cover only the 500 to 600 nm region, such that only part of the visible domain can be exploited for the radial-velocity measurement. The "self-calibration" technique is considered to be theoretically very precise, since the stellar and the reference spectrum are recorded perfectly simultaneously. The two spectra are hold by the same photons which follow the same optical path. Therefore, possible instrumental effects, such as changes of the IP, affect the spectra in the same manner. In practise, however, the stellar spectrum is "polluted" by the reference spectrum. The self-calibration technique is therefore firmly linked to spectral disentangling and deconvolution processes. Both require much higher signal-to-noise-ratios (SNR) in order to achieve a given Doppler precision, and in any case, a degradation of the spectroscopic quality of the spectrum must be taken into account.

Although the nature of these calibration sources is very different, the two solutions actually deliver similar radial-velocity precisions. With the iodine technique precisions down to 2 to $3\,\text{ms}^{-1}$ are routinely achieved, while using the thorium even sub-meter-per-second precision could be demonstrated. Some limitations start to be identified at that precision level, however. The main problem was, to our surprise, the input atlas of the thorium lines. For the construction of the wavelength solution we have been using the Palmer & Engleman [5] atlas. This atlas is very complete and recorded with very high resolution, typically of $R = 600,000$. It resulted from actual measurements of ThAr lamps with an FTS. Due to this, the determined wavelengths are affected by photon noise, which depends on the lines' intensity. Assuming these input wavelengths of the atlas lines as fixed, the original photon-noise error will be finally converted into a systematic error during the wavelength calibration process. Indeed, when reducing and wavelength-calibrating several hundreds of HARPS calibration spectra, we discovered that a given line had always a (within photon noise) fixed residual in line position with respect to the best-fit wavelength solution. As expected, the systematic deviation from the wavelength solution was well proportional to the photon noise given for the respective line in the Palmer input catalog. Thanks to the great stability of HARPS, we have actually been able to go in the reverse direction, namely to determine the systematic errors on the HARPS spectra and correct the wavelength value in the input atlas. In addition, and thanks to the great sensitivity of HARPS, we have been able to extend the number of usable lines in the visible wavelength range from the original 3000 to about 6000. The new atlas has been published in [3].

Another important limitation may arise from the reference-line wavelength itself. The electronic transitions depend indeed on environmental parameters, such as temperature and pressure, for instance. It can therefore not be guaranteed that the theoretical λ_i of a given line is constant with time. Laboratory measurements we made with the HARPS instrument actually show that, for example, the average thorium line position can change by about 50 of $cm\,s^{-1}$ when the lamp current is increased from 7 to 10 mA. Also pressure changes may affect the line position, in particular the argon lines of the ThAr lamp. Finally, aging effects of the lamp, acting mainly via pressure variations, may also produce changes in the reference wavelengths. Fortunately, these effects have been proven to be at the one ms^{-1} level or below on thorium lines over more than 4 years.

3.3 Uncertainties on the Line-position Measurement

As mentioned before, we have to determine the position of a spectral line on the detector, e.g. by applying a Gaussian fit or by measuring the centroid of the line. Whatever method is used, the measurement will be affected by measurement errors. In the following, we shall provide a list of error sources and their magnitude for the case of the HARPS instrument. We shall recall at this occasion that $1\,ms^{-1}$ corresponds to a relative wavelength precision of 3×10^{-9}.

Photon noise The photon noise on a single line contributes by $1\,ms^{-1}$ (almost saturated line) to $150\,ms^{-1}$ (weak line). The average over all used lines is $8\,cm\,s^{-1}$.

Systematics due to blends The principle limitation arises from the dense and blended spectral lines (see Figs. 1 and 2), in particular if this feature is coupled with varying line-intesities: Since the lines are not resolved, they may be seen as one single line by the calibration process. A relative intensity variation between the two lines, will then shift the measured line position, e.g. the centroid, and influence negatively the value x_i in the calibration data pair (λ_i, x_i) and consequently the whole wavelength solution. By comparing the residuals to the obtained wavelength solution for spectral lines which are present in two neighboring echelle orders of the same spectrum, we have been able to estimate that the residual error due to blends is of the order of $8\,ms^{-1}$ on a single line and about $10\,cm\,s^{-1}$ in average over all lines ("zero point").

Systematics due to "pixelisation" The problem arises from the fact that the pixel geometry is not perfectly square and that the distance between the pixels is not absolutely constant. In addition, we cannot assume that the response of a pixel to photons is completely flat across the pixel. Therefore, the pixel space does not correspond perfectly to an equidistant space sampling with integer value, but rather one should assign real-valued coordinates to the pixel. When ignoring this fact, as usually done,

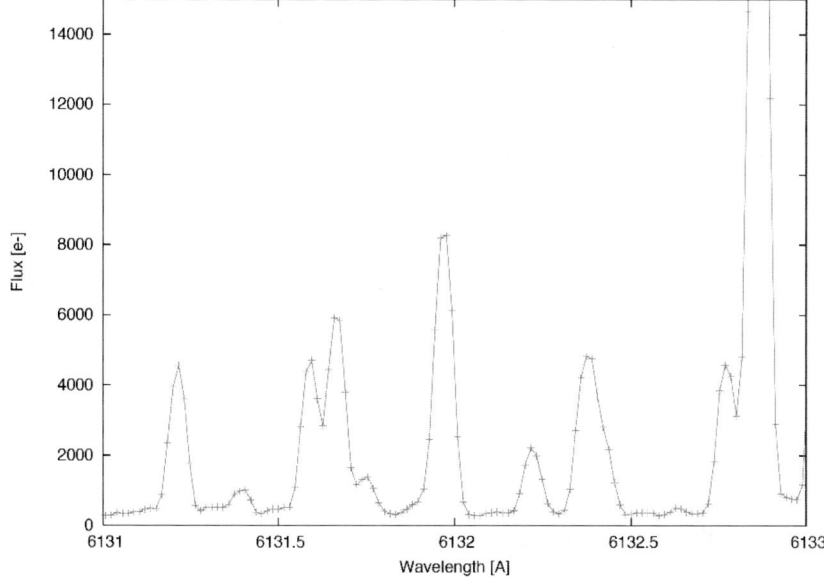

Fig. 1. Portion of the ThAr-lamp spectrum showing the high dynamic range in line intensity and the high line density causing blends

an additional error is introduced on the measurement of x_i. We have been able to estimate the residual error due to pixelisation by comparing, for the same lines observed on the respective spectra of the two fibers (object and reference fiber), the residuals of the line position to the value given by the fitted wavelength solution. Again we obtain an effect of the order of $8\,\mathrm{m\,s^{-1}}$ on a single line and about $10\,\mathrm{cm\,s^{-1}}$ in average over all calibration lines.

3.4 Errors Due to the Finite Accuracy of the Calibration Algorithm

The first task of the algorithm is to determine the line position x_i of each line of the input atlas. As mentioned above, the measurement of the line position is affected by errors. Appropriate algorithms allow to minimize the systematic errors and to stick as close as possible to the photon noise. Big improvements have been made recently in this process [3], especially with respect to the blend problem, by treating the spectral lines not individually but instead making multiple line fitting for those line groups at close distance to each other. Doing so, it could be avoided that changes in relative line intensity affect the centroid of a group of lines and thus the respective calibration data

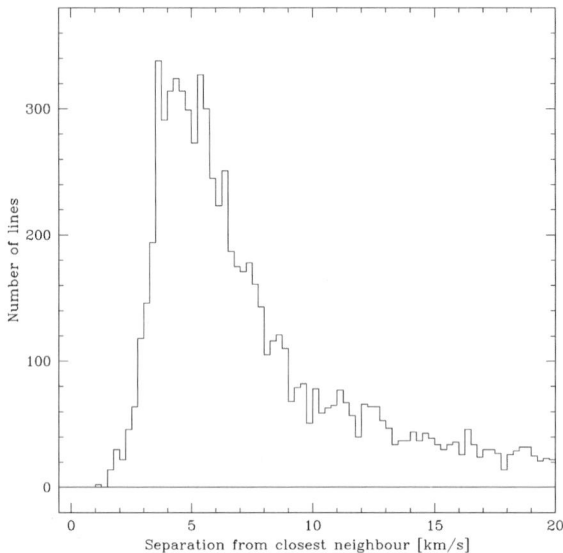

Fig. 2. Histogram of the separation between two neighbouring lines in a thorium spectrum. The majority of the lines have a neighbour at a distance corresponding to about the resolution of the spectrograph, meaning that most of the lines are blended and cannot be treated as single independent lines, but require multiple line fitting instead

pair. A blended line (λ_i, x_i) was thus transformed into a group of individual lines $(\lambda_{i,j}, x_{i,j}, I_{i,j})$. with independent intensities $I_{i,j}$. Another important step has been to couple different echelle orders in the wavelength solution, by the fact that many spectral lines are present on two neighboring orders simultaneously. Besides delivering information on common systematic errors of the same spectral line distributed over two different orders, this procedure allowed to improve the wavelength solution robustness, though without over-constraining the solution. Finally, the improvements made on the input thorium atlas greatly contributed to make the wavelength solution converge rapidly and to make it robust against small variations, e.g. due to simple photon noise.

In the case of HARPS, the mentioned improvements on the atlas and the algorithms helped to reach a much better precision and stability on the wavelength solution. The previously obtained $80\,\mathrm{cm\,s^{-1}}$ *rms* on the zero-point of the wavelength solution, could be reduced to finally $20\,\mathrm{cm\,s^{-1}}$ *rms*, a value which corresponds much better to the expected photon-noise precision of about $10\,\mathrm{cm\,s^{-1}}$. We refer here to [3] for details concerning the new thorium atlas, the used algorithms and solutions, and the obtained performances.

4 Outlook on a "Perfect Calibrator"

4.1 Requirements to the "Perfect Calibrator"

Present calibration techniques have allowed us to reach $\mathrm{ms^{-1}}$-precision level. From our experience it becomes however evident that the used wavelength references such as thorium or iodine gas cells may not be adapted for the $\mathrm{cm\,s^{-1}}$ level, i.e relative precisions up to 10^{-11}. The main limitations arise from the source stability, the dynamical range in line intensities, blends, and intrinsic stability. In accordance with this experience, we shall therefore try to define the characteristics of an ideal calibration source:

1. The calibration source must cover the full wavelength domain of the spectrograph.
2. The source must possibly deliver a set of equal-intensity spectral lines.
3. The distance between different lines is perfectly stable at the level of 10^{-11} and ideally constant in the $\Delta\lambda/\lambda$ space.
4. The shape of single lines must be stable at a level which would correspond to a centroid shift of less than 10^{-11}.
5. The zero-point of the source, defined as the average over all spectral lines with respect to some "zero velocity", must be stable to better than 10^{-11}.
6. The zero-point of the source, or the wavelength of the lines, must be known with an absolute accuracy of better than 10^{-7}.
7. The line distance must be minimum 2 and maximum 3 FWHMs of the spectrograph IP, in order to be optimum with respect to photon noise.
8. The spectral lines must be narrow, i.e unresolved by the spectrograph, in order to be optimum with respect to photon noise.
9. No blends are present in the spectrum (separation between any two lines > 2 FWHM of the IP).

4.2 Possible Solutions

We foresee two possible solutions for the "perfect" calibrator. The first would consist in a simple Fabry-Perot illuminated in white light. The second solution would consist in using a Laser comb (see corresponding paper in the present proceedings). Both solution deliver a nice spectrum which would be very close to the requirements spelled out in the previous section. Technological challenges remain, however. In the former, the question is how to obtain long-term stability on the absolute "zero point". This problem is solved in the latter possibility, since the laser is coupled to an atomic clock providing eventually wavelength stability up to 10^{-14}. The laser comb, however, shows too high line density and poor extendibility into the UV wavelength region. In both cases developments are being carried out and positive results are expected for the coming years.

5 Conclusions

The most challenging application for the calibration of high-resolution optical spectrograph is probably the search for extra-solar planets, where precisions of typically 10^{-9} are reached nowadays. In order to find even smaller exoplanets, e.g. of Earth's masses, the precision is pushed even further down to the 10^{-11} level. If this level of precision becomes available, however, new possibilities arise also in other domains of astrophysics. Here we shall refer to the CODEX feasibility study [1], a spectrograph proposed for the E-ELT. This study shows that new and interesting applications can be defined in other fields, such as the direct measurement of acceleration of the Universe, or the determination of a possible variability of physical constants. All these science cases push toward high wavelength accuracy combined with luminous instrument. These goals are possibly within reach, although new developments are required. The wavelength calibration will however remain one of the most critical aspects.

References

1. L. Pasquini, S. Cristiani, H. Dekker, et al.: The Messenger **122**, 10 (2005)
2. C. Lovis, M. Mayor, F. Pepe, et al.: Nature **441**, 305 (2006)
3. C. Lovis, F. Pepe: A&A **468**, 1115 (2007)
4. M. Mayor, F. Pepe, D. Queloz: The Messenger **114**, 20 (2003)
5. B.A. Palmer, R. Engleman Jr: Los Alamos National Laboratory LA 9615, USA (1983)

Discussion

S. Moehler: If you have equally spaced wavelengths in your calibration "lamp", how would you detect a shift in wavelength?
G. Lo Curto: You do not expect a drift as large as the spacing between two lines. The spacing is ~ 3 resolution elements corresponding to $\sim 10\,\mathrm{km/s}$. We are talking about a drift of $\sim 1\,\mathrm{m/s}$.
C. Izzo: What fraction of a pixel is $\sim 1\,\mathrm{m/s}$?
F. Pepe: $\sim 1/3000$ pixel.
S. Stefl: Your plots may indicate that you assume a linear wavelength solution. Is it so or what is the optimal degree for your polynomial fit?
F. Pepe: In HARPS a polynomial of degree 3 is used. Increasing the degree does not significantly improve the solution rms.

Laser Comb: A Novel Calibration System for High Resolution Spectrographs

C. Araujo-Hauck[1], L. Pasquini[1], A. Manescau[1], Th. Udem[2], T. W. Hänsch[2], R. Holzwarth[2,3], A. Sizmann[3], H. Dekker[1], S. D'Odorico[1], and M. T. Murphy[4]

[1] ESO, Karl-Schwarzschild-Strasse 2, 85748 Garching, Germany; caraujo@eso.org
[2] Max-Planck-Institut für Quantenoptik, Hans-Kopfermann-Str. 1, 85748 Garching, Germany
[3] Menlo Systems GmbH, Am Klopferspitz 19, 82152 Martinsried, Germany
[4] Institute of Astronomy, University of Cambridge, Madingley Road, Cambridge CB3 0HA, UK

Abstract. With HARPS, the ESO community has established a new standard in the wavelength accuracy and stability of Doppler shift measurements of astrophysical sources. To meet new scientific challenges, like the detection of Earth-like planets and the measurement of the expansion of the Universe, one crucial subsystem is the wavelength calibration source. ESO and the Max-Planck-Institut für Quantenoptik (MPQ) have studied the feasibility, and are now starting the development of a laboratory demonstrator, for a wavelength calibration system based on laser frequency comb techniques. Such a calibration system will provide superior wavelength accuracy and stability by producing perfectly equidistant emission lines, covering a large wavelength range in the visible perfectly stabilized to the 10^{-11} to 10^{-15} level. The absolute reference is provided by an atomic clock, which guarantees the long term stability over the years required for some of the measurements.

1 Introduction

High precision spectroscopy is a field which has recently experienced a tremendous boost from a number of astrophysical disciplines ranging from exo-planet search, to measurements of varying physical constants, to element isotopes ratio determination and primordial nucleosynthesis, to the direct measurements of the expansion of the universe (cf. e.g. [1]). High precision spectroscopy requires, on the other hand, very accurate calibrations. The most commonly used wavelength calibration light sources are the hollow-cathode Th-Ar lamps, which are reasonably cheap and easy to find, however they have a number of disadvantages [2]: the appearance of the spectrum changes depending on the particular characteristics of the lamp used, the Ar spectral lines are sensible to pressure changes, which implies shifts in the wavelength with changing lamp pressure or current, the lines differ in intensity and spacing... In addition the presence of line blends affect the centre of gravity of the spectral lines, and aging of the lamp varies its performance.

The ideal reference source should eliminate the limitations of the Th-Ar lamp. The perfect calibration source should cover a wide wavelength range, have a very large number of reference lines in the desired range, with wavelength known to a very high accuracy and with similar intensity. It should guarantee wavelength stability ($< 10^{-11}$) over time frames up to many years. As we can now use old plates for astrometry, fully calibrated spectra will represent a potential legacy for the future generations of astronomers.

A very promising route to obtain a higher stability wavelength reference has been identified in the so-called laser frequency comb system [3]. A frequency comb consists of thousands of equally spaced frequencies over a bandwidth of several THz. It is based on the properties of mode locked lasers and brings a direct link from optical to radio frequencies. With respect to the Th-Ar lamps, the laser frequency comb will have a number of advantages: it guarantees long term stability over the years which are required for some of the measurements, the absolute wavelength of each line in the comb is known a priori (i.e. without the need of previous laboratory measurements) to a very high precision, the precision is only limited by the reference signal, which could be an atomic clock or a GPS, depending on the required stability. Figure 1 shows an example of a small spectral region around 500 nm from a Th-Ar lamp. Figure 2 shows a simulated comb spectrum covering about the same wavelength interval as shown in Fig. 1. The advantages of the comb generated spectrum are evident.

Another interesting feature of the laser frequency comb is the high density and equidistant emission lines; this allows to trace and to model the wavelength solution in the focal plane of the spectrographs with a very high

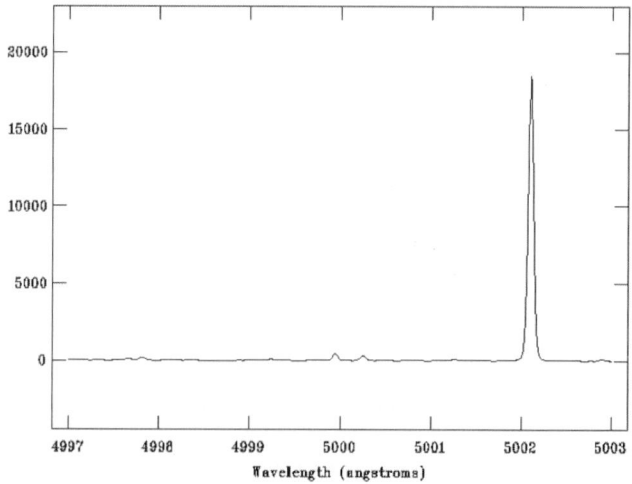

Fig. 1. Recorded Th-Ar lamp spectrum in the range from 499.7 to 500.3 nm

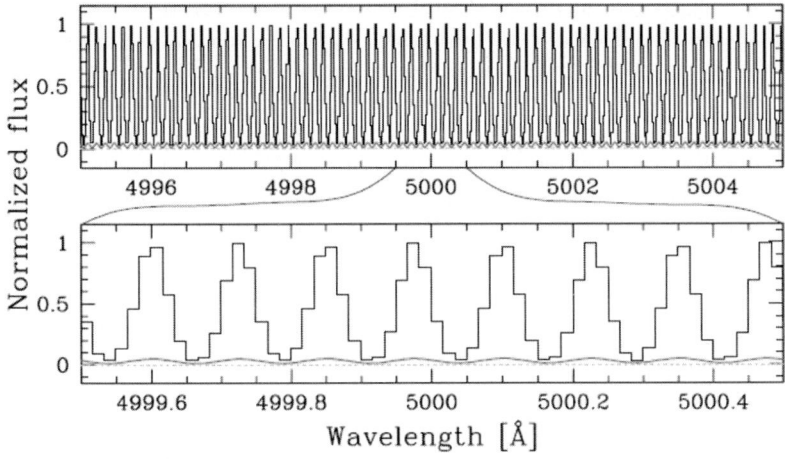

Fig. 2. Simulated comb spectrum in the same spectral region of Fig. 1 [4]

accuracy. Finally, once the single lines of the comb are resolved, blends do not exist.

Although the laser frequency comb is a very promising solution, its implementation as calibration source for an astronomical spectrograph is challenging. To further investigate the technological status and feasibility of a new calibration source, ESO has commissioned a study to Menlo Systems GmbH. Following the recommendations of this study, which has shown that it should be possible to fulfill our requirements, ESO and MPQ have started a collaboration aiming at the development of a laser comb prototype, suitable for spectroscopic applications. This development is of general interest for both, high resolution astrophysical spectroscopy and as a laboratory standard.

2 Requirements and Specifications

The specifications and requirements for a wavelength calibration system based on a laser frequency comb were discussed in the framework of the CODEX project, an ultra-high accuracy spectrometer for the ESO ELT telescope [5]. The aim of this instrument is to reach an accuracy of $1\,\mathrm{cm\,s^{-1}}$ over timescales of at least 10 years:

- The calibration system should have minimum wavelength coverage from 380 to 680 nm with a goal from 350 to 1000 nm. Ideally it should be obtained from a single-comb system. For the wider range two combs may be combined. To fulfill this requirement, high power requirements for non-linear conversion have to be considered.

- The wavelength accuracy should be of the order of 10^{-11} or better. It can be achieved with a self-referenced optical frequency comb locked to a precision radio frequency oscillator like a GPS or a cesium clock.
- For the stability required, the accuracy should remain at the same level for decades, or beyond the experiment / instrument lifetime.
- The minimum required photon flux is 10^{-11} W per line, in order to have an integration time as short as a few seconds. The intensity stability of each line shall be better than 10% or traceable over a period from 30 s to 24 h.
- Due to the limited dynamic range of the CCD (full well 100,000 electrons), within the spectral range, the flux (detected photoelectrons/second/comb element) should not vary more than a factor ≈ 5 to have for all the lines a reasonable flux in a single exposure.
- The line distribution in the wavelength range shall be almost continuous with no gaps to well sample the detector.
- The comb line separation is one of the most challenging requirements. Simulations show, that the optimum mode spacing should be of the order of 13 GHz [4]. In the ideal case that a 13 GHz femtosecond laser source was available, the problem would be solved, but to our knowledge there is no demonstration of such a source.
- In order to combine high repetition rate, short pulses and a large spectral width, the laser beam will be spectrally filtered outside the laser oscillator, which will be the final step to achieve the desired 13 GHz mode spacing.
- The other challenging requirement is the need to attenuate the supermodes or side modes, which need to be suppressed to a level of 10^{-5} with respect to the main modes, or in other words, below the dynamic rage of the detector. This factor has to be accommodated to choose the proper mode filter cavity finesse, which defines the losses in the cavity (a high finesse means low losses but might be difficult to achieve).

3 Status and Perspectives

The delivered feasibility study has identified the most critical technological challenges: increase of the comb frequency, large wavelength coverage, and has suggested two potential technical solutions. ESO and MPQ have agreed to launch a second phase which shall bring to a fully functional laboratory prototype in the timeframe of 3 years, with a possible third phase, which will bring to an easy to operate prototype to test on the telescope in 4 years from now.

Figure 3 shows a schematic representation of the solution to be developed, which is composed of three subsystems: a frequency stabilized comb high repetition rate femtosecond source with power amplifier in the near infrared, a non linear frequency conversion chain to generate the desired wavelength range and a mode filter cavity, needed for generating the desired mode spacing

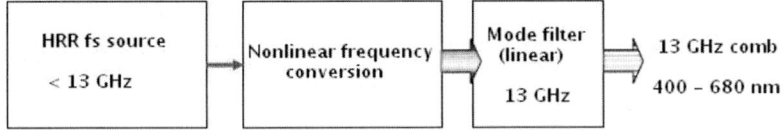

Fig. 3. Basic scheme of the Laser Frequency Comb calibration system to be developed. The architecture of the system is based on three building blocks: a frequency stabilized comb high repetition rate (HRR) femtosecond source, a non linear frequency conversion chain to generate the desired wavelength range and a mode filter cavity

of 13 GHz. We are therefore confident that in a few years we will be able to offer to our community a novel and very reliable wavelength calibration system. The wavelength calibration source based on laser frequency comb will offer the possibility of substantially improved wavelength calibrations, capable of meeting the increasing demands from various areas of astrophysical research.

References

1. N. Santos, L. Pasquini, M. Romaniello, M. eds. *Precision Spectroscopy in Astrophysics*, Proceedings of the Aveiro Conference (Springer) 2007
2. C. Lovis, F. Pepe, et al. Proc. SPIE **6269**, (2006)
3. Th. Udem, R. Holzwarth, T.W. Hänsch: Nature **416**, 233 (2002)
4. M.T. Murphy, et al.: MNRAS **380**, 839 (2007)
5. L. Pasquini et al. (CODEX Team) In: *The scientific requirements for ELTs* Proceedings of IAU Symposium 232, P. Whitelock, M. Dennefeld, B. Leibundgut eds. p. 193 (Cambridge)

Discussion

S. Kimeswenger: Why are the lines so wide after mode remowement?
C. Araujo: The simulations presented includes the "pixelation" on the detector.
S. Kimeswenger: How do you solve the problem addressed by S. Moehler in her talk concerning the pipeline reduction – pattern matching?
C. Araujo: The laser comb based calibration system is thought for high resolution echelle spectra. While pattern matching is successfully applied to MOS low resolution spectra, it is not clear that it will be suitable for high precision echelle data. In any case, with the comb calibration all echelle orders will be 'identical'. Of course a zero point additional signal is necessary, and this will also solve the problem of pattern recognition.
D. Baade: How much will such a system cost?

C. Araujo: Realistic costs will be known after the demonstrator has been built and re-engineered for operations at the observatory, a process which will last about 3 years. After this period we could make a realistic estimation of cost for a unit suitable to be used on the telescope.

Instrumental Line Shape Function for High Resolution Fourier Transform Molecular and Atmospheric Spectroscopy

M. Badaoui[1], F. Schreier[2], G. Wagner[2], and M. Birk[2]

[1] IAV Hassan II, Physics Unit, B.P.6240, Madinat Al-Irfane, 10101 Rabat, Morocco; nouralhakim@yahoo.com
[2] DLR – Remote Sensing Technology Institute, Oberpfaffenhofen, 82234 Wessling, Germany

1 Introduction

Remote sensing by spectroscopic techniques is well established in atmospheric sciences. It has contributed significantly to our knowledge of planetary atmospheres, and is clearly indispensable for investigation of astronomical objects.

The improving quality of current instruments and the ever increasing number of data sets has also increased the demands on analysis of the observed spectra. First, accurate retrieval of the atmospheric state (e.g., temperature, composition, ...) requires precise knowledge of molecular spectroscopic parameters. Furthermore, quantitative analysis of spectra relies on efficient yet accurate forward modeling (i.e., radiative transfer and instrumental response) along with inverse problem solvers.

Fourier transform (FT) spectroscopy is a powerful technique for both molecular laboratory spectroscopy and atmospheric remote sensing. Here we discuss atmospheric spectroscopy requirements and some molecular spectroscopy analysis tools with an emphasis on instrumental line shape.

2 Atmospheric Remote Sensing Requirements

A quantity of central importance to radiative transfer is the transmission describing the attenuation of radiation along its path. In the infrared and microwave it is usually possible to neglect extinction due to scattering, and the monochromatic transmission $\mathcal{T}$ is given according to Beer's law [6]

$$\mathcal{T}(\nu) = \exp\left(-\int_{\text{path}} \mathrm{d}s \sum_m n_m(s)\, k_m\bigl(\nu, p(s), T(s)\bigr)\right) \qquad (1)$$

where ν is wavenumber, p and T are atmospheric pressure and temperature, and k_m and n_m are the absorption cross section and number density of molecule m. The cross section is given by a superposition of many lines at positions $\hat{\nu}_l$, each described by the product of line strength S and a normalized line shape function g describing the broadening mechanism,

$$k(\nu, p, T) = \sum_l S_l(T)\, g\left(\nu, \hat{\nu}_l, \gamma_l^{(L)}(p,T), \gamma_l^{(D)}(T)\right) \qquad (2)$$

The combined effect of pressure broadening (Lorentzian line shape with width proportional to pressure, $\gamma^{(L)} \sim p$) and Doppler broadening (Gaussian with $\gamma^{(D)} \sim \hat{\nu}\sqrt{T}$) is modelled by a Voigt profile $g_V(\nu) = g_L(\nu) \otimes g_D(\nu)$

Obviously radiative transfer is depending only on the product $S \cdot n$ of line strength and number density; thus any uncertainty in the line strength immediately maps into corresponding uncertainties in the molecular densities to be retrieved in remote sensing. Likewise, accurate broadening parameters are crucial to derive altitude information from remote sensing.

Several comprehensive datasets have been compiled in the past decades, nb. HITRAN [11], GEISA [7], and JPL [8]. Despite significant advances further improvements are mandatory, especially with respect to the accuracy of line strength and pressure broadening.

3 Least Squares Line Parameter Fitting

The standard approach to retrieve molecular line parameters from spectroscopic measurements is least squares fitting, i.e., the minimization of the norm of the residual vector between the observation y and model f by appropriate adjustment of the model parameters x, $\min_x \|y^{(\text{obs})} - f(x)\|$. Note that the model is a nonlinear function of the parameters comprised in the vector x.

The FT spectrometer will give rise to a smoothing of the spectrum that can be described by a convolution of the monochromatic spectrum (1) with an instrumental line shape (ILS, response or apparatus function),

$$f(x) \longrightarrow \mathcal{T}_{\text{fts}}(\nu) = \mathcal{T} \otimes \mathcal{R} = \int \mathcal{T}(\nu')\, \mathcal{R}(\nu - \nu')\, d\nu', \qquad (3)$$

The ideal response function (normalized to one) accounting for the maximum optical path difference L of the interferogram is — in a first approximation

$$\mathcal{R}(\nu) \equiv 2L\,\text{sinc}\,(2\pi L\nu) = \frac{\sin(2\pi L\nu)}{\pi \nu} \qquad (4)$$

For high resolution spectroscopy further instrumental effects have to be considered, nb. the finite aperture (also known as optical apodization or through-put effect), vignetting, phase errors due to misalignment, ... The effect of the finite input beam is weighting the interferogram by the central arch of the function $P(\Delta) = |\text{sinc}\tfrac{1}{2}\nu_0\Omega\Delta|$ for $0 \leq \Delta \leq L$ [5, 4]. The solid angle $\Omega = \pi(\tfrac{R}{F})^2$ of the light beam is defined by the iris radius R and by the focal length of the collimator F.

3.1 FitMAS — Fit Molecular Absorption Spectra

The FitMAS code [13] determines molecular line parameters using nonlinear least squares solved iteratively with Gauss–Newton– or trust region algorithms. Collisional narrowing effects can be modelled with, e.g., Rautian profiles [12]. The ILS is either (4) or a convolution of sinc and box

$$\mathcal{R}(\Delta\nu) = \frac{2}{\pi\alpha^2\bar{\nu}} \left\{ \text{Si}\left[2\pi L\left(\Delta\nu + \tfrac{1}{2}\alpha^2\bar{\nu}\right)\right] - \text{Si}\left[2\pi L\Delta\nu\right] \right\} \quad (5)$$

(with $\bar{\nu} = \tfrac{1}{2}(\nu_{\min} + \nu_{\max})$ and $\Delta\nu = \nu - \nu'$) accounting for the finite input iris ($\alpha = R/F$). Baseline effects are considered by an appropriate polynomial.

3.2 The "Chain of Programs"

Developed independently, the nonlinear least squares problem is solved using a "DUD — Doesn't use Derivatives" technique [9]. The instrumental response $\mathcal{R}(\nu - \nu_0)$ is taking into account the throughput effect, phase errors, and internal non-multiplying channel (if any) [1, 2].

Phase errors due to misalignment of the moveable mirror of the interferometer inevitably occurs. The path difference is $(\Delta \pm \epsilon)$, and the adjustable phase error (or asymmetry) parameter is then $\Phi = \pm 2\pi\nu_0\epsilon$. A "realistic" ILS is the FT of a monochromatic line's interferogram truncated by the boxcar function $\Pi(\Delta)$, optically apodized by $P(\Delta)$ and affected by phase error:

$$\mathcal{R}(\nu - \nu_0) = \text{FT}[I_{\nu_0}(\Delta + \epsilon)\,\Pi(\Delta)\,P(\Delta)]. \quad (6)$$

4 Analysis of Pure Rotational Spectra of Ozone

Due to its importance for atmospheric chemistry and its impact on climate ozone is one of the best investigated molecules in spectroscopy and is a key species of databases [10]. For astronomy ozone is important as a marker for extraterrestrial life. Whereas line positions and strengths of pure rotational and ro–vibrational transitions are known quite well, there is considerable need to improve knowledge of pressure–broadening parameters.

Pure rotation spectra of ozone (with various mixtures of nitrogen and oxygen) in the range $10-50\,\text{cm}^{-1}$ have been recorded within the framework of an ESA study [3]. Line parameter retrieval is under process, cf. Fig. 1. Agreement within noise level was obtained between FitMAS and "Chain of Programs", and also with data in HITRAN.

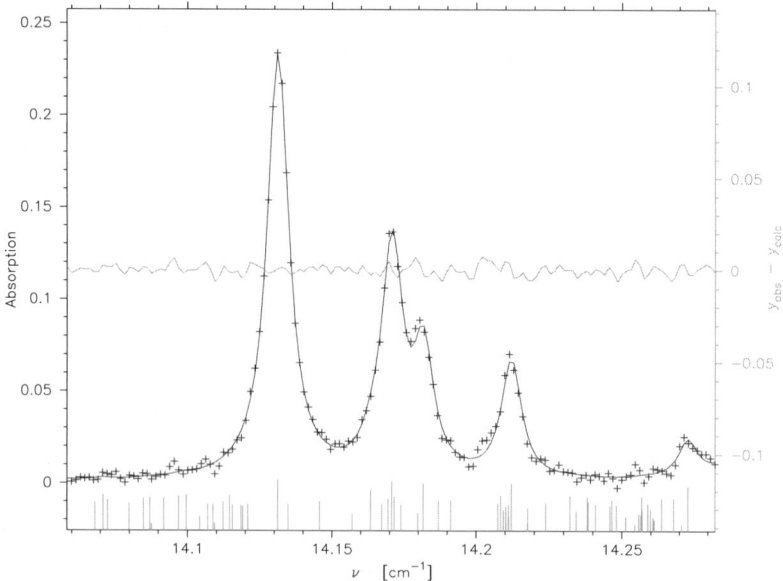

Fig. 1. FitMAS example: Ozone pure rotational spectrum (*Crosses* (+) and *solid curve* are observed and modelled absorption spectrum ($\mathcal{A}(\nu) = 1 - \mathcal{T}(\nu)$); y-axis for the residual (difference) spectrum on the *right*; *Vertical bars* indicate positions and strengths of ozone lines)

Acknowledgement. Financial support from DFG (Germany) and CNR (Morocco) is gratefully appreciated.

References

1. M.Y. Allout, et al.: J. Quant. Spect. & Rad. Trans. **60**, 979–988 (1998)
2. Mohammed Badauoi, PhD Thesis, Université de Paris VI, Paris (1977)
3. M. Birk, G. Wagner: Study on Spectroscopic Database for Millimeter and Submillimeter Wavelength, Final Report, Contract No. 11581/95/NL/CN; ed by A. Bauer, ESA/ESTEC, 1997
4. J.W. Brault: Fourier Transform Spectroscopy. In: *High Resolution in Astronomy, Proceedings of the 15th Advanced Course* ed by R.S. Booth and others (Swiss Society of Astronomy and Astrophysics, Saas-Fe, Geneva Observatory, 1985)
5. J. Connes: Recherches sur la spectroscopie par transformation de Fourier, Revue d'Optique, **40**, 45 (1961)
6. R.M. Goody and Y.L. Yung: *Atmospheric Radiation — Theoretical Basis*, 2nd edn (Oxford University Press 1989)
7. N. Jacquinet–Husson et al: The 2003 edition of the GEISA/IASI spectroscopic database, J. Quant. Spect. & Rad. Trans. **62**, 205–254 (2005)
8. H.M. Pickett, et al.: J. Quant. Spect. & Rad. Trans. **60**, 883–890 (1998)

9. M.L. Ralston, R.I. Jennrich: DUD, a derivative–free algorithm for nonlinear least squares, Technometrics, **20**, 7 (1978)
10. C.P. Rinsland, et al.: J. Quant. Spect. & Rad. Trans. **82**, 207–218 (2003)
11. L.S. Rothman, et al.: The HITRAN 2004 molecular spectroscopic database J. Quant. Spect. & Rad. Trans. **96**, 139–2104 (2005)
12. P.L. Varghese, R.K. Hanson: Appl. Opt. **23**, 2376–2385 (1984)
13. G. Wagner, M. Birk, F. Schreier, J.-M. Flaud: J. Geophys. Res. **107**, D22, 4626 (2002)

Calibration Sources for CRIRES

F. Kerber[1], A. Seifahrt[1], P. Bristow[1], G. Nave[2], H. U. Käufl[1], M. R. Rosa[3], C. J. Sansonetti[2], R. Siebenmorgen[1], and A. Smette[4]

[1] ESO, Karl-Schwarzschild-Strasse 2, 85748 Garching, Germany;
   fkerber@eso.org
[2] National Institute of Standards and Technology, 100 Bureau Dr., Gaithersburg, MD 20899-8422, USA
[3] Space Telescope European Coordinating Facility, Karl-Schwarzschild-Strasse 2, D-85748 Garching, Germany
[4] ESO, Alonso de Cordóva 3107, Vitacura, Santiago, Chile

Abstract. We report on our efforts to provide high accuracy wavelength calibration to the scientific observations with the Cryogenic High-Resolution IR Echelle Spectrograph (CRIRES), ESO's new high resolution (R≈100,000) infrared (IR) spectrograph at the VLT. In order to provide reliable and accurate wavelength standards for CRIRES the European Southern Observatory (ESO), in collaboration with the Space Telescope European Co-ordinating Facility (ST-ECF) and the National Institute of Standards and Technology (NIST), embarked on a project to establish Th–Ar wavelength standards in the 950 nm to 5000 nm operating range of CRIRES. In addition, and for use at wavelengths larger than 2200 nm, we have established a gas cell filled with N_2O as a calibration source based on existing data from NIST. Both sources were extensively tested during CRIRES commissioning runs and both are now being used for routine operations of the instrument.

1 CRIRES Wavelength Calibration

Traditionally, astronomical spectroscopy in the near infra-red (IR) has relied on atmospheric features of the night sky for wavelength calibration [8]. The lines from rotation-vibration levels of the hydroxyl radical OH (Meinel bands), which account for the night time OH airglow emission, are routinely used since they are very numerous, cover a wide wavelength range, and have been studied in detail at high resolution in the laboratory [1]. Originally, the concept for CRIRES wavelength calibration was also based on the use of features provided by the night sky. In the course of CRIRES commissioning it was found that at CRIRES high resolution only few features are evident in most wavelength settings. These were obviously not appropriate for high accuracy calibration purposes. We therefore intensified efforts to establish external source wavelength standards in the near-IR for use with CRIRES. Available sources are discharge lamps and gas cells. Based on considerations of scientific potential, practicability and availability we selected Th–Ar hollow cathode lamps (HCLs) and N_2O gas cells as prime candidates.

2 CRIRES and Its Calibration Sources

CRIRES [5] is a cryogenic, infrared echelle spectrograph designed to provide a resolving power $\lambda/\Delta\lambda$ of 100,000 between about 950 and 5000 nm. A ZnSe prism is used as pre-disperser. A multiple applications curvature adaptive optics (MACAO) feed is employed to minimize slit losses and to provide diffraction limited spatial resolution along the slit. An echelle grating with a blaze angle of 63.5° and a groove density of 31.6 lines/mm provides dispersion in the main spectrograph. CRIRES (http://www.eso.org/instruments/crires/) has been available to the community since April 2007.

2.1 Th–Ar Hollow Cathode Lamps

The Th spectrum was studied from 278 to about 1000 nm at high resolution more than 20 years ago by Palmer and Engleman [7]. Its emission lines are very narrow and the spectrum is rich over a wide wavelength range. In nature Th has only one isotope, ^{232}Th, which has zero nuclear spin. Thus the use of Th for calibration lines avoids complex and asymmetric line profiles attributable to isotopic or hyperfine structure. Two valuable studies of the Th–Ar spectrum in the near IR have recently been published, but neither is directly applicable to the operation of CRIRES:

Hinkle et al. [4] produced an atlas of the Th–Ar spectrum covering the range 1000–2500 nm. Their list of about 500 lines contains significant gaps in wavelength coverage for technical reasons.

A fundamental analysis of the Th–Ar spectrum was provided by Engleman et al. [3]. Their list contains more than 5000 lines derived from observations of a high current Th–Ar source which is not well-suited for operation at an astronomical facility. Hence ESO and NIST jointly investigated the spectra of commercial low current Th–Ar HCLs. Details of this effort and the operations of HCLs are described in Kerber et al. [6].

2.2 Gas Cells

The CRIRES gas cell is a sealed tube of about 18 cm length which contains N_2O (nitrous oxide or "laughing gas") at a well defined pressure. During operations we use a halogen lamp as background illumination. The lines from the transition of the gas appear in absorption superimposed on the bright continuum of the lamp. N_2O was chosen for practical reasons and because it is a primary calibration molecule from Heterodyne frequency measurements at NIST (Maki and Wells, http://physics.nist.gov/PhysRefData/wavenum/html/spect.html). Hence wavelengths for a large number of lines of its fundamental mode have been established covering the range 523–2845 cm^{-1} (1912–3515 nm). In addition a large data set on N_2O has been measured

with the MARK IV FTS at NASA's Jet Propulsion Laboratory (JPL) and
N_2O (http://mark4sun.jpl.nasa.gov/n2o.html) is also included in the
HITRAN 2004 (http://cfa-www.harvard.edu/hitran) database based on
a list by R.A. Toth. Model spectra (see Fig. 2) can be calculated as a function of pressure and temperature using a spectral synthesis code (FASCODE
in a PCLNWIN/IDL wrapper). The use of gas cells was originally foreseen
as a somewhat specialized application in e.g. radial velocity studies. During
commissioning we have established that they can also be used as a standard
source for routine wavelength calibration.

3 Laboratory Work, Results and Outlook

Spectra (Fig. 1) of the Th–Ar lamps operated at 20 mA were recorded on the
NIST 2 m FTS. Wavelength calibration of the spectra was derived by using
laser measurements of seven Th lines [2] as internal standards. The relative
uncertainty of the calibration (one standard deviation) is 1.4×10^{-8} × wavelength. For a detailed description of the experimental procedures and the
results we refer to Kerber et al. [6]. A total of about 2500 lines in the range
900–4800 nm has been established as wavelength standards and is available
for CRIRES wavelength calibration. In addition we have used published data
from NIST and JPL to investigate the use of N_2O gas cells as calibration
source. Both Th–Ar hollow cathode lamps and N_2O gas cells have been suc-

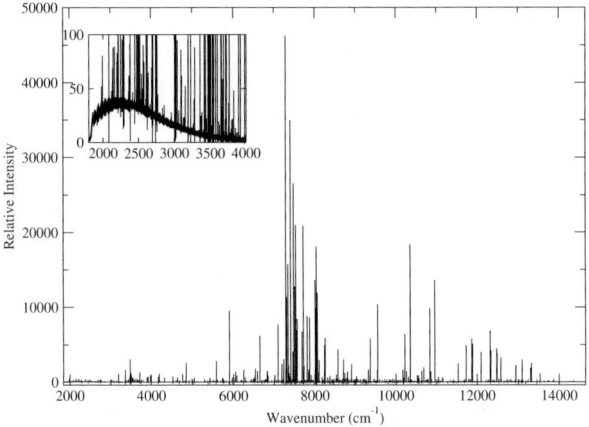

Fig. 1. Overview spectrum of a Th–Ar lamp. Wavelength range is 710–5000 nm
(14,000–2000 cm^{-1}). The line intensity is given in arbitrary units. The inset shows
that longwards of 2500 nm (<4000 cm^{-1}), thermal emission from the hot cathode
produces a continuum that underlies the pure emission line spectrum

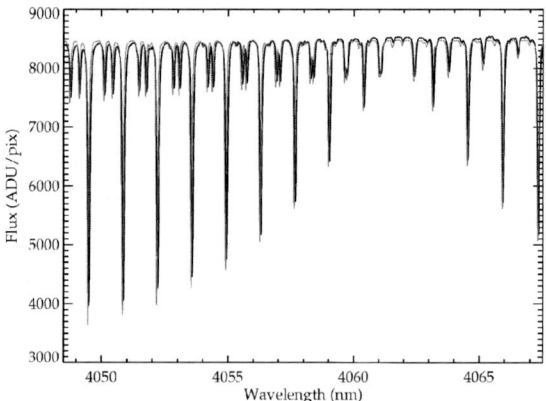

Fig. 2. 1-D representation of the spectrum of an N_2O gas cell recorded during commissioning III (October 2006). The spectrum, taken with black body illumination and a 2 s exposure time, was flatfield corrected and smoothed to a resolution of 65,000 using a preliminary wavelength solution. Overplotted (*grey line*) is a simulated HITRAN/FASCODE spectrum of N_2O for a 18 cm gas cell at a pressure of 50 mb

cessfully tested during commissioning and both are now available for routine use on CRIRES.

A number of steps are planned to further enhance the value of both sources. In particular we are making additional Th–Ar lines from the literature [3] available for use in wavelength calibration. For N_2O we will characterize the spectrum of the CRIRES gas cell as a function of temperature and pressure with the ESO FTS and we will investigate additional fill gases such as carbonyl sulfide (OCS).

With this development, wavelength calibration in the near IR will become very similar to the UV-visible region. It will be possible to support high accuracy absolute wavelength calibration without having to rely on atmospheric features. This is an essential step towards supporting the fundamental studies planned for the next generation of extremely large telescopes.

References

1. M.C. Abrams, S.P. Davis, M.L.P. Rao, et al.: ApJS **93**, 351 (1994)
2. W. DeGraffenreid, C.J. Sansonetti: J. Opt. Soc. Am. **B19** 1719 (2002)
3. R. Engleman Jr., K.H. Hinkle, L. Wallace: JQSRT **78**, 1 (2003)
4. K.H. Hinkle, R.R Joyce, A. Hedden, et al.: PASP **113**, 548 (2001)
5. H.-U. Käufl, P. Ballester, P. Biereichel, et al.: In: *Proceedings of the SPIE*, vol *5492*, ed by A.F.M. Moorwood, M. Iye, pp 1218–1227 (2004)

6. F. Kerber, G. Nave, C.J. Sansonetti, et al.: In: *ASP Conference Series vol 364*, ed by C. Sterken, pp 461–478 (2007)
7. B.A. Palmer, R. Engleman Jr.: Los Alamos Report, LA 9615, USA (1983)
8. P. Rousselot, C. Lidman, J.-G. Cuby, et al.: A&A **354**, 1134 (2000)

Recent Developments in the Optimal Extraction of UVES Spectra

J. M. Larsen[1], P. Ballester[2], V. D'Odorico[3], C. Ledoux[4], A. Modigliani[2], J. Pritchard[2], H. Sana[4], and A. Smette[4]

[1] Rovsing A/S, Dyregaardsvej 2, DK-2740 Skovlunde, Denmark; jmlarsen@eso.org
[2] ESO, Karl-Schwarzschild-Strasse 2, 85748 Garching, Germany
[3] INAF – Osservatorio Astronomico di Trieste, via G.B. Tiepolo 11, Trieste I-34131, Italy
[4] ESO, Alonso de Cordóva 3107, Vitacura, Santiago, Chile

1 Introduction

The UVES data reduction pipeline [1] uses optimal extraction to achieve higher signal-to-noise (S/N) of faint objects, corresponding to an increase in effective exposure time up to 70% compared with a simple aperture extraction (see Horne [2] and the further developments by Marsh [3] and Mukai [4] for an introduction). Initial releases of the UVES pipeline had limitations in the extraction quality at certain S/N ranges. We describe the implementation in version 3 of the pipeline, working in the common pipeline library CPL [5] context, which shows significant improvements with respect to earlier versions. The new implementation is used operationally since the beginning of P79.

2 Algorithm

In order to measure the spatial profile a preliminary estimate and subtraction of the sky is carried out by taking the median of all pixels (after masking out the object using a rough object localization). The algorithm then follows Horne's scheme but with the following differences:

- The spatial profile is measured either using an analytical (Gaussian or Moffat) profile (as described in [4]), or by resampling the empirical profile to a grid with a resolution of 0.2 pixels in the spatial direction, and fitting a low order polynomial to the spatial profile at each resampled position.[1] See also Fig. 3.

---

[1] While resampling the data is often avoided because it introduces resampling noise [3], this resampling noise is smoothed when fitting a low-degree polynomial to the spatial profile. However, when the model profile is later used to extract the data, it is important that the model is rebinned to the sampling of the data rather than the other way around. Mukai dubbed this "virtual resampling".

- In order to fully exploit the peculiarities of the echelle format the free parameters of the respective models (analytical profile centroid and width, or virtually resampled profile at each spatial position) are modelled as 2D polynomials in wavelength *and* order number. In this way, regions (which may span entire orders) where the profile cannot be accurately determined due to very low signal are interpolated from neighbouring regions having presumably higher S/N.
- Horne's formula for the optimally extracted flux (which is equivalent to profile fitting at every wavelength [2]) assumes that the sky background has been already subtracted, and furthermore that the interpolated sky level is effectively noise-free. Because of the short slits typically used in echelle spectrography (to ensure order separation), the assumption of a noise-free sky determination may not be valid; we therefore generalized the method to give combined optimal sky and object flux estimates by minimization of

$$\chi^2 = \sum_i \frac{(f_i - (S_i + Fp_i))^2}{\sigma_i^2} \quad (1)$$

where f_i and σ_i^2 are the flux and variance at the i-th pixel, p_i is the normalized spatial profile, and F and S_i are the object flux and sky levels to be determined. Assuming a simple model where the sky background is constant, $S_i = S$, a two-parameter minimization of (1) yields

$$F = \frac{(\sum_i 1/\sigma_i^2) \sum_i p_i f_i/\sigma_i^2 - (\sum_i p_i/\sigma_i^2) \sum_i f_i/\sigma_i^2}{D} \quad (2)$$

$$S = \frac{(\sum_i p_i^2/\sigma_i^2) \sum_i f_i/\sigma_i^2 - (\sum_i p_i/\sigma_i^2) \sum_i p_i f_i/\sigma_i^2}{D} \quad (3)$$

with variances

$$V(F) = \frac{\sum_i 1/\sigma_i^2}{D}; \qquad V(S) = \frac{\sum_i p_i^2/\sigma_i^2}{D}, \quad (4)$$

where $D = (\sum_i 1/\sigma_i^2) \sum_i p_i^2/\sigma_i^2 - (\sum_i p_i/\sigma_i^2)^2$, and where the error bars of p_i and σ_i^2 are not propagated because the final variances are dominated by the contribution from f_i (following Horne). The object and sky spectra are extracted by applying (2) and (3) to the non sky-subtracted image. In comparison with a separate sky subtraction, this method improves the final object S/N by a few percent (see Fig. 1) in the low to intermediate S/N range. While this is not a dramatic improvement, we consider it worth the effort because the additional computational cost is practically zero.

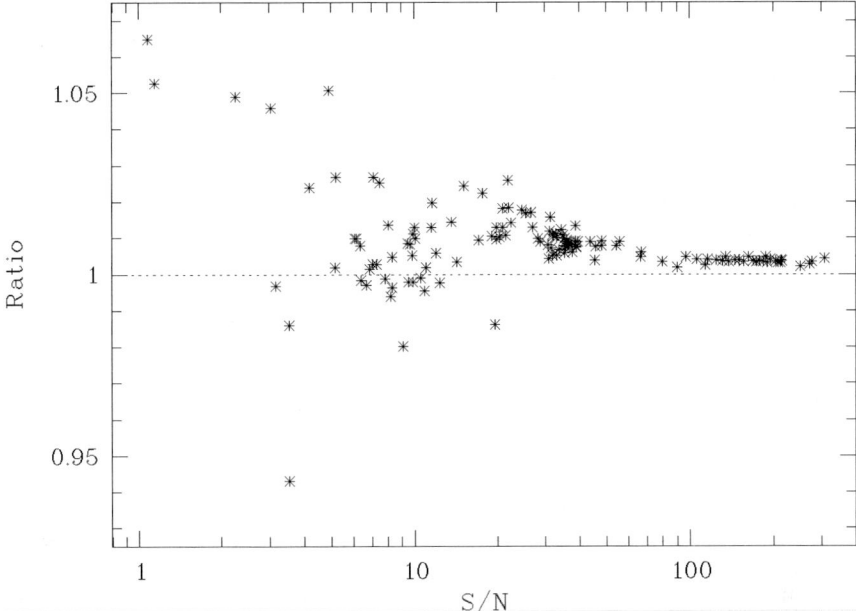

Fig. 1. Ratio of the S/N obtained with the optimal sky subtraction and the S/N obtained with a separate, initial sky subtraction. Each point in the figure corresponds to the extraction of one echelle order in various data sets with different count levels

3 Robust Automatic Data Reduction

Considering the default resolution of 0.2 pixels, a 30 pixel slit length and a two-dimensional 2nd degree polynomial, the virtual method needs to determine 1350 polynomial coefficients from the science image. In contrast, the analytical methods use only 18 coefficients for the two 2nd degree polynomial fits to the centroid and width of the spatial profile. For this reason, the analytical methods are more robust at very low S/N (see Fig. 4). However, at higher S/N the often significant mismatch between the analytical profile and the empirical profile (see Fig. 2) is known to bias the extracted flux [3] and make the rejection of cosmic rays unreliable.

In order to have a fully automatic data reduction with always good quality science results the appropriate profile measuring method is selected at runtime, depending on an initial estimate of the object S/N. If it is less than 10, an analytical method is used, otherwise the virtual method is selected. The Gauss and Moffat methods usually give very similar results, and the Gauss method is chosen as default because it makes the data reduction slightly faster.

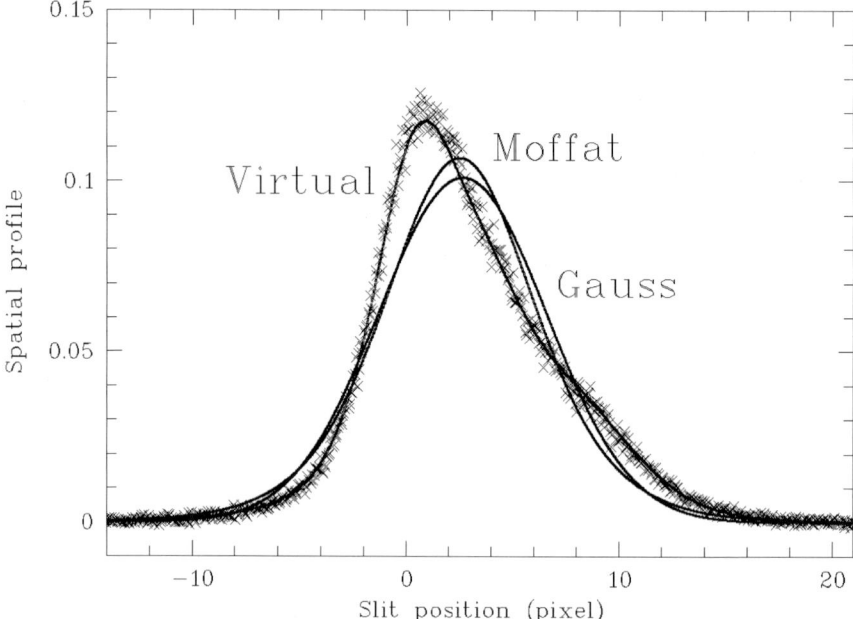

Fig. 2. Empirical spatial profile (*crosses*) and three models (*lines*). At intermediate or high S/N the analytical methods usually cannot fit the object spatial profile accurately. (The UVES instrumental PSF is known to be slightly asymmetric, but the high degree of asymmetry shown here for illustration purposes is due to the source being double)

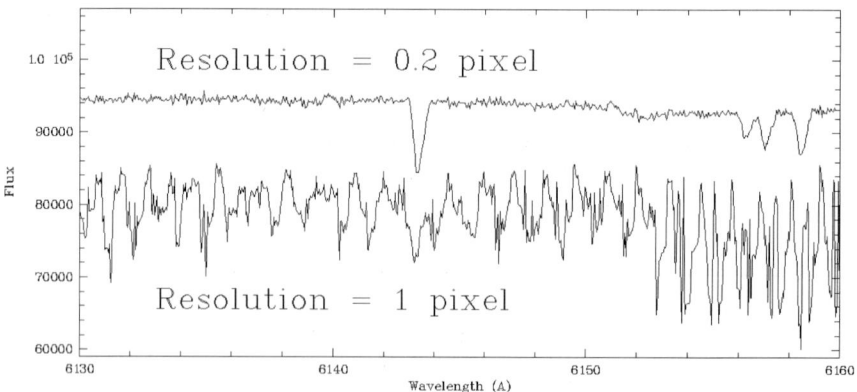

Fig. 3. Zoom of spectrum with S/N ≈ 290. Using a model profile of too low resolution may cause a quasi-periodic pattern in the extracted spectrum

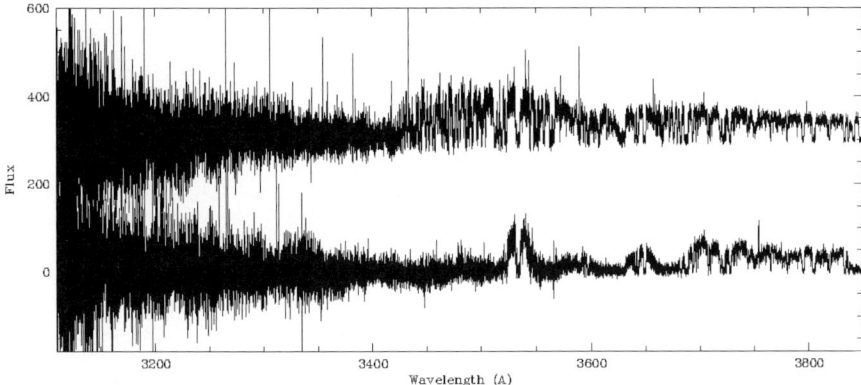

Fig. 4. Comparison of an extracted spectrum obtained with the new algorithm, analytical method (*upper spectrum*, shifted +300 units for clarity) with the same spectrum as produced using the previous algorithm (*lower spectrum*). The latter presents regions of missing signal because the object tracing fails in some orders. The global model allows to locate and extract the object in every order in a robust manner

Also, the degrees of all two-dimensional, low-order polynomials involved are determined at runtime by starting from (0, 0) and increasing the degrees in steps of one or two as long as the residuals decrease significantly. All polynomial fits are weighted, and outliers are rejected iteratively using a robust kappa-sigma clipping. Although this method of determination of the optimal polynomial degrees involves some ad hoc heuristics, it has proven to consistently give better results than setting the degrees manually.

References

1. P. Ballester, et al.: ESO Messenger **101**, 31 (2000)
2. K. Horne: PASP **98**, 609 (1986)
3. T.R. Marsh: PASP **101**, 1032 (1989)
4. K. Mukai: PASP **102**, 183 (1990)
5. http://www.eso.org/cpl

A new Generation of Spectrometer Calibration Techniques Based on Optical Frequency Combs

P. O. Schmidt[1], S. Kimeswenger[2], and H. U. Käufl[3]

[1] Institute of Experimental Physics, Technikerstr. 25, A-6020 Innsbruck, Austria; Piet.Schmidt@uibk.ac.at
[2] Institute of Astro- and Particle Physics, Technikerstr. 25, A-6020 Innsbruck, Austria
[3] ESO, Karl-Schwarzschild-Strasse 2, 85748 Garching, Germany

Abstract. Typical astronomical spectrographs [1, 2] have a resolution $\lambda/\Delta\lambda$ ranging between a few hundred to 200,000. Deconvolution and correlation techniques are being employed with a significance down to 1/1000th of a pixel. HeAr and ThAr lamps are usually used for calibration in low and high resolution spectroscopy, respectively. Unfortunately, the emitted lines typically cover only a small fraction of the spectrometer's spectral range. Furthermore, their exact position depends strongly on environmental conditions. A problem is the strong intensity variation between different lines[1] (intensity ratios > 300). In addition, the brightness of the lamps is insufficient to illuminate a spectrograph via an integrating sphere, which in turn is important to calibrate a long-slit spectrograph, as this is the only way to assure a uniform illumination of the spectrograph pupil.

Laboratory precision laser spectroscopy has experienced a major advance with the development of optical frequency combs generated by pulsed femto-second lasers. These lasers emit a broad spectrum (several hundred nanometers in the visible and near infra-red) of equally-spaced "comb" lines with almost uniform intensity (intensity ratios typically < 10). Self-referencing of the laser establishes a precise ruler in frequency space that can be stabilized to the 10^{-18} uncertainty level [3, 4], reaching absolute frequency inaccuracies at the 10^{-12} level per day when using the Global Positioning System's (GPS) time signal as the reference. The exploration of the merits of this new technology holds the promise for broadband, highly accurate and reproducible calibration required for reliable operation of current and next generation astronomic spectrometers. Similar techniques are also proposed in [5, 6].

1 Calibration of High Resolution Spectrometers

We will consider optical frequency combs based on fiber lasers that have a repetition frequency $f_{\text{rep}} \sim 250\,\text{MHz}$, therefore producing an equally spaced spectrum with lines separated by 250 MHz. These systems have the advantage

---

[1]see http://www.eso.org/instruments/fors/inst/arc_lines_MIT/atlas_GRIS_1400V+18.jpeg – 587.6 nm He line overexposes while nearly no other line is visible between 400 and 690 nm.

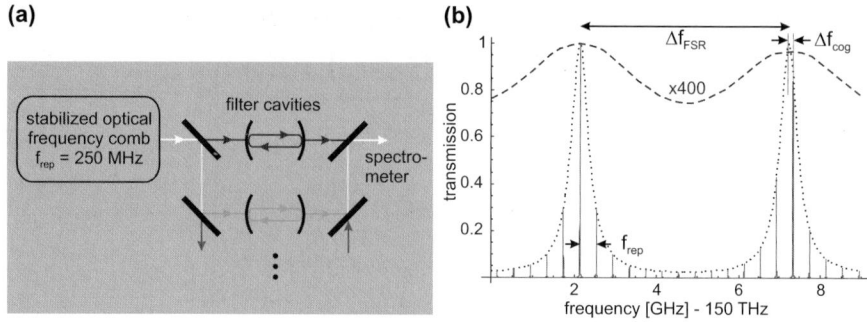

Fig. 1. (a) Schematic setup of the calibration source: the frequency comb provides a stabilized ruler in frequency space that is filtered by an external cavity. The distance between transmission maxima of the cavity can be adjusted to match the resolution of the spectrometer. (b) Cavity-filtered output spectrum of a frequency comb (exaggerated for visibility). *Solid line*: filtered comb spectrum; *dotted line*: transmission curve of the cavity; *dashed line*: spectrum seen by the spectrometer (magnified by 400)

over Ti:Sapphire based frequency combs to be more reliable and require less maintenance. To resolve individual lines of the frequency comb, a resolution of more than 2×10^{-6}, would be required. Therefore, astronomical spectrographs will see these devices as white light. For the calibration of high resolution spectrometers, we propose to filter the output of a frequency comb generator with external cavities as shown in Fig. 1(a). Interference inside the cavity leads to a frequency dependent transmission. The separation of transmission maxima (free spectral range: $\Delta f_{\rm FSR}$) can be chosen via the length of the cavity, whereas the width of the maxima is determined by the reflectivity of the optical coating applied to the surfaces of the cavity mirrors.

By matching the repetition frequency of the optical comb to be an integer multiple of the cavity's free spectral range ($\Delta f_{\rm FSR} = n f_{\rm rep}$), the effective repetition frequency of the laser is increased to $\Delta f_{\rm FSR}$. If the transmission maxima have a spacing well exceeding the spectrometer's resolution, and the frequency of one of the observed lines can be identified unambiguously, one can assign a precise frequency to all other observed lines simply by counting. This identification can be achieved by overlapping a cw laser (referenced to the frequency comb via one of the transmitted comb lines) with the comb spectrum before the filter cavity and observing the light of this laser on the spectrometer. At the same time, this laser serves as a reference to actively stabilize the length of the filter cavity and therefore its spectral properties. Due to limitations in the optical coatings, it is unlikely that a single filter cavity can cover the entire spectral range. A solution would be to spectrally split the output of the frequency comb laser into several wavelength regions, each filtered by an optimized cavity. The filtered output of the frequency comb will be similar to the solid line in Fig. 1(b). The recorded spectrum is a convolution between the filtered spectrum of the frequency comb and the spectrometer's resolution (dashed line in Fig. 1(b)). The achievable quality of the optical coating (dispersion compensated bandwidth vs. reflectivity/absorption) determines the width of the transmission resonances. This may result in insufficient suppression of neighboring comb lines and thus shift the center of gravity of the line observed by the spectrometer. It will be difficult to exactly match the filter cavity's free spectral range to the repetition frequency of the optical comb over the whole spectral range due to residual dispersion effects. This will result in an imperfect match of the comb lines to the transmission maxima of the filter cavity and thus induce a shift in the observed line center by Δf_{cog} as shown in Fig. 1(b)[2].

In the following, we will give an estimate of the shift in line center calibration due to uncompensated dispersion in the filter cavity based on CRIRES (2 pixel Nyquist sampling resolution $\lambda/\Delta\lambda = 10^5 \equiv 1.5\,\text{GHz}$ at $\lambda = 2\,\mu\text{m}$). Emission line centers can currently be determined to within 0.05 pixels (0.001 pixels $\equiv 0.75\,\text{MHz}$ anticipated in future experiments) [7]. To achieve accurate fitting of the center of gravity of the maxima, a separation of ≈ 27 pixels is required. This corresponds to $\Delta f_{FSR} = 20\,\text{GHz}$.

Figure 2 shows the effect of a frequency shift between a comb line and a filter cavity resonance. The intensity of a single transmission maximum as seen by the spectrometer is periodic in f_{rep} (several comb lines contribute to each maximum with decreasing intensity as their distance from the maximum increases). The observed shift in the center of gravity exhibits plateaus as comb lines approach the transmission maximum of the cavity. From the inset in Fig. 2(b) we see that a maximum frequency shift between comb line and cavity resonance of 18 MHz can be tolerated to maintain the required line center accuracy of 0.75 MHz. Figure 2(a) shows that at this shift, the inten-

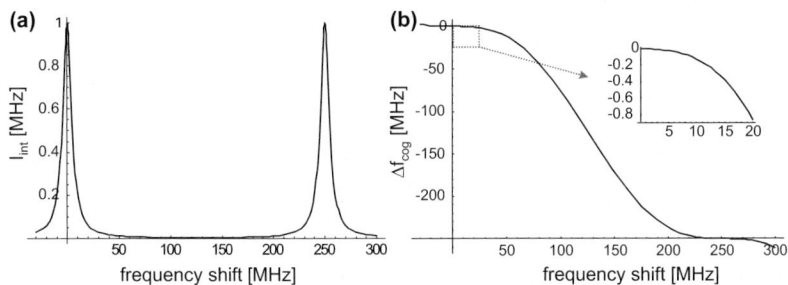

Fig. 2. Effect of frequency shift between filter cavity resonance and frequency comb lines. (a) Intensity as seen by the spectrometer, integrated over one cavity transmission maximum. (b) Shift in the center of gravity of the observed line (simulation parameters: $f_{rep} = 250\,\text{MHz}$, $\Delta f_{FSR} = 20\,\text{GHz}$, Finesse = 2000)

---

[2] Such a shift may not be an issue as long as it is reproducible and absolute frequency accuracy is not required.

sity has reached 7% of its maximum value. This defines a threshold intensity below which the observed line is discarded for calibration purposes. Since the spectrum of the frequency comb can not be assumed uniform in intensity to that level, a scan of the frequency shift between cavity resonances and frequency comb teeth is required to calibrate the magnitude of the transmission maxima for each cavity resonance [8].

Even more stringent requirements in terms of reproducibility and resolution apply to e.g. HARPS and CODEX: At similar resolutions the required stability over a few months to several years has to be $10^{-3} \ldots 10^{-5}$ pixels [9].

2 Calibration of Mid Resolution Spectrometers

For medium resolution spectrometers with an effective resolution below 100,000, the stability of the frequency comb is not required. Instead, it can be replaced by a fiber laser-based high-brightness white light source that is then filtered by the cavities. In this case, the cavity transmission maxima provide the ruler required for spectrometer calibration. The dispersion properties of the cavity can be calibrated using a frequency comb as described in [8]. We propose to use a cw laser locked to a stable reference (gas cells or a GPS-referenced frequency comb) to stabilize the length of the filter cavity to sub-MHz precision. A similar technique has been successfully implemented previously using an unstabilized cavity [10, 11].

References

1. I. Appenzeller, K. Fricke, W. Furtig, et al.: The Messenger **94**, pp. 1–6 (1998)
2. H.U. Käufl, P. Ballester, P. Biereichel, et al.: Ground-based Instrumentation for Astronomy. In: *Proceedings of the SPIE*, vol 5492, ed. by A.F.M. Moorwood and M. Iye (SPIE, Glasgow, Scotland, United Kingdom 2004) pp. 1218–1227
3. J. Stenger, H. Schnatz, C. Tamm, H.R. Telle: Phys. Rev. Lett. **88**, 073601-1-4 (2002)
4. M. Zimmermann, C. Gohle, R. Holzwarth, et al.: Opt. Lett. **29**, 310 (2003)
5. C. Araujo-Hauck, L. Pasquini, A. Manescau, et al.: The 2007 ESO Instrument Calibration Workshop (2008)
6. M.T. Murphy, Th. Udem, R. Holzwarth, et al.: MNRAS **380**, 839 (2007)
7. H.U. Käufl, et al.: The 2007 ESO Instrument Calibration Workshop (2008)
8. A. Schliesser, C. Gohle, T. Udem, T.W. Hänsch: Opt. Express **14**, 5975 (2006)
9. C. Lovis, F. Pepe: The 2007 ESO Instrument Calibration Workshop (2008)
10. R. Bacon, Y. Georgelin, G. Monnet: Bull. CFHT, **23** (1990)
11. C.B. Foltz, F.H. Chaffee, D.B. Ouelette, et al.: MMT Tech. Mem. **85**–4, 1 (1985)

Part IX

Session 8: Atmospheric Effects

Photometric Calibration of LSST Data

D. L. Burke and The LSST Collaboration

Kavli Institute for Particle Astrophysics and Cosmology Stanford Linear Accelerator Center Stanford University, Stanford, CA 94309, USA;
daveb@slac.stanford.edu

Abstract. Science studies made by the Large Synoptic Survey Telescope (LSST) will reach systematic limits in nearly all cases. Requirements for accurate photometric measurements are particularly challenging. Advantage will be taken of the rapid cadence and pace of the LSST survey to use celestial sources to monitor stability and uniformity of photometric data. A new technique using a tunable laser is being developed to calibrate the wavelength dependence of the total telescope and camera system throughput. Spectroscopic measurements of atmospheric extinction and emission will be made continuously to allow the broad-band optical flux observed in the instrument to be corrected to flux at the top of the atmosphere. Calibrations with celestial sources will be compared to instrumental and atmospheric calibrations.

1 Introduction

The Large Synoptic Survey Telescope (LSST)[1] is designed to carry out scientific investigations that would require a century to complete with existing telescopes. Its mission to rapidly and repeatedly survey one half the sky with high-quality deep imagery and photometry will address a broad and diverse range of scientific goals. The LSST will detect 250,000 Type Ia supernovae ($z \leq 0.8$) in each year of observing. Prompt alerts will be issued, and followup carried out in selected LSST deep fields ($z \leq 1.2$). Surface brightness shapes of over 3 billion galaxies ($z \leq 3$) will be measured during the ten-year LSST mission.

The LSST 8.4 m primary aperture, combined with its 65 cm diameter focal plane, will produce a large 3.5 deg field of view (FOV). The optics is a three-mirror design that provides good control of the image PSF over the full FOV, and a compact physical profile that will allow rapid slew and quick settling to the next pointing on the sky. The focal-plane detector will be a 3.2 Gigapixel mosaic with 10 µm pixels in 4k×4k sensor packages with 6400 channels of electronics to achieve fast read-out (2 s) following each 15 s exposure. The short focal length of the telescope (f/1.2) delivers a narrow depth of focus (10 µm confocal length) that imposes stringent requirements on the flatness of the focal plane sensors. The LSST optical photometric bands (u,g,r,i,z,y)

---

[1] See http://www.lsst.org.

will be similar to those used in the Sloan Digital Sky Survey (SDSS) [1]. The telescope active optics, wave-front sensing, guiding, and observational monitoring systems are modern technologies with counterparts working in the field today.

The optical through-put of the LSST (étendue = aperture × FOV = 319 m^{2}deg^2) will be more than an order of magnitude greater than any existing facility. Such a large step in survey power will make possible a ten-year mission that will produce 2000 observations of each 10 square degree patch of the sky and spanning the six photometric bands. Science investigations will in nearly all cases be limited by systematic rather than statistical errors. Precise determination of the PSF across each image, accurate photometric calibration, and continuous monitoring of system performance and observing conditions will be needed to reach the full potential of the mission.

2 Calibration Requirements and Goals

The science goals for the LSST mission pose stringent requirements on the stability and accuracy of photometric measurements. Specifications are given for four quantities that are defined in terms of broad-band photometry of isolated stars. These are summarized in Table 1. These requirements do not directly address measurement of photometric quantities of extended objects, nor specific requirements for separation of blended objects in crowded fields. It is anticipated that specifications for stellar photometry will properly fix the performance of the photometric system.

The first three specifications given in Table 1 are relative quantities defined within the native LSST photometric system in approximate AB magnitudes. These specifications are for bright stars not limited by photon statistics in a single exposure (e.g. r < 21 will assure that photon statistics are below 0.5%). Repeated measurements of the flux of a single object should be distributed approximately normally with 0.5% rms width, the rms width of the zero-point of objects measured across the sky no greater than 1%, and zero-point transformations between bands known with rms accuracy of 0.5% (this is a specification on the accuracy of measured colors). These requirements are of central importance to accurate broad-band photometric determination of redshifts of galaxies, and combined with good stability, are important for studies of supernovae and other transients. The fourth column in the table

Table 1. LSST goals for stellar photometry

Repeatability (rms)	Spatial Uniformity (rms)	Color Accuracy (rms)	Absolute (rms)
0.005	0.010	0.005	0.020

states a goal to understand transformations from the internal LSST photometric system to a definable physical scale to 2%.

3 Calibration Issues and Strategy

Photometric measurements in present-day large surveys with facilities such as SDSS and the Canadian-France-Hawaii Telescope (CFHT) are approaching the accuracies specified above for the LSST [2, 3, 4]. Selected data taken in good photometric conditions, have indeed, reached these numbers [5]. But more typically, ground-based telescopes produce measurements with errors a factor of two or so larger than the goals set out in Table 1.

There are many sources of calibration error that contribute at the fraction-of-percent level, and that can become worse if not carefully monitored and corrected. These include errors in the instrumental PSF and uncertainty in the effects of atmospheric turbulence on the image quality, errors in flat-fielding instrument response, scattering of light into view of the camera, errors in corrections for atmospheric extinction and refraction, uncertainties in processing background subtraction and de-blending objects in crowded fields, and ultimate uncertainties in calibration standards that are matched to spectral densities of celestial science targets.

It is becoming apparent [6] that the cadence and pace of sky surveys are important factors in achieving good performance. Rapidly paced surveys with good control over the stability of instrumental performance and minimal time between exposures allow identification and analyses of varying observing conditions and of changes in instrumental calibrations. Multiple epochs of observation of the surveyed sky enable parsing of data for analysis of systematic errors and unbiased elimination of the poorest quality data without significant loss of statistical accuracy. Additionally, observing the sky in multiple epochs provides the ability to define calibration standards with data taken during the best observing conditions. The LSST intends to take full advantage of its quick cadence and rapid return to each point on the sky. (Each epoch of survey of the half of the sky seeable from the LSST site will be completed approximately every 4 days.)

The LSST calibration plan builds on experience gained from the SDSS survey. The process decouples the establishment of a stable and uniform internal relative calibration from the task of assigning absolute optical flux to celestial objects. The latter task requires determination of a relatively few number of factors appropriate for the multi-epoch data set accumulated at any given time. Celestial sources will be used to define the internal photometric system and to monitor stability and uniformity of photometric data. Standardization of photometric scales will be done through direct observation of stars with well-understood spectral energy distributions. It is anticipated this method will achieve the photometric design specifications for the LSST mission.

3.1 Celestial Calibration of LSST Photometry

For purposes here, the broadband magnitude of a science target in a particular filter can be written as,

$$m = -2.5 \cdot \log[(\int F(\lambda) \cdot A(\lambda) \cdot T(\lambda) \cdot \lambda d\lambda)/(\int F_{cal}(\lambda) \cdot A(\lambda) \cdot T(\lambda) \cdot \lambda d\lambda)] + m_0$$

where, F and F_{cal} = flux (ergs $\times$ cm$^{-2}\times$ s$^{-1}\times$ nm^{-1}) at the top of the atmosphere from the target and calibration sources, $A(\lambda)$ = transmission through the atmosphere, and $T(\lambda)$ = transmission through the telescope, filter, and camera.

The value of m_0 is chosen to fix the zero-point of the photometric scale in a convenient fashion. The bright star Vega (α Lyr) has historically been used as the primary calibration source, and the zero-point chosen to fix its magnitude in a specific band-pass. The UBVRI system of Johnson, Morgan, and Cousins [7, 8] chooses the visible magnitude V = 0.03 for Vega. Alternate monochromatic AB systems [9] lead to definition of broad-band magnitudes with F_{cal} fixed to a convenient constant value. The photometric magnitude scale is defined by a value for the zero-point at a specified wavelength, and in practice is done in a way that keeps Vega near zero broad-band magnitude in the visible (V = 0.019 in SDSS AB95). Advantage has also been taken of several cool and hot hydrogen-dominated white dwarf stars with simple spectra to define calibrations at fainter magnitudes.

It is not practical for the LSST to continually observe a small set of calibration stars, nor is it possible for a deep survey like the LSST to observe bright objects such as Vega. It is necessary to establish a network of well-understood sentinel stars with magnitudes within the dynamic range of the telescope and that span the surveyed sky. These will define the internal LSST relative photometric system, and transfer and monitor the calibration from the reference standards to science observations. Precursor surveys to the LSST, particularly the SDSS but also others, will provide adequate starting points for this task. It is desired there be sufficient overlap between the dynamic range of the LSST and these precursors to avoid the need for intermediate transfer of calibrations via third-party instruments. Objects in images taken by the LSST will saturate the full-well potential of pixels when point-sources fluxes reach about 17 AB magnitude in the r-band, and somewhat fainter in other bands. On the other hand and for example, images in the SDSS telescope become noise limited at r < 20, and deep-field archives focused on the Southern sky reach r < 22 in full stacks of images. This provides a range of 3–5 magnitudes in which the SDSS and LSST overlap, and in that range we expect to find $\geq$ 100 main sequence stars on every chip (180 square arc-minutes) in LSST images taken of the same fields. A full internal network of sentinel stars will be established in the early months of LSST commissioning, and improved as the first epochs of the survey are completed.

The growing data archive of the multi-epoch survey will allow this network to be periodically optimized with increasing precision and definition.

The LSST will be able to observe in galactic equatorial fields where SDSS has spectroscopic confirmation [10] of ~ 4000 white dwarf stars with magnitudes $17 < g < 20$. Sub-classes of these well understood stars will be used as reference standards for the LSST photometric scale. It is planned to extend the SDSS catalog across the Southern sky, and a precursor campaign will be pursued as necessary to complete a suitable LSST standards catalog. The density of acceptable standards is expected to be great enough that, even at high galactic latitudes, one will be in a large fraction of LSST science images.

3.2 Toward Flat ABs

Standardization of LSST photometry will also follow a new and parallel approach [11]. Goals for this approach are yet more aggressive than stated above, and R&D now underway must be successful to realize the necessary hardware and software. The strategy is to determine independently the instrumental throughput of the telescope and camera ($T(\lambda)$ in the above expression for the broad-band magnitude), and corrections for extinction ($A(\lambda)$ above), emission, and refraction of light in the atmosphere. The goal is to construct from a monochromatic diffuse-screen illumination of the LSST a narrow-band "flat cube" indexed by the coordinates of the pixels in the camera and optical wavelength (i,j,λ). Precision photodiode detectors, whose responsivity as a function of optical wavelength can be calibrated at the U.S. National Institute of Standards and Technology to $\sim 0.1\%$ accuracy (in g,r,i),[12] will be used to provide normalization of the wavelength dependence of the illumination of the telescope and camera. The spectra of bright stars and cloud cover will be continuously monitored with independent instruments at the LSST site. Combined with atmospheric models (e.g. MODTRAN4 [13]), these measurements will be used to determine corrections for atmospheric effects in LSST broad-band observations. The instrumental and atmospheric measurements and calculations can be used to reconstruct the total system throughput of light from the top of the atmosphere to the signals read from the detector electronics.

There are several possible ways to use the narrow-band information described above to do broad-band photometry. The "flat cube" itself will be a data product, just as other flat-field measurements. This would allow any hypothetical spectral energy distribution (SED), even a rather badly behaved one, to be fitted to the observed broad-band data (perhaps combined with other observations, priors, or nuisance parameters). An effective broad-band flat-field could be constructed by averaging a "calibration SED" over the wavelength dependence for each pixel - the particularly simple case of a constant SED would produce a "Flat ABs" LSST photometric system. If successful, this process will provide LSST instrumental photometric parameters pegged to an absolute energy scale via calibrated precision detectors. To be most useful, it would be cross referenced to standard DA white dwarf stars that have been observed from above the atmosphere.

3.3 Cross Calibration and Systematic Errors

The two approaches sketched above will allow important cross checks to be made from comparison of quantities that can be derived independently from each approach. For example, dithered images of sentinel stars can be used to check the instrumental flat-fielding. And the rapid pace of the LSST cadence will allow gray extinction due to cirrus clouds or ice in the upper atmosphere to be monitored as coherent variations from exposure to exposure in the average magnitudes of sentinel stars. Similar information will be gathered from a co-bore sighted thermal ($9\,\mu$m) IR camera on the main telescope. Agreement between comparisons such as these will provide confidence that systematic errors are understood, and will allow good estimates to be made of errors in quantities that are derived from the directly measured data.

4 Instrumental Calibration

The LSST design includes subsystems dedicated to calibration and monitor of the primary telescope and camera functions. Additionally, the LSST project plan includes test and calibration of the telescope and camera hardware and software components and systems as they are completed and assembled. These initial measurements, and data generated by in-situ systems during operations, will yield products for on-line assurance of data quality and off-line science analyses.

The mirrors of the telescope and the refractive lenses of the camera will be individually tested in the laboratory with additional optics specifically designed for the purpose. Special diagnostic hardware will be built to test and calibrate the reflective optics as it is assembled into the telescope. This will include a small camera to be used in early commissioning to study light propagation through the mirror system, and an in-situ laser tracker (or equivalent) to measure alignments and monitor stability of the telescope mirrors and camera. This system combined with optical wave-front sensors located in the focal plane of the camera will calibrate and monitor the instrumentation and control of the active optics system.

Careful measurement of the optical throughput of the camera filters will be made over the full range of incident angles encountered in the LSST optics. The camera shutter timing and exposure pattern will be calibrated, and an in-situ monitor of the shutter timing read out for each exposure.

Calibration of the focal plane sensors is of particular importance. Quantities of interest include some that are not expected to vary significantly from one sensor to the next and to be stable over time. These include the full-well potential of the pixels, cross-talk between pixels and between electronic channels, and the charge transfer efficiency between pixels. These "operational" parameters can be determined from measurements made on prototype and first production sensors. Others vary more from chip to chip or depend more strongly on environmental changes. Acceptance testing will be done during production by the manufacturer and by LSST personnel to reject units that fail to meet specifications. Full calibrations of subassemblies (rafts) of nine sensors will be done with a combination of diffuse-screen and laser optical instrumentation, and with X-ray sources. The combination of optical and X-ray exposures will provide calibrations of quantum efficiencies, electrical gains and offsets, read-out noise, and physical cosmetic features (electron traps, truncated columns, dead electronic channels, etc) over a range of temperature and operating conditions. The goal of the photometric calibrations on the raft subassembly will be to obtain all needed operational data and a 1% relative calibration of the sensor pixel responses. Precision optical metrology will be done on sensors and rafts to check the pitch and alignment of the sensor coordinates, and to establish flatness of the surface of the focal plane to within the specified 6 micrometer peak-to-peak depth demanded by the f/1.2 optics of the telescope. The mechanical design of the rafts provides kinematical positioning of each sensor.

Tests and calibrations will also be done of the fully assembled camera before it is delivered to the mountain observatory to be mounted on the telescope. Diffuse-screen and laser optical calibrations will be carried out; these can be done both with and without the full compliment of refractive optics that are part of the camera structure. The design of the camera includes in-situ systems to monitor the flatness of the focal plane array, and to provide X-ray calibrations and monitor of the sensors and read-out electronics. These will be fully exercised and used to establish baseline operating calibrations. The goal of the photometric calibrations will be to establish a 0.5% relative calibration of the responsivity of the detector pixel-by-pixel. It is anticipated that the equipment used for these tests will accompany the camera to the LSST observatory site so that calibrations can be repeated on the mountain top whenever necessary.

4.1 In-situ Monochromatic Diffuse-screen System

As discussed in Sect. 3.2 above, the LSST facility plan includes an in-situ system designed to calibrate the optical and electronic response of the telescope and camera. This system is being developed and tested in collaboration with the ESSENCE[2] and Pan-STARRS[3] surveys. The concept is to use a tunable monochromatic laser to back-light a diffuse screen mounted in the dome of

---

[2] See http://www.ctio.noao.edu/essence/.
[3] See http://pan-starrs.ifa.hawaii.edu/public/home.html.

the LSST enclosure. The physical size of the screen, and the angular distribution of the light emitted from it, must fill the étendue of the telescope. In this case, every point in the pupil plane of the telescope will illuminate every point in the focal plane of the camera. The requirements for spatial uniformity of the light emitted by the screen are not very stringent (10% variations can be tolerated), but the angular distribution of light in the pupil plane must be uniform (Lambertian), and scattering of light from outside the telescope beam into view of the camera must be eliminated (or accurately subtracted). The intensity of light emitted from the screen and passing through the telescope optical path is monitored by photodiodes. This system is intended to provide the efficiency of the optical transport through the telescope and camera combined with the electronic response of the detector as a function of two-dimensional position in the focal plane (pixel coordinates) and optical wavelength. In a modification of the traditional terminology of astronomy, this is a "flat cube" that will allow reconstruction of optical flux in the telescope pupil from measured signals in the camera. First tests of such a system made at the CTIO Blanco telescope look promising [14].

The monochromatic dome screen can be combined with the in-situ X-ray source in the LSST camera to enable a complete characterization of the optical and electronic instrumental response. The quantum efficiency of the detector can be separated from the purely (wavelength independent) electronic response using the statistical behavior of the data during illumination by the monochromatic screen. And the electronic response can be directly measured with the X-ray calibration. Agreement between these two provides a powerful check and constraint on the overall calibration. It will also be possible to bias the electronics with the optical system in steps through the dynamic range of the electronics. This will allow a measure of the linearity of the system across its dynamic range from measurement of the constant X-ray signal atop the optical signal.

5 Monitoring Observing Conditions

Correction of the flux of light measured by the LSST for extinction and emission in the atmosphere is discussed in Sect. 3.2 above. Correction of photometric measurements for atmospheric effects must be done with accuracy better than 0.5% to meet the specifications in Table 1. So it will be essential to make accurate contemporaneous measurements of atmospheric conditions through which science data are being taken. The LSST facility plan includes auxiliary instrumentation to do this job.

Presence of cloud cover will be monitored by a thermal IR camera co-bore sighted with the main LSST telescope. This camera ($9\,\mu$m wavelength) will be sensitive to emission from water vapor, and will need to be able to resolve cloud structure on arc-minute angular scales. An auxiliary spectroscopic telescope is planned to measure extinction and emission at optical wavelengths.

This telescope need not have high resolution (R ~ 100), and can have a limited FOV. But it needs to be able to monitor the spectrum of a bright star that is in the FOV of the main telescope as science observations are made. The aperture of the auxiliary telescope must be large enough, and the readout and slew times rapid enough, to keep up with the cadence of the main telescope. This will require a 1.5 meter auxiliary telescope.

Accurate estimation of the PSF of each image is essential, and auxiliary equipment is planned to monitor turbulence in the atmosphere through which the LSST is observing. This includes a MASS-DIMM monitor as well as general weather station information. The site that has been selected for the LSST observatory on Cerro Pachon in Chile is well known, and satellite information on upper atmospheric conditions readily available.

Acknowledgement. The LSST design and development activity is supported by the National Science Foundation under Scientific Program Order No. 9 (AST-0551161) through Cooperative Agreement AST-0132798. Portions of this work were performed in part under Department of Energy contracts DE-AC02-76SF00515, DE-AC02-98CH10886, DE-FG02-91ER40677 and W-7405-Eng-48. Additional funding comes from private donations, in-kind support at Department of Energy laboratories and other LSSTC Institutional Members.

References

1. M. Fukugita, et al.: AJ **111**, 1748 (1996)
2. C. Stoughton, et al.: AJ **123**, 485 (2002)
3. Z. Ivezic, et al.: Astron. Nachr./AN **325** (2004)
4. E. A. Magnier, J.-C. Cuillandre: ASP **116**, 449 (2004)
5. N. Padmanabhan, et al.: ApJ 674, 1217 (2008)
6. Z. Ivezic, et al.: AJ **134**, 973 (2007)
7. H.L. Johnson, W.W. Morgan: ApJ **117**, 313 (1953)
8. A.W. Cousins: MNASSA **37**, 8 (1978)
9. J.B. Oke, J.E. Gunn: ApJ **266**, 713 (1983)
10. D.J. Eisenstein, J. Liebert, H.C. Harris, et al.: ApJS **167**, 40 (2006)
11. C. Stubbs, J. Tonry: ApJ, **646**, 1436 (2006)
12. National Institute of Standards and Technology, NIST Special Publication 250–41
13. U.S. Air Force Research Laboratory, 29 Randolph Road, Hanscom AFB, MA 01731
14. C. W. Stubbs PASP 199. 1163 (2007)

The GTC Photometric Calibration Programme

J. M. Rodríguez Espinosa, A. Di Césare, and P. L. Hammersley

Instituto de Astrofísica de Canarias, La Laguna, Tenerife, Spain

Abstract. The Gran Telescopio Canarias (GTC) is a 10.4 m segmented telescope that is about to have first light. Science operation will begin in about one year time. In order to ensure that the data coming from the GTC have the required quality and homogeneity and to foster the re-utilization of the data archive, a strategy for an adequate calibration of the data is being implemented. As part of this strategy, a catalogue of standard stars suitable for the photometric calibration of the GTC facility instruments is being put together. The GTC facility instruments are quite versatile and include both broad and narrowband filters plus some tuneable filters, and span the range from the near UV to the mid IR. The GTC Catalogue of Standard Stars that we are producing consists of a number of calibrated stellar fields, which include stars suitable for a 10 m telescope, and are not variable, non-binary and not having infrared excesses if they are to be used in the infrared. The catalogue consists of 30 star fields evenly distributed across the sky. The fields contain sources over the range 12 to 22 magnitude, spanning a wide range of spectral types (A to M) for the visible and near infrared. The stars are being spectro-photometrically calibrated, through the use of spectral templates. This allows selecting the bandpasses and thus the calibration magnitudes appropriate for the variety of filters in the GTC facility instruments.

1 Introduction

As telescopes as well as science instruments become large and expensive, an accurate and traceable photometric calibration of the data is mandatory. The Gran Telescopio Canarias (GTC) has recognised this aspect and there are currently two programmes underway to ensure that suitable calibration standards are in place for instruments operating in the optical, near and mid IR. The strategy is to adopt the methodology used for the development of the set of photometric standards for the Infrared Satellite Observatory (ISO). This approach has the advantage that once the set of photometric standards has been properly characterized and their full spectral energy distributions obtained, the magnitudes in any photometric system can be determined using synthetic photometry from the standard templates.

2 The Importance of a Proper Set of Standard Stars for GTC

Accurate and traceable photometric calibration standards are important for several reasons, including the obvious one that scientific observations can only be as accurate as the calibrations available for them. Besides, accurate calibrations are important because

- Astronomers are increasingly using and combining data from many different instruments covering a very wide wavelength range.
- Astronomers are increasingly using observations from archives, and/or mixing archival data with data from other telescopes and instruments. Astronomers, in general have no control over how the archival data is taken. Hence the calibration of data that is to be archived has to take special care in their accuracy and traceability. Indeed, without a proper calibration for all data all kind of systematic effects will contaminate the final results.

The current catalogues of calibration standards pose a number of problems when trying to use them on large telescopes. Problems due to either the size of the telescopes or to how the standard data were taken in the first place. Indeed,

- most of the commonly used standard stars will saturate the detectors on any 8–10 m class telescope very rapidly. For visible and near IR imaging a 12 magnitude source will saturate in one second.
- Large telescopes take some non-negligible time to move from a direction to another in the sky. Calibration sources are then required well distributed over the sky, so that the time to reach a standard star is minimised.
- Instruments on large telescopes are complex with many observing configurations. It is important to reduce the time required for calibrations hence the traditional method of measuring a number of stars in each configuration must be streamlined.

3 The GTC

Just for completeness, a few notes on the GTC, a 10.4 m segmented primary mirror telescope (Fig. 1), now being completed at the *Observatorio del Roque de Los Muchachos*, in the island of La Palma, Spain. The GTC is being prepared for having first light in Spring this year. Currently there are six segments installed in their primary mirror cells 1, and both the Secondary and Tertiary mirrors are in their respective mounts in the telescope. The instrument rotators are also mounted and soon the first Acquisition and Guiding box will be mounted on one of the rotators. The plan is to have first light with these six segments. All 36 segments are at the observatory, however

Fig. 1. Top down view of the GTC primary mirror cell. Note the six segments installed. Photo credit: IAC

only six will be on line for First Light in order to simplify the stacking of the images onto the 1 arc-minute A & G array. More segments will be added as soon as we master the behaviour of the primary mirror control system.

After First Light, we have foreseen one year of test and optimization of the telescope plus two first light instruments. One of these instruments is OSIRIS. OSIRIS is a low dispersion multi-object spectrograph and wide field imager, employing both broad and narrow band filters as well as tuneable filters. The large number of possible configurations of OSIRIS is one of the drivers of this calibration programme. The other First Light instrument is CanariCam, a thermal infrared imager, spectrograph, polarimeter and coronagraph. Calibration standards for CanariCam are also being characterised in a parallel programme.

4 The GTC Calibration Approach for the Optical and Near IR

A calibration plan based on techniques similar to those used for space observatories [2] is being implemented for the GTC. The methodology is based on the use of spectral templates, which combined with the footprint of the filters used for the various instrument/detector, allows recovering the magnitudes of the standards for the said combination of filter/detector. The requirements for the GTC calibration standards are as follows:

- Calibration fields will be used, not single stars, distributed all over the sky.
- Zero points accuracy of 0.01 magnitude in all filters. Vega (HR7001) is taken as reference standard.
- Zero Point changes from field to field of less than 0.005 magnitudes in all filters.
- Linearity errors under 0.001 magnitudes for up to magnitudes beyond magnitude 20.
- Colour errors less than 0.015 magnitudes.

Note that the objective is not to improve the accuracy of the flux to magnitude conversion, as this conversion is taken as a definition, thus no error is introduced. The important point is that it must be clearly established what this conversion is based on.

4.1 The Calibration Fields

The strategy is thus to obtain both photometry and spectroscopy of a set of calibration sources in a number of calibration fields. These data will be used to produce spectral energy distributions (SED) for the selected sources over the entire spectral range. Once these SED have been characterised, the isophotal flux in each combination of instrument/filter is obtained by convolving the instrumental transmission in each filter with the SED for each standard source. 180 stellar fields have been pre-selected that have been studied for over 3 years in an intensive observational campaign. From within each field, sources are selected such that they are sufficiently bright to give a high S/N in a few seconds, however they should not saturate the GTC instrument detectors. The stars are chosen to be single stars, and non variable, with spectral types in the range A0–M0.

The observations are being performed at the Canary Island Observatories (OT and ORM). More than 120 nights have been devoted so far to this programme, and more observations are ongoing. To date, we have collected photometry in 6 filters (V, R, I in the visible and J, H, K in the Mauna Kea filters set). The visible data have been acquired with the CCD on the IAC80 telescope. The near IR data have been obtained with the *Cámara Infrarroja* (CAIN) on the TCS. We have also acquired spectra of the some selected sources using the red and blue arms of the intermediate dispersion spectrograph ISIS on the William Herschell Telescope. The observations of the programme stars were always done at airmasses between 1.0 and 1.3. Observation of classical Standard stars [3, 1] were used to calibrate the atmospheric transmission and the instrumental response.

4.2 Photometry: Data Processing and Analysis

The data reduction is being performed using IRAF (Image Reduction and Analysis Facility), and several of its NOAO packages. Bias and flat-field corrections are carried out to calculate instrumental magnitudes as following:

$$m_{instr} = m_0 + k1 + k2 * \text{Air mass} + k3 * \text{Colour index} \qquad (1)$$

where m_{instr} are the instrumental magnitudes, m_0 are the standard calibration stars magnitudes, and $k1, k2, k3$ are parameters to be determined. Likewise, the stars apparent magnitudes can be calculated.

After reducing the images of various nights, we identify variable stars by comparing the star magnitudes of the same fields for the different nights. We thus identify all variable stars, which are discarded. We then compute new statistical standard deviations, obtaining representative values for each field. Our photometry, for various objects from the Landolt and UKIRT catalogues included in our calibration fields, has been cross checked in order to detect possible inconsistencies in the photometry. To this effect, we use several stars that were used for calibrations for several observing nights. Figure 2 shows a comparison between the Landolt and UKIRT photometry and our own photometry as a function of the optical (V-I) and ner IR (J-K) colours of the sources.

These plots show that we are achieving the expected accuracy in the calibration. Moreover this holds for both the optical and the near IR, even if in this last case the number of sources we have used is smaller than in the optical. The precision in the photometry is however still increasing with the number of observations.

4.3 The Spectroscopy

The aim of this programme is to be able to obtain in-band magnitudes for any instrument/filter configuration of the GTC instruments, including the OSIRIS tuneable filters. To this end we need to reconstruct Spectral Energy Distributions (SED) for our calibration sources. The next step is therefore to obtain spectra of pre-selected, i.e. photometrically well characterised, sources. We have therefore undertaken an observing campaign with the ISIS spectrograph on the WHT to secure spectra of these sources. The spectra were calibrated with observations of A0-A2 stars at the same air-mass as the actual sources.

Figure 3 shows a subsample of 4 spectra. The blue and red parts of the spectra have been joined through interpolation, after reducing each separate part of the spectrum. Note that these spectra are not yet absolute flux calibrated. This will be done by using the accurate photometry we already have for all our sources.

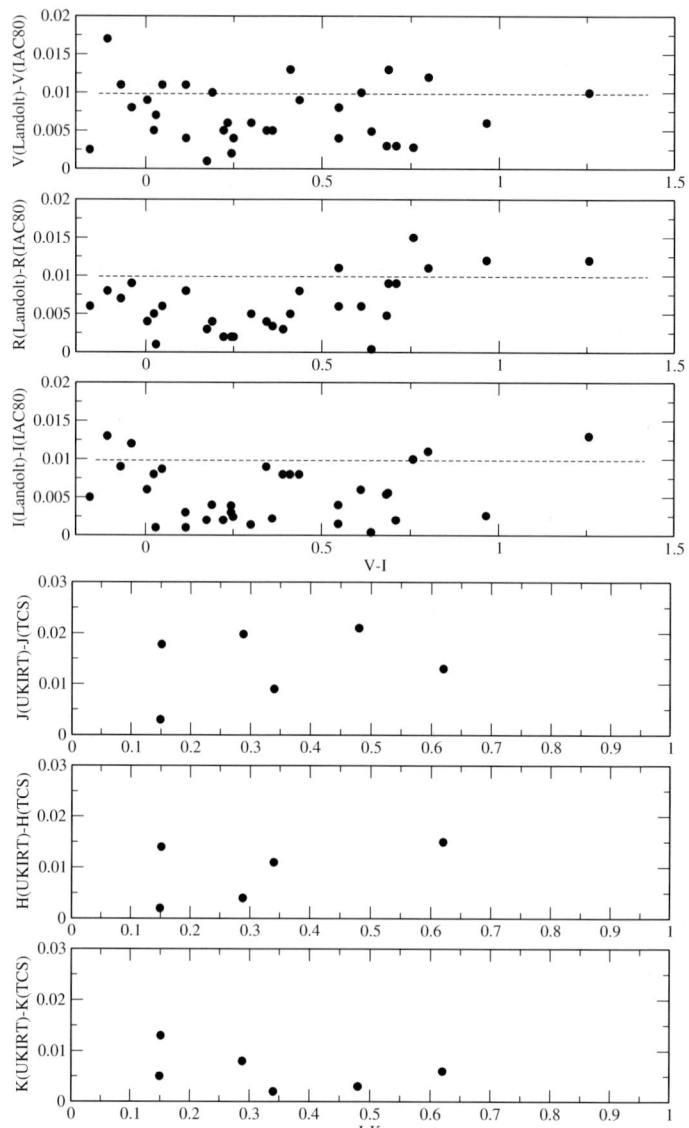

Fig. 2. Comparison between Landolt and UKIRT photometry with IAC80 and TCS photometry as a function of colour

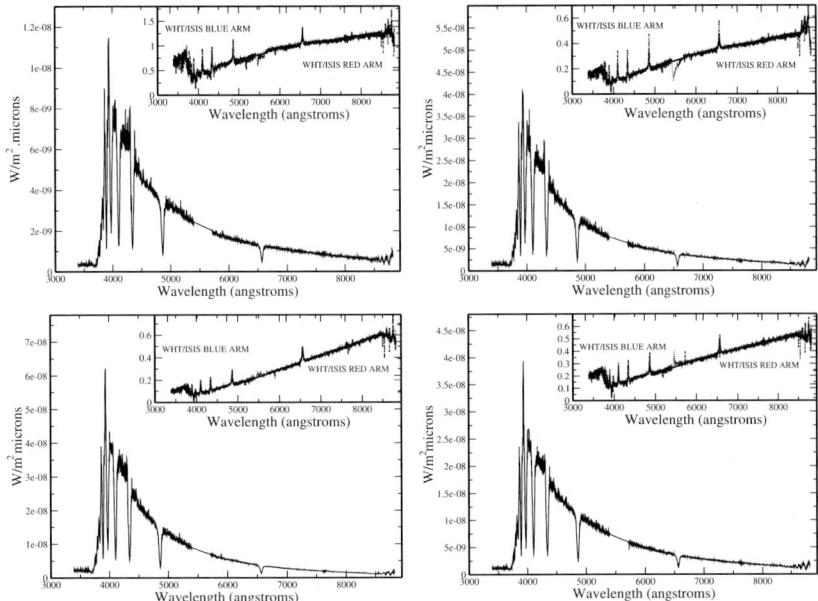

Fig. 3. Spectra obtained with ISIS (WHT) for 4 pre-selected objects

We are now in the process of establishing a methodology that using the SEDs of the GTC calibration sources accurately predict the magnitude of any of these stars in any GTC instrument/filter configuration. This we expect will save substantial calibration overhead time during the operation of the GTC.

5 Conclusions

The limited and costly observing time available together with the very complex and large number of instrument/modes to be calibrated on large telescopes means that new approaches to calibration must be explored. Indeed the traditional methods used for developing calibration standards are not time efficient. Therefore, for the GTC calibration standards we are undertaken a calibration programme based on the use of SEDs of all of the standard stars over the wavelength range of interest. Then we use these SEDs to determine, using synthetic photometry, the magnitudes for any instrument/filter. This approach was successfully used for the calibration of ISO.

We have so far shown that it is possible to reach the required accuracy in the photometry. We thus expect to have soon a catalogue of appropriate calibration standards meeting the GTC calibration requirements.

References

1. M.M. Casali, T.G. Hawarden: *JCMT-UKIRT Newsletter*, No. 3, August 1992, p. 33
2. M. Cohen, R.G. Walker, B. Carter, P.L. Hammersley: Astron. J. **117**, 1864 (1999)
3. A.U. Landolt: Astron. J. **88**, 853 (1983)

Discussion

D. Osip: Since you are beginning first light with six segments, will you be able to carry out calibration observations with your science instruments prior to facing the saturation problems that occur for the full aperture?

J.M. Rodriguez Espinosa: We will have the full mirror installed, i.e., 36 segments, by the time we bring the first science instrument. In any case, the suggestion is valid for the filters of the A+G box, for instance.

Measuring the Amount of Precipitable Water Vapour with VISIR

A. Smette, H. Horst, and J. Navarrete

ESO, Alonso de Córdova 3107, Vitacura, Santiago, Chile; asmette@eso.org

Abstract. We describe a method to determine the amount of precipitable water vapour (PWV) in the atmosphere above the Paranal observatory based on short exposure 19.5 µm VISIR spectra. We compare the values obtained by this method with the ones based on satellite data. We also show the dependence of the sensitivity and conversion factors on the PWV for different filters.

1 Introduction

Water is one of the main constituents of the atmosphere. It is mainly concentrated in its lower layers. When meteorological conditions allow an optical or infrared telescope to observe, water in the atmosphere can be found either in condensed form (liquid: droplets, in most clouds or solid: ice crystals in cirrus clouds) or as water vapour. The condensed forms will significantly degrade the quality of optical and Near Infra-Red (NIR) observations, while a significant amount of water vapour will negatively affect NIR and, more importantly, Mid-Infra-Red (MIR) observations, as can be seen http://www-atm.physics.ox.ac.uk/group/mipas/atlas/.

To quantify the amount of water vapour along a line of sight across the atmosphere, the commonly used quantity is the amount of precipitable water vapour (PWV). It is defined as the equivalent to the liquid precipitation that would result if all the water vapour in the column is condensed, and is expressed in mm.

Variations of the PWV during the night will influence the sensitivity and conversion factor (the relation between the number of detected photons and the flux above the atmosphere) of observations in the MIR, in a way that depends on the passband of the filter used. Therefore, time expensive observations of standard stars are necessary if accurate estimations of these quantities are required. In addition, before starting observations of a science target, it would be advantageous to know the sensitivity that can be reached. In service mode observations, this allows the night astronomer to select which instruments to use or which observation blocks (OBs) are most suited to the current conditions. In visitor mode, the visiting astronomer can best schedule the time to execute a given observation. Therefore, it would be useful to assess the relationship between the sensitivity and conversion factor, on one hand,

and the observing conditions, such as the PWV, on the other hand. Hence, a reliable and fast method to determine the PWV is needed.

At ESO, the WWW page developed by Marc Sarazin and Andre Erasmus (http://www.eso.org/gen-fac/pubs/astclim/forecast/meteo/ERASMUS/l_p_f0.html) provides an estimate of the PWV over the ESO observatories. This value is based on Upper Tropospheric Humidity derived from 6.7 µm images obtained by the GOES satellite, surface relative humidity and on the European Centre for Medium-Range Weather Forecasting model.

Other methods include radiometers [1], in particular, as support to submillimeter observatories (APEX, ALMA [2]), as a high-accuracy ($\approx 10^{-6}$ mm) is required; sky dips [1] that are also expensive in terms of execution time; or measurement of absorption lines in stellar spectra (see Osip et al., these proceedings). Here we describe a method based on short exposure VISIR sky spectra.

2 PWV Based on VISIR Sky Spectra

Since December 2005, we have regularly obtained 10 s, 19.5 µm medium-resolution spectra of the sky using VISIR when it is in use. The wavelength coverage corresponds to the Q3 filter. The execution time required to produce such spectra is actually dominated by the set-up of the instrument.

The data are obtained in burst mode (see the VISIR user manual). The planes of the produced cube are averaged. The resulting frame is divided by a flat-field obtained with the telescope at zenith. The sky spectrum is obtained by averaging 37 columns in order to obtain a high S/N while not being too affected by the spectrograph distortion.

These spectra are then fitted with a model spectrum based on the Reference Forward Model (RFM), a line-by-line modeling Fortran code, developed by Anu Dudhia (Oxford) to analyze data from MIPAS on-board ENVISAT (see http://www.atm.ox.ac.uk/RFM). It uses the HITRAN'2004 Database (Version 12.0) [3], which includes individual line parameters for 1,734,469 spectral lines for 37 different molecules. Other components of RFM include an 'Atmospheric profile' describing mean pressure, temperature, and concentration of H_2O, CO_2, etc. for typically 50 layers of the atmosphere.

The 19.5 µm VISIR spectra are however dominated only by the H_2O lines. Corrections to the 'tropical' atmospheric profile are made based on the atmospheric pressure and temperature at the observatory at the time of the observation. The data are then fitted by a least square method by adjusting the amount of water vapour with a unique factor for all atmospheric layers. In practice, this only affects the lower layers. Figure 1 shows two examples and their corresponding best fitted models.

As the number of measurements increases, comparison with satellite data become possible. As can be seen in the left graph of Fig. 2 the overall trend is similar between the values derived by the two methods. However, it is clear

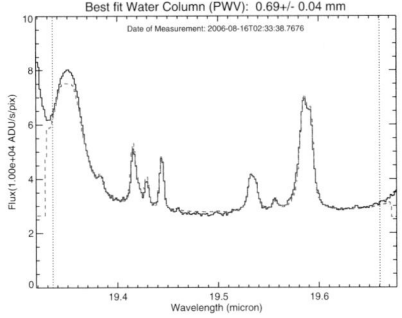
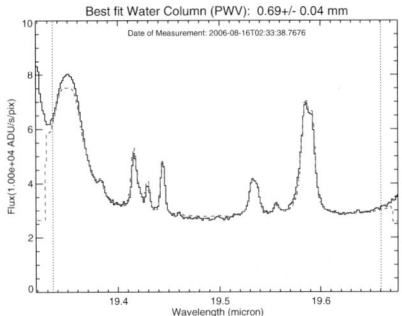

Fig. 1. Examples of sky spectra (*solid line*) and their best fit (*dashed line*). The spectral range used for the fit is delimited by the vertical dotted lines

that in a large number of cases, the satellite data provides smaller PWV estimates compared to the VISIR ones. In particular, some short time scale events are just not seen by the satellite: the right graph shows a rapid decrease of PWV from the beginning to the end of the night started on February 26, 2006.

Preliminary graphs showing the sensitivities and conversion factors as a function of the PWV are shown in Fig. 3 for PAH1, Fig. 4 for PAH2 and Fig. 5 for Q2. For PAH1, a significant dependence is put in evidence: the sensitivity is getting worse for larger value of PWV, while at the same time the conversion factor decreases. Note that for PAH1, the graph shows the

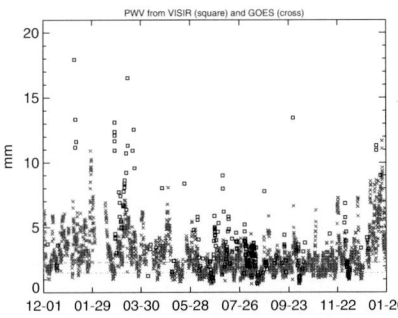
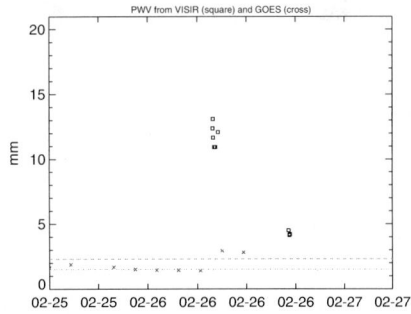

Fig. 2. Comparison between the GOES-based estimates of PWV (*crosses*) and the ones derived from the VISIR sky spectra (*open squares*). The left graph covers the whole period for which VISIR spectra have been taken. Overall trends are identical, although significant differences are present most of the time. On the other hand, the satellite based data are unable to detect short time scale events, such as the one displayed on the right graph

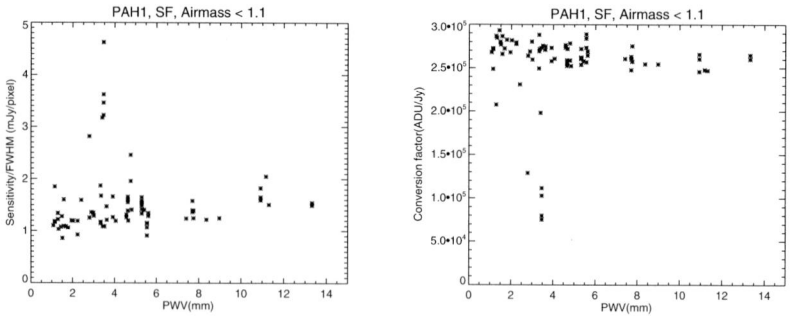

Fig. 3. Dependence of sensitivities and conversion factors on the PWV, for the PAH1 filter. In this case, the sensitivity has been divided by the FWHM in order to cancel its dependence on the seeing

sensitivity divided by the FWHM of the PSF, since one expects that the sensitivity linearly depends on the FWHM. This dependence also exists for other filters, but is less important as most of the observations are diffraction limited.

As the PAH2 filter bandpass avoids any strong water vapour line, one expects only a weak dependence of the sensitivities and conversion factors on the PWV due to the larger continuum opacity for large PWV. Figure 4 confirms these theoretical predictions. On the contrary, for Q2 (Fig. 5), one expects a strong dependence, which is confirmed by the measurements.

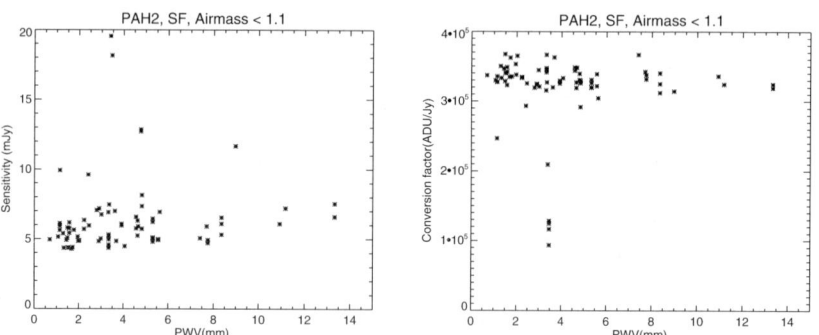

Fig. 4. Idem, for PAH2 filter

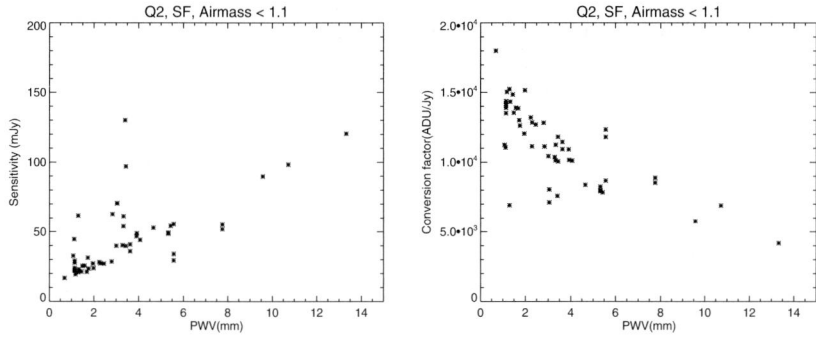

Fig. 5. Idem, for Q2 filter

3 Conclusion

We have demonstrated that the water vapour content of the atmosphere can be quickly estimated thanks to a 10 s sky spectrum. Significant differences are seen between a top-down (satellite-based) and the bottom-up estimates.

Clear dependences of the sensitivities and conversion factors on the PWV are put in evidence. Therefore, time expensive observations of photometric standards to determine the quality of the sky for MIR observations can be avoided. A paper describing these results as well as numerical relations for a larger set of filters is in preparation.

For the future, one could consider the possibility that the PWV becomes a constraint for observations in the MIR. This could be relevant for imaging as well as for spectroscopy, where some observations could critically depend on the width of some water vapour lines.

Accurate determination of the PWV as well as behavior of the PWV on a short time scale (minutes) to long time scale (months) are also important in the context of the selection of a site for the E-ELT.

The authors wish to thank Marc Sarazin for numerous discussions, as well as the constant support from Jason Spyromilio and Andreas Kaufer.

References

1. I.M. Chapman, D.A. Naylor, R.R. Phillips: MNRAS **354**, 612 (2004)
2. J.R. Pardo, M.C. Wiedner, E. Serabyn, C.D. Wilson, C. Cunningham, R.E. Hills, J. Cernicharo: ApJ **153**, 363 (2004)
3. L.S. Rothman, et al.: JQSRT **96**, 139 (2005)

Discussion

D. Osip: If the PWV calculation can be derived readily from 1 min of executed observation – and – it appears that these measurements can be indicative of sensitivity variations, would it be a good idea and would it be possible to implement such a measurement for *all* mid-IR science target and standard start observations?

A. Smette: The measurements I have described were meant to demonstrate the feasibility of the method. It is now up to the observatory to implement such a scheme, which I believe would be very useful.

NIR Extinction Coefficients in Paranal

E. Mason[1], G. Lombardi[2], C. Lidman[1], and A. O. Jaunsen[3]

[1] ESO, Alonso de Cordóva 3107, Vitacura, Santiago, Chile; emason@eso.org
[2] University of Bologna, Italy
[3] Institute of Theoretical Astrophysics, University of Oslo, Norway

1 Introduction

The atmosphere attenuates the starlight. The actual degree of attenuation (i.e. extinction in mag/airmass) varies with the wavelength of the observations and the astronomical site. It is possible to measure the atmospheric extinction of a site by fitting the Bouguer curve.

ISAAC at UT1 is a NIR imager and spectrograph and has been operating since 1998. ISAAC calibration plan foresees the observation of a photometric standard star during the evening twilight whenever the instrument is in use. The ESO archive (http://archive.eso.org) contains a large amount of photometric standard stars which can be used to characterize the NIR extinction of the Paranal site. We have reduced and analysed the dataset and here we present our results.

2 The Sample

We retrieved, from the ESO archive, all the Persson standard stars [2] that have been observed with ISAAC SW arm in the broadband filters J, Js, H and Ks during photometric or clear nights. This resulted in data from 494 different nights between Jan 2000 and Jun 2004, 40 different stars and 578 data points in the J band, 606 in Js, 607 in H and 678 in Ks.

The data were reduced with the *zpoint* recipe in the eclipse data reduction package (v 5.0.0). In Fig. 1, we plot the zero-point, $ZP(i)$, which is computed as the difference between the catalog magnitude (mag_i) and the instrumental magnitude, i.e.: $ZP(i) = mag_i + 2.5 \times log(\text{flux(ADU)/DIT})$, with no correction for extinction.

Figure 1 shows that the data points can be divided in subsamples or periods, each delimited by a mirror re-coating and/or an instrument intervention. We analysed (Sect. 3) each period separately.

3 Data Analysis

We started by removing any obvious trend of the data points within each period. The ZPs are expected to change after an instrument intervention. They also decrease in time due to the progressively reduced reflectivity of

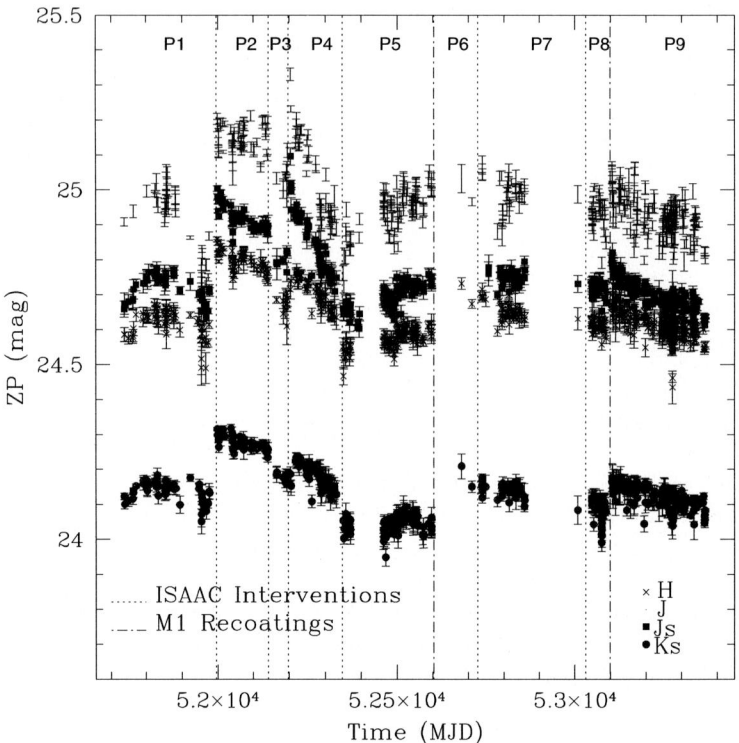

Fig. 1. The J, Js, H, and Ks zero-points as computed by the eclipse *zpoint* recipe. They are not corrected for extinction. Note that each subsample/period is identified by the label P_i, with i running from 1 to 9

the mirror. Therefore, we applied a linear fit to the $ZP(i)$ vs time, t, and removed the time trend by adding the amount $ZP(t=0) - ZP(t_i)$ to each zero point $ZP(i)$. Note that we identified a total of nine periods, but we applied the above correction to just five of them. The remaining four periods either do not show a clear decrease of the ZP versus time or contain too few data points.

We then checked for possible dependency of the computed ZP on the precipitable water vapor (PWV) content in the atmosphere. Indeed, the ZP is sensitive to the H_2O column abundance above the site [1]. Figure 2 (left panel) shows that there is a seasonal variation of the PWV above Paranal; the amount of measured PWV being larger during the so called Bolivian winter (i.e., the trimester Jan-Mar). Figure 2 (right panel) also shows how the ZPs decrease for larger PWV values. We computed the median PWV value of 2.279 mm for the period 2000–2004 and rescaled all the ZP(i) (already corrected for time dependence) to such median value.

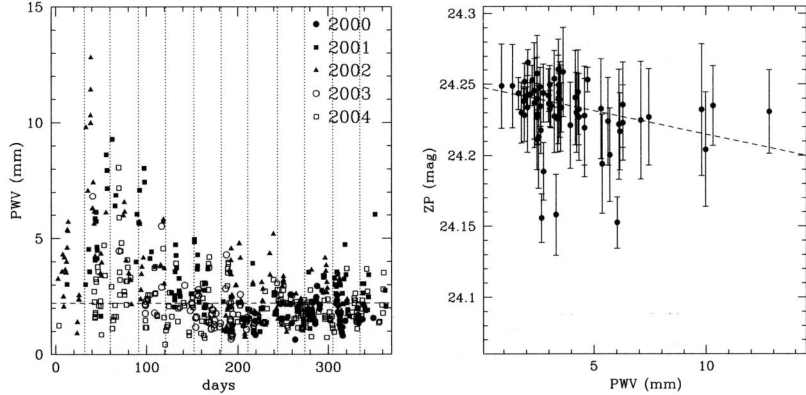

Fig. 2. *Left*: seasonal variation of the PWV (different symbols refer to different years). The *dotted vertical lines* delimit different months, while the *horizontal dashed* one represents the computed median value. *Right*: an example of the dependency of the zero points on the PWV value (Ks filter and period 4 data points)

Finally we plot the "corrected" $ZP(i)$ versus airmass and fit the Bouguer curve to each period and each filter. Figure 3 plots the period 9 data points and their fit (Bouguer curve) as an example. The extinction coefficients (i.e.

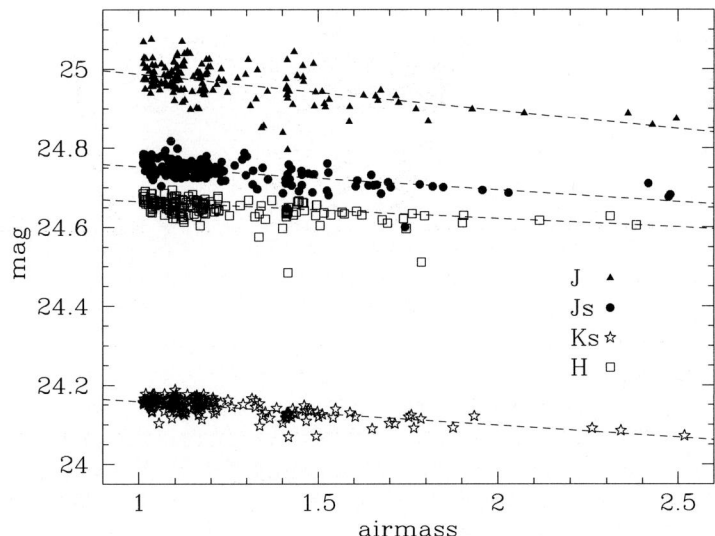

Fig. 3. Corrected ZPs vs airmass and Bouguer curve for each filter in period 9

Table 1. Extinction coefficient for each broad band filter J Js H and Ks and for each subsample or period (raw 1 to 9). The last row (10) reports their weighted average (WA)

subsample	K_J	K_{Js}	K_H	K_{Ks}
P1	0.06±0.07	0.04±0.03	0.07±0.03	0.05±0.03
P2	0.04±0.04	0.00±0.04	0.02±0.03	0.02±0.01
P3	0.17±0.19	0.03±0.49	0.16±0.10	0.09±0.05
P4	0.10±0.04	0.06±0.01	0.03±0.01	0.02±0.09
P5	0.08±0.03	0.05±0.03	0.03±0.02	0.05±0.01
P6	–	–	–	–
P7	0.04±0.06	0.02±0.03	0.03±0.02	0.07±0.01
P8	0.15±0.06	0.03±0.03	0.04±0.03	0.06±0.01
P9	0.09±0.01	0.06±0.01	0.04±0.01	0.06±0.01
WA	0.09±0.01	0.05±0.01	0.04±0.01	0.06±0.01

the slopes of the linear fit) for each period and filter are reported in Table 1, together with their weighted average.

References

1. A. Manduca, R.A. Bell: PASP **91**, 848 (1979)
2. S.E. Persson, D.C. Murphy, W. Krzeminski, et al.: AJ **116**, 2475 (1998)

Improvements in the Residual OH Emission Removal in SINFONI Pipeline Spectra

A. Modigliani[1], R. Davies[2], C. Dumas[3], and M. Neeser[1]

[1] ESO, Karl-Schwarzschild-Strasse 2, 85748 Garching, Germany;
amodigli@eso.org
[2] Max-Planck-Institut für extraterrestrische Physik, Giessenbachstraße, D-85748 Garching, Germany
[3] ESO, Alonso de Cordova 3107, Vitacura, Santiago, Chile

1 Introduction

The strong and variable OH airglow emission lines in the 1–2.5 µm spectra impose severe restrictions on the observing strategy. This is particularly true for instruments using Integral Field Spectroscopy techniques to sample the sky, as they are characterized by a narrow field of view.

SINFONI is the Spectrograph for INtegral Field Observations in the Near Infrared (1.1–2.45 µm) at the ESO-VLT. SINFONI was developed and built by ESO and MPE in collaboration with NOVA. It is an integral field spectrograph which combines Near Infrared spectroscopy and adaptive optics in a field of view that is image sliced [1, 2]. It is mounted at Yepun, the fourth Unit Telescope (UT4) of the ESO-VLT and has been operational since April 1st, 2005. SINFONI operations are supported by a data reduction pipeline [3], as part of the Data Flow Operations [4].

The standard SINFONI observational strategy is to do a sequence of object-sky observations to sample the object and the sky at regular intervals. This observing technique is limited temporally by changes in the flux of the OH lines on time scales of 2–3 minutes, spectrally by variations in the flux among individual OH lines, and by instrument flexures which may result in spectral format shifts, which can lead to P-Cygni type residuals.

These effects are indeed present in some spectra generated by early pipeline releases and triggered the development of an improved algorithm that can correct them. As part of a collaboration between ESO and MPE, we have implemented in the SINFONI pipeline an algorithm originally developed by the MPE to solve these problems. In this chapter we present the first results of the new algorithm.

2 Sky Emission

Near infrared air-glow emission originates in OH radicals which are created by reactions between ozone and hydrogen high in the atmosphere. Removing the emission lines which result from the subsequent radiative cascade is a crucial part of processing near infrared (1–2.5 µm) spectra.

This problem is relevant as it has been proven that the strongest OH lines, which lie in the H band, have fluxes of the order of 400 photons m^{-2} arcsec^{-2}. This contrasts strongly with the background continuum measurable between lines of only 590 photons m^{-2} arcsec^{-2}µm^{-1}. This means that even at moderate spectral resolutions (R $\approx$ 3000), like the one of SINFONI, the background level on an OH line can be more than three orders of magnitude higher than the level between the lines.

Due to temporal changes in the absolute flux in OH sky lines exposure times are usually limited to 2–3 min. This sets a lower limit to the statistical photon noise. An additional source of problems is the peculiar change of absolute flux in OH lines, in which the (vibrational/rotational) bands of lines vary with respect to one another. Finally the effects of instrumental flexures, which are typical of instrument rotation in a Cassegrain or Nasmyth telescope configuration, may lead to P-Cygni type residuals.

While long-slit spectrographs, using a slit length much longer than the object of interest may prevent such problems by allowing an appropriate fit of the residual sky background, integral field spectrographs like SINFONI, with very limited FOV, impose constraints on the observations and require more sophisticated data reduction techniques.

3 Algorithm

More details on the algorithm to remove residual OH emission from near infrared spectra can be found in [5] or in the SINFONI pipeline User Manual at http://www.eso.org/pipelines.

In a nutshell the algorithm allows one to find a scaling as a function of wavelength that can be applied to a spectrum from a sky cube, in order to match it optimally to the sky background in an object cube. This scaling function is then applied separately to the spectrum at each wavelength position of the sky cube, creating a modified sky cube. This modified cube is then subtracted from the object cube.

The scaling function may be found by taking into account that the main contribution to the OH lines is coming from vibrational and rotational transitions, with the former contributing the most. Transitions between vibrational bands lie within well defined wavelength limits, with only a small amount of overlap between different vibrational transitions. Thus, to a first approximation one may divide the spectrum into sections corresponding to specific vibrational transitions and treat these separately.

The sky residual correction algorithm is outlined in the following steps:

– The wavelength ranges which correspond to the different vibrational and rotational transitions of the OH emission lines are defined.

- The noise associated with the object frame is calculated.
- Object intensities more than two times the object's noise above the background are flagged.
- Sky pixels are identified, flagged, and selected as good sky pixels when they make up at least a given fraction (this can be defined by the user and is set to 80% by default) of the spectrum.
- The average object and sky spectra are calculated. For each plane in the cube (both object and sky) an average is computed by excluding pixels which are more than three sigmas above the median. This defines the object and sky intensity.
- The thermal contribution to the sky background is estimated and subtracted from the sky spectrum. To improve accuracy, as the sky spectrum contains many emission lines, we first smooth this spectrum with a running box and then fit a Boltzmann function (see Fig. 1, left panel).
- To properly compute the shift between object and sky only wavelengths corresponding to vibrational or rotational transitions are considered. This removes the effect from improper spectrum features.
- The pixel shift between the sky and the object spectra is computed by cross correlating the two spectra and the sky spectrum and the sky cube are correspondingly shifted.
- The scaling factor that is applied to the sky emission lines in order to remove them from the object spectrum is computed.
  The object and sky sub-spectra are extracted over wavelength ranges corresponding to each vibrational transition. A scaling factor is computed for each range. As a first approximation, this factor is derived from the line contributions only due to the vibrational transitions. Then the recipe

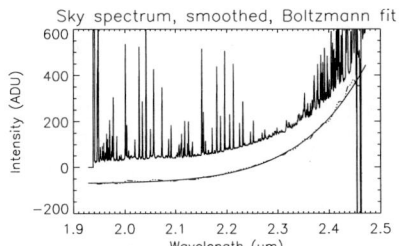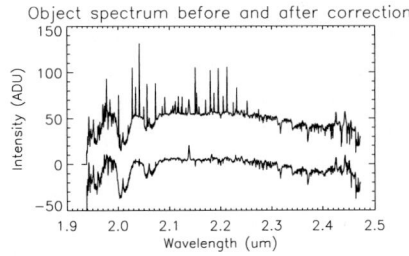

Fig. 1. On the *left panel* are shown the sky background, its smoothed profile and the fit by a Boltzmann function, for clarity shifted down by 50 ADU. On the *right panel* we compare un-corrected and corrected sky subtracted object spectra which for clarity are shifted down by 50 ADU. Rotational transitions affect the spectrum quality by only 1–5%. Not only can one see 1-0S(1) at $2.12\,\mu m$ and Brγ at $2.17\,\mu m$, but now the H_2 1-0Q(1) and 1-0Q(3) emission lines are clearly seen longward of $2.4\,\mu m$, and several CO band-heads are also visible at 2.3–$2.4\,\mu m$

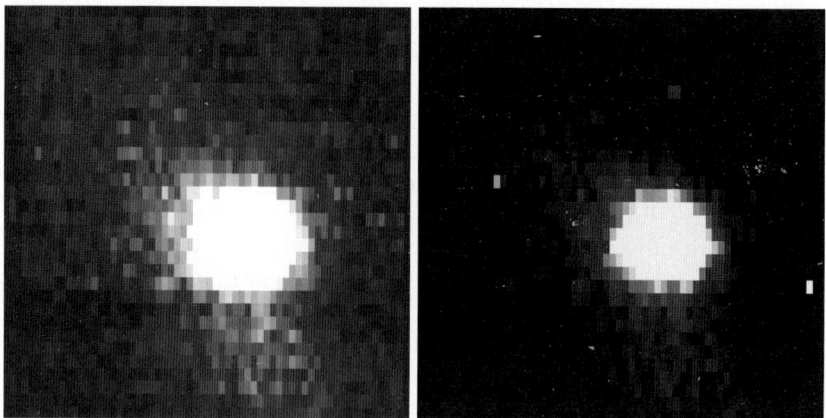

Fig. 2. Un-corrected (*left*) and corrected (*right*) object cubes. The cube plane corresponding to 2.151 μm is shown. Input data included a dedicated sky frame. The image on the *left* has the same intensity scale as the image on the *right*

computes a scaling factor based on the contribution of the rotational transitions. Later the two factors are combined and the shifted sky spectrum is accordingly rescaled and subtracted from the object spectrum, to obtain an object spectrum cleaned of sky lines (see Fig. 1, right panel).
– The shifted and scaled sky cube is finally subtracted from the object cube. The resulting cube is thus corrected for the sky residuals (see Fig. 2, right panel).

Summary

We have described an algorithm in the SINFONI pipeline which allows the correction of residual sky lines, possibly present in the extracted object. For a large set of data the accuracy in the estimation of the thermal background (which affects the definition of the continuum level) is 1%, and the sky line residuals, after correction, improve by a factor of 40 in comparison to their values before the application of this new algorithm.

References

1. F. Eisenhauer, et al.: SPIE **4841**, 1548 (2003)
2. H. Bonnet, et al.: The Messenger **117**, 17 (2004)
3. A. Modigliani, et al.: astro-ph/0701297 (2007)
4. D. Silva, M. Peron: The Messenger **118**, 2 (2004)
5. R. Davies: MNRAS **375**, 1099 (2007)

Calibration of the Relationship Between Precipitable Water Vapor and 225 GHz Atmospheric Opacity via Optical Echelle Spectroscopy at Las Campanas Observatory

J. Thomas-Osip[1], A. McWilliam[2], M. Phillips[1], and D. Osip[1]

[1] The Observatories of the Carnegie Institution of Washington, Las Campanas Observatory, Le Serena, Chile; jet@lco.cl
[2] The Observatories of the Carnegie Institution of Washington, Washington, DC, USA

Abstract. We report on the calibration of the relationship between precipitable water vapor and opacity at 225 GHz at Las Campanas Observatory as measured by a Tipping Radiometer. This relationship is a function of altitude and temperature and thus is highly dependent on location. We determine the relationship applicable at Las Campanas Observatory by using high-resolution Magellan Echelle spectra to measure the precipitable water vapor independently and absolutely. Temperature insensitive (between 220–300 K) lines allow the use of a single temperature atmospheric model as long as the lines are unsaturated. Absolute calibration was achieved by measuring the humidity in the path length of the McMath Solar telescope with a psycrometer [1]. We have expanded the method presented by Brault et al. (1975) with improved partition functions and additional lines. Based on this calibration, we present Southern hemisphere winter-time precipitable water vapor statistics for Las Campanas Observatory as measured during a two month campaign. We find that the median winter value of 2.8±0.3 mm is consistent with that measured at the nearby La Silla Observatory during the VLT site survey [2] and inconsistent (lower by a factor of approximately two) with estimates, also for La Silla, derived from GOES-8 satellite imagery and the European Centre for Medium-Range Weather Forecasting (ECMWF) meteorological numerical model [3]. Furthermore, in the Southern hemisphere winter months, we can expect good conditions for infrared observing ($\lesssim$1.5 mm) at the tenth percentile level. Further details can be found in Thomas-Osip et al. [4].

References

1. J.W. Brault, J.S. Fender, D.N.B. Hall: JQRST **15**, 549 (1975)
2. D. Morse, F. Gillett: 1982, AURA Eng. Report 73
3. http://www.eso.org/gen-fac/pubs/astclim/lasilla/h2o/
4. J. Thomas-Osip, A. McWilliam, M.M. Phillips, N. Morell, I. Thompson, T. Folkers, F.C. Adams, M. Lopez-Morales: PASP **119**, 697 (2007)

Part X

Session 9: Interferometric Instrumentation

IOT Overview: Calibrations of the VLTI Instruments (MIDI and AMBER)

S. Morel[1], F. Rantakyrö[1], T. Rivinius[1], S. Stefl[1], C. Hummel[1], S. Brillant[1], M. Schöller[1], I. Percheron[2], M. Wittkowski[2], A. Richichi[2], and P. Ballester[2]

[1] ESO, Alonso de Cordova 3107, Vitacura, Santiago, Chile; `smorel@eso.org`
[2] ESO, Karl-Schwarzschild-Strasse 2, 85748 Garching, Germany

Abstract. We present here a short review of the calibration processes that are currently applied to the instruments AMBER and MIDI of the VLTI (Very Large Telescope Interferometer) at Paranal. We first introduce the general principles to calibrate the raw data (the "visibilities") that have been measured by long-baseline optical interferometry. Then, we focus on the specific case of the scientific operation of the VLTI instruments. We explain the criteria that have been used to select calibrator stars for the observations with the VLTI instruments, as well as the routine internal calibration techniques. Among these techniques, the "P2VM" (Pixel-to-Visibility Matrix) in the case of AMBER is explained. Also, the daily monitoring of AMBER and MIDI, that has recently been implemented, is shortly introduced.

1 Introduction

Long-baseline optical interferometry (LBOI) is a technique which aims to reconstruct images of celestial objects at very high angular resolution. Although imagined at the end of the 19th century by Fizeau [1], and attempted by Michelson with some successful results [2], it really has emerged at the end of the 20th century with the works of Labeyrie [3].

LBOI consists in coherently recombining the signal (i.e., the photons from the celestial source of interest) collected by several apertures (i.e., telescopes). The vector spanning between two of the apertures is called a "baseline". If we consider $I(\alpha, \delta)$ as the angular distribution of the source on the sky (at a given wavelength λ), and we note (B_x, B_y) the components along (α, δ) of the projected baseline (i.e., the projection of the baseline vector onto the plane which is perpendicular to the line-of-sight of the source), the interferometer will measure:

$$\gamma = \hat{I}(u,v)/\hat{I}(0,0), \qquad (1)$$

where $\hat{I}$ is the Fourier transform of I, $u = B_x/\lambda$, $v = B_y/\lambda$. What an interferometer measures can be therefore summarized by Fig. 1. The "visibility" is defined by:

$$V = |\gamma| \qquad (2)$$

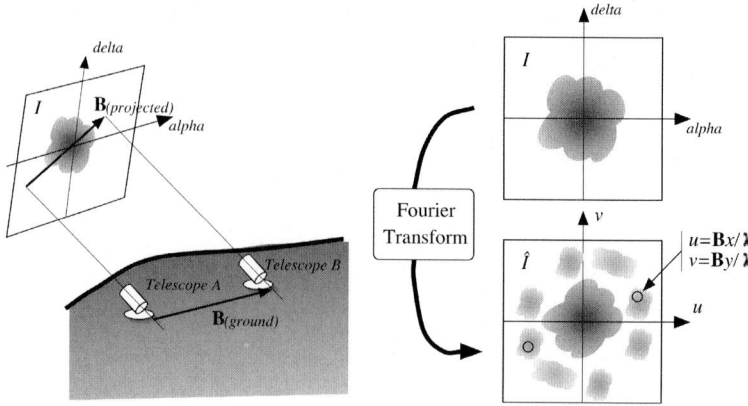

Fig. 1. Basic principle of ground-based long-baseline optical interferometry. The sample of $\hat{I}$ for the given projected baseline and wavelength is given by the small *circle* (the graphical representation of $\hat{I}$ is fictive)

It is common in literature to encounter V^2 instead of V because the estimators of V^2 are most robust. The phase φ of γ cannot be directly recovered by a baseline. Using three baselines (i.e, three telescopes A, B and C) allows to compute the "closure phase" term by summing the phases measured on the three baselines AB, BC and CA: $\varphi_{AB} + \varphi_{BC} + \varphi_{CA}$. This term is constant and allows to partly recover the phase information. Another technique for one-baseline interferometry consists in using "phase-referenced imaging" by simultaneously observing another object giving a phase reference (see [4]).

Ultimately, the idea of LBOI would be to imitate what already exists in radio-astronomy: cophase a large number of apertures (e.g., 27) and make direct imaging at high angular resolution [5].

Physically, V, for a given baseline, is given by the contrast of the interference fringes that are formed by combining the two beams coming from the apertures. The general expression of the detected fringe signal is:

$$I(\Delta L) = I_1 + I_2 + 2V\sqrt{I_1 I_2}\cos(2\pi \Delta L/\lambda + \varphi), \tag{3}$$

where ΔL is the optical pathlength difference (OPD) between the two beams, I_1 and I_2 are the intensities of the beams. The knowledge of these values is therefore required to get V from I_1 and I_2 that are measured either by "tapping" into each beam before combination to get simultaneous "photometric" measurements, or after fringe measurement by feeding the interferometric instrument with the light coming from only one aperture (and then from the other aperture).

For various reasons (see Sect. 3.1), the measurement of V is not as simple as it may seem. It requires a complex calibration process that is described from Sect. 3.

2 Interferometric Instrumentation at Paranal

The Very Large Telescope Interferometer (VLTI) at Cerro Paranal currently features two scientific instruments: MIDI and AMBER. The characteristics of these instruments are the following.

2.1 MIDI

MIDI has been the first VLTI instrument to be commissioned. It is a two-beam pupil-plane combiner which operates in the mid-infrared N-band (8 to 13 µm). The beams are dispersed after combination, allowing to measure $V(\lambda)$ at two possible resolutions (either $R = 30$ or $R = 230$). For a complete description of MIDI, see [6].

The detector image of MIDI (Fig. 2, left) can show up to four channels that are spectrally dispersed in the horizontal direction: two interferometric channels containing fringes (at the center of the detector), and two optional "photometric beams" which sample the photometric level (at a given λ) of each beam coming from a telescope. A spatial distortion appears on the detector when the highest spectral resolution mode is used.

2.2 AMBER

AMBER is a three-beam image-plane combiner which operates in J-, H- and K-bands (from 1.3 to 2.3 µm). Before combination by focusing the parallel

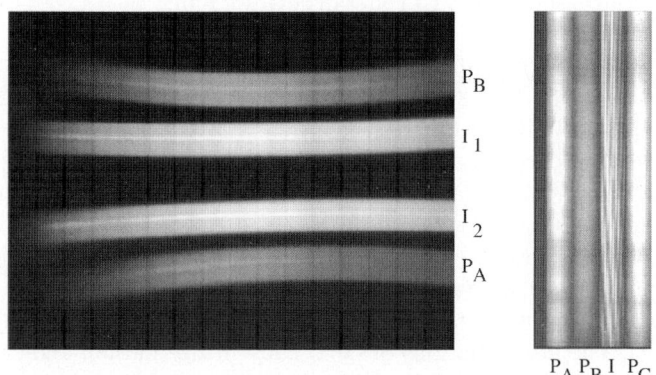

Fig. 2. MIDI (*left*) and AMBER (*right*) detector frame. The photometric channels are indicated by P... and the interferometric channels by I... The fringes can be seen in the interferometric channel of AMBER

beams onto the detector plane, the beams are dispersed. AMBER has three possible spectral resolutions ($R = 35$, $R = 1500$ and $R = 12,000$). With the lowest resolution, it is possible to simultaneously observe the J-, H- and K-bands. With the two other resolutions, observations are currently in K-band only. AMBER features a set of monomode fibers into which the incoming light is injected in order to have "clean" plane wavefronts before combination (at the expense of increased flux fluctuations). A description of AMBER can be found, for example, in [7].

On the detector plane of AMBER, the light is dispersed along the vertical axis. Four channels (columns) can be seen on (Fig. 2, right): three photometric channels (similar to the ones of MIDI), and a sole interferometric channels where three systems of dispersed fringes are imaged.

3 Interferometric Calibration Requirements

A general calibration plan for interferometric instruments should include:

- Calibration of the visibility V.
- Flux calibration to remove the thermal background from the signal.
- Internal instrument calibrations (characterization of the optics and of the detector).
- Instrument health-checks and monitoring of important fluctuating parameters.

It is also important to notice that the data quality provided by an interferometer and its instruments strongly depends on factors like optical alignment, pupil re-imaging, adaptive-optics performances, vibration control, fringe-tracking performances, etc... However, these issues are not *stricto sensu* within the scope of instrumental calibration and will not be presented in this chapter. In the case of the VLTI, details can be found in [8].

3.1 Principles of Visibility Calibration

The measured visibility μ differs from the expected visibility V because of several sources of perturbation. Amongst these sources are the wavefront errors (caused by the external atmospheric turbulence and by the turbulence in the interferometer itself) from each aperture and in the OPD between two apertures, the mechanical vibrations, the polarization effects in the instrument. Wavefront errors cause fluctuations of the beam overlap (in the case of MIDI) or of the flux that is injected in the monomode fibers (in the case of AMBER). The fast variations of OPD cause fringe motion and smears the fringe image. The result of these perturbation is a random underestimation of the measured visibility. In general, $\mu = 0.4V$ to $0.9V$.

In order to recover V from μ, the method is to obtain a measured visibility μ_0 of a celestial source for which the visibility V_0 is known a priori. This source is called a "calibrator". Getting μ_0 and V_0 allows to calculate the "interferometric transfer function" $T = \mu_0/V_0$. Therefore, the estimation of the visibility $\tilde{V}$ of the source of interest (the "science target") will be: $\tilde{V} = \mu/T$. It is important that the calibrator is observed with the same instrumental setup (including detector integration time) as for the science target. Also, the calibrator should be selected to be observed as close as possible to the science target, both in time and in space (i.e., in the observed sky region) This method has been used for a long time in LBOI (see for example [9]). Since calibrators and science targets must be alternatively observed, at least half of the observing time in interferometry is devoted to calibrations, which partly explains why, at Paranal, the open-shutter time statistics of interferometric instruments are low, compared to classical single-telescope instruments. It is important to notice that, in spite of an observation of the calibrator right before or after the science target observation, the transfer function cannot be the same: $T(\mu) \neq T(\mu_0)$. Therefore, T must be interpolated in order to estimate its value at the time when the science target was observed [10]. For this reason, all the calibrator data are public.

The choice of a suitable visibility calibrator is therefore of paramount importance in interferometry. The absence of variability of its shape (i.e., its intensity distribution $I(\alpha, \delta)$) and the simplicity of that shape are critical for a calibrator. One would wish that all calibrators could be considered as point sources ($V_0 = 1$ always), but in this case the flux of the calibrators would probably be too faint to measure μ_0 correctly. Actually, the point source model can be used for very short baselines only. Therefore, it is more realistic to use as model a uniform disk (UD) function having as parameter the angular diameter θ of the calibrator: $I(r) = I_0$ if $r < \theta/2$; $I(r) = 0$ otherwise. The knowledge of θ is required. It can be derived from photometric observations of the calibrator:

$$\theta = \frac{\sqrt{F_0}}{K T_{eff}^2}, \qquad (4)$$

where F_0 is the bolometric flux from the calibrator, K is a constant term based on the Stefan-Boltzmann constant σ and T_{eff} is the effective temperature of the stars. The problem to measure θ comes from the uncertainty on T_{eff}. In some cases [10], T_{eff} is actually determined from a direct measurement of θ.

The uniform disk model can be refined by using the "limb-darkened" (LD) disk model. The radial intensity distribution can be, for example: $I(r) = I_0(1 - 4r^2/\theta^2)^{\alpha/2}$, where α is the limb-darkening factor [11]. A knowledge of the physics of the atmosphere of the calibrator, or interferometric observations in the second lobe of the visibility function are required, in order to know α. Physically, the LD model is really useful at long baselines, where V_0 becomes small. At θ equal, and for $V_0^2 < 0.4$, the maximum difference of V_0^2 between the UD and the LD models is 0.1% [12], while the best achieved visibility accuracy is 5% with MIDI, and 1% with AMBER.

4 VLTI Calibrator Catalogs

4.1 The MIDI Calibrator Catalog (MCC)

This catalog [13] has been established by the MIDI consortium. The authors used the following process: first, calibrator candidates were picked from the IRAS and MSX point source catalogs (PSC) with criteria of flux (5 Jy at $\lambda = 12\,\mu\mathrm{m}$, the limiting magnitude of MIDI with the 8 m unit telescopes being 1 Jy), of declination ($\delta \leq +35°$) and of color temperature (> 4000 K). This yielded a list of 511 candidates. Then, for these candidates, photometric data at different spectral bands were collected. For the UV and the visible, the Geneva photometric catalog was used. For the near-IR, observations with the MkII infrared photometer at the 0.75 m telescope of the South-African Astronomical Observatory were carried out. For the mid-IR, IRAS photometric data were used. For each candidate, the best Kurucz SED model fitting the photometric data (after corrections) was determined, using T_{eff}, θ and the average interstellar extinction constant A_v as free parameters. From these fits, it has been possible to identify candidates showing an infrared excess by computing the χ^2 between the mid-IR flux given by the model and the actual flux. A filtered list of 178 calibrators having $\chi^2 < 5$ could be established.

4.2 The Calibrator Catalog for AMBER (Mérand et al.)

No calibrator catalog was purposely established for AMBER, but a catalog of interferometric calibrator for near-IR observations with baselines up to 200 m [12] is well suited for AMBER. This catalog is an update of a previous one for shortest baselines [14]. It has been established from the Cohen catalog [15]. This catalog (C99) was a selection of IRAS PSC objects with criteria of flux (> 1 Jy at 25 µm, infrared color (quadrant of giant stars), IR environment, spectral types and luminosity classes (A0-G9 and II-IV; K0-M0 and III-V). Emission-line, carbon, and variable stars were rejected. From the C99, some constraints on IRAS PSC were relaxed to get more candidates (flux ≤ 1 Jy at 25 µm ; acceptance of stars associated with small extended sources), though a new criterion on the sphericity of the shape was introduced (by discarding spectroscopic binaries and fast rotators). Also, all the candidates required an entry in SIMBAD. As for the MCC, photometric data were collected. This time, the 2MASS catalog was used for the near-IR, and the MSX catalog for the mid-IR. These data were used to fit Cohen templates in order to get the limb-darkened diameter for each candidate. The result is a catalog of 1320 calibrators.

4.3 The CHARM Catalog

CHARM (Catalog of High Angular Resolution Measurements) is a database of the published direct measurements of stellar angular diameters obtained either by lunar occultation or by LBOI techniques [16]. Some of the entries in this catalog can be used as calibrators. A first version of CHARM was released in 2002 [17]. The current version includes the results obtained with VINCI (the K-band commissioning instrument of VLTI) and during the MIDI commissioning. CHARM also includes photometric data (2MASS and DENIS K-magnitudes, IRAS PSC and MSX 12 μm fluxes) on the sources. In its latest version, CHARM contains 1596 objects which can be used as calibrators.

Between the MCC, the Mérand et al. and the CHARM catalogs, many entries are indeed common. It is interesting to notice, as the authors of the MCC have pointed out, that there is no real discrepancy between the angular diameters given by the MCC and the ones given by CHARM (see Fig. 3).

4.4 Tools for Searching the VLTI Calibrator Catalogs

The data of the previously described catalogs are accessible with the ESO "CalVin" web-tool. In CalVin, the user enters information about the science target (coordinates) and the observing setup (instrument, baseline,...). Then, CalVin returns a list of suitable calibrators for that target. It is important to notice that the users are responsible for choosing the calibrators of their programs. Although they are advised to use CalVin, they can use any calibrators of their choice which are not in CalVin.

There exist other calibrator selection tools, but they are not supported by ESO: the "search calibrator" tool integrated in the ASPRO software issued by the Jean-Marie Mariotti Center [18] and the "getCal" tool issued by the Michelson Science Center. These tools are software which query existing databases (unlike CalVin which has its own database).

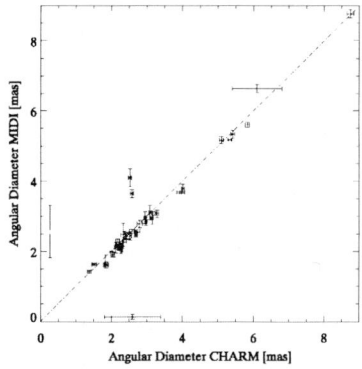

Fig. 3. Correlation between the angular diameters given by CHARM (x-axis) and by the MCC (y-axis). Courtesy of R. van Boekel, B. Stecklum, T. Verhoelst

5 Flux Calibrations

The calculation of the raw visibility μ relies on the measurement of the correlated flux (i.e., the fringe amplitude) and on the measurement of the uncorrelated flux I_1 and I_2 (see (3)). The thermal background is a problem to measure these values. The VLTI instruments use different methods that are used to remove the thermal background from the measured signal.

5.1 MIDI

On MIDI, the uncorrelated flux level is corrected by using telescope chopping, usually at a 0.5 Hz frequency. For the correlated flux, the two interferometric channels are subtracted: since the thermal background is highly correlated in both channels and the fringes are in phase opposition because of the optical design of MIDI, the background is removed but the fringe signal enhanced by channel subtraction.

The spatial distortion of the photometric channels when the grism is used (see Fig. 2, left) implies to take one-beam-only exposures in this mode to correct the data, although it should not be necessary in theory.

5.2 AMBER

On AMBER, a dark exposure with the shutter closed is taken before the fringe exposure starts. After the fringe exposure, a "flat" exposure on the sky (by offsetting the telescopes) is taken.

6 Internal Calibrations and Daily Monitoring

Like all the VLT instruments, MIDI and AMBER feature in their respective calibration plans daily internal calibrations that are executed after the night. The data are reduced by a pipeline and the results displayed as trending plots on an internal website, in order for the Paranal astronomers to monitor the "health" of the instruments.

6.1 MIDI

When the "SCI_PHOT" setup allowing to simultaneously observe photometry and fringes is used (as in Fig. 2), It important to measure the "κ-matrix" of the beam combiner. This matrix gives the relation, at a given λ between the flux in the interferometric channels and the flux in the photometric channels:

$$\begin{pmatrix} I_1 \\ I_2 \end{pmatrix} = \begin{pmatrix} \kappa_{11} & \kappa_{12} \\ \kappa_{21} & \kappa_{22} \end{pmatrix} \begin{pmatrix} P_A \\ P_B \end{pmatrix} \quad (5)$$

The κ-matrix is computed once a night from one-beam-only exposures taken on a bright calibrator, for each dispersive element (prism and grism).

The detector readout noise, the detector linearity, the throughput of the cryogenic dispersive elements (the prism and the grism), the stability of the spectral dispersion, and the internal alignment of MIDI are monitored daily. After several months of monitoring, it has been noticed that MIDI is a very stable instrument. For example, the readout noise is always around 12 ADUs, and the dispersion of the prism is 90% at $\lambda = 10.5\,\mu$m.

6.2 AMBER

The main AMBER internal calibration is the "P2VM" (pixel-to-visibility matrix) which gives the internal transfer function of the instrument for each pixel of the detector. The P2VM is computed during the night after each setup of AMBER, before the observation starts. It consists of several exposures using the internal source of AMBER (for which $V = 1$). An exposure is taken with one shutter open to determine the profile of the "continuum" of the interferometric channel. Then, another exposure is taken with an etched filter to determine the offset in wavelength between the interferometric and the photometric channels. Finally, other exposures are taken with two shutters open (therefore, one system of fringes) at OPD = 0 and OPD = $\lambda/2$. A series of P2VM is also taken after the night with different setups to characterize the instrument.

Another critical point of AMBER is the alignment of the dichroics that split the three incoming beams into J-, H- and K-bands before injection in the monomode fibers; and of the dichroics at the exit of these fibers which recombine the bands for each beam to relay them to the spectrograph. This alignment, as well as the internal OPD of AMBER, is monitored everyday, and corrected if needed.

Finally, the calibration plan includes the establishment of a bad-pixel detector map and of a flat-field map that are measured every 6 to 12 months. It is foreseen however to measure these maps more often and to deliver them to the users.

7 Conclusion and Perspectives

We have shown that the calibration of the interferometric visibility relies on a careful selection of the possible calibrator stars. However, it is important that an effort is made to have more calibrators available. For phase-referenced imaging with PRIMA at VLTI [4], a list of phase calibrators will have to be built, and the internal OPD variations of MIDI and AMBER be monitored. The implementation of daily monitoring of MIDI and AMBER is recent, and should, in the near future, include a monitoring of the on-the-sky performances of the instruments.

References

1. H. Fizeau: C. R. Acad. Sci. **66**, 932 (1868)
2. A.A. Michelson, F.G. Pease: Astrophys. J. **52**, 249 (1921)
3. A. Labeyrie: Astrophys. J. **196**, L71 (1975)
4. F. Delplancke, et al.: Astrophys. Space Sci. **286**, 99 (2003)
5. A. Labeyrie: Astron. Astrophys. Suppl. Ser. **118**, 517 (1996)
6. C. Leinert, et al.: The Messenger **112**, 13 (2003)
7. R. Petrov, et al.: Astrophys. Space Sci. **286**, 57 (2003)
8. H. Bonnet, et al.: The Messenger **126**, 37 (2006)
9. M. Shao, et al.: Astrophys. J. **327**, 905 (1988)
10. G. Perrin, et al.: Astron. Astrophys. **331**, 619 (1998)
11. D. Hestroffer: Astron. Astrophys. **327**, 199 (1997)
12. A. Mérand, P. Bordé, V. Coudé du Foresto: A&A **433**, 1155 (2005)
13. T. Verhoelst: Evolved stars: a combined view from interferometry and spectroscopy, PhD Thesis, Catholic University Leuven, Belgium (2005)
14. P. Bordé, et al.: Astron. Astrophys. **393**, 183 (2002)
15. M. Cohen, G.W. Russel, B. Carter: Astron. J. **117**, 1864 (1999)
16. A. Richichi, I. Percheron, M. Khristoforova: A&A **431**, 773 (2005)
17. A. Richichi, I. Percheron: Astron. Astrophys. **386**, 492 (2002)
18. D. Bonneau, et al.: Proc. SPIE **5491**, 1160 (2004)

Discussion

U. Hopp: Using the limb darkening of a star assumes that the star is not a binary, has no large spots, and no disk! How can you be sure about that? Polarimetry might point out disks, photometric monitoring star spots.

S. Morel: For the Mérand et al., catalog used for AMBER, binaries were eliminated from the beginning. I have no idea about large spots, but I don't think it might have an effect on the limb darkened diameter, considering the precision on visibility we currently achieve with interferometric instruments. Disks can be revealed by a discrepancy between measured mid-IR fluxes and synthetic SEDs (IR-excess). Since I am not a calibrator expert, please check [13] for MIDI calibrations. The paper is in preparation. Contact B. Stecklum (stecklum@tls-tautenburg.de) for more details.

AMBER on the VLTI: Data Processing and Calibration Issues

F. Millour[1], R. Petrov[2], F. Malbet[3], E. Tatulli[4], G. Duvert[3], G. Zins[3], E. Altariba[3], M. Vannier[5], O. Hernandez[3], and G. Li Causi[6]

[1] Max-Planck-Institut für Radioastronomie, Auf dem Hügel 69, 53121 Bonn, Germany; `fmillour@mpifr-bonn.mpg.de`
[2] Laboratoire Universitaire d'Astrophysique de Nice, Parc Valrose, 06108 Nice, France
[3] Laboratoire d'AstrOphysique de Grenoble, 414 rue de la piscine, Domaine Universitaire, 38400 Saint-Martin d'Hères, France
[4] Osservatorio di Arcetri, L.go E. Fermi, 5, 50125 Firenze, Italy
[5] ESO, Alonso de Cordova 3107, Vitacura, Santiago, Chile
[6] Rome Astronomical Observatory, Via di Frascati, 33, 00040 Monte Porzio, Catone, Rome, Italy

1 Introduction

We present here the current performances of the AMBER/VLTI instrument for standard use and compare these with the offered modes of the instrument. We show that the instrument is able to reach its specified precision only for medium and high spectral resolution modes, differential observables and bright objects.

For absolute observables, the current achievable accuracy is strongly limited by the vibrations of the Unit Telescopes, and also by the observing procedure which does not take into account the night-long transfer function monitoring.

For low-resolution mode, the current limitation is more in the data reduction side, since several effects negligible at medium spectral resolution are not taken into account in the current pipeline.

Finally, for faint objects (SNR around 1 per spectral channel), electromagnetic interferences in the VLTI interferometric laboratory with the detector electronics prevents currently to get unbiased measurements. Ideas are under study to correct in the data processing side this effect, but a hardware fix should be investigated seriously since this limits seriously the effective limiting magnitude of the instrument.

2 The AMBER Instrument

2.1 Short Description

AMBER is the near-infrared interferometric re-combiner of the VLTI. Its general and technical descriptions are contained in the two articles by Petrov

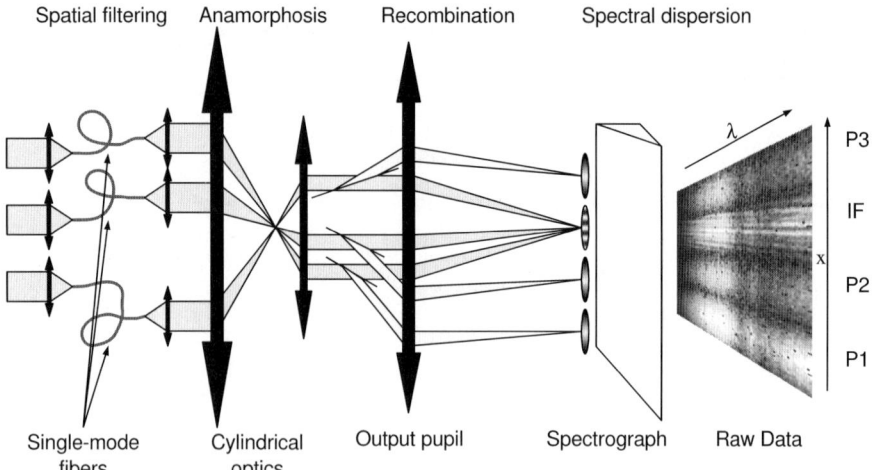

Fig. 1. The AMBER optical schematics, showing the principal elements of the instrument: the spatial filtering is made with optical fibers, then an anamorphosis optics shrinks the beam in one direction to feed it into a long slit spectrograph where the spectrally dispersed fringes are finally imaged on the detector

et al. [13] and by Robbe-Dubois et al. [14]. In short, it features simultaneous observations in J, H and K bands observations, low (R = 35), moderate (R = 1500) and high (R = 12000) spectral resolutions, and 3 telescopes operation. The use of optical fibers to improve calibration and the multiaxial scheme adopted complete this short view of the instrument (see Fig. 1). The AMBER instrument features also a limited number of pixels in the fringes leading to the use of very specific algorithms for data reduction.

2.2 AMBER Data Processing: Principle

The AMBER data processing is based on the fitting in the image plane of the fringe pattern. One can find a complete description of the process in the article by Tatulli et al. [17]. The resulting basic information is a measurement of the coherent flux (instantaneous complex coherence factor multiplied by the flux) for each single frame. Three time-averaged squared visibility, a closure phase and three differential phases can be extracted from these measurements, using respectively specific techniques such as quadratic estimator [12], bi-spectrum estimator [4] and inter-spectrum estimator [1]. Added to these interferometric observables, AMBER provides the object spectrum which adds simultaneous velocimetric measurements to the purely geometric measurements brought by the interferometric observables.

3 AMBER and the VLTI

3.1 How Does AMBER Behave on VLTI?

The AMBER instrument was installed and tested at the Paranal observatory (ESO, Chile) during the month of May 2004 [15]. Since then, a series of commissionings were performed to check the performances of the AMBER instrument together with the VLTI infrastructure. These first tests showed that the UTs have vibrations that strongly affect the AMBER signal (see Fig. 2) with a drastic average instrumental visibility decrease with regards to specifications (20% instead of 80%, leading to a loss of a factor 4 in signal to noise ratio). These vibrations come mainly from the Coudé train of the UTs and are in the process of being damped by ESO. However, one has to consider using the AMBER instrument in the today limited state of VLTI, taking into account these vibrations.

Therefore, changes in observing and data processing strategies were needed for AMBER to successfully be opened to the community:

- A longer exposure time (i.e. more individual short exposure frames) is needed to maximize the chances to get some "useful" frames for the data processing.

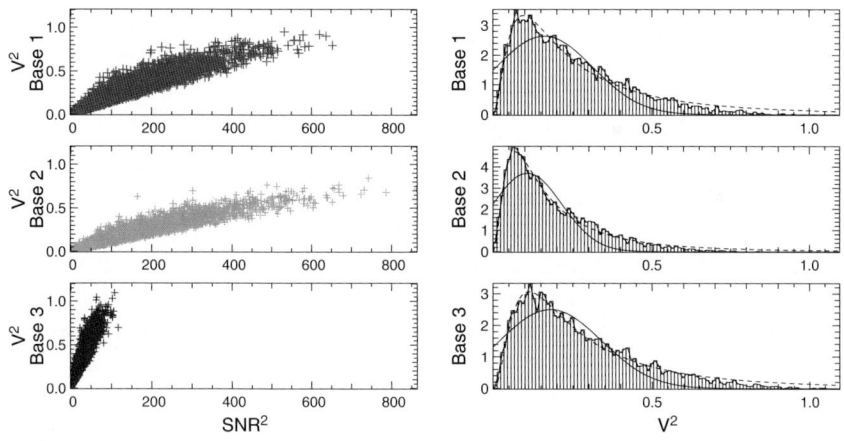

Fig. 2. *Left:* A correlation plot between fringe SNR and visibilities shows a strong correlation, linked to the domination of the jitter effect (blurring of the fringes by their move during integration time). *Right:* AMBER squared visibilities histogram on the same star, showing the highly non-symmetric effect induced by UT vibrations. The resulting histogram looks like a log-normal distribution (*dashed line*) and not like a Gaussian distribution (*solid line*), which makes it difficult to extract an average and an error

- A frame-selection process, i.e. a removal of spurious frames where there is either no fringes, no flux or a too high piston has been added to the data processing software.

3.2 The Standard Operating Mode Performances

These observing and data processing strategies are used today on the AMBER/VLTI instrument, allowing a better visibilities histogram (closer to a Gaussian and therefore easier to compute realistic error bars), but are quite observing time-consuming (about 50–90% of the shutter-opened observing time is lost in practice, added to the already large overheads linked to optical interferometry). This allows an internal precision (i.e. visibilities scatter inside an exposure time) of roughly 0.01–0.05 for a bright star ($K \leq 4$ at medium spectral resolution), depending on the observing conditions (wind, vibrations, seeing). However, in practice, the AMBER visibilities precision cannot be better than 0.05 due to several strong limitations coming from the infrastructure itself and from the observing strategy used at Paranal.

- The time between two measurements cannot be less than 30 min (for star and calibrators, i.e. 1 h between 2 science measurements), leading to very large time gaps in the instrumental/atmospheric/vibrations transfer function (see Fig. 3). This prevents today from interpolating such transfer function and gain in precision and stability of the measurement.
- The individual frames exposure times changes a lot during the night (see Fig. 3), leading to miscomparisons between full-night data sets, which provokes a typical night-long visibilities scatter of 0.05. This translates into typical calibrated visibilities errors of about 0.07.

The Table 1 summarizes the current situation with typical observing conditions and the AMBER/VLTI instrument. Please note that this is an indicative table and does not represent all the conditions, which can change strongly depending on the air-mass, seeing, vibrations conditions, etc. Therefore, improvements can be expected for AMBER in standard mode, following several tracks for improving operation:

- Restrain strongly the number of available and effectively used exposure times during the observing nights. This would probably enhance the final transfer function scatter and therefore the calibration accuracy.
- Accelerate the time between observations by working on the telescopes overheads: from the beginning of AMBER operation to today, huge improvements have been achieved, and the arrival of FINITO will improve again these overheads.
- Improve the data processing software, in terms of accuracy, so it takes into account the identified problems and proposed solutions. An effort is also needed in terms of ergonomy and documentation.

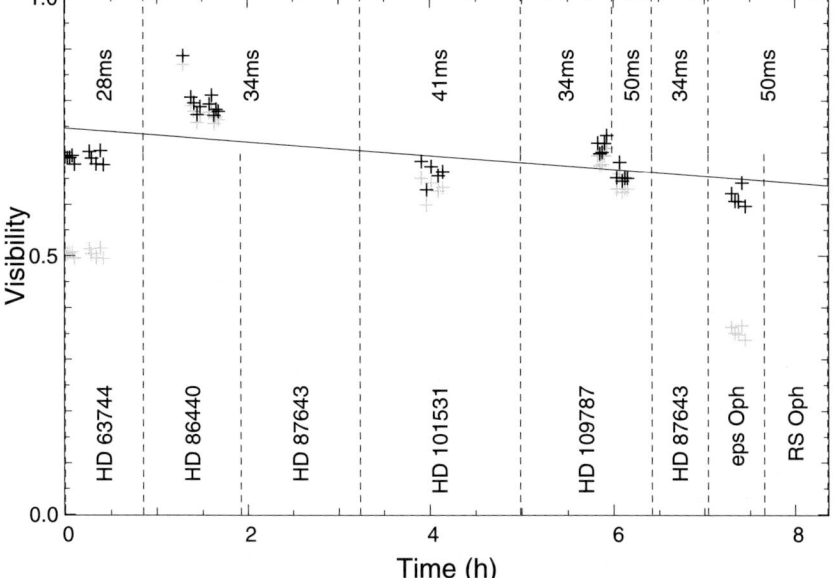

Fig. 3. Instrument+atmosphere transfer function during the night of 17/02/2006. In gray are the original visibilities and in black the visibilities corrected from the intrinsic calibration star's visibilities. The stars where no visibilities are plotted are the science stars. One can see the frame exposure time dependence of the transfer function for the star HD 109787

Table 1. Error bars order of magnitude one can expect from the AMBER instrument in the current status using UTs for calibrated measurements. Visibility errors are dominated by the transfer function error and not by internal visibilities scatter. LR differential data reduction is highly biased by the atmospheric phase bias and the errors given here take into account this bias as an error. The figures given here can be seen as the result of a hard work data processing and very careful calibration process and not as a pipeline black-box output error estimate

Observable	LR ($R = 35$) Bright Star ($K = 5$)	MR ($R = 1500$) Bright Star ($K = 3.5$)
V	0.07	0.07
V_{diff}	0.1	0.01
ϕ_{diff} (rad.)	0.1	0.01
ψ_{123} (rad.)	0.01	0.05

3.3 Closure Phase and Differential Phases

Due to the low number of frames where all three fringes patterns from the three baselines are present together, the closure phases are very much affected by the current state of the VLTI. Therefore, the best achievable closure phase accuracy on a bright star is of the order of 10^{-2} rad, i.e. ~ 1 deg, in low spectral resolution, and about 10^{-1} rad, i.e. ~ 10 deg in medium spectral resolution. For "standard" applications, this is in general sufficient, but for high dynamics or high accuracy measurements, this low precision is very much killing the use of such observable. Added to that, the specific behaviour of phases obliges one to compute very specifically the error bars (see Fig. 4), resulting in under-evaluated error bars in the current data processing software and very noisy data (but this will be solved in the next releases).

The situation on differential phases is much better, with already reached 10^{-2} rad accuracy at medium spectral resolution [9] and 10^{-3} rad accuracy (but with a 10^{-1} rad bias) at low spectral resolution [11]. However, systematic biases related to the amount of water vapour are still under investigation and the calibration of such biases for low spectral resolution is still under development.

3.4 The Low Flux Issue

For low number of photons, the AMBER signal switches from a photon-noise driven behaviour to a detector-noise driven one. In case the detector behaviour is well known, this signal with few photons is workable with a lower signal-to-noise regime than with high flux. However, for AMBER, the detector noise behaviour has changed between the integration laboratory in Grenoble in 2003 and the Paranal interferometric laboratory where it is installed today:

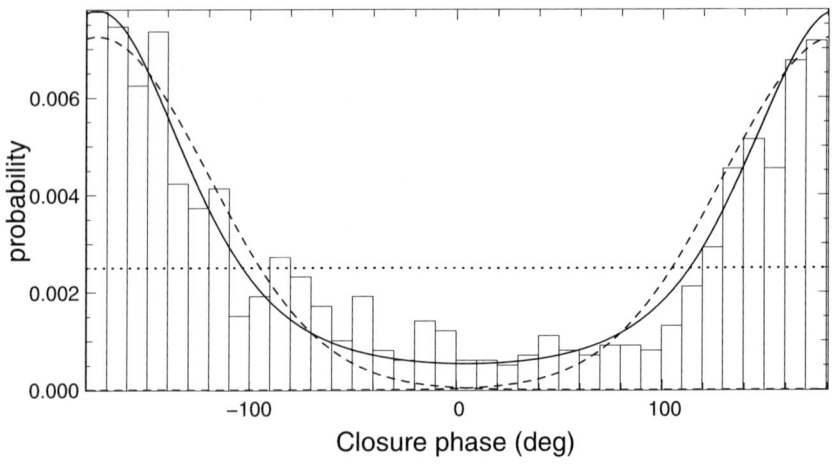

Fig. 4. Example of a closure phase histogram on an observed star (ϵ Sco, courtesy of O. Chesneau), illustrating the difference between a Gaussian distribution (*dashed line*) and a wrapped phase distribution (*solid line*): the phase distribution is in-between a Gaussian-like (*dashed line*) and a white noise distribution (*dotted line*), leading to a problematic estimation of the error bars

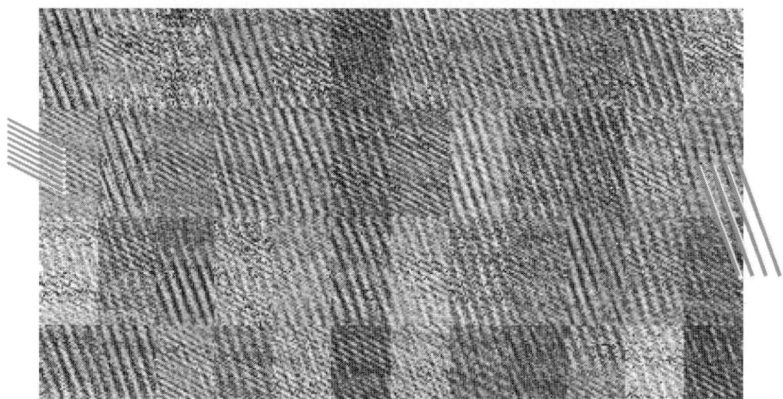

Fig. 5. AMBER detector fringes induced by electromagnetic interferences [5]

electromagnetic interferences from an unknown source occurs and creates a correlated noise which appears as "detector fringes" (see Fig. 5). Therefore in the current data reduction scheme and knowledge of the detector put in the AMBER software, low flux data reduction results in non reliable results.

Therefore, while ESO puts manpower to solve this problem in the hardware side [7], this problem is also under investigations in the data reduction software side [5], in order to achieve a workable data reduction solution for already observed targets. The idea is there to evaluate the detector correlated noise from exposures without fringes (dark or photometric beams) and to try to subtract it from the interferometric beam where the pattern affects the fringes. A prototype version of this algorithm can be found on the AMDC[1] web page.

However, this study is only for the already-acquired data and this problem affects also the on-site real time acquisition of the targets, preventing AMBER to reach its goal limiting magnitude without the external fringe tracker FINITO.

3.5 The Low Resolution Issues

AMBER is facing a series of data processing issues very specific to low resolution observations, and which prevents it from working in an optimal way. We present here a series of identified points to enhance significantly the data processing in this mode:

- The jitter effect affecting the visibilities (Fig. 2) is very important in the way that it introduces an "artificial" slope to the visibilities, very hard to calibrate since this jitter effect cannot be calibrated efficiently in the current state of the infrastructure.

---

[1] AMBER detector cleaner, http://www.mporzio.astro.it/~licausi/AMDC/

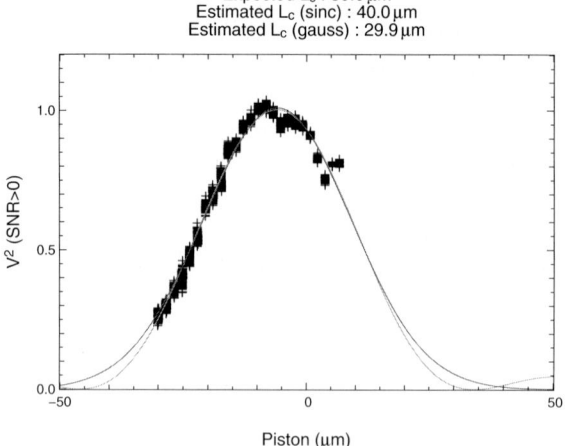

Fig. 6. Coherence length (L_c) dependence of the visibilities of AMBER measured in lab using the internal piezo mirrors of the instrument. It corresponds to a Gaussian of FWHM L_c

- The coherence length visibility decrease plays also an important role (see Fig. 6), and is difficult to calibrate.
- Problems of bias removal in the squared visibilities introduce a visibility flux dependence, which is highly problematic for low resolution, since between the center and the edges of a band, a difference of flux of up to 100 can be found.

All these issues except the first one have an associated solution, included hopefully in the next version of the AMBER data reduction software. The jitter issue is still under investigation and has not reached up to now a level of comprehension allowing us to present a solution. For this issue, the use of FINITO (available first with ATs in P80 and probably soon thereafter with the UTs) is expected to improve a lot the problem.

4 Conclusion

We presented here an overview of what AMBER can do in the current state of the instrument, data reduction pipeline and infrastructure. Table 1 gives typical figures (taken from the personal experience of the authors, and that must be taken as indicative values and not as specified performances) of what can reach in practice the AMBER instrument using the Unit Telescopes of VLTI. With the use of Auxiliary Telescopes and FINITO in a near future, these figures are expected to improve a lot.

As one can see, the instrument does not work in optimal conditions and there is room for near future improvements as well as long term instrumental study to correct all the expected and unexpected effects on the AMBER signal. However, the huge potential of the AMBER instrument has been already proved by the numerous first articles published in a special feature of A&A [6, 16, 9, 8, 18, 10, 2, 3], and many other new successful observing programs can be expected for the future, even with the infrastructure-limited performances of the instrument.

Acknowledgement. This paper is based on data taken at the Paranal observatory, ESO, Chile, and during the AMBER integration in Grenoble in 2003.

The AMBER project (The structure and members of the AMBER Consortium can be found in the AMBER website[2]) has been founded by the French *Centre National de la Recherche Scientifique* (CNRS), the *Max Planck Institut für Radioastronomie* (MPIfR) in Bonn, the *Osservatorio Astrofisico di Arcetri* (OAA) in Firenze, the French Region *Provence Alpes Côte D'Azur* and the *European Southern Observatory* (ESO).

The AMBER data reduction software `amdlib` is freely available on the AMBER website. It has been linked to the open source software Yorick[3] to provide the user friendly interface `ammyorick`.

References

1. J.M. Beckers: Differential speckle interferometry, a new tool for double star research, LowOB **9**, 165 (1981),
2. O. Chesneau, N. Nardetto, F. Millour, et al.: A&A **464**, 119 (2007)
3. A. Domiciano de Souza, T. Driebe, O. Chesneau, et al.: AMBER/VLTI and MIDI/VLTI spectro-interferometric observations of the B[e] supergiant CPD-57 2874, A&A **Special AMBER issue 4134**, (2007)
4. K.-H. Hofmann, G. Weigelt: Image reconstruction from the bispectrum using an iterative algorithm and applications of the method to astronomical objects, In *Digital Image Synthesis and Inverse Optics*; *Proceedings of the Meeting*, San Diego, CA, July 9–13, 1990 (A92-18776 05-74). Bellingham, WA, *Society of Photo-Optical Instrumentation Engineers*, ed by Gmitro, A.F., Idell, P.S. and Lahaie, I.J., pp. 522–525 (1990)
5. G. Li Causi, E. Antoniucci, E. Tatulli: Detector interference pattern, A&A **479**, 589 (2008)
6. F. Malbet, M. Benisty, W.-J. de Wit, et al.: A&A **464**, 43 (2007)
7. P. Mardones: AMBER Image Interferences Report, Chille: ESO (2007)
8. A. Meilland, F. Millour, P. Stee, et al.: A&A **464**, 73 (2007)
9. A. Meilland, P. Stee, M. Vannier, et al.: A&A **464**, 59 (2007)
10. F. Millour, R.G. Petrov, O. Chesneau, et al.: A&A **464**, 107 (2007)
11. F. Millour, M. Vannier, R.G. Petrov, B. Lopez, F. Rantakyrö: Extrasolar Planets with AMBER/VLTI, What can we expect from current performances? In: *Direct Imaging of Exoplanets: Science and Techniques*, IAU Colloq 200, ed by Aime, C. and Vakili, F., pp. 291–296 (2006)

---

[2] http://amber.obs.ujf-grenoble.fr
[3] http://yorick.sourceforge.net

12. G. Perrin: A&A **400**, 1173 (2003)
13. R.G. Petrov, F. Malbet, G. Weigelt, et al.: A&A **464**, 1 (2007)
14. S. Robbe-Dubois, S. Lagarde, R.G. Petrov, et al.: A&A **464**, 13 (2007)
15. S. Robbe-Dubois, R.G. Petrov, S. Lagarde, et al.: The VLTI focal instrument AMBER: results of the first phase of the alignment, integration on Paranal. In: *New Frontiers in Stellar Interferometry, Proceedings of SPIE*, Volume 5491, ed by W.A. Traub. Bellingham, WA: The International Society for Optical Engineering, pp. 1089 (2004)
16. E. Tatulli, A. Isella, A. Natta, et al.: A&A **464**, 55 (2007)
17. E. Tatulli, F. Millour, A Chelli, et al.: A&A **464**, 29 (2007)
18. G. Weigelt, S. Kraus, T. Driebe, et al.: A&A **464**, 87 (2007)

Discussion

A. Grado: Can you give some more details about the vibrations of VLTI?
F. Millour: About the vibration peaks near 50 Hz, this is not expected to be an effect from the power supply since it was observed already by VINCI and there is no peak at 50 Hz.
A. Richichi: An atlas of vibrations and their main sources has been compiled (work by interferometric task force, ITF). Worst offender is a 18 Hz vibration of the VLT primary mirror cells. Other vibrations are coming for example from cooling elements, electronic racks, etc.

QC and Analysis of MIDI Data Using *mymidigui* and OYSTER

C. A. Hummel

ESO, Alonso de Cordova 3107, Vitacura, Santiago, Chile; chummel@eso.org

1 Introduction

Long Baseline Optical/IR Interferometry (LBI) is a niche technique used to answer specific scientific questions requiring the highest spatial resolution available. By including the VLT Interferometer in its service mode (SM) operations offered to a wide astronomical community [4], ESO is now subjecting LBI to an entirely new perspective, namely that of non-specialized scientists evaluating this technique for their research, and of operators and observatory users alike scrutinizing its feasibility, reliability, and performance.

In the ideal world of Paranal Science Operations (PSO) and Quality Control (QC) Garching, data reduction pipelines ensure rapid feedback on the quality of the data obtained, and provide the means for long term monitoring of instrument parameters. In reality, the same process comes in handy if a more thorough understanding of the actual instrument performance beyond the mere specifications and predictions is desired, which is especially true for complex techniques such as LBI. Input to this process is derived from data taken on calibrators and internal calibrations. We would like to "put the rubber to the road" with LBI, exploiting the increasing public archive of calibrator observations. This chapter is a status report of this effort.

2 Tools

MIDI [1] was the first interferometric instrument offered in SM worldwide [2]. The data reduction tools as well as the pipeline are based on software developed by the consortia who built the instrument. The MIDI pipeline has been integrated into ESO's Data Flow System (DFS) and produces on Paranal in real time quality control parameters to help classify the observations during science operations. In Garching, an identical pipeline is run in semi-automatic mode with various sets of input parameters and is used to verify the quality of the data and for instrument trending and health checks. The resulting products are also provided in the SM packages provided to the principal investigators. A description of the Garching Quality control and instrument trending operations is available at http://www.eso.org/observing/dfo/quality/.

For a more in-depth look at intermediate results of the data reduction such as acquisition images and spectra extracted from the chopped frames, a tool has been written as a front-end to the MIA+EWS software (MIA+EWS-1.5, Köhler et al., `http://www.strw.leidenuniv.nl/~nevec/MIDI/index.html`), with some of its own processing capabilities. *mymidigui* is based on the IDL scripting language (ITT Industries, Inc.), and is described at `http://www.sc.eso.org/~chummel/midi/mymidigui/mymidigui.html`. This tool is integrated into the OYSTER software so that configuration and results can be automatically checked into OYSTER data structures for analysis and display. The integration with a general interferometry package such as OYSTER also enables the implementation of a pipeline reducing entire nights of data for the study of correlations and instrumental stability.
(For OYSTER, see `http://www.sc.eso.org/~chummel/oyster/oyster.html`.)

3 Acquisition

Interferometers are notorious for bad throughput (on the order of 2%), caused by the many reflections typically necessary to bring the light into the lab for recombination. Therefore, it is a good idea to establish reference values for the detector signal as a function of target flux. Within PSO, work is in progress to use `chop_nod_phot` (part of MIA+EWS) to integrate the detector ADUs within the PSF in the chopped acquisition images, and to use Table 1 to estimate 12 µm fluxes. This scheme is approximative since we relate flux measured in the N8.7 filter to 12 µm IRAS fluxes (using a conversion $N = 4 - 2.5 \log F[\text{Jy}]/\log 10$), making it dependent on the color of the object. Significant deviations of the fluxes predicted in this way from the expectations would indicate vignetting or detector problems.

Furthermore, the acquisition GUI allows to plot flux and relative jitter of individual exposures.

4 Photometry

In HIGH_SENS mode, where all the light is used for the interferometric outputs of the beam combiner, the uncorrelated fluxes have to be measured

Table 1. Table of values for N_0, where $N = N_0 - 2.5 \log C$, with N, the IRAS magnitude, and C, the integrated flux from `chop_nod_phot` (part of MIA+EWS). Valid only for the N8.7 filter

UT1	UT2	UT3	UT4	AT
16.3	16.4	16.9	16.9	12.9

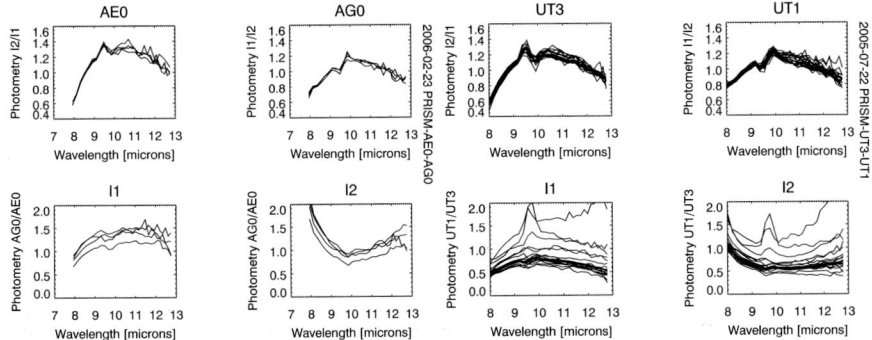

Fig. 1. Photometric ratios of the two interferometric channels (*upper panels*) for each shutter position (beam A or B), and ratio in each interferometric channel between beams A and B. Observations with the ATs on the *left*, UTs on the *right*. The scatter in the lower *right panel* for the photometric ratios between telescopes is due to faint targets, while the brighter ones define the locus of the mean ratio

separately with only one beam shutter open at a time. The quality of the photometry thus obtained has been shown to have a critical impact on the accuracy of the normalized visibility amplitudes, and is therefore usually the first focus of attention if something seems wrong with the results.

Aside from the absolute throughput or sensitivity calibration which can be accomplished with the acquisition data, the spectra extracted from the photometric observations can be compared between the two interferometric channels and also between the two input channels. The former is related to the splitting ratio of the beam combiner, which of course should be quite stable. In Fig. 1 we show photometric ratios obtained in two nights with the auxiliary telescopes (ATs) and with the unit telescopes (UTs).

5 Transfer Function

The visibility amplitude measured on an unresolved calibrator (or on a star with known diameter and corrected for its effect) is used to calibrate the visibility amplitude measured on the science target. This calibration factor has also been called transfer function, system visibility, or interferometric efficiency. It is an important measure of the quality of the interferometer including the effects of the atmosphere. For VLTI/MIDI, it is usually between 0.6 and 0.8 (with 1.0 being the maximum theoretical limit).

In 1 arcsecond seeing, the focus of the 8 m aperture of an UT is almost but not quite free of speckles, while in 0.5 arcseond seeing it is (not unusual for Paranal). With the MACAO adaptive optics system, the first Airy ring pattern is easily produced under these conditions. Only in seeing worse than one arcsecond is a drop in visibility expected, and that seems to be confirmed

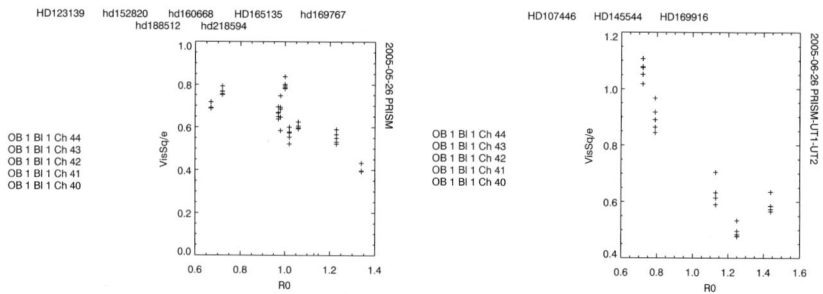

Fig. 2. Squared visibility amplitude measured on calibrators as a function of r_0 (in arcseconds), a measure of the seeing. Results shown for two different nights. Corrections for the finite calibrator diameters are negligible. Five channels between 8 and 9 μm are plotted. For other nights, this correlation is often not obvious

when looking at Fig. 2. For these QC reasons, some of the MACAO system parameters while in closed loop are recorded [3].

The question of correlations between visibility amplitude and ambient parameters such as seeing, coherence time, and beam jitter (measured by IRIS) is an important one because of its impact on the calibration. If correlations could be found and are well understood, they could increase the scope of a single calibrator to a much wider range in location, time, and seeing conditions. We have used gaps in the service mode schedule for the ATs to observe calibrators to study these effects. In Fig. 3 we show a visibility variation apparently correlated with a change in flux level, albeit just at the highest zenith angles.

In another test observing the same calibrator repeatedly, a sharp drop of the visibility amplitude apparently does not correlate with total flux, but with a drop of the seeing, as characterized by r_0 (see Figs. 4 and 5). Using EWS, we are able to compute the RMS of the group delay as another measure of the atmospheric conditions.

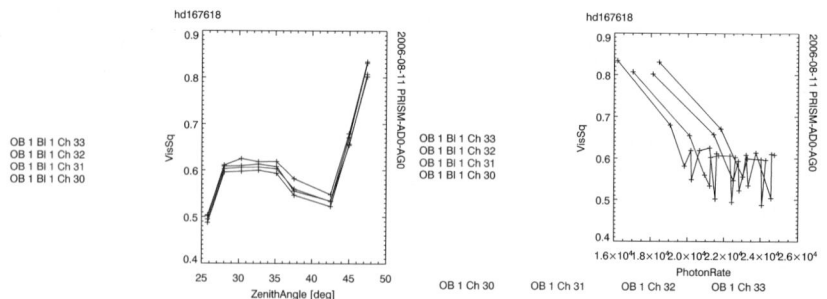

Fig. 3. Systematic variation of the visibility due to photometric variations, observed with the ATs on a bright calibrator in HIGH_SENS mode

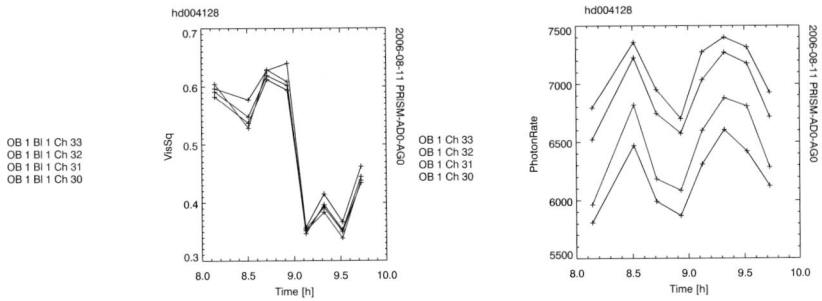

Fig. 4. A sharp drop of visibility observed with time, and the photometry at the same time. The two are not correlated. Data of four wavelength channels are displayed

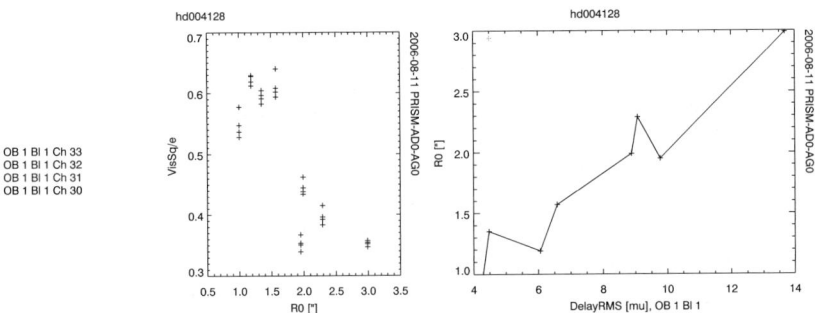

Fig. 5. The same observations shown in Fig. 6, but plotted versus seeing r_0. The seeing correlates with the RMS of the group delay (*right*)

6 SCI_PHOT Mode

The SCI_PHOT mode was designed to provide the highest possible calibration accuracy by performing photometry at the same time as the fringes are measured. This is done by splitting off light for two photometric channels recorded alongside the interferometric channels. This will ensure that systematic opacity variations caused by clouds, airmass differences between fringe and photometry exposures, etc. will not introduce systematic errors in the visibility normalization. The splitting ratios (the four coefficients of the so-called κ-matrix) are measured once on a bright target, and are then used to determine the normalizing uncorrelated flux for the other observations. In theory, the κ-matrix needs to be only determined once a night, and Fig. 6 shows that the variation during a night is on the order of 4%. More frequent measurements could be performed for instrument trending, but they would also increase the duty cycle.

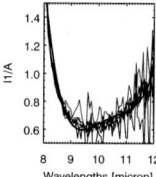 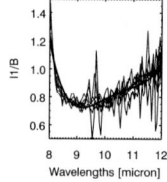 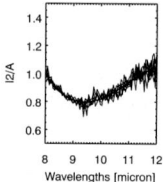 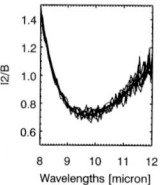

Fig. 6. The four elements of the κ-matrix as a function of wavelength, measured six times on HD 120323 over an interval of about 4 hours. Observations were performed with the ATs (D0–G0) on Feb. 26, 2006. These data are in the public archive and can be used to verify the data reduction methods

7 Conclusions

A lot is known on how to turn interferometric observations into results of scientific value. Much less is known on how to turn interferometric operations into papers in referred journals. We have a unique opportunity to integrate a complex technique such as long baseline interferometry into mainstream astronomy.

References

1. Ch. Leinert, et al.: "Ten-micron instrument MIDI: getting ready for observations on the VLTI" in *Interferometry for Optical Astronomy II*, W. A. Traub, ed., pp. 893–904, SPIE, Bellingham, 2003
2. S. Morel, et al.: "Preparing MIDI science operation at VLTI" in *New Frontiers in Stellar Interferometry*, W. A. Traub, ed., p. 1666, SPIE, Bellingham, 2004
3. I. Percheron, et al.: "Adaptive optics quality metrics and user constraints set for VLTI" in *Advances in Stellar Interferometry*, J. Monnier, M. Schöller, W. Danchi, eds., p. 626848, SPIE, Bellingham, 2006
4. M. Wittkowski, et al.: "Observing with the VLT interferometer" in *New Frontiers in Stellar Interferometry*, W. A. Traub, ed., p. 617, SPIE, Bellingham, 2004

Discussion

R. Hanuschik: Your motivation for developing this software was the lack of QC1 functionality in the MIDI pipeline. I guess the pipeline has been developed further in the meantime. Is there still functionality missing, and if so, shouldn't there better be an effort to close this gap with the pipeline?

C. Hummel: The current MIDI pipeline produces unreliable results for the SCIPHOT mode. Some results from the photometry (normalized flux, PSF FWHM, etc.) should also be made available.

A. Richichi: How long would it take to get a detailed insight into the whole database of MIDI combined using the mymidigui approach?

C. Hummel: Just to crunch through all the data, several nights worth of data could be processed per day. However, it is equally important to have a good understanding of the effects before running everything through the pipeline to make sure all necessary ancillary data are processed.

D. Osip: Given 178 MIDI calibrators are a subset (brighter) of the 1320 AMBER calibrators with the difference being one set uses stellar atmosphere models while the other adopts Cohen templates, can anyone comment on the comparison between these two options? Are the "standards" really adequate to the 1% goals of the instruments?

C. Hummel: To calibrate a visibility to 1% it is not necessary to know the diameter to 1%. It really depends on the diameter of the calibrator. If they are nearly unresolved then a 10% accuracy could be sufficient. The predictions are usually better than 5%.

A. Smette: Michael Sterzik noticed last year that in a number of cases the visibility appears larger over the O_3 band. Why is this?

C. Hummel: Visibilities in the O_3 Band are more difficult to normalize due to the sensitivity of the photometry in that region. For faint targets this is often seen.

U. Hopp: You said you can "predict" the stellar radii? Is this prediction based on stellar atmosphere and its theory?

C. Hummel: The predictions are based on calibrations of, e.g., the surface brightness method or measurements of bolometric absolute fluxes and effective temperatures from spectrophotometry.

Calibration of AMBER Visibilities at Low Spectral Resolution

P. Cruzalèbes, A. Spang, and S. Sacuto

Observatoire de la Côte d'Azur, Laboratoire Gemini, UMR-CNRS 6203, av. Copernic, F-06130 Grasse, France; pierre.cruzalebes@obs-azur.fr

1 AMBER Observations

AMBER (Astronomical Multi BEam Recombiner) is the VLTI beam combiner, operating in the near-infrared, using optical fibers as spatial filters [8]. Anamorphic optics compress the interferometric beams perpendicularly to the fringe coding, before injecting them into the slit of a spectrograph (left part of Fig. 1). The instrument can operate at a spectral resolution up to 10^4 and delivers wavelength-dispersed visibilities. The giant star π^1 Gru and its reference δ^2 Gru were observed on 2005/07/14 (MJD=53565.3) during the "Paranalization" run of the AMBER instrument with the single AT-VLTI baseline E0-G0 (B=15.9 m; $\theta=71°$). AMBER observations were made with frame exposure time of 29 ms, in the LR spectral mode (spectral resolution R=35). They were carried out under non-optimal conditions since biases related to injection problems may exist due to the lack of the Variable Curvature Mirror, responsible for wrong pupil superposition and limited field of view.

2 Raw Data Pre-Processing

The AMBER instrument follows the standard data flow system implemented at ESO/VLT during the data acquisition. The software simultaneously records the images of the spectrally dispersed fringes (interferometric output) and those of the telescope beams (photometric outputs). Each data file consists of one thousand of frames provided by the detector subsystem (right part of Fig. 1). The standard data reduction method developed and optimized for AMBER is called P2VM for Pixel-To-Visibilities Matrix [4, 10]. Twelve observation files by groups of three are recorded alternatively with the reference and with the science target. Standardized OI-FITS files [6] are produced using standard reduction routines with *amdlib 1.21* and *ammyorick 0.56*.

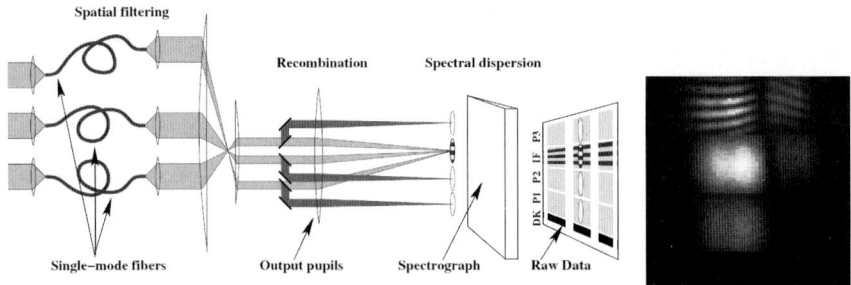

Fig. 1. *Left*: Sketch of the AMBER instrument with 3 beams [10]. *Right*: Example of an AMBER raw short-exposure frame obtained with the internal calibration source and 2 beams. Each frame is composed of 4 horizontal windows with (from *top to bottom*): interferometric output; photometric outputs 1 and 2; dark (masked pixels). Wavelength decreases in the horizontal direction from left to right (K-band at *left*; H-band at *right*)

3 Aberrant Data Removal

To estimate the instantaneous visibility in each spectral channel of the K and H bands, a fringe sinusoidal function is fitted on the interferometric signal along the spatial direction (on 32 pixels). In case of poor signal to noise ratio (specially near the bandwidth limits of the H and K filters), the reduction process may produce aberrant data which must be removed. Three selection criteria are successively applied in order to reject instantaneous measurements with negative flux ratios, with negative total fluxes, and with V^2 outside the range between 0 and 1 (left panel of Fig. 2). Insofar as the mean visibilities (for the science target and its reference) lie between 0.3 and 0.7, this sorting process, although severe, does not change the V^2-statistics, and is sufficient

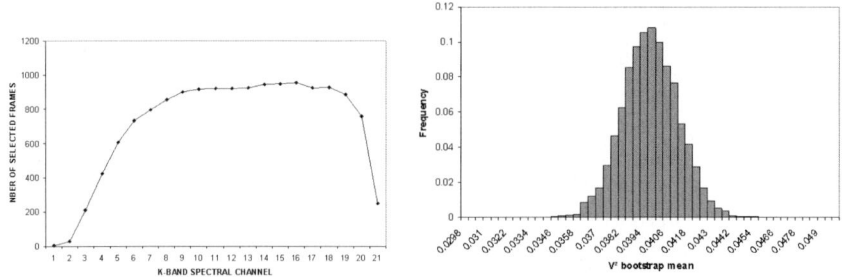

Fig. 2. *Left panel*: Number of remaining frames in one reference-star file after aberrant-data removal for each spectral channel of the K-band. *Right panel*: Histogram of the V^2 means after bootstrapping at the K-band central wavelength for the same data file

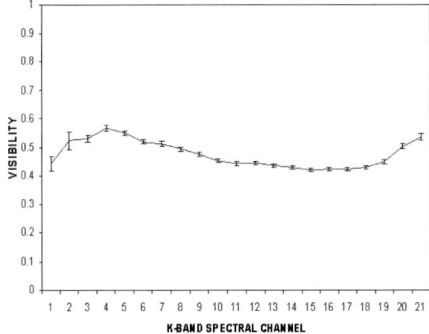

Fig. 3. Measured visibility of the reference after aberrant-data removal for each spectral channel of the K-band (decrease of wavelength from *left* to *right*)

to insure a "relative" reliability of the visibility measurements. Then, the mean visibility and its error are calculated in each spectral channel using the bootstrap analysis (right panel of Fig. 2), a method successfully used with the VINCI instrument [3]. Figure 3 shows the measured visibility of the reference for the K-band (with the 15.9 m baselength).

4 Visibility Calibration

Because the reference star δ^2 Gru is also a cool giant with a predicted angular radius near 5 mas, it clearly cannot be considered as a point-source and the calibration process of the visibility must take this size into account. Interferometric observations provide measurements of the source visibility equal to the product of the true source visibility with the system response, a combination of the atmospheric and the instrumental responses. The system response can be estimated thanks to the ratio between the measured visibility of the reference and its true visibility [11, 1], which needs to be all the better known the more the reference is resolved. The photospheric diameter of the reference is estimated using the MARCS photospheric model [9] fitted on spectral energy distribution measurements in the infrared obtained from the ground and with IRAS-LRS (left panel of Fig. 4).

5 Visibility Model-Fitting

To extract the physicochemical parameters of the circumstellar material, we finally fit the true (calibrated) AMBER visibility (right panel of Fig. 4) with parametric models [2]. As a preliminary result, we find that a two-shell model, with a cool molecular H_2O layer located between 2.0 and 2.3 R_* and a thin,

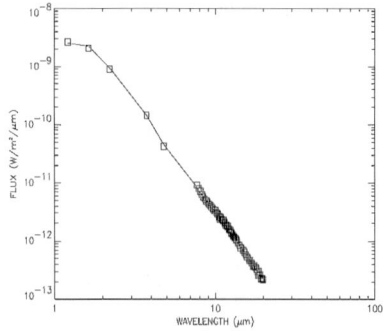 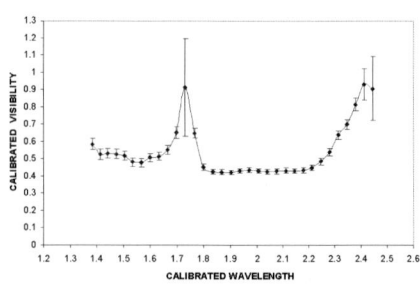

Fig. 4. *Left panel*: Best χ^2-fit of the MARCS model with T_{eff}=3600 K on the spectro-photometric measurements of δ^2 Gru in the infrared, giving an infrared photometric diameter of the reference of 8.8±0.2 mas. *Right panel*: Final true (calibrated) visibility of π^1 Gru in H and K (B=15.9 m). The increase of the visibility around 1.7 µm and after 2.25 µm results from residual calibration errors due to low signal to noise ratios at these wavelengths

dusty envelope, can reproduce the AMBER measurements of π^1 Gru in the near-infrared, with final parameters in good agreement with the values proposed in previous works [12, 5, 7].

References

1. D. Bonneau, et al.: A&A **456**, 789 (2006)
2. P. Cruzalèbes, S. Sacuto: A&A **452**, 885 (2006)
3. P. Kervella, D. Ségransan, V. Coudé du Foresto: A&A **425**, 1161 (2004)
4. F. Millour, et al.: In: *New Frontier in Stellar Interferometry*, vol 5491, ed by W.A. Traub (Bellingham, WA : SPIE), 2004, pp 1222–1230
5. K. Ohnaka, et al.: A&A **424**, 1011 (2004)
6. T.A. Pauls, et al.: PASP **117**, 1255 (2005)
7. G. Perrin, et al.: A&A **426**, 279 (2004)
8. R. Petrov, et al.: In: *Interferometry for Optical Astronomy II*, vol 4838, ed by W.A. Traub (SPIE), 2003, pp 924–933
9. B. Plez, et al.: In: *Modelling of Stellar Atmospheres*, vol 5491, ed by N. Piskunov, W.W. Weiss, and D.F. Gray (IAU-ASP), 2004, pp A2–A11
10. E. Tatulli, et al.: A&A **464**, 29 (2007)
11. G.T. van Belle, G. van Belle: PASP **117**, 1263 (2005)
12. I. Yamamura, et al.: A&A **348**, L55 (1999)

VLTI Instruments: From J to N Band Instrumental Calibrations, from Short to Long Baseline Astronomical Calibrations

I. Percheron

ESO, Karl-Schwarzschild-Strasse 2, 85748 Garching bei München, Germany; ipercher@eso.org

1 VLTI Layout, Instruments and Sub-systems

The VLT Interferometer (VLTI) has been operating on Paranal since 2001 with first fringes obtained at 2 µm with VINCI, the VLTI commissioning instrument. Since October 2005, two VLTI instruments are offered for Service Mode (SM) observations, MIDI (offered since April 2004) is operating in the N Band between 8 to 12 µm [3] and AMBER (near-infrared/red focal instrument for the VLTI) operates in the J, H and K bands [4]. MIDI observations are performed using two UT (Unit Telescopes 8.2 m) or two ATs (Auxiliary Telescopes: 1.8 m), while AMBER has been operating until now with three UTs and will be offered from April 2007 also with three ATs [1, 2, 5]. The ATs can be located on different stations with baselines range from 8 to 200 m with different orientations.

Several sub-systems such as the telescopes (UTs or ATs), the AO systems (MACAO), the fringe tracker (FINITO and in the future PRIMA), the delay lines, are also part of the VLTI. The status of all these sub-systems has to be understood. They are monitored either regularly with the daily calibrations or at the time of the science observation with the attached calibrations. We are also monitoring the correlation between data taken by external instruments (such as the DIMM installed on Paranal to record real time seeing and meteorology data).

2 VLTI Daily Calibrations and Reference Frames

VLTI like the other VLT instruments follows a calibration plan which includes daily calibrations. The detector as well as some of the most important elements of the instrument are monitored. In the case of MIDI, the detector Read Out Noise and its linearity are measured daily and trended. The position of the reference beams (see Fig. 1), the transmission of the dispersive elements and data for wavelength calibration are also taken daily.

For AMBER, the calibration plan includes daily calibration of the internal status of the instrument as well as a monitoring of the alignment of the instrument. Other calibrations such as the Pixel to Visibility Map are taken

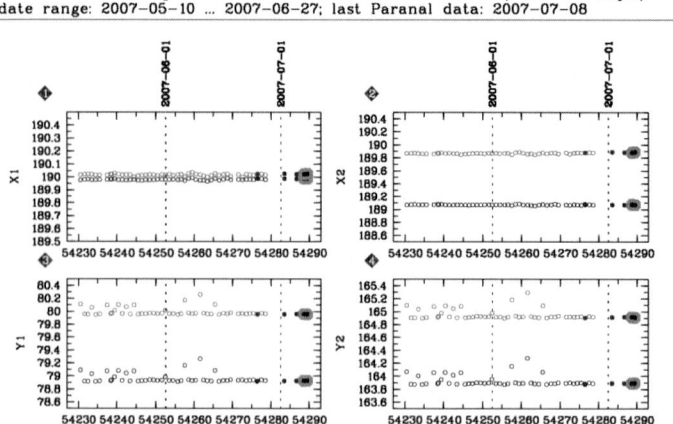

Fig. 1. Instrument trending for MIDI. In this trending plot, the position of the two beams is shown for one of the MIDI instrument setting, there is sometimes a jump of both beams in the Y position, this problem is under investigation

together with the observations. During the alignment and the calibrations of the VLTI instruments, reference frames are available to the observer for comparison.

3 Monitoring of the VLTI Sub-systems

The VLTI is a complicated system with two main instruments MIDI and AMBER and several sub-systems (Telescopes, AO systems, fringe tracker, Delay lines...). Since the beginning of VLTI operations, studies have been made to understand better the effects of the turbulence and of the sub-sytems on the data [6]. Data from the different sub-systems are recorded during the daily calibrations and during the observations.

3.1 Monitoring of the Different MACAO Sub-systems

To get diffraction limited images with MIDI and AMBER on the UTs, MACAO is used during the observations. The MACAO systems [7] are installed on the UT Coude foci and are feeding the VLTI delay lines with a corrected IR beam from 1000–13000 nm with up to 50% Strehl at 2.2 μm. The operational conditions for good quality operations with MACAO are a seeing less than 1.5 arcsec, a coherence time larger than 1.5 ms, an air mass smaller than 2 and the distance between the science target and the AO target less than 57.5 arcsec. The plots (Fig. 2) show the comparison between the different MACAO units during AMBER observations with 3 telescopes.

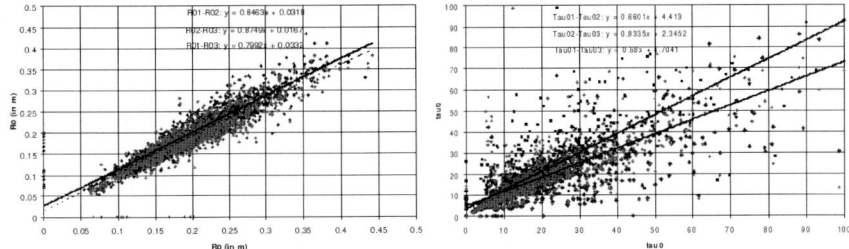

Fig. 2. The *left* plot shows the Fried parameter (R0) in m, and the *right* plot the coherence time Tau0 in ms, both are calculated for 500 nm. The data for each telescope is represented by a different shape. These data were taken both under SM operations and under test where the atmosphere did not allow a good correction. The differences seen between the MACAO systems on the 3 telescopes are under analysis

3.2 Comparison Between MACAO and the DIMM Measurements

The VLTI is part of the Paranal Science Operations Team (PSO) and the performances of the instruments and of the different sub-systems are also compared to other Paranal instruments. When VISIR (operating also in the N band) is operating at the same time as MIDI, some data (such as the sky background) taken on both instruments could be compared. The differential image motion monitor (DIMM) records seeing parameters such as the Fried parameter R0 or the coherence time (Tau0). We are for example studying the correlation between the data measured by the DIMM, the different MACAO units and the atmospheric conditions.

Future studies involve the comparison between the Strehl Ratio recorded by the MACAO sub-systems, the distance from the source and other atmospheric parameters such as the isoplanatic angle with the quality of the MIDI acquisition image.

3.3 Photometric and Instrumental Visibility Trending

We implemented a calibration plan for the MIDI and the AMBER instruments, to observe regularly some calibrator and photometric standard stars. For MIDI photometric observations are taken on spectro-photometric objects (not known as variable), some of them are also observed regularly with the VISIR instrument. A list of potential bright standard calibrators is also available. They are extracted from the VLTI list of potential calibrators CalVin. Regular observations of these objects are done on the different baselines for instrument monitoring.

3.4 Astronomical Calibrators and Observations of Standard Stars

In addition to the data taken to monitor the instrument, astronomical calibrators are measured during the night to calibrate the science data. Depending of the night (Service Mode (SM), Visitor Mode (VM) or technical nights), the astronomical calibrators are either defined by the SM or the VM PI or taken from CalVin (for technical nights). The instrumental Transfer function is trended using these data.

Each night, some photometric standard stars are observed, the flux of these objects ranges from a few Jansky to more than 100 Jy. We study the variation of the flux on the detector as a function of different parameters (station, telescope, delay line, airmass, atmospheric parameters). This is still under study.

4 Conclusions

It is essential to monitor and trend all the important sub-systems to insure a good quality of the science data. Knowing which part of system to calibrate, monitor and trend is important and is in constant evolution and discussed between members of the VLTI team. This knowledge is also transferred to the community with the instruments Health Check pages maintained by the Quality Control team at ESO.

References

1. S. Morel, et al.: Preparing MIDI science operation at VLTI, *New Frontiers in Stellar Interferometry, SPIE* 1666 (2004)
2. F. Rantakyro, et al.: Experiences from the first AMBER open time observations, *Advances in Stellar Interferometry, SPIE* 6268, p. 6268 IN (2006)
3. C. Leinert, et al.: Ten-micron instrument MIDI: getting ready for observations on the VLTI, *Interferometry for Optical Astronomy II, SPIE* 893–904 (2003)
4. R. Petrov, et al.: Using the near infrared VLTI instrument AMBER, *Interferometry for Optical Astronomy II, SPIE* 924–933 (2003)
5. M. Wittkowski, et al.: Observing with the ESO VLT Interferometer, *ESO Messenger* 119, 14 (2005)
6. E. Di Folco: Atmospheric and internal turbulence measured on the Very Large Telescope Interferometer with VINCI, *Interferometry for Optical Astronomy II, SPIE* 4838 1115–1126 (2003)
7. R. Arsenault, et al.: MACAO-VLTI Adaptive Optics system performance, *Advancements in Adaptive optics, SPIE* 5490 47 (2004)

Part XI

Session 10: Polarimetry

IOT Overview: Polarimetry

N. Ageorges

ESO, Alonso de Cordóva 3107, Vitacura, Casilla 19001, Santiago 19, Chile;
nageorge@eso.org

Abstract. This contribution concentrates on the polarimetric modes offered by different instruments at ESO. In the introduction, I will demonstrate the importance of polarimetry, the kind of science it permits to achieve and list the instruments which offer these modes. Sects. 2 and 3 will present the involved modes in more details as well as the currently related calibrations, as part of the calibration plans.

ESO does not offer any pure polarimetric instrument. As a consequence the polarimetric modes are just one (or more) mode(s) of the given instruments. Polarimetric modes might be mentioned in the related IOT but are not followed up thoroughly as is e.g. spectroscopy.

1 Introduction

This section concentrates on showing why polarimetry is important and how it can be measured. The latter should help understand which calibrations are needed.

1.1 Why Polarimetry?

To some extent all sources of radiation are partially polarised. Polarization can be produced in many different ways, e.g. differential (dichroic) absorption of radiation passing through the interstellar medium or scattering of radiation, which is the most frequent cause of polarisation, at visible and near-infrared wavelength. Polarimetric observations allow one to derive a wealth of information, e.g. what is the direction of the plane of oscillation of photons, study the magnetic field of a cloud, create 3D maps of dusty environments ...

Because of the large number of application, polarimetric techniques are used for a large variety of astronomical research. The most recent publications based on the use of the polarimetric mode of ESO instruments go from studies of trans-neptunian objects (see e.g. [14]) to gravitational lensing (see e.g. [6]), but encompass as well supernovae (e.g. [10]), open clusters (e.g. [3]), ULIRGs (e.g. [13]), planetary nebula (e.g. [2]), young molecular nebula (e.g. [5]) and the measurements of magnetic fields (e.g. [9]).

Polarimetric measurements can either be based on imaging or spectroscopy. Both methods can be used to measure either the linear or circular polarisation of a source, thus delivering different information.

1.2 How to Measure Polarisation?

To measure the polarisation of a source one needs a polarising (and analysing) optical element. In most instruments, this is a Wollaston prism, which divides the incoming light beam in two orthogonal polarisation beams; but this could as well be achieved with a wire-grid. The polarisation status of a source is fully described by the four Stokes parameters, I, Q, U & V. To determine them one needs at least a measurement for 4 different polarisation angles; typically 0, 45, 90 & 135 degree. I represents the full intensity of the source and V the circular polarisation. Q & U are needed to calculate the linear polarisation of the source and are defined, in their normalised form, as:

$$Q = I_0 - I_{90} \quad \& \quad U = I_{45} - I_{135} \tag{1}$$

From there one can thus derive the polarisation degree P and polarisation angle θ as

$$P = \sqrt{(Q^2 + U^2)} \quad \& \quad \theta = \frac{1}{2}\arctan(\frac{U}{Q}) \tag{2}$$

Please refer to [11] for a detailed definition of the Stokes parameters.

At ESO, only the optical instruments allow to measure both the linear and circular polarisation. They also deliver both imaging and spectroscopic polarimetric measurements. These instruments are EFOSC2 in La Silla and FORS1 in Paranal. They are presented in more detail in Sect. 3. The ESO infrared instruments equipped with a polarimetric mode have been presented in [1]. With the exception of CRIRES, which is not considered further in this contribution since its mode to measure circular spectroscopic polarimetry has not been commissioned yet, all other ESO near-infrared instruments equipped with a polarimetric mode (SofI, ISAAC & NaCo) only allow to perform linear imaging polarimetric measurements. The calibration plan of the polarimetric mode of these instruments is discussed in Sect. 2.

In an ideal case, the instrumental polarisation is well known and documented and can be removed easily from the science data. Moreover it is re-measured regularly to check the instrument stability. In a less ideal world, the instrumental polarisation is badly known but measured together with each science observation to allow a good calibration of the data.

2 Near-IR Polarimetric Instruments

2.1 Introduction

All near-infrared instruments equipped with a polarimetric mode are installed at a Nasmyth focus. As a consequence, the instrumental polarisation depends on the parallactic angle and is thus variable. For SofI, ISAAC & NaCo, the polarimetric mode is a last minute addendum; not an integral part of the instrument from the early design.

In CRIRES the polarising element is a cryogenic Wollaston prism, while the analysing element is composed of a retarder plate located close to the telescope focal plane. For SofI & ISAAC, the polarising element is a Wollaston prism. One such prism is also installed in NaCo, which is however also equipped with 4 wire-grid analysers and a half-wave plate. With a Wollaston prism, one gets simultaneous measurements of two orthogonal polarisation parameters and thus saves on telescope time. However in average half of the field is masked out so two observations are needed to cover the entire field of view. To get the measurements at different angles, the intrument needs to be rotated except if, as in NaCo, a half-wave plate is installed in the beam and allows to rotate the polarisation.

Due to the non constant instrumental polarisation, it is important to be able to calibrate properly each set of observations in order to derive absolute polarisation (and not only relative values). The problem of the polarimetric standards is thus raised. Very few near-infrared polarimetric standard are known and most of them are too bright for 4–10 m class telescope. This explains why some ESO internal project has been started with the aim to determine secondary polarimetric standards for the near-infrared.

Some calibration of the SofI polarimetric mode has been provided by Wolff, Vanzi & Ageorges (2003) and can be found in the SofI official webpages at http://www.ls.eso.org/lasilla/sciops/ntt/sofi/archive/pol/tech_rep_polarimetry.ps

2.2 Calibration for Near-IR Polarimetric Instruments

The polarimetric mode is poorly supported by any calibration plan. There is neither an Exposure Time Calculator nor pipeline recipes for it. If the user wants to obtain absolute polarimetric values, he must take into account the time for observing the standard star into his requested telescope time, and must also provide the observing blocks.

The only thing that the respective calibration plan foresees is to provide flat fields. For ISAAC, normal twilight flats are taken without the Wollaston prism. For NaCo both imaging and polarimetric lamp flats are provided. In service mode, SofI offers flat fields as well. Due to the polarisation of the polarimetric dome flats, imaging flats, without polarimetric optical elements, are provided. Of course in visitor mode, one can request to take polarimetric twilight flats but this is not optimum since the sky light is polarised. In general flatfielding of polarimetric data is an issue, independently of the observing wavelength. Indeed, flats without the polarising elements, e.g. the Wollaston prism, do not correspond fully to the flat field on the polarimetric data since light does not cross the same optical elements. However due to the sky polarisation, ideally the flat field with the full instrumental set-up should be taken over different polarisation angles in order to average out the sky contribution. One way to do this would be to rotate e.g. an half wave plate while integrating the data. In principle this would be possible with NaCo, which is equipped with such a rotating element. In the two other cases, the full instrument would need to be rotated while taking the flat field.

2.3 Polarimetric Mode & NIR Instruments: Conclusion

The existing polarimetric modes on the near-infrared instruments at ESO are "under-used". It is not clear if this is a consequence of the poor calibration of these modes or its cause. Indeed a higher pressure on these modes would encourage ESO to spend more time for proper calibrations and to characterise these modes better.

It is interesting to note that most users do not take polarimetric standards. They either are interested in differential polarisation or can boot strap on the data themselves to calibrate, e.g. knowing the instrinsic polarisation of some of the sources present in the field of view.

It has to be mentioned that common users polarimetric facilities are available to the astronomical community since the 80s. But as J. Hough [8] nicely states in his review of polarimetric techniques at optical and infrared wavelengths: "Despite this development, and the huge benefits in virtually all areas of astronomy, polarimetry has rarely been included in the baseline design of either telescopes or instruments, at least for night time astronomy, and adding polarimeters to existing systems often leads to what should have been unnecessary compromises and perhaps less user-friendly instruments for the non specialist."

3 "Optical" Instruments

Only two ESO instruments working at optical wavelength provide polarimetry: EFOSC2 at La Silla and FORS1 at Paranal. They are described separately hereafter. Contrarily to their infrared counter part, these instruments are placed at a Cassegrain focus and a polarimetric mode was planned from the beginning of the instrument design. For FORS1, one can even say that the instrument has been build around the polarimetric mode. In both cases, the percentage of observing time reserved for polarimetry is substantial, as well as the resulting publications. One direct consequence is the installation of an additional polarimetric mode for EFOSC2 as of April 2007.

3.1 EFOSC2

Introduction

EFOSC2 is a focal reducer multi-mode instrument, which is mounted part-time at the 3.6 m telescope in La Silla. It works from 305 to 1100 nm and has a field of view of 5.2' × 5.2'. Aside imaging, coronography and spectroscopy, its polarimetric observing modes are:

- IPOL imaging (linear) polarimetry
- SPOL spectropolarimetry

The instrument has two Wollaston prisms providing different separation between the two orthogonal polarisation beams 10" (for imaging only) and 20" for both imaging and spectropolarimetry. It also has a half-wave plate. As of P79, a new polarization unit, to measure circular polarimetry, is offered. This unit is a copy of the previously existing one, but holding a new super-achromatic quarter-wave retarder plate.

Calibration Plan

The calibration plan of this instrument specifies that for imaging polarimetry, 10 BIAS frames are provided as well as three dome or sky flats taken with the rotating half-wave plate, which is the best possible for a polarimetric flatfield, and an image of the Wollaston mask. The observatory does not provide any standard star observation.

For spectropolarimetry, on top of the bias and flats with rotating half-wave plate, an arc with the HeAr lamp is provided. Depending on the science, a spectropolarimetric and/or spectrophotometric standard will be observed as part of the calibration plan.

3.2 FORS1

Introduction

FORS1 stands for FOcal Reducer/low dispersion Spectrograph 1. It is a multi mode optical (330–1100 nm) instrument installed at the Cassegrain focus of the second unit telescope in Paranal. The image arrives on the 2048×2048 detector with two different magnifications. With the high resolution collimator, the pixel scale is of 0.1" thus providing a field of view of 3.4'×3.4'; while the pixel scale with the low (or "standard") resolution (SR) collimator is of 0.2". One has to calibrate each magnification independently. On top of imaging and spectroscopy, the offered polarimetric observing modes are:

- IPOL imaging polarimetry
- PMOS multi-object spectropolarimetry (for SR collimator only)

Polarimetric Mode

The polarimetric mode uses a remotely controlled rotating half-wave plate or quarter wave plate installed in front of a Wollaston prism. The retarder

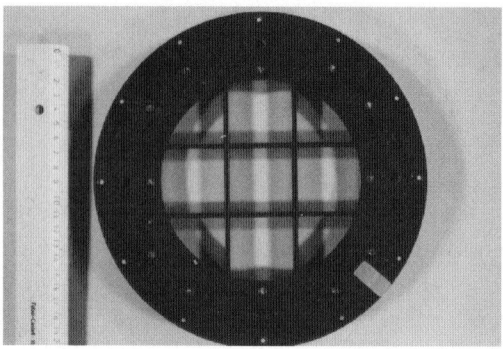

Fig. 1. Photography of the FORS1 retarder plate

plate (Fig. 1) is of superachromatic type. It however has been impossible to procure waveplates of sufficient size; therefore mosaics of 3×3 plates of 45.5×45.5 mm each are used (with an inter-plate gap of 3mm) providing a resulting free mosaic diameter of 138 mm. It has a positioning accuracy of 0.1 degree.

Linear polarimetric observations obtained with the retarder plate are affected by its wavelength dependence and require to be corrected for the chromatic zero angles (Fig. 2). This is explained in the user manual in more detail. From the FORS webpage, one can download a fits table with the tabulated values measured with a Glan Thomsen Prism, which was mounted at the M2 spider and aligned in a very tricky procedure.

For imaging polarimetry the Mag-limit of R=23.0 (given for linear/circular polarisation) is the one which allow a 1% accuracy in the determination of the degree of polarisation. This limit is calculated for a point source of zero

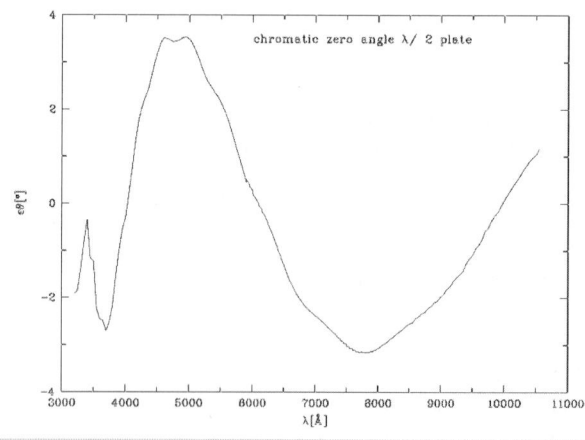

Fig. 2. Plot illustrating the required correction for the half-wave plate chromatic term

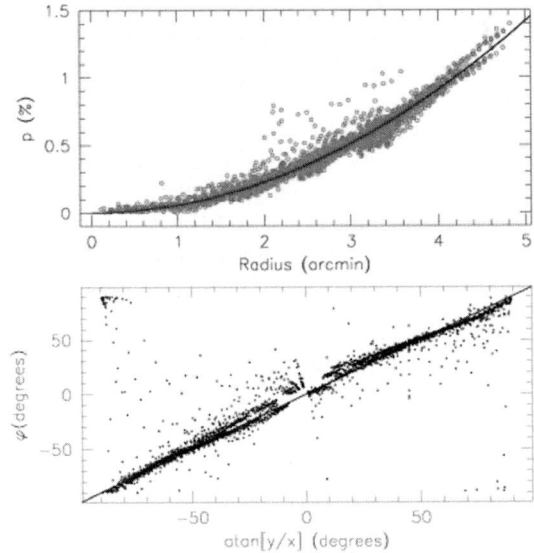

Fig. 3. Illustration of the radial dependency of the linear instrumental polarization

colour (A0V star) which would give a S/N of 5 in one hour with dark sky, clear conditions, a seeing FWHM of 0.8" and an airmass of 1.2.

A field-dependent instrumental polarization pattern was discovered in the FORS1 linear polarization mode. This spurious polarization field shows a high degree of axial symmetry, as illustrated in Fig. 3, and smoothly increases, in V band, from less than 3×10^{-4} on the optical axis to 7×10^{-3} at a distance of 3 arcmin from it. In case of the V band the spurious polarization can be modeled with a radial function:

$$p(r) = 0.057 \times r^2 \text{ in } [\%] \quad \& \quad \theta(x,y) = \arctan(\frac{(y-yc)}{(x-xc)}) \quad (3)$$

depending on the radial distance from the center of the field of view (xc,yc) and the position angle PA(x,y). The origin of the problem is still under investigation.

The spectropolarimetric Mag-limit, is the R-band magnitudes of a point source of zero colour which would give a S/N of 5 per pixel at 650 nm in the continuum in one hour with dark sky, clear conditions, a seeing FWHM of 0.8", an airmass of 1.2, and using a 1.0" slit and the SR-collimator. The limits, R=19.2 & 17.2, are respectively for the resolution of 260 & 1700, i.e. with grism 150I and 600R, respectively. The limits on spectropolarimetry are again those which for linear/circular polarisation allow a 1% accuracy in determination of degree of polarisation for one hour of total integration time.

Calibration Plan

The calibration plan foresees the observation of one polarized standard star every night polarimetric science observations are performed. For linear polarisation, the same mode and set-up is used for the science observations and the standard. The idea is to check that the polarimetric optics is correctly positioned and to improve the characterisation of the achromatism of the retarder plate. Observation of one unpolarised standard star is planned once a year to measure the instrumental polarisation.

It has been found however that the accuracy reached with the IPOL mode is not sufficient to measure the instrumental polarisation with high precision.

What is known to be missing in the calibration plan is:

– to monitor the instrumental polarization as derived from the calibration plan
– to measure with higher precision the circular polarization
   The instrumental circular polarization is around zero [4]. However results of a survey of magnetic stars in open cluster indicates there might be some cross-talk. Sterzik & Bagnulo have been granted technical time to investigate this on lunar polarization.
– a further characterisation of the instrumental polarization, which is needed as shown by Patat & Romaniello [12] in their analysis of errors related to dual beam optical liner polarimetric measurements.
   Fossati et al. [7] report on further analysis of the linear polarisation of FORS1 based on observations obtained within the calibration plan.

4 Conclusion

It is not by coincidence that this presentation has been divided between near-infrared and optical instruments. It does not only reflect the difference of observational techniques but also highlight the trench between these two types of instruments.

For optical instruments the polarimetric mode has been part of the instrumental design from the beginning. As a consequence, the instrumental polarisation is low and well characterised, altough not well enough for detection of low polarisation levels. Some follow-up is provided by the calibration plan but it is largely insufficient: not all polarimetric modes are well covered.

In the infrared, the polarimetric mode has been added more or less at the last minute of the instrument's design, so the instrumental polarisation can be high. It is moreover variable due to the installation of the instrument at a Nasmyth focus. Knowing that science targets typically show lower degree of polarisation at these wavelengths, it is easy to understand the difficulty of such observations. On top of that the number of suitable standard stars for telescopes with a diameter bigger than 4 m is small. All this explains the small number of publication output from these modes.

At ESO we are aware of our weakness in regards of the calibration plan for the polarimetric modes. We are working on improving this but the feedback of users is very valuable and expected to avoid overlooking some issues. Moreover it is important to understand if there is a true interest e.g. in near-infrared polarimetry and if so what is needed. For all instruments, the question is: "How can we help you performing good observations leading to publications?" My conclusion is: **ESO needs your feedback!**

In these proceedings, the case of polarimetry and science is further discussed in the paper by S. Katajainen. More information about calibration sources in the NIR can be found in Barrena et al. Complementary discussion about instrument and polarimetry is treated both by H.-M. Schmid (these proceedings) and S. Hubrig's contribution.

Acknowledgement. Many thanks go to S. Bagnulo for his active help in providing very useful detailed input as far as optical polarimetric instruments are concerned. F. Patat and M. Romaniello are warmly thanked for their constant productive feedback for the FORS1 polarimetric calibration plan. To finish I would like to express my gratitude towards A. Smette for his careful reading and commenting of this contribution.

References

1. N. Ageorges, A. Moorwood: ASP Conf. Series **343**, 15 (2005)
2. N. Ageorges, J.R. Walsh: A&A **357**, 801 (2000)
   I. Belskaya, M.A. Barucci: A&A **450**, 1239 (2006)
3. S. Bagnulo, J.D. Landstreet, E. Mason, V. Andretta, J. Silaj, G. Wade: A&A **405**, 777 (2006)
4. S. Bagnulo, T. Szeifert, G.A. Wade, J.D. Landstreet, G. Mathys: A&A **389**, 191 (2002)
5. A. Burkert, B. Stecklum, Th. Henning, O. Fischer: A&A **353**, 153 (2000)
6. R. Falomo, J. Melnick, E.G. Tanzi: A&A **255**, L17 (1992)
7. L. Fossati, S. Bagnulo, E. Mason, L. Degl'Innocenti: ASP Conf. Series **364**, 503 (2007)
8. J.H. Hough: ASP Conf. Series **343**, 3 (2005)
9. S. Hubrig, P. North, M. Schoeller, G. Mathys: AN **327**, 289 (2006)
10. D. Kasen, P. Nugent, L. Wang, D.A. Howell, J.C. Wheeler, P. Hoeflich, D. Baade, E. Baron, P.H. Hauschildt: ApJ **593**, 788 (2003)
11. E. Landi Degl'Innocenti, S. Bagnulo, L. Fossati: ASP Conf. Series **364**, 495 (2007)
12. F. Patat, M. Romaniello: PASP **118**, 146 (2006)
13. C. Pernechele, S. Berta, A. Marconi, C. Bonoli, A. Bressan, A. Franceschini, J. Fritz, E. Giro: MNRAS **338**, L13 (2003)
14. P. Rousselot, A C. Levasseur-Regourd, K. Muinonen, J.-M. Petit: Earth, Moon, and Planets **97**, 353 (1999)

Polarimetry with ESO Instruments

H. M. Schmid

Institute of Astronomy, ETH Zurich, CH-8092 Zurich, Switzerland;
schmid@astro.phys.ethz.ch

Abstract. A small literature survey is presented which demonstrates that FORS1 and EFOSC2 are scientifically very successful polarimetric instruments, while the near-IR instruments SOFI, ISAAC and NACO are rarely used in polarimetric mode.

First the measuring principle and the calibrations for FORS1 and EFOSC2 are discussed. These Cassegrain instruments are well suited for polarimetry and a high polarimetric precision is routinely achieved. Contrary to this, the near-IR instruments SOFI, ISAAC and NACO are not ideal for polarimetry because they were not designed as high precision polarimeters. It is outlined how the performance of the polarimetric modes of the near-IR instruments could be improved with a dedicated calibration program in combination with models for the instrument polarization.

Finally, an outlook is given on some general calibration issues related to the polarimetric mode foreseen for SPHERE, the future VLT planet finder instrument. In SPHERE similar polarimetric issues have to be solved like for the polarimetric modes of the near-IR instruments SOFI, ISAAC and NACO.

1 Scientific Use of ESO Polarimetry

Different instruments at ESO telescopes offer polarimetric observing modes. Characteristic properties of these polarimetric modes are summarized in Table 1. All these instruments offer imaging polarimetry for linear polarization. The visual instruments FORS1 and EFOSC2 offer in addition modes for circular polarimetry and low resolution spectropolarimetry. Not offered at ESO telescopes is low resolution spectropolarimetry in the near-IR and high resolution spectropolarimetry in the visual. CRIRES will provide in the future high resolution spectropolarimetry (circular polarization) in the near-IR.

A small literature survey was made in order to get an idea about the scientific use of the different polarimetric modes. For this I searched the refereed publications for the years 2000–2006 in the astronomical data system (ADS) for keywords like "polarimetric", "polarization", or "magnetic" combined with instrument names like "FORS1", "ISAAC", etc., or with "VLT", "NTT", etc., and "ESO". In addition I scanned the list of refereed publications in the ESO annual reports for titles indicating polarimetric observations or authors who are known to publish often polarimetric papers. For all selected papers it was checked whether they include indeed new polarimetric

Table 1. Properties of polarimetric modes of ESO instruments

	FORS1	EFOSC2	ISAAC	NACO	SOFI
telescope	VLT	ESO 3.6 m	VLT	VLT	NTT
wavelength range	visual	visual	near-IR	near-IR	near-IR
focal station	Cassegrain	Cassegrain	Nasmyth	Nasmyth	Nasmyth
imaging polarimetry	yes	yes	yes	yes	yes
spectropolarimetry	yes	yes	–	–	–
linear polarization	yes	yes	yes	yes	yes
circular polarization	yes	yes	–	–	–
straight beam[a]	yes	yes	–	–	yes
Wollaston	yes	yes	yes	yes[b]	yes
retarder plates	$\lambda/2, \lambda/4$	$\lambda/2, \lambda/4$	–	$\lambda/2$	–

[a] No reflection from inclined surfaces in the instrument.
[b] A set of wire-grid polarizers can be inserted.

data gained with ESO telescopes. This publication survey has not the quality of a careful scientific study. However, I guess that more than 80% of the papers were included, which seems to be adequate for the present purpose.

In this survey I found 58 papers for the years 2000–2006; 4 in 2000, 4 in 2001, 7 in 2002, 9 in 2003, 14 in 2004, 8 in 2005, and 12 in 2006. Very interesting is the distribution of the papers with respect to scientific topic and instrument used, which is shown in Fig. 1.

1.1 Scientific Topic

Three science topics dominate in the polarimetric papers:

- Measurements of stellar magnetic fields with circular spectropolarimetry of the Zeeman effect (e.g. [3, 2]).

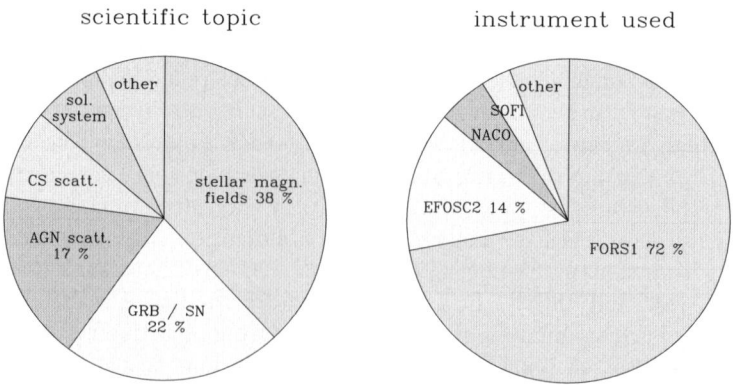

Fig. 1. Distribution of scientific papers based on ESO polarimetry with respect to scientific topic (*left*) and instrument used (*right*)

- Polarimetry of supernovae (SN) and gamma ray bursts (GRB); for SN mainly the asymmetry of explosions is explored with spectropolarimetric observations (e.g. [15]), while for GRB the physics of the relativistic shocks or the presence of interstellar polarization from the host galaxy are investigated (e.g. [9]).
- For active galactic nuclei (AGN) most polarimetric studies investigate the AGN geometry via the scattering polarization using spectropolarimetry, aperture polarimetry, or imaging polarimetry (e.g. [13, 14]).

Further science topics are the scattering polarization from circumstellar regions (CS) or the scattering polarization of solar system bodies.

1.2 Instrument Used

The VLT instrument FORS1 dominates strongly the output for ESO polarimetry (Fig. 1), followed by EFOSC2, which is an equivalent instrument, but at the ESO 3.6 m telescope. One can say that ESO polarimetry is essentially the polarimetric modes of FORS1 and EFOSC2.

It is important to understand for the future planning of ESO polarimetry why FORS1 and EFOSC2 are so successful, and why the polarimetric modes of the near-IR instruments yield much less scientific results.

The most important aspect is that FORS1 and EFOSC2 are ideal polarimetric instrument because they are attached to the Cassegrain focus and they have no inclined folding mirrors in the beam. The whole optical setup is rotationally symmetric and all instrumental polarization effects cancel essentially to zero, at least near the optical axis. In Sect. 2 the polarimetric modes of FORS1 and EFOSC2 are discussed in more detail.

The near-IR instruments SOFI, ISAAC, and NACO are Nasmyth instruments and already the telescope mirror M3 introduces substantial instrument polarization. Clearly, these instruments were not designed for high precision polarimetry and the polarimetric modes were just included as add-on. For this reason there are features which are not ideal for polarimetry, like inclined mirrors in the instrument (ISAAC, NACO), or the lack of retarder plates (ISAAC, SOFI). Therefore, SOFI, ISAAC, and NACO are difficult instruments for polarimetric measurements. Up to now they provided published results for a few special targets, for example highly polarized sources where a low polarimetric precision is acceptable [1, 5]. However, it is well possible that the polarimetric performance of these instruments can be substantially enhanced with a dedicated calibration effort as discussed in Sect. 3.

2 Polarimetry with FORS1 and EFOSC2

FORS1 and EFOSC2 are multi-mode instruments with exchangeable focal plane apertures and slits and a collimated beam section where different analyzing components can be inserted for filter imaging and grism spectroscopy.

In addition, also rotatable retarder plates and a Wollaston prism can be inserted for imaging polarimetry and spectropolarimetry in combination with special focal plane apertures.

2.1 Basic Measuring Principle

In polarimetry one measures the Stokes parameters Q, U, and V, which are the differential intensity signals between two opposite linear polarization modes, e.g. $I_0 - I_{90} = Q$ and $I_{45} - I_{135} = U$ (where $0, 90, ...$ stand for the polarization direction on the sky), or for the two circular polarization modes $I_l - I_r = V$ (l, r stand for left and right). For a high polarimetric precision it is important that the two opposite modes are measured simultaneously in order to avoid errors due to atmospheric seeing and transmission variations. The basic principle of the polarization analysis in FORS1 and EFOSC2 is illustrated in Fig. 2. The first polarimetric component is a rotatable retarder, either a half wave plate (HWP or $\lambda/2$), which rotates the linear polarization into a particular orientation or a quarter wave plate (QWP or $\lambda/4$) which converts circular polarization into measurable linear polarization. The Wollaston splits then the $I_\perp$ and $I_\parallel$ polarization direction as defined by the orientation of the Wollaston into two images separated on the CCD by about $10''$ or $20''$. Thanks to the special aperture masks the signals from the two Wollaston beams do not overlap. Rotating the retarder between the two exposures, e.g. from $0°$ to $45°$ for Stokes Q, allows to swap the two polarization images, so that differential effects in the two Wollaston beams cancel out in the polarization signal (including the individual pixel efficiencies of the CCD). Thus two images taken with different retarder plate orientations yield one normalized Stokes parameter, for example for Q/I according to:

$$Q/I = \frac{R-1}{R+1} \quad \text{where} \quad R^2 = \frac{(I_\perp/I_\parallel)_{0°}}{(I_\perp/I_\parallel)_{45°}} \tag{1}$$

and similar for U/I (HWP $22.5°$ and $67.5°$), or V/I (QWP $-45°$ and $+45°$). In some ESO manuals one can also find the formula:

$$Q/I = 0.5 \left(\left(\frac{I_\perp - I_\parallel}{I_\perp + I_\parallel}\right)_{0°} - \left(\frac{I_\perp - I_\parallel}{I_\perp + I_\parallel}\right)_{45°} \right)$$

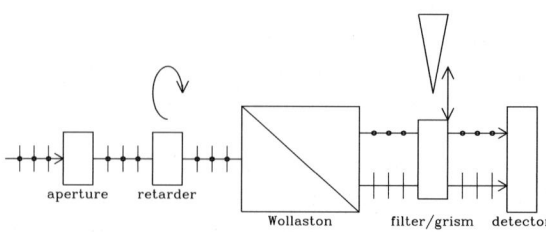

Fig. 2. Schematic concept for the FORS1 and EFOSC2 polarimetry

This formula is not identical to (1). It represents an approximation which provides good results for many but not all (!) cases. In particular it is important that the $I_\perp$ and $I_\parallel$ fluxes are calibrated with respect to each other.

Data Reduction Issues

For unresolved sources one can measure the signal $I_\perp$ and $I_\parallel$ within adequate apertures and just apply the formula of (1) to the net counts in the apertures.

For spatially resolved data (extended sources) the reduction is applied to 2-dimensional pixel arrays. It is of importance to achieve a good alignment of the signal in the $I_\perp$ and $I_\parallel$ beams. This can be difficult, because the Wollaston introduces some differential chromatic aberrations, so that the two images are not exactly identical.

For spectropolarimetry (1) can be applied to the one-dimensional (extracted) $I_\perp$ and $I_\parallel$ spectra. Important is that the wavelength calibration is accurate (done separately for the two beams). For Zeeman spectropolarimetry of spectral lines a bad wavelength calibration results in a spurious detection of a magnetic field. Particular care is required for this type of data analysis.

Calibrations for Linear Polarization Data

Required calibrations are the determination of the amount of instrument polarization $p_{\rm inst}$ with observations of zero-polarization standard stars, and the measurement of the zero point of the polarization angle $\theta_{\rm inst}$ with observations of polarized standard stars. Polarized standard stars yield also the polarization efficiency of the instrument, which should be close to 1. For FORS1 and EFOSC2 the instrument polarization is $p_{\rm inst} \cong 0.1\%$ near the optical axis. It has been noticed that there exists a centro-symmetric instrument polarization for off-axis positions. For FORS1 $p_{\rm inst}$ increases steadily from zero to about $p_{\rm inst} \approx 1.5\%$ at the edge of the field of view at $d = 5'$ [11]. For FORS1 and EFOSC2 the zero point angle $\theta_{\rm inst}$ varies slightly with wavelength due to the color dependent orientation of the optical axis of the HWP (see FORS web page). The polarization of all science data must be corrected for these instrumental effects. The correction is best carried out in the Q/I and U/I plane where the instrument polarization is equivalent to a shift of the data points and the position angle zero point calibration is equivalent to a rotation. With standard star observations it is easily possible to calibrate FORS1 and EFOSC2 data (taken on axis) to a precision of $\Delta p = \pm 0.2\%$ and $\Delta\theta = \pm 1°$–$2°$. A much better calibration accuracy is difficult to achieve mainly due to the lack of well measured polarization standard stars which are not too bright for the VLT. Fortunately, efforts are undertaken to improve the quality of the standard stars (e.g. [10, 6]). For extended objects or objects measured off-axis ($> 30''$) the field dependent instrument polarization has to be taken carefully into account.

Calibrations for Circular Polarization Data

For circular polarization one should check with a zero polarization standard star the instrument polarization and with a linearly polarized standard star the cross talk effect, the conversion from $Q, U \rightarrow V$. Circularly polarized standard stars (which deserve the name standard star) are essentially not existing. For Zeeman spectropolarimetry, it is recommended to measure always a comparison star with known magnetic field orientation in order to be sure about the sign $+/-$ of the measured V-signal.

2.2 Calibration Needs for FORS1 and EFOSC2

For FORS1 it would be useful to investigate the standard star instrument calibrations from archive data. This would clarify how many standard star observations are required to verify the instrument polarization for a particular run. I guess that quite some VLT time is spent unnecessarily on standard star observations which are repeated for each polarimetric run.

For EFOSC2, where the polarimetric components are specially introduced for a polarimetric run, at least two zero polarization standard stars and two polarized standard stars should be observed, and checked at the telescope to ensure that everything is working as expected.

Faint Targets

The precision of polarimetric observations of faint targets, e.g. AGN, SN, GRB, is dictated by the photon statistics. Thus the predominant requirement for better polarimetric data is simply: *more photons*.

Bright Targets

The huge light collecting power of the VLT allows to achieve a much higher polarimetric precision for bright targets than previous instruments. One main problem is, as discussed above, the limited precision of standard star measurements. Thus it is important to have a list of good standard stars. The overhead, mainly the data read-out, is quite long for bright star polarimetry. There are in principle enough photons with FORS1 and the VLT to reach a polarimetric precision of $\Delta p = 0.01\%$ within 1 h of integration time for imaging polarimetry of a 13^m-object or low resolution spectroscopy of a 10^m-object (see [12]). But this requires that on the order 10–100 frames are taken and the overhead due to the frame read-out may become a real issue. Thus, it would be useful if measuring strategies are investigated and tested which reduce the overhead (super-fast readout mode, extreme windowing, defocusing the image in order to collect more photons per frame, etc.).

Extended Targets

The field dependent instrument polarization is a real concern for the high precision measurements. For this reason, a pioneering program should be undertaken which investigates the achievable calibration accuracy of imaging polarimetry for a wide field.

3 Polarimetry with SOFI, ISAAC and NACO

The polarimetric modes of the near-IR instruments SOFI, ISAAC and NACO are not often used (see Fig. 1). The design of these instruments was not driven by polarimetry and therefore there exist several draw-backs for the polarimetry with these instruments:

- The location at the Nasmyth focus is not ideal because the inclined telescope mirror M3 introduces substantial instrument polarization which rotates with respect to the field.
- SOFI and ISAAC are not equipped with half wave plates for the selection of the polarization direction, so that the whole instrument must be rotated to select the Stokes Q or U direction.
- ISAAC and NACO have inclined mirrors in the beam which introduce polarization effects which have not yet been determined accurately.

A further important problem, which is not instrumental, is the severe lack of well measured polarized standard stars for the near-IR.

A dedicated polarimetric calibration program and the modeling of the polarimetric properties of the instruments is required to overcome these problems. Such an initiative would significantly improve the polarimetric performance of the near-IR instruments.

3.1 Modeling the Telescope and Instrument Polarization

An accurate polarimetric calibration of the near-IR instrument requires a polarimetric modeling of the instrument with Stokes vectors and Mueller matrices. The Stokes vector describes the polarization of the light:

$$\mathbf{I} = \begin{pmatrix} I \\ Q \\ U \\ V \end{pmatrix} = \begin{pmatrix} I \\ I_0 - I_{90} \\ I_{45} - I_{135} \\ I_l - I_r \end{pmatrix} = \begin{pmatrix} \text{intensity} \\ \text{linear polarization } 0°/90° \text{ direction} \\ \text{linear polarization } 45°/135° \text{ direction} \\ \text{circular polarization} \end{pmatrix}$$

Mueller matrices $\mathbf{M}$ describe the effect of optical components on light described as Stokes vector $\mathbf{I}$. The resulting Stokes vector is then $\mathbf{I'} = \mathbf{M} \cdot \mathbf{I}$.

The polarization effects of the instrument can be characterized with different elements of the matrix $\mathbf{M}$:

$$\mathbf{M} = \begin{pmatrix} m_{11} & m_{12} & m_{13} & m_{14} \\ m_{21} & m_{22} & m_{23} & m_{24} \\ m_{31} & m_{32} & m_{33} & m_{34} \\ m_{41} & m_{42} & m_{43} & m_{44} \end{pmatrix} = \begin{pmatrix} I \to I & Q \to I & U \to I & V \to I \\ I \to Q & Q \to Q & U \to Q & V \to Q \\ I \to U & Q \to U & U \to U & V \to U \\ I \to V & Q \to V & U \to V & V \to V \end{pmatrix}.$$

For example the elements $I \to Q, U, V$ describe the instrument polarization introduced by the telescope and instrument, and $Q \leftrightarrow U$, $Q \leftrightarrow V$, $U \leftrightarrow V$ are the polarization cross talks. The Mueller matrix for a perfect instrument would be just the unit matrix $\mathbf{M}_{i,t} \cong c_{i,t} \cdot \mathbf{1}$ multiplied with the transmission loss $c_{i,t}$. The Cassegrain instruments FORS1 and EFOSC1 are close to this ideal case.

Reflections from inclined surfaces are often the cause for instrument polarization. The effect depends on the component's surface coating and on incidence angle and wavelength. For an Al-coated mirror with an incidence angle of 45° (Nasmyth mirror) the Mueller matrix is for the I-band (800 nm) and H-band (1.6 µm):

$$\mathbf{M}_{t(\mathrm{I})} = \begin{pmatrix} 0.95 & 0.05 & 0 & 0 \\ 0.05 & 0.95 & 0 & 0 \\ 0 & 0 & -0.94 & -0.14 \\ 0 & 0 & 0.14 & -0.94 \end{pmatrix}, \quad \mathbf{M}_{t(\mathrm{H})} = \begin{pmatrix} 0.99 & 0.01 & 0 & 0 \\ 0.01 & 0.99 & 0 & 0 \\ 0 & 0 & -0.99 & -0.08 \\ 0 & 0 & 0.08 & -0.99 \end{pmatrix}.$$

For Nasmyth instruments we have to consider the polarization introduced by the inclined mirror M3 and the field rotation as described by Giro et al. [7]. Thus rotation matrices $\mathbf{R}(\theta)$ have to be introduced in the Mueller calculus:

$$\mathbf{R}(\theta) = \begin{pmatrix} 1 & 0 & 0 & 0 \\ 0 & \cos 2\theta & \sin 2\theta & 0 \\ 0 & -\sin 2\theta & \cos 2\theta & 0 \\ 0 & 0 & 0 & 1 \end{pmatrix}.$$

The full Mueller matrix for the telescope and instrument polarization for a Nasmyth instrument is given by:

$$\mathbf{M} = \mathbf{M}_i \cdot \mathbf{R}_2^{Q,U}(p,z) \cdot \mathbf{R}_1(z) \cdot \mathbf{M}_t \cdot \mathbf{R}_0(p) \qquad (2)$$

The different matrices stand for:
- $\mathbf{R}_0(p)$: rotation for the parallactic angle p (telescope orientation),
- $\mathbf{M}_t$: Mueller matrix for telescope (essentially mirror M3),
- $\mathbf{R}_1(z)$: rotation for the zenith angle,
- $\mathbf{R}_2^{Q,U}(p,z)$: rotator position for Nasmyth instrument,
- $\mathbf{M}_i$: rotation matrix for Nasmyth instrument.

The calibration has to be achieved with measurements of polarization standard stars. The telescope and instrument matrices, $\mathbf{M}_t$ and $\mathbf{M}_i$, have to be determined, while the rotation matrices are defined by the telescope pointing direction. Because SOFI has no folding mirror in the beam we may assume as starting point $\mathbf{M}_i(\text{SOFI}) \cong \mathbf{1}$. ISAAC and NACO have due to the folding mirrors certainly some instrument polarization and it remains to be investigated how difficult it is to disentangle with standard star measurements the effects of the telescope and the instrument.

4 Polarimetric Calibration of ZIMPOL/SPHERE

SPHERE is the future VLT planet finder instrument for the search and investigation of extra-solar planets [4]. SPHERE is a Nasmyth instrument which contains a polarimetric focal plane instrument, called ZIMPOL (Zurich Imaging Polarimeter). A schematic concept of the instrument is shown in Fig. 3.

ZIMPOL is a high precision imaging polarimeter based on a fast modulation/demodulation technique using a fast switching (1 kHz) ferroelectric liquid crystal polarization modulator and a special, demodulating CCD detector [8]. ZIMPOL is located after strongly inclined mirrors of a derotator, an AO system, and a coronagraph. The components drawn with bold lines in Fig. 3, the VLT and the derotator, are most critical in terms of instrument

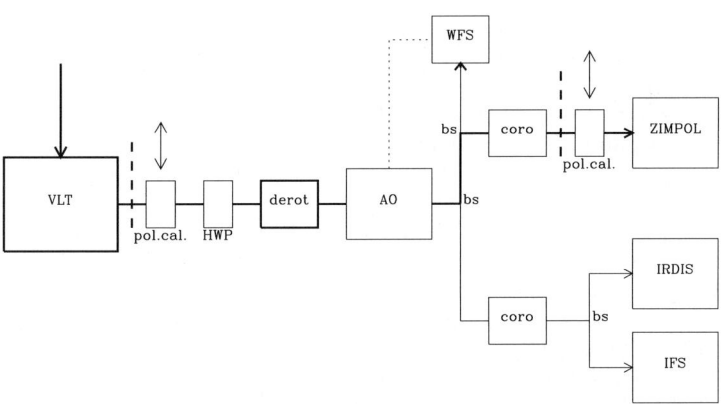

Fig. 3. Schematic concept of the SPHERE/ZIMPOL polarimeter. The rotatable half wave plate (HWP) is in this instrument close to the telescope while the high sensitivity polarimeter is located after derotator (derot), AO system, and visual coronagraph (coro). WFS stands for wave front sensor, IRDIS for infrared dual-beam imaging spectrometer, IFS for integral field spectrograph, and "bs" for beam splitter. Polarization calibration components (pol.cal.) can be inserted at two positions for the polarimetric calibration of the instrument

polarization. Thus, we face for ZIMPOL similar problems with instrumental polarization effects as described for the near-IR Nasmyth instruments.

For the polarimetric calibrations, the instrument is divided into three sections which are indicated in Fig. 3 with dashed lines: (a) the telescope, (b) the AO/coronagraph part, and (c) ZIMPOL. Each of these sections can be calibrated separately. The AO/coronagraph part and ZIMPOL have their own insertable calibration components which can be used with an artificial light source for daytime calibrations. The calibration of the telescope polarization has to be carried out with night time observations of polarization standard stars.

In order to minimize the calibration efforts a detailed polarization model of the instrument will be developed. Further it is foreseen to create a map of spurious polarization features from the instrument. All this will then be incorporated into the data reduction pipeline to allow a fast and reliable polarimetric data reduction.

The development of a polarimetric model and a calibration strategy for SPHERE/ZIMPOL may be adapted, at least partly, for the polarimetric modes of the near-IR Nasmyth instruments SOFI, ISAAC and NACO.

References

1. N. Ageorges, J.R. Walsh: A&A **357**, 255 (2000)
2. R. Aznar Cuadrado, S. Jordan, R. Napiwotzki, et al.: A&A **423**, 1081 (2004)
3. S. Bagnulo, T. Szeifert, G.A. Wade, et al.: A&A **389**, 191 (2002)
4. J.-L. Beuzit, M. Feldt, K. Dohlen, et al.: ESO Messenger **125**, 29 (2006)
5. A. Eckart, R. Schödel, L. Meyer, et al.: A&A **455**, 1 (2006)
6. L. Fossati, S. Bagnulo, E. Mason, E. Landi Degl'Innocenti: "The Future of Photometric, Spectophotometric and Polarimetric Standardization", ASP Conference Series **364**, p. 503 (2007)
7. E. Giro, C. Bonoli, F. Leone, et al.: in: "Polarimetry in Astronomy", SPIE Conference **4843**, 456 (2003)
8. D. Gisler, H.M. Schmid, C. Thalmann: in: "Ground-based Instrumentation for Astronomy", SPIE Conference **5492**, p. 463 (2004)
9. J. Greiner, S. Klose, K. Reinsch, et al.: Nature **426**, 157 (2003)
10. E. Mason, S. Bagnulo, T. Szeifert, et al.: "Astronomical Polarimetry", ASP Conference Series **343**, p. 85
11. F. Patat, M. Romaniello: PASP **118**, 146 (2006)
12. H.M. Schmid, I. Appenzeller, J.O. Stenflo, A. Kaufer: "Scientific Drivers for ESO Future VLT/VLTI Instrumentation", ESO Astrophybics Symposium, p. 231 (2002)
13. H.M. Schmid, I. Appenzeller, M. Camenzind, et al.: A&A **372**, 59 (2001)
14. D. Sluse, D. Hutsemekers, H. Lamy, et al.: A&A **433**, 757 (2005)
15. L. Wang, D. Baade, P. Höflich, et al.: ApJ **653**, 490 (2006)

Discussion

R. Barrena: Given the high dependence of the polarization on the position angle (PA), how do you correct the data of faint targets which need long exposure times? This correction is specially important for spectropolarimetry modes, where the targets usually need longer exposure time. If this correction is not possible, what is the limiting magnitude you can reach and what is the tolerance in the PA variation?

H.M. Schmid: For Cassegrain instruments, there is essentially no instrument polarization which depends on the pointing direction. Thus, the problem addressed by you concerns mainly Nasmyth instruments. There exists a complex relation between integration time, field/pupil rotator speeds and required polarimetric precision. Thus I cannot give numbers but just state that your point should certainly be considered for programs with long integrations.

F. Patat: Great talk! Just a comment on the polarimetry in Service Mode. This is already in place and, actually, it is how SNe and GRBs have been observed. Also, I do not reckon the current calibration plan is a limitation for attracting people to polarimetry. I rather think the problem is that polarimetry is not a well known technique, even though it is a really powerful tool, giving us geometric info that no other technique can provide us. Just give us a larger telescope!

Faint NIR Polarimetric Standards

R. Barrena, J. A. Acosta-Pulido, and A. Manchado

Instituto de Astrofísica de Canarias, Vía Láctea s/n, E-38205 La Laguna, Canary Islands, Spain; rbarrena@iac.es

Abstract. We report on the preparation of a catalog of faint polarized standards for near infrared bands. The catalog will only contain faint targets (below $J \simeq 14$) which should be a reference list for observations in 10 m class telescopes. We expect to have a list with more than 50 targets contained in several fields distributed in a wide right ascension range. The data are being collected using the polarization mode of LIRIS instrument (mounted at the Cassegrain focus of the WHT telescope). Here we show some examples of the use of LIRIS polarimetry and a first sample of the catalog.

1 The Polarimetry Mode of LIRIS

LIRIS is a near-infrared (0.9–2.4 µm) intermediate resolution spectrograph [1, 2], conceived as a common user instrument for the WHT at the Observatorio del Roque de los Muchachos (ORM, La Palma). The detector is a Hawaii 1024×1024 pixel array that covers a $4' \times 4'$ field of view. LIRIS also offers the polarimetric observing mode thanks to a Wedged double Wollaston device (WedoWo; [4]).

Polarization measurements can be performed with any of the available filters. Four polarized beams ($\theta=0$, 45, 90 and 135 degrees) are registered simultaneously thanks to the WeDoWo prism, that produce a 2D field of $4' \times 1'$ FOV (see Fig. 1). The normalized Stokes parameters (q, u) are computed from differences in orthogonal polarization planes:

$$q = (I_0 - I_{90})/I_1 \qquad I_1 = (I_0 + I_{90})/2$$
$$u = (I_{45} - I_{135})/I_2 \qquad I_2 = (I_{45} + I_{135})/2$$

The degree of linear polarization is obtained as $P = \sqrt{q^2 + u^2}$. The position angle of polarization vector as $\theta = atan(u/q)/2$.

1.1 Instrumental polarization

We used the unpolarized standard WD1344+106 (J=14.407, H=14.139, Ks=14.235) of ISAAC polarimetric standards and UKIRT polarized and unpolarized standards catalogs [6]. From these measurements we derive the instrumental polarization of LIRIS during the observations (see Table 1).

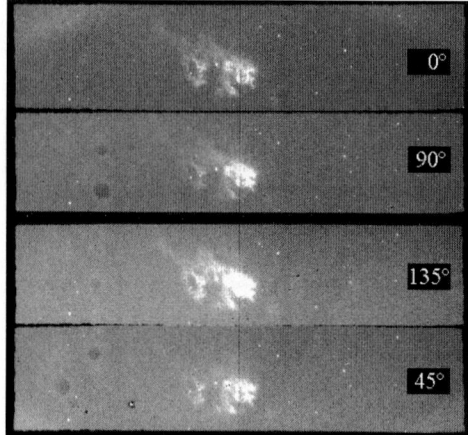

Fig. 1. Raw polarization frame of Cep-A nebula

Table 1. Instrumental polarization

	(J)	(H)	(Ks)
Q(%)	0.321	−0.098	0.640
U(%)	−0.350	−0.436	−0.548
P(%)	0.419	0.446	0.843

The errors of this determination are around 0.2% for J and H bands, and $\sim 0.35\%$ for Ks.

2 The Case for Faint Polarimetric Standards in the NIR?

Polarimetry is nowadays applied to a wide field of branches in Astronomy, from cometary physics or planetary envelopes to accretion discs or AGNs. In particular, NIR polarimetry is specially interesting on the interstellar medium studies. In fact, an appropriate analysis of the polarization state of the light crossing interstellar clouds can give information on the composition, size, forms and possible alignments of the dust in the interstellar medium [3].

The dependence of the linear polarization P with λ in the interstellar dust is well known. This dependence follows the Serkowsky law [5, 7] $P/P_{max} = \exp-(0.01 + 1.66\lambda_{max})\ln\lambda_{max}/\lambda$; where P_{max} depends on the column density, and λ_{max} is the wavelength where P_{max} occurs.

Two types of polarimetric standards are commonly used to calibrate polarimeters: unpolarized stars and stars with well established non zero polarization. However, there are two important problems with the current catalogs of polarimetric standards: (1) There are few catalogs which contain a poor number of standards, and most of them can only be used from the southern hemisphere; (2) Only a small subset of the sample is apropriate to be observed in 10 m class telescopes. In fact, only very few catalogued standards show magnitudes $J > 12$.

Therefore, our main goal is to construct a list of polarimetric standards, including unpolarized targets, in J, H and K_s infrared bands, appropriate to be observed in large aperture telescopes located in the northern hemisphere. This means to catalog targets with $J, H, K_s > 14$.

Since the origin of polarization is dichroic absorption by the intervening interstellar dust, the best zones to get standards with appreciable polarization are likely the edges of molecular clouds at low Galactic latitudes where the extinction is high. These fields contain stars behind (polarized sources) and in front of the interstellar cloud (not extinguished and so unpolarized stars). The Barnard clouds meet these requirements. We have chosen Barnard 64, 68, 163, 346 and 352 clouds as targets for our observations in July 2006.

3 The First Standards

We have identified six polarized standards in these fields. The degree of polarization P in J band is about 2–3% and decreases until 0.5–1% for K_s band following the Serkowski law (see Fig. 2). Table 2 summarizes the magnitudes, linear polarization P and polarizarion angle θ in each band.

In addition, we have found some unpolarized standards. In fact, the field Barnard 352 only contains stars with zero-polarization at our sensitivity limit. Table 3 compiles some of these targets.

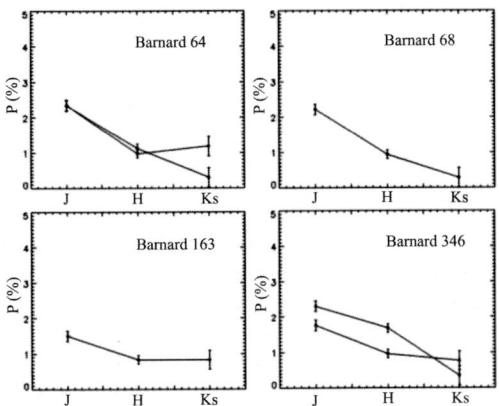

Fig. 2. Linear polarization P of some targets in the observed Barnard fields

Table 2. Preliminary list of polarization standard stars

ID	RA & Dec (J2000)	J	J-H	J-Ks	P (%) (J)	(H)	(Ks)	θ (deg) (J)	(H)	(Ks)
B64$_1$	17:17:14.5 −18:29:28	15.94	1.00	1.00	2.35	0.97	1.18	6.4	3.5	−10.8
B64$_2$	17:17:13.6 −18:29:27	16.56	0.66	0.81	2.33	1.13	0.30	−12.2	−13.5	0.6
B68$_1$	17:22:37.2 −23:51:24	16.52	0.78	0.96	2.22	0.94	0.29	12.1	18.3	21.6
B163$_1$	21:35:57.9 +57:26:08	14.76	0.43	0.47	1.50	0.83	0.83	21.8	9.8	11.4
B346$_1$	20:27:04.5 +43:39:01	15.87	1.01	1.34	2.30	1.68	0.35	43.3	42.2	32.2
B346$_2$	20:27:03.7 +43:38:49	13.64	0.90	1.20	1.76	0.96	0.77	37.5	28.3	1.7

Table 3. Zero-polarization standards. In all cases $P(J)$ is below 0.2%

ID	RA & Dec (J2000)	J	J − H	J − Ks
B352$_1$	20:57:06.2 +45:49:58	14.76	0.47	0.73
B352$_2$	20:57:06.5 +45:49:51	16.50	0.35	0.35
B352$_3$	20:57:07.2 +45:50:13	17.20	0.25	0.27
B352$_4$	20:57:04.8 +45:50:18	16.52	0.32	0.34
B352$_5$	20:57:06.7 +45:50:19	15.46	0.37	0.40

4 For the Future ...

Our aim is to continue the observations of Barnard fields to make a larger list of standards, in particular for winter time, using LIRIS at the WHT. We also plan to follow a parallel strategy in order to confirm the above presented standards: (1) to reject targets with possible variabilities we will measure the polarization of above presented standards once again; (2) We will also perform optical polarimetry of these targets. Optical observations will be carried out using the TURPOL polarimeter in the NOT next August 2007. If the polarization follows the Serkowski law these candidates will be confirmed as actual polarization standards.

References

1. A. Manchado, M. Barreto, J.A. Acosta-Pulido, et al.: SPIE, **4008**, 1162 (2000)
2. A. Manchado, J. Acosta-Pulido, E. Cadavid, et al.: SPIE, **5492**, 1510 (2004)
3. P. Martin: IAU Symp. **135**, 375 (1989)
4. E. Oliva: A&AS **123**, 589 (1997)
5. K. Serkowski, D.L. Mathewson, V.L. Ford: ApJ **196**, 261 (1975)
6. UKIRT polarized and unpolarized standards at http://www.jach.hawaii.edu/UKIRT/instruments/irpol/irpol_stds.html
7. D.C.B. Whittet, P.G. Martin, J.H. Hough, et al.: ApJ **386**, 562 (1992)

Developmental Aspects of a Multi-Slit Spectro-Polarimeter

K. George[1], K. Sankarasubramanian[1], R. Bayanna[2], and H. Lin[3], and P. Venkatakrishnan[2]

[1] ISRO Satellite Centre, Bangalore, India; koshy@isac.gov.in
[2] Udaipur Solar Observatory, Udaipur, India
[3] University of Hawaii, HI, USA

Abstract. We report the development aspects of an integral field unit, multi-slit spectro-polarimeter (MSSP) optimized for optical to near infrared regime, which can be used to derive simultaneous spectral and vector magnetic field information at high spatial, spectral and temporal resolution of any extended astronomical object like the Sun, with limited spectral coverage of few Angstrom. The instrument will be first developed and tested in laboratory which in a later stage will be used as a focal plane instrument for the Multi Application Solar Telescope (MAST). The major technological challenges involved in setting up and calibration of the instrument are discussed. The scientific motivation for the system is highlighted, with special emphasis on science limitations imposed by similar existing instruments elsewhere.

1 Introduction

Multi-slit spectrographs had been a reality in astronomical community for quite some time, but a multi-slit integral field unit based spectro-polarimeter where one can achieve high polarimetric accuracy for any given source at high spectral and spatial resolution is a relatively new concept [1, 2]. The concept involves the accurate measurement of polarization signals across spectral line profiles at high spatial, spectral and temporal resolutions of any extended astronomical object like the sun optimized for a wavelength range of 4000 Å to 1.6 μm. The major science drivers for MSSP are in understanding the dynamics of small scale magnetic fields, estimation of turbulent magnetic fields, study of evolution of penumbral filaments and its connection to the evershed flows, observe and understand the small-scale magnetic elements in sunspots and small scale magneto-hydrodynamic waves on the solar surface.

The simultaneous measurement of vector magnetic field information at high spatial, spectral and temporal resolution is a major limitation imposed by existing instruments. Conventional filter based systems compromise for high spectral resolution resulting in inaccurate measurements of physical parameters like vector magnetic field where as tunable filter based systems (like Fabry-Perot etalon) compromise on the temporal resolution so as to tune through the temporal line profile and also posing a serious problem of velocity cross talks in magnetic field estimates [3, 4]. Spectrograph based systems need to compromise on temporal resolution for covering the field-of-view (FOV) of interest but accurately estimate the physical parameters [5]. Solar spectro-polarimetric observations in contrast to the stellar analogue concentrate on a single line or on a multiplet separated by few angstroms, and measure stokes profiles at high spatial, spectral, and temporal resolution along with a good polarimetric accuracy. This chapter deals with the developmental aspects of one such instrument, the concept of which can be used to study any extended astronomical object at high spectral, spatial, and temporal resolution with good polarimetric accuracy but with limited spectral coverage.

2 New Technologies

The existing astronomical instrumentation had its own limitations which is a major set back for the above defined science goals. This calls for new innovations taking into account the recent technological advancements. The major limitation of existing instruments are in attaining simultaneous spatial and temporal coverage with good spectral resolution. In most of the solar spectro-polarimetric observations a wavelength coverage of few Å is good enough. The excess CCD pixels along the wavelength axis can then be used to accommodate more slit positions and hence a multi-slit spectrograph will be a better option than a single slit spectrograph particularly when the spectrograph is used in scanning mode to cover a large FOV. However, a conventional filter of few angstrom will not work since the signal-to-noise (SNR) will vary across the line profile which will then limit the accuracies achieved in the derived physical parameters. The bandwidth of the filter need to be at least five times larger than the required bandwidth to attain a uniform SNR across the line profile. Dense wave division multiplexing (DWDM) filters used in telecommunication industry is a good alternative for conventional filters. DWDM filters have narrow band and also rectangular profile (Fig. 1) [2]. This will enable to accommodate more slit positions within a single CCD chip.

Even though the multi-slit concept reduces the amount of time required to scan a FOV of interest, it still requires scanning to cover the field in between the slits. Also, it imposes the observations of larger FOV irrespective of the user requirements. An integral field unit will avoid the above difficulty. The integral field unit (IFU), consisting of an array of optical fibers aligned inside a fiber bundle, helps in scrambling the 2D FOV of interest into several 1D slits [2]. The single slit of conventional spectro-polarimeters can be replaced by a multi-slit configuration there by multiplexing the temporal and spatial resolution, which when coupled to the IFU will further improve the performance in terms of spatial and temporal resolution without the requirement of scanning. The limitation in the FOV of this instrument comes from the available detector size, fiber-bundle, and the plate scale at the input to the fiber.

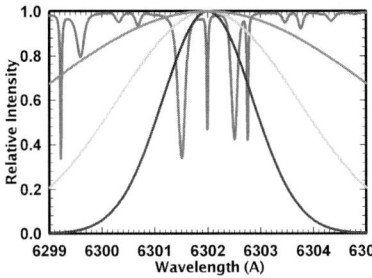 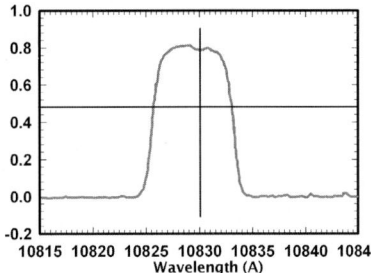

Fig. 1. Conventional filter profiles (two, three and four FWHM) overlaid over a solar spectrum centered at 6302 Å and the measured DWDM filter profile of an actual filter with central wavelength at 10830 Å

3 System Requirements

The system requirements are listed in Table 1. A study on the availability of fiber bundles and DWDM filters need to be undertaken before finalizing the design. Depending on the availability of the system components, a trade-off may need to be done to realise the system. The requirement suggests that the individual fibers should preserve polarization information and need to have small cladding and large fill fraction so as to have good image quality and spatial resolution. A fiber bundle similar to the one used for VisIRIS can be used, however practically achieving 12 μm diameter may be a limitation [2]. The most challenging part of the instrument development will be the calibration of IFU and its coupling with the multi-slit unit. Spectral mask will be used at the output of spectrograph in order to limit the wavelength coverage to one full line profile with good sampling. The orthogonal polarization states produced by the polarising beam splitter will be placed in between adjacent slit spectra for optimum use of the CCD real estate. The polarimeter package will consist of two ferro-electric liquid crystals (FeLCs), a half-wave retarder with a switching angle of 45°, another one with a retardance of 102.2° and a switching angle of 32.2° allowing for a balanced modulation where all stokes

Table 1. System requirements

Parameter	Requirement	Parameter	Requirement
Spectral resolution	≤ 30 mÅ	Dispersion	15 mÅ/pixel
Wavelength coverage	2 Å–4 Å (6302 Å & 1.56 μm)	SNR	≥ 1000
Expected Cadence	10 s	F/#	10
Detector	2k×2k CCD	IFU	100×100 fibers
FOV	16"×16"	Spatial sampling	0.16"/pixel

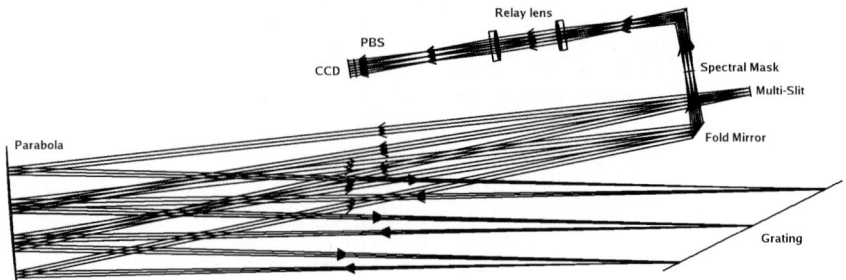

Fig. 2. Multi-slit Spectro-polarimeter optical design

profiles can be measured with equal weightage and hence minimising seeing induced polarization.

4 Optical Design & Current Status

The desired design of multi-slit consists of five identical slits (12 μm wide, 25 mm high) each separated by 2.4 mm. The collimator is an off-axis parabola of focal length 1277 mm with a 5° tilt. The dispersing element is a large format grating with 79 lines/mm, blazed at 63° for an operating wavelength of 6302 Å. The five spectral masks of width 1.2 mm are separated by 2.4 mm, and is having a one to one correspondence with the multi-slit arrangement. The relay system consists of two convex lenses of focal length 300 mm, which reimages the spectral mask onto a CCD through a polarization beam splitter (PBS), a modified savart plate with a beam separation of 1.2 mm. The DWDM filter will need a band width of 150/300 GHz at a peak wavelength of 6302 Å. The complete design of the spectrograph is optimised for a f/10 system, taking into account the focal ratio degradation due to fiber (Fig. 2).

The optical design of the instrument is presently undergoing rigorous iterations and more changes are expected in the course of development. The feasibility study on the DWDM and IFU unit is underway and depending on the availability the design may be modified. We expect to deliver the instrument as a focal plane user system for the MAST at Udaipur solar observatory optimistically in about three years time. The polarization calibration unit planned for MAST will be used to attain the required polarimetric accuracy.

References

1. N. Srivastava, S.K. Mathew: SoPh **185**, 61 (1999)
2. H. Lin, A. Versteegh: SPIE **6269**, 18 (2006)
3. B. Lites, et al.: SoPh **155**, 1 (1994)
4. S. Gosain, et al.: JAA **27**, 285 (2006)
5. K. Sankarsubramanian, et al.: ASP Conf. Proc. **358**, 201 (2006)

Current and Future Instruments Providing a Spectro-Polarimetric Mode

S. Hubrig[1] and I. Ilyin[2]

[1] ESO, Alonso de Cordova 3107, Vitacura, Santiago, Chile; shubrig@eso.org
[2] Astrophysikalisches Institut Potsdam, An der Sternwarte 16, 14482 Potsdam, Germany

Abstract. We present the potentials and advances of existing and planned facilities which have a spectropolarimetric mode.

1 Introduction

Magnetic fields are ubiquitous in the universe and play an important role in various branches of astrophysics. Virtually all measurements of magnetic fields make use of the Zeeman effect. Generally, the methods to measure the polarisation status of light are based on $\lambda/2$ and $\lambda/4$ retarders, which can be inserted along the optical path and rotated with reference to a beam displacer to measure the linear and the circular polarisation respectively. Alternatively, the polarisation is measured by the dual-waveplate method, which is based on a combination of two independently rotatable quarter-wave plates with a Wollaston polariser as a beam splitter inserted along the optical beam.

In spite of the challenging character of polarisation measurements (due to e.g., the generally very low level of intrinsic polarisation, presence of instrumental polarisation, crosstalk), an increasing number of instruments built for 8 to 10 m class telescopes (e.g., Gemini, Keck, Subaru, UT2) offer spectropolarimetric modes for observations at high or low spectral resolution. A number of smaller telescopes with 2–4 m diameters have been upgraded in the past or are awaiting an upgrade with high resolution spectrographs with spectropolarimetric capabilities (e.g., CFHT, WHT, TNG, TBL, AAT).

In the following, we present the most recent developments and advances of existing and planned facilities outside of ESO, which have a spectropolarimetric mode.

2 Existing Spectropolarimetric Facilities

SOFIN at the 2.6 m Nordic Optical Telescope

The Stokesmeter (Q, U, V, I) in SOFIN has been upgraded in 2006. A new cross-dispersion prism with higher refraction index for a larger separation of Echelle orders and a calcite beam splitter allow to record two polarised

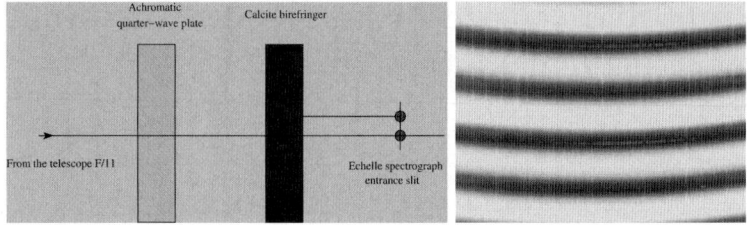

Fig. 1. *Left panel*: A sketch of the Stokesmeter used in SOFIN. *Right panel*: Echelle orders of left and right polarised light in γ Equ around 5000 Å

beams simultaneously (Fig. 1). Before 2006 it was only possible to make two subsequent exposures. The observations can be carried out at spectral resolutions of 30,000 and 80,000. In Fig. 1 we show an Echelle image of two double spectral orders (out of 25 in one image) of γ Equ around 5000 Å.

SARG at the 3.55 m Italian Telescopio Nationale Galileo

The spectrograph with a maximum resolution of $R=164,000$ works in the 3700–10,200 Å wavelength range and is equipped with a polarisation analyser. This analyser consists of a $\lambda/2$ and a $\lambda/4$ retarder rotating with respect to a beam displacer (Savart plate). The SARG is fed by means of the tertiary telescope mirror and an additional folding mirror, so that the instrumental polarisation has to be determined as a function of the target position in the sky. Usually it is removed by means of the Müller calculus. The accuracy of the measurement of polarisation depends on the accuracy of the alignment of the optical axes of the retarders with respect to the Savart plate. As the efficiency of the Echelle grating depends on the polarisation state of the incident light, a $\lambda/4$ retarder usually has to be added to convert the linear polarised beam into a circularly polarised beam.

ESPaDOnS at the 3.6 m Canada-France-Hawaii Telescope

The high-resolution spectrograph (R up to 70,000) is bench-mounted and fed by low-OH H-treated Ceram-Optec optical fibers from a Cassegrain module containing all calibration and polarimetric facilities, making it possible to have extremely good wavelength stability and minimal instrumental polarisation. The achromatic polarimeter includes one quarter-wave and two half-wave Fresnel rhombs coupled to a removable Wollaston prism. Unfortunately, the efficiency of the spectrograph is rather low in the spectral region around the Ca II H and K lines. These lines are frequently used for the analysis of circumstellar and photospheric magnetic fields in hot Be stars or in PMS stars. In Fig. 2 we show the low resolution FORS 1 spectra ($R = 4000$) of the Herbig star HD 190073. The Ca II H and K profiles in the spectra of this

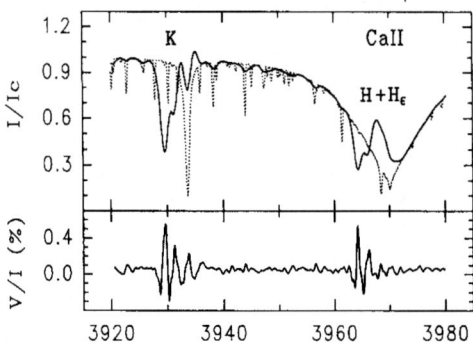

Fig. 2. Multi-component structure of Ca II H and K profiles in Stokes I and V spectra of HD 190073. The *dotted line* in the *upper panel* presents the synthetic photospheric spectrum

star exhibit a multi-component structure of polarisation features indicating the presence of a magnetic field in the magnetic circumstellar components of the Ca II H and K lines. Here we emphasise the importance of future magnetic field measurements using high resolution spectropolarimeters to be able to measure the magnetic field separately for lines of different elements in order to study both the magnetic field configuration in Herbig Ae/Be stars and the interaction of the circumstellar material with the magnetic field.

MUSICOS and NARVAL at the Telescope Bernard Lyot

The planned instrument NARVAL for the TBL is supposed to be an adapted copy of ESPaDOnS. It is expected that this spectrograph will be 10 to 20 times more efficient than MUSICOS, currently in use at TBL.

ISIS at the William Herschel Telescope

The setup of WHT/ISIS with the polarimetry module allows to measure both linear and circular polarisation at a resolution of up to 14,000.

Main Stellar Spectrograph at the 6 m telescope of the Special Astrophysical Observatory

The MSS is equipped with several circular polarisation analysers which are located in the Nasmyth focus of the telescope. The instrumental polarisation introduced by the oblique reflection on the mirror that sends the light to the Nasmyth focus of the 6 m telescope produces a 5% decrease in the measured longitudinal magnetic field.

3 Planned Spectropolarimetric Facilities on Large Telescopes

PEPSI at the Large Binocular Telescope

The Potsdam Echelle Polarimetric and Spectroscopic Instrument (PEPSI) is currently under construction at the Astrophysikalisches Institut Potsdam and will use the unique feature of the LBT (double mirror configuration) to provide high resolution full-Stokes four-vector spectra in the wavelength range 4500–11,000 Å. The accuracy of linear polarisation measurements depends on the alignment of a reference element of the polarimeter with the parallactic axis of the AGW (LBT wavefront sensing and guiding unit) platform. All other components of the polarimeter and its calibration unit are aligned with respect to this reference element. The accuracy of circular polarisation measurements depends on the uncertainty in the quarter-wave plate position and the retardation error, which is wavelength dependent. The other source of systematic errors is the relative transmission of the ordinary and extraordinary beams having the absolute transmissions T_1 and T_2. The estimate of the parameters (at least T_2) can be done measuring the residual intensity of linearly polarised light through the Wollaston prism.

PFIS at the Southern African Large Telescope

The University of Wisconsin - Madison's Department of Astronomy, together with Rutgers University and the South African Astronomical Observatory, is building a Prime Focus Imaging Spectrograph (PFIS) for the SALT telescope. The PFIS will specialize in very high throughput low- and medium-resolution ($R = 500$–$10,000$) spectroscopy, high time resolution spectroscopy, and spectropolarimetry from 3200 to 9000 Å.

4 Future Instruments at ESO

Currently, X-shooter does not foresee a spectropolarimetric option although both linear and circular polarisation could be supported by the insertion of a polarisation analyser into the beam.

It is not clear yet whether the ELT will host an instrument with spectropolarimetric capabilities. In any case, both high spectral resolution and high time resolution spectropolarimetric observations will certainly be of great scientific interest.

Circular Polarization Observations at ESO Using UT2 and FORS1

S. Katajainen[1], V. Piirola[1], H. Lehto[1,2], and A. Berdyugin[1]

[1] Tuorla Observatory, University of Turku Vaisalantie 20, FI-21500 Piikkio, Turku, Finland; sekataja@utu.fi
[2] Physics Department, University of Turku, Turku, Finland

Abstract. We present results from circular polarimetry made by using UT2 and FORS1. In our ESO Programs 076.D-0608A, 076.D-0608B and 077.D-0499A between Dec 12, 2005 and Apr 20, 2006, we have observed a sample of Intermediate Polars (IPs, magnetic cataclysmic variables) by using FORS1 in circular imaging polarimetry mode. The aim of studies were to detect possible polarization effects in Intermediate Polars (IPs) and to search how many of them are polarized and via polarisation data to calculate their magnetic field strengths and to explain accretion behaviour and their evolution.

The measured polarization found in our sample of soft X-ray Intermediate Polars has varied from nearly zero percent up to few percent, and thus high S/N is needed. As the polarization in these objects is due to cyclotron cooling effects near the surface of the white dwarf (WD), and the WD has typically a spin period between 10 and 20 min (in some systems the spin of the WD can be only few minutes) we needed high time resolution to cover rapid circular polarization variations well. These requirements all together emphasize needs of the large telescopes and their light collecting power, such as VLT.

1 Intermediate Polars

Intermediate Polars are a subclass of magnetic Cataclysmic Variables (CV). They are binary stars, where a low mass main sequence star (the secondary) transfers matter via Roche lobe overflow onto a compact white dwarf (WD) primary (see below). They form a subclass among CVs, and they typically have magnetic field strength B near the surface of the WD from few Mega-Gauss to 10–20 MegaGauss. WDs in IPs typically have spin periods of order of 1000 s, orbital periods are often between 3–5 h. Due to their high magnetic field strength B, accreted matter is channelled along the magnetic field lines towards the magnetic poles near the surface of the WD, emitting hard X-rays (often kT=10–30 keV) and soft X-rays as bremsstrahlung as well as blackbody radiation and cyclotron emission in the optical and infrared.

1.1 Polarisation in Intermediate Polars

Only five IPs have been detected to emit polarised light yet. All these IPs are so called "soft X-ray emitters" to some extent and thus they resemble polars (a strongly magnetic CVs) which have magnetic field strength up to hundreds of MG. There is overlap between the magnetic field strengths in these two subclasses. Important question here is: do the IPs evolve into the polars in their final stages of evolution?

It is not clear yet how the evolution of IPs and polars (strongly magnetic CVs) is connected. It is suggested [3] that the relatively high accretion rates in IPs may effectively bury the WD magnetic field, so making them appear less magnetic than they really are. IPs might have so much unpolarised light that polarised emission is not seen in most cases at all.

2 Circular Polarimetric Observations of IPs, Using FORS1

At the ESO the only instrument to obtain circular polarimetry has been (until the period P79) FORS1 (=Focal Reducer Optical Range Spectrograph). FORS1 is installed at the Kueyen (UT2) on Cerro Paranal, and it is a multi-mode instrument with polarisation optics comprising super-achromatic half- and quarter-wave phase retarder plates (for the linear and circular polarisation observations, respectively) and a Wollaston prism. FORS1 is suitable for observations for the wavelength range from 330 to 1100 nm and it has an image scale of 0".2/pixel (or 0".1/pixel with the high resolution collimator) with the 2048 × 2046 pixels CCD (pixel size of 24×24 µm). In its imaging-polarimetry mode (IPOL-mode) a strip mask is produced in the focal area of FORS1 to prevent overlapping of the two light beams (ordinary and extra-ordinary) of polarised light on the CCD chip. Displacement of the star images produced by the polarisation optics is 22", and thus only half of the full field of FORS1 is imaged on the chip. The small, reduced field size (3.4 arcmin) could be a problem in some cases. In our studies (polarimetry of mCVs) we have always only one polarised target in the CCD frame, however.

2.1 Program 076.D – 0608(A)

In our ESO program 0.76.D – 0608A we observed in Dec 11/12, 2005 and Mar 18/19, 2006 an Intermediate Polar UU Col (RX J0512.2-3141) by using FORS1 and circular polarimetry. Our hypothesis was that soft X-ray IPs, such like an Intermediate Polar UU Col, are all emitting circularly polarised light to some extent, as it is already observed with previously found 5 soft X-ray IPs. UU Col is a relatively faint target, even for polarimetry at the VLT, B=18.0 and V=17.6 [4, 2], and its WD spin period is short, only 863.5 s. As the possible polarisation level in some IPs might be as low as few tens of percents, and the detection of the polarisation with 0.3% errors requires S/N = 470, we applied for time on VLT and FORS1.

FORS1 was used in its ABCD 4-port (1×1, 1.4 e⁻/adu) mode, and a full CCD frame was read. This resulted in quite long reading times, but as the reading of a small sub-frame in FORS1 does not really improve the reading time (it would still take almost 40 s to read a small subframe) we have chosen reading a full frame. Bias and -flatfield calibrations were done by using several bias-images and sky-flats. Two different retarded angles were used: −45 deg, and +45 deg to achieve highest accuracy (see FORS1 manual) these were varied as a sequence of −45 deg, +45 deg, −45 deg, +45 deg, −45 deg and +45 deg...etc.

The amount of the circular polarisation V can then be measured (FORS1 manual, p.38 the equation 4.1.) eliminating effectively instrumental cross-talk effects (if the flat-fielding has been done properly) by using measurements of two different retarder angle positions and ordinary and extraordinary fluxes measured from those positions.

The star was always put in the same position in the CCD frame (almost in the center of the field) and this same position was kept for all targets were circular polarisation was measured, to avoid any calibration problems due to spurious instrumental polarisation found near the edges of the CCD in FORS1 (see [1]). The sign of the circular polarisation was accurately calibrated by observing well known circular polarisation standard LP790-29, and in addition to that, also polar VY For was observed as a calibration target (the circular polarisation of VY For is almost constant and not variable over orbital period) and a magnetic white dwarf HE 0330−0002. Zero polarisation level was checked by observing a zero (linear) polarisation target WD 0310−688.

3 Preliminary Results

Some calibration results from FORS1 circular polarimetry: High polarisation star LP790-29: 6.2 per cent (B) and 7.2 per cent (I), AM Her star (a polar) VY For: −5.3 per cent (I), Zero polarisation star WD 0310−688: 0.04 per cent (B) and 0.03 per cent (I), which are all in agreement with the previous observations. In Fig. 1 preliminary results from the target star UU Col are presented. Our results show that Intermediate Polar UU Columbae emits circularly polarized radiation, both negative polarization (about −1 per cent in both in B and I) and positive polarization (+1 per cent in B and I), and this is modulated according to WD spin period of 863.5 s. Errors in measurements are about 0.08 per cent both in B and I, exposure times of 230 s were used for B and I observations and the seeing was very poor and variable during our observations, between 1 and 2.3 arcsec (Dec 11/12, 2005).

These polarimetric variations of UU Col are in agreement with the X-ray observations [4] of UU Col, i.e. that there are two active accretion region active in UU Col, where positive pole correlates with the soft X-ray maximum.

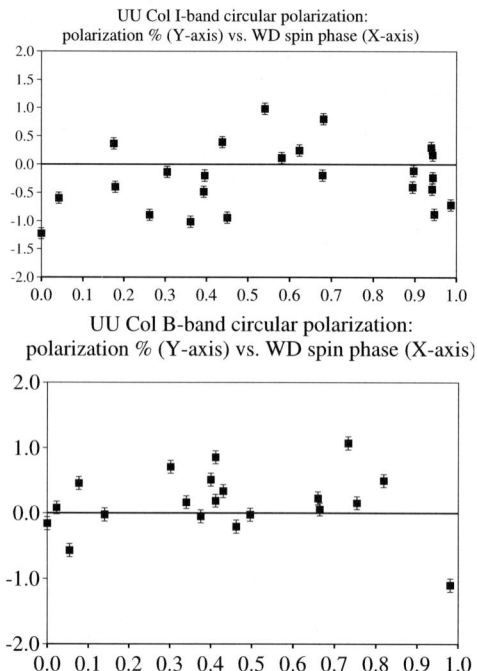

Fig. 1. Circular polarimetry for the Intermediate Polar UU Col, FORS1 (UT2). The WD spin period is only 863.5 s, and the brightness of UU Col is $B = 18.0$, and $V = 17.6$. The errors are about 0.08 per cent, and 230 s integration times were used

More results from FORS1 polarimetry for UU Col will be presented in a paper by Katajainen et al. [5].

References

1. S. Bagnulo, et al.: A&A **389**, 191 (2002)
2. V. Burwitz, et al.: A&A **310**, L25 (1996)
3. A. Cumming: ASPC **315**, 58 (2004)
4. D. De Martino, et al.: A&A **454**, 287 (2006)
5. S. Katajainen et al.: A&A **475**, 1011 (2007)

Part XII

Session 11: Wide Field Imagers

IOT Overview: Wide-Field Imaging

F. J. Selman

ESO, Alonso de Cordóva 3107, Vitacura, Santiago, Chile; fselman@eso.org

Abstract. The Wide Field Imager (WFI) instrument at La Silla has been the workhorse of wide-field imaging instruments at ESO for several years. In this contribution I will summarize the issues relating to its productivity for the community both in terms of the quality and quantity of data that has come out of it. Although only surveys of limited scope have been completed using WFI, it is ESO's stepping-stone to the new generation of survey telescopes.

1 The Science

The Wide-Field Imager at the 2.2 m MPG telescope at La Silla has been in operations since 1999 [3]. It has produced approximately 160 refereed papers during that time (up to Jan 2007). The topics researched range from the study of large scale structure and cosmology all the way down to the detection of satellites of Uranus and the search of extra-solar planets. Table 1 shows the number of papers per topic (with some duplicity as some papers address more than one topic). Up to January 2007 WFI had a healthy Hirsch-index [11] of 30.

Approximately 50% of the papers used WFI as an auxiliary instrument providing accessory information to the data obtained with one of the VLT instruments. In this regard WFI fills an important role as a support for research which is principally carried out with other instruments and in other bands. Nevertheless, WFI has been used as the main research tool for many programmes. High redshift galaxies, up to $z=5.65$, have been discovered using its unique set of medium band filters [27]; two new quadruple lensed quasars have been also discovered by doing a shallow wide-angle survey of high galactic latitude fields [18]. The large set of medium band filters had been used to determine photometric redshift and study the large scale structure of the universe with the COMBO17 survey [28]. The search of dark galaxies with weak lensing is another area where WFI has been extensively used [21]. The WFI has contributed to the study of the extragalactic distance scale through the Chilean Araucaria project observing Cepheid variables in several galaxies [8]. High risk projects have also been started with WFI and completed with VLT and HST: CSL-1 was a gravitational lens by a cosmic string candidate which was later shown to be a chance projection of two very similar elliptical

Table 1. WFI science 1999–2006

Topic	Number of Refereed Papers
Clusters and cosmology	48
Galaxies outside LG	22
Local Group	22
Galactic structure	4
Star formation	21
Stellar astrophysics	41
Extra-solar planets	1
Slitless spectroscopy	1
Mini-surveys	32
Total	192

galaxies [20]. Another important work is the search for intra-cluster planetary nebulae in Virgo [2]. Closer to home it has been used for studies of dwarf galaxies in the Local Group such as that of the Sagittarius dwarf spheroidal [17] and those of the putative Canis Major dwarf/overdensity [16, 4]. Even closer to home WFI has been used to study the rotational/variability properties of young stars in clusters of the Milky Way [10, 22]; the Initial Mass Function in 30 Doradus [23]; and even to discover satellites of Uranus [9]. WFI has also been part of the Planet collaboration searching for extrasolar planets using microlensing [13].

The WFI has also been used succesfully for a variety of surveys. It participated in the Great Observatories Origins Deep Survey with images of the CDF-S [6, 7]. It has done follow up work for the ISO survey (e.g. [19]). And it has been an active participant for ESO's Imaging Survey, EIS [15]. It has provided the wonderful images of 30 Doradus presented by P. Rosati as examples of ESO's Advanced Data Product project.

All these different observing programmes use and need different capabilities of WFI and are affected differently by its virtues and shortcomings. In this contribution I will briefly describe the instrument, and its operation. I will focus on what are the shortcomings of WFI specially regarding its photometric properties giving an explanation of the origin of that anathema of focal reducer instruments: light concentration. Finally I will briefly discuss how some of these issues might be absent from ESO's next generation survey instrument: OmegaCam.

2 WFI@2.2 m MPG Telescope at La Silla: An Overview

WFI is mounted at the Cassegrain focus of the 2.2 m MPG telescope at La Silla. The 2.2 m telescope is a classic Ritchey-Chretien design mounted in a fork design equatorial mount. The combined telescope instrument throughput approaches 80% at 500 nm, and it is useful between 350–1000 nm.

Table 2. WFI in a nutshell

Field of view	34'×33'
Pixel scale	0.238 arcsec/pixel
Detectors	4×2 mosaic of 2k×4k CCDs
Read-out time	27 s
Read-out noise	4.5 e$^-$/pixel
(Inverse) gain	2.0 e$^-$/ADU
Dynamic range	16 bit
Full-well capacity	>200,000 e$^-$
Telescope aperture	2.2 m
Intrinsic image quality	0.4 arcsec
Geometrical distortions	≤0.08%
RAW data format	FITS (with extensions) 142 MBytes/file

A detailed diagram of the instrument can be found at http://www.ls.eso.org/lasilla/sciops/2p2/E2p2M/WFI/docs/Widefield_Imager.jpg. Table 2 gives its characteristics in a nutshell. It has the capability to reach limiting magnitudes of 23.8, 26.0, 25.5, 25.3, and 24.3 in UBVRI respectively (assuming observations at seeing of 0.8" and airmass 1.2 of an A0V star, S/N = 5 in a 1 h integration).

Perhaps one of the most notable characteristics of WFI is its large number of filters which convert it into a redshift machine. Figure 1 illustrates all the passbands available for WFI together with a graph of the sky emission. Note how the infrared windows have been sampled by medium-band filters.

Table 3 gives a rough view of how the different ESO instruments compare between them and with other currently available and planned widefield instruments worldwide. Notice that the figure-of-merit, FOM, is just

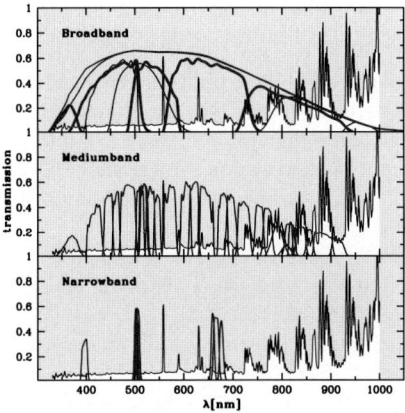

Fig. 1. The WFI filter set

Table 3. WFI in context

Instrument	Field-of-View	Pixel Scale	FOM (AΩ)
WFI	34'×33'	0.238	1.2
SUSI2	5.5'×5.5'	0.161	0.08
FORS2	6.8'×6.8'	0.250	0.67
VIMOS	4x7'×8'	0.205	3.3
SDSS	3.9 deg²	0.396	6
SUPRIME	34'×27'	0.202	13
Megaprime	1 deg²	0.187	10
OmegaCam	1 deg²	0.220	5.5
SkyMapper	5.7 deg²	0.500	7.5
PanSTARRS	7 deg²	0.300	60
LSST	9.6 deg²	0.200	300

the multiplication of telescope mirror area (m²) and field-of-view (square degrees), also called Etendue. An instrument with a FOM 10 times larger could in principle complete a survey in a factor 1/10 of the time for the smaller instrument. Notice that a 1 deg² camera at one of the VLT UT's prime focus would have a FOM of 50 making it competitive with PanSTARRS.

3 WFI Operations

WFI has been in fully flexible queue operations during the period 2002–2005. The observing programmes have been assigned priorities by ESO's OPC. Medium term queues are created for each of the priorities defined by the OPC, and each night Observation Block (OBs) are executed according to observing condition constraints provided by the Principal Investigators (PI). The observations are carried out by an astronomer together with an Telescope Instrument Operator (TIO). The following conditions are recognized:

- Transparency. This can be PHO for photometric conditions, CLR for clear skies; THN for thin cirrus covering, and THK, for almost any other condition in which some light arrives at the telescope. This last condition is usually reserved for time-critical, one-chance observations. Notice that PHO implies clear conditions (less than 2% transparency variations over the standard zero points) for at least half a night and over the whole sky.
- Seeing. This is defined as the FWHM in arcseconds, at the wavelength of observations, measured at the focal plane of the detector. Most WFI programmes request seeing to be less than 1.2".
- Airmass. This is the maximum airmass allowed for the observations. Notice that airmass and seeing, as defined above are interrelated constraints: because of differential atmospheric refraction we have elongated images when observing with broadband filters. This elongation could easily be interpreted as poor seeing or image quality problems of the telescope (if the observer is not experienced).

– Moon. Most of the programmes for WFI request dark time. To keep the telescope in operation most of the time we first installed FEROS and then GROND at the 2.2 m. These instrument can produce useful science under any Moon conditions. Regrettably, the installation of hardware around the light path complicates the operations of a photometric instrument (see below).

This has been a learning period in the intricacies of queue observing. Perhaps where we most have to learn is in the proper assessment/prediction of the available time for a given set of conditions. The completion of some programmes has been very slow, in part due to period of poor weather at La Silla, but also because of too optimistic time allocation.

3.1 Instrument Monitoring and Calibration Plan

The instrument is managed by an Instrument Operation Team, IOT, consisting of an Instrument Scientist and a couple of instrument fellows, all of them astronomers. They work together with instruments "friends" which are usually engineers and TIOs. They are supported by a group at Garching from the User Support Department, USD. That group is in charge of interfacing with the users during Phase I and II of proposal preparation and submission, and of maintenance of the queue. Starting 2007 there will probably be a pipeline working for WFI.

The current calibration plan of WFI is separated into three different stages: instrument monitoring, instrument calibration, and scientific calibration. The objective of the instrument monitoring is to asses the *health* of the instrument, and to check that all system are ready to go for the night. The health checks are done during the afternoon when there is still time to fix a problem if any is found. There is a short term, daily health monitoring that we use to measure readout noise, RON, gain, flat level, and bias level. During this phase several kind of problems have been detected through the years: improper instrument interventions which have left the instrument in an unknown state; pick-up noise from faulty electronic boards; instrument startup problems; etc. The long term health checks are used to determine the proper functioning of the detector. We monitor the shutter delay, the detector nonlinearity, etc. Notice that this is done separately for each of the 8 WFI detectors. No health check is done on its tracker camera. The monthly health check parameters from the start of operations can be seen at the http://www.ls.eso.org/lasilla/sciops/2p2/CCDs/WFI. We started performing these tests on a weekly basis but we later switched to a monthly one after assessing the high stability of the system. The variation of the shutter from values near zero to values near -0.04 s is due to an initially improperly configured subroutine. The jump in the linearity after 2004 is due to modification of the beta-light intensity (with a neutral density (ND) filter) and subsequent modification of integration times to sample the whole dynamic range more completely. The decay seen in the amplitude before 2004, and also present in the trend after 2004 is due to the progressive dimming of the radioactive beta-light which slowly reduces the dynamic range of the linearity curve. There is an increase of the RON in the first half of 2005. This occurred after an "upgrade" of an embedded computer which is currently under investigation.

The instrument and scientific calibration for WFI are quite straightforward:

- 10 daily biases
- 5 screen flats on demand
- 5 twilight sky flats per observed filter
- STD star observations at at least three airmasses during PHO nights
- Dark frames
- Shutter error frames
- zero-point variation map frames
- STD field in all chips

Note that the operation plan calls for the observation of these frames, and that apart from the health checks described above no further processing of the frames is performed. In this respect the operation is different to that of Paranal Observatory where there is a pipeline processing of all the calibration frames and performing what is called QC1, or quality control level one. With WFI we only do QC0 which involves the visual inspection of the frames.

Although we do not have support for a formal QC1 step we do a bit of QC1 through the use of MIDAS scripts embedded in Python or bash scripts. We do monitor the screen flat and twilight flat variations and we also monitor the photometric zero point variations in all chips. We do this last step using the raw, unprocessed science frames of Landolt fields. It is remarkable that when using the same field night after night we recover the same zero point to within 1% accuracy. The photometric zero points for all chips for a period of a couple of years can be seen at the http://www.ls.eso.org/lasilla/sciops/ 2p2/E2p2M/WFI/zeropoints/Nightly/. Notice that the plot includes data from photometric and non-photometric nights. No attempt to determine the atmospheric extinction coefficients is made and the data reduction relies on the standard La Silla values.

We have been keeping track of the dome and sky flat field variations using an automatic reduction script. The reader can look at a plot of the ratio of daily sky flats over a fiducial one at http://www.ls.eso.org/lasilla/ sciops/2p2/CCDs/WFI/QC1/. That figure shows three sets of points: large dots are the 4 chip corners, smaller dots are areas near the centers of each of the 8 frames, and even smaller dots come from 4 areas near the center of the mosaic. A total of 16 points should be plotted. With the exception of the corner points (and the U filters) most variations are within 1–2% for periods of weeks.

4 What We Could Have Done Better

Although the operation of WFI has been highly successful there are several issues that we believe could have been done better. I will summarize those here.

4.1 Data Reduction Support

For some time the only available reduction environment for WFI data was MIDAS and IRAF which required to split the images into its individual extensions. This lack of proper support after the user gets its data could explain the initial lag in the publication of papers based in WFI data. The situation was somewhat ameliorated by the arrival of IRAF's MSCRED package, and specially the ESOWFI sub-package. This was partly due to the work by Jacoby and De Marco [12] to search for PNe in the SMC using WFI which lead to the necessary adaptation of IRAF's MSCRED [25]. The situation now is much better and is probably going to improve in the future. Users can choose between the MVM pipeline designed for the ESO EIS project [26], or the Theli processing pipeline ([5]; and Erben and Schirmer, these proceedings). Furthermore, ESO is adapting/testing the OmegaCam pipeline with WFI. Once completed that pipeline could be distributed to users to reduce WFI data. *Lesson number one: we should have the end-to-end operation in mind from the beginning. Great data is of no use if users do not have the tools to extract the science from them.*

4.2 Configuration Control/Improvement

Sometimes better is not good. For astronomical photometry to be viable it is important to ensure the constancy of conditions during the execution of a scientific project. In this context an improvement to the telescope/instrument configuration which results in uncalibratable data is no improvement at all. Figure 2a shows an image of the 2.2 m telescope taken pointing to the dome-flat field screen through a pin-hole. The image shows what you could see looking up from the focal plane. The bright doughnut at the center is the pupil image, and it is all that we should see. All other sources of light come from scattered light through the refracting optics, or through reflections. This extra light reaching a pixel can strongly contaminate the flat field frames, and through the use of these contaminated frames it can affect the photometry. All instruments are affected by this to larger and lesser degrees. What is important is to reduce it to a minimum and then to ensure that proper calibration frames are obtained (see below). In our case we could be criticized by not reducing it soon enough. Figure 2b shows the change after the baffle of the 2.2 m telescope was changed during a campaign to improve the straight light characteristics of the telescope so that FEROS and WFI could co-exist. So *lesson number two: work hard to improve and characterize the initial configuration, and then freeze it (if at all possible).*

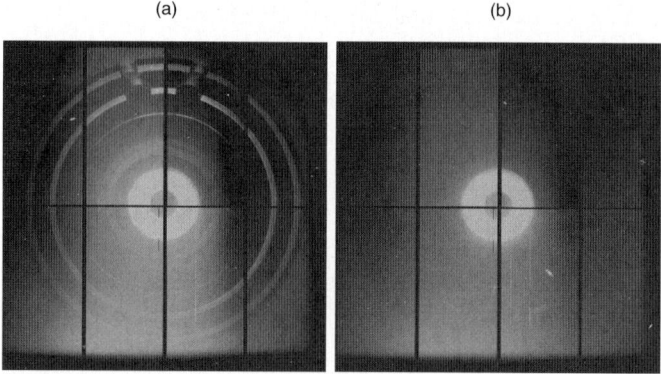

Fig. 2. Pinhole image of the 2.2 m telescope. (**a**) prior to September 2002, and (**b**) after the September 2002 baffle improvement

4.3 Proper Filter Passbands

To have an instrumental system which can be converted to a standard system is a very important property. In general one can always choose to work in the instrumental system and to determine physical parameters directly from there. But in this case the work looses value as people will not be able to use such photometry for inter-comparison. This is the case for the B and to some extent the U filters of WFI. Although both filters were replaced at least once the replacement were not better than the originals. Figure 3a show the simulated colour equations for the WFI U and B filters, with reddenings of $E(B-V) = 0$ and 1. We can see that the errors in (U-B) can be up to 15% when observing highly reddened stars. The author noticed the problem when doing stellar photometry of highly reddened massive stars in the 30 Doradus region of the LMC. WFI is not the only ESO instrument suffering from this disease. Figure 3b shows the reddening trajectories for NACO magnitudes of the stars in the Arches cluster. Improper transformation to Johnson JHK is shown as the data points. The continuous thin line shows the 2 Myr isochrone, and the dashed line show the standard extinction law reddening vector. The left figure shows the data in the instrumental magnitude plane where we have simulated the reddening vector in the standard NACO system. It is tempting to conclude that the reddening law is different when the only thing that happened is that the conversion to the standard system is faulty (standard stars are usually not very much reddened). So *lesson number three: match the instrumental photometric system as closely as possible to the standard one (and/or properly define the new one).*

4.4 Zero-Point Variation Map

ESO commit itself to provide data that can be calibrated to a 5% absolute level. The decision for such a low precision photometry has turned out to

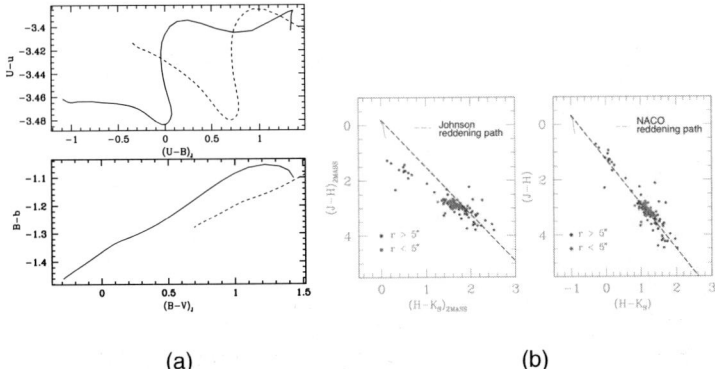

Fig. 3. (a) Simulated colour equations for the U and B WFI filters. (b) Reddening trajectories for the NACO JHK converted to Johnson, left, and instrumental magnitudes, right

be quite wise as it is not easy to go below this level, and almost impossible to go below a 1% level. In the case of WFI even this conservative goal is violated because the instrument suffer from variations of zero points across the field that can be as high as 30% peak-to-valley in some filters. We have not measured the narrow band filters but the author suspects that for some of them the effect could be even larger.

The origin of this effect, which was first reported for the case of DFOSC at the Danish telescope by Andersen et al. [1], is quite subtle (and nasty). When one obtains the flat field frame light is rearranged through the optical elements of the telescope resulting in a frame which is not really proportional to the gain-variation map as all proper flat-field frame should be, but it has been contaminated. It is as if there is a little gnome in the detector unevenly illuminating the CCDs with a flashlight while the exposure proceeds. The nasty part is that this little gnome arrange things so that every science frame is illuminated in the same proportion as the flat field. Thus, the end result after dividing by the flat-field is a perfectly flat sky level. But this perfect flat background level hides a problem for the photometry: stars measured in those areas where the illumination was high will have underestimated flux. With a proper gain map we would have been able to see the uneven illumination pattern. We have designed a method that allow us to correct for this uneven illumination down to a 2–3% level. The method is still experimental and ESO is not providing the tools to easily correct the data for this effect. The reader is directed to the many references in the literature that address this issue [1, 14, 23, 24] *Lesson number four: it is important to have astronomers doing science with the data of the instrument they are supporting.*

4.5 Detector Linearity

It was standard knowledge that ESO's detector were linear to ±0.5%. Nevertheless the linearity test performed for WFI every week showed a peak-to-valley variation larger than 2% in some cases. It took some time to figure out that what was called a detector linearity curve was not that, but that it included a non-linear shutter effect.[1] A problem arose when trying to disentangle both effects: the exposure times for the detector linearity determination where in the same range as those times for which the shutter showed non-linear behaviour. Once we understood this we designed a shutter linearity test which obtained exposures with the same total number of counts but with different shutter times. We need to use the multiple exposure mode of the CCD controller. We thus compare a single 100 s exposure with ten 10 s exposures which are only read out after the last one of the ten has completed. We are currently only obtaining the data so that anyone who is interested in applying a linearity correction can do so and attempt to perform photometry at or below the 1% level (after dealing with the effects previously described!). *Lesson number five: understand your instrument!*

4.6 Scheduling of Service Observations

It turned out that in actual terms the amount of time assigned to programmes was for the most time overestimated by approximately a factor of two. A period of particular poor weather at La Silla did have an influence on this, but only partly. There was strong competition in RA and for good seeing. Several PIs were requested to increase their seeing constraint. The large programmes dedicated for surveys, by having the largest priority of all did have a detrimental effect in the completion of normal programmes. We believe that this could have a negative impact on the perception of completion of service mode programmes as viewed by the community. At some point the WFI queue, including programmes carryover from previous periods, had twice the amount of OBs that could be executed in the available time (and the oversubscription rate was a factor of 4 in terms of what could reasonably be predicted!). *Lesson number six: forecast conservatively the availability of time for each constraint set (seeing, transparency, Moon) and try to ensure that all class A and B programmes can be completed within the available time.*

5 Conclusions

The work of a support astronomer at ESO has been marked by a tension, unhealthy at times, between the need to keep operating, the need to characterize

---

[1] Careful reading of the WFI manual showed that the non-linear shutter effect was detected from the beginning of operations and that ESO's Optical Detector Team, ODT, was well aware of this. Nevertheless the operation group at La Silla, and this instrument scientist in particular, had not properly digested this info.

the instrument at a high level, and the need to produce science with the instrument. It is most important for instrument scientists to be producing the higher standards of science, pushing the envelope of what can be accomplished with an instrument. It is only then that one can see the ugly bugs crawling in all directions inside the instrument and a proper understanding begin. Nevertheless, no matter how close is the relation of an instrument scientist with the instrument it is impossible to explore the complete parameter space. It is thus also important to keep close ties with the community of users as it is in the community where a large part of the knowledge of the instrument is. So I thank the organizers of this workshop to give us this wonderful opportunity to interact.

References

1. M.I. Andersen, L. Freyhammer, J. Storm: Gain calibration of array detectors by shifted and rotated exposures. In: *Proceedings of an ESO/ST-ECF Workshop on Calibrating and Understanding HST and ESO Instruments*, ed by P. Benvenuti (ESO Conference and Workshop Procedings, Garching, Germany), p. 87 (1995)
2. M. Arnaboldi, J. Aguerri, L. Alfonso, et al.: AJ **123**, 760 (2002)
3. D. Baade, K. Meisenheimer, O. Iwert, et al.: ESO Messenger **95**, 15 (1999)
4. D.J. Butler, D. Martinez-Delgado, H.-W. Rix, et al.: AJ **133**, 2274 (2007)
5. T. Erben, M. Schirmer, J. Dietrich, et al.: AN **326**, 432 (2005)
6. R. Giacconi, P. Rosati, P. Tozzi, et al.: ApJ **551**, 624 (2001)
7. M. Giavalisco, H.C. Ferguson, A.M. Koekemoer, et al.: ApJL **600**, L93 (2004)
8. W. Gieren, G. Pietrzyński, A. Walker, et al.: AJ **128**, 1167 (2004)
9. B. Gladman, et al.: Nature **412**, 163 (2001)
10. W. Herbst, C. Bailer-Jones, R. Mundt, et al.: A&A **396**, 513 (2002)
11. J.E. Hirsch: PNAS USA **102(46)**, 16569 (2005)
12. G.H. Jacoby, O. De Marco: AJ **123**, 269 (2002)
13. D. Kubas, A. Cassan, J.-P. Beaulieu, et al.: **435**, 941 (2005)
14. J. Manfroid, F. Selman, H. Jones: ESO Messenger **104**, 16 (2001)
15. A. Mignano, J.-M. Miralles, L. da Costa, et al.: A&A **462**, 553 (2007)
16. Y. Momany, S. Zaggia, G. Gilmore, et al.: A&A **451**, 515 (2006)
17. L. Monaco, E. Pancino, F.R. Ferraro, et al.: MNRAS **349**, 1278 (2004)
18. N.D. Morgan, J.A.R. Caldwell, P.L. Schechter, et al.: AJ **127**, 2617 (2004)
19. F. Pozzi, P. Ciliegi, C. Grupioni, et al.: MNRAS **343**, 134 (2003)
20. M. Sazhin, G. Longo, M. Capaccioli, et al.: MNRAS **343**, 353 (2003)
21. M. Schirmer, T. Erben, M. Hetterscheidt, et al.: A&A **462**, 875 (2007)
22. A. Scholz, J. Eislöffel: A&A **429**, 1007 (2005)
23. F. Selman: The IMF of the 30 Doradus super-association. PhD Thesis, Caltech, California (2004)
24. F. Selman: SPIE **5493**, 453 (2004)
25. F.G. Valdes: In: The IRAF Mosaic Data Reduction Package ASP Conference Series 145, Astronomical Data Analysis Software and Systems VII, ed. R. Albrecht, R.N. Hook, and H.A. Bushouse (San Francisco: ASP), p. 53 (1998)

26. B. Vandame: Traitments d'images a grand-champs et multi-longeurs d'ondes. PhD Thesis Universite de Nice-Sophia Antipolis, France (2004)
27. E. Westra, D.H. Jones, C.D. Lidman, et al.: A&A **455**, 61 (2006)
28. C. Wolf, K. Meisenheimer, H.-W. Rix, et al.: A&A **401**, 73 (2003)

Discussion

U. Hopp: The largest survey we have established has a standard system: SDSS. VST/OmegaCam tried hard to match this system. In addition, it offers further filters like Stroemgren-v to support stellar astronomy, Johnson B and V and a few narrow bands.

P. Bonifacio: When you have an instrument like WIFI or VST, which provides an enormous amount of data, it is really worth the effort to define the instrument's "natural" photometric system. It is a big effort, but the payback is large.

F. Selman: I totally agree with your statement. Furthermore, there should be calibration proposals for this, as the task is just too large. When you work on a standard system, you have all the information from a large amount of past research. We cannot redo all this. But some of it should be done, e.g. stellar properties, reddening law, etc.

Processing Wide Field Imaging Data

M. J. Irwin

Cambridge Astronomy Survey Unit, Institute of Astronomy, Madingley Road, Cambridge CB3 0HA, UK; mike@ast.cam.ac.uk

Abstract. Optical and near-infrared wide field cameras on 2 m-, 4 m- and 8 m-class telescopes are increasingly being used to conduct large scale imaging surveys. With total data rates approaching 1 Tbyte/night automated pipeline processing and data management facilities are crucial. Although processing near-infrared data is in some ways technically more challenging than the optical, the majority of the processing requirements and the adopted solutions are common to both. In this contribution I describe the general princples required for an end-to-end system capable of delivering science-grade, quality-controlled calibrated images and catalogues. This is illustrated with some practical examples from current optical and near-infrared surveys.

1 Introduction

Typical average data rates from modern optical and near-infrared (NIR) mosaic cameras lead to hundreds, if not thousands, of on-sky images each night. Manual processing using conventional software packages is generally impractical. However, automating what used to be a labour-intensive processing and interpretation task is also non-trivial. The natural solution is to provide an end-to-end integrated data flow system where all aspects from data taking, through data processing, to data archiving are linked in a coherent whole. Imposing constraints on the overall system architecture enables a unified approach to converting raw observations to quality-controlled, calibrated and science-ready data products. This is still a challenging problem but if achieved opens up new avenues for exploiting panoramic wide field imaging surveys, thereby enabling exciting scientific opportunities across a broad spectrum of contemporary astronomical research.

In this overview I discuss the general approach that we at the Cambridge Astronomical Survey Unit have developed to address this problem and in particular focus on the requirements and processing strategy adopted to convert multi-terabyte raw data volumes to useful science-ready products.

2 Overall Data Structure

Our general philosophy is that all fundamental data products are FITS multi-extension files (MEFs) with headers describing the data taking protocols in sufficient detail to trigger the appropriate pipeline processing components. If this is not the case, as in, for example, data taken using WFCAM on UKIRT, a pre-processing stage is used to convert the raw data to this data structure. Within this format it is natural to require all generic information about the observations, such as telescope parameters, weather details and so on, to be present solely in the primary FITS header unit. Each data extension can then contain detector-specific information such as detector format and initial World Coordinate System (WCS) parameters.

To reduce the data storage, I/O overheads and transport requirements, we use, as much as possible, the lossless Rice tile compression scheme e.g. [5] as used transparently, for example, in CFITSIO [4]. For NIR data which is typically 32-bit integer, Rice compression gives an overall factor of 3–4 reduction in raw file size, with the added advantage that the CFITSIO implementation also maintains the overall FITS file structure by making use of FITS binary tables to encode the compression. A further benefit is that by tiling images prior to compression, efficient fast access to subsets of the images, without decompressing them in there entirety, is also possible.

A common problem is that reliable transfer of raw data from observatory to processing centre is often a bottleneck. The three main methods we use in order of preference are: Internet transfers – for example all Isaac Newton Group (ING) data from La Palma in the Canary Islands are automatically transferred to on-line disks in Cambridge in almost real time; tape transfers – we use LTO tapes to transfer all WFCAM data from Hawaii and have found the tapes extremely reliable, to date 200+ tapes have been used to transfer around 50 Tbytes of raw data via this route; disk transfers – these are generally IDE disks either in a standalone enclosure or with a suitable mounting "caddy" attached, unsurprisingly these either work fine or fail completely.

One added advantage of tape transfers is that they provide an off-line archive store once they have been read. For both other transfer methods we write LTO tapes to provide a guaranteed off-line data store.

3 Overall Processing Philosophy

Our overarching approach is that all derived information, quality control measures, photometric and astrometric calibration and processing details, are also incorporated within the FITS headers and that generated object catalogues are also stored as multi-extension FITS binary tables. These FITS headers provide a simple basis for ingest into databases for archiving and databases for real time monitoring of survey progress and survey planning (e.g. `http://casu.ast.cam.ac.uk/surveys-projects/wfcam/data-processing/`).

In principle, the data-processing pipelines should be automatically driven by appropriate (agreed) FITS keywords and by the overall observing structure. In practice because we have to deal with observations taken on a wide variety of telescope systems, primarily in our case from: Subaru-Suprimecam

8 m; CTIO-mosaic 4 m, KPNO-mosaic 4 m; CFH-megacam 3.6 m; INT-WFC 2.5 m; ESO-WFI 2.2 m and UKIRT-WFCAM 3.8 m; this is not always the case and various pre-processing stages are often needed. Over the years this led to our developing a modular processing system whereby the pipeline processing is controlled from scripts and individual software modules are used for specific processing tasks. For example, not all data suffers from cross-talk or fringing and therefore each new telescope system added to the canon requires a specific pipeline procedure "commissioning".

Occasionally this requires un-anticipated software modules to be developed (for example see the contribution by Lewis et al. these proceedings). However, in general, the science requirements are well-defined and the majority of science deliverables hinge on a series of fundamental processing stages: removal of two-dimensional instrumental signature; astrometric and photometric calibration; quality control assessment; deep image stacking; production of detected object catalogues; and merging object information across passbands or in time series.

We briefly discuss some of these stages in the next sections and illustrate the overall processing structure in Fig. 1.

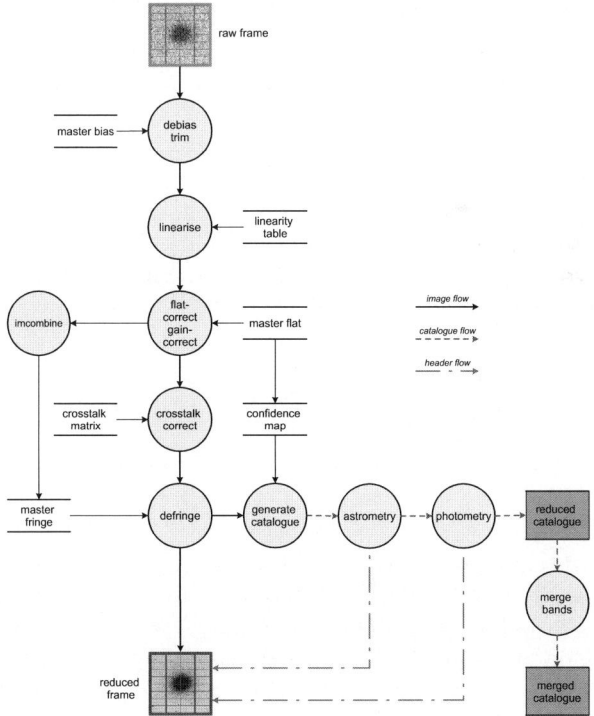

Fig. 1. A summary of the typical processing operations required to go from raw data to science products

4 Image Processing

For NIR data from WFCAM we process each night of data "independently" using master calibration twilight flats (updated at least monthly) and a series of nightly generated dark frames covering the range of exposure times and readout modes used during that night. For all the optical cameras we generally group data into observing blocks or runs, of typical duration ≈1 week and derive overall master calibration twilight flats and bias frames using the entire run. The typical data volume for a processing "run" is then similar (≈100 Gbytes), convenient to manage, and contains enough independent pointings to derive dark sky calibration frames such as sky frames in the NIR, and fringe frames for the redder optical bands.

We find generally that twilight sky flats give better results than dome flats due to better uniformity of illumination, and normally, in the optical at least, do not then require additional flatfielding stages using dark sky stacks. NIR data on the other hand invariably requires a dark sky correction stage. In both cases we find subsequent photometric calibration is greatly simplified if the flatfielding stage includes gain correction to place all the detectors on the same internal photometric system.

The image processing problem for both optical and NIR can be encapsulated in the following way:

$$D(x,y) = f(x,y) \left[S(x,y) + F(x,y) + O(x,y) + T(x,y) \right] + d(x,y) \quad (1)$$

where $D(x,y)$ is observed, $f(x,y)$ is the flatfield function, $S(x,y)$ is the sky illumination, $F(x,y)$ is the fringe contribution, $O(x,y)$ is the object contribution, $T(x,y)$ is the thermal contribution, $d(x,y)$ is the dark or bias frame structure, and without loss of generality we have excluded any explicit wavelength- and time-dependence for clarity. There are two obvious approaches to solving for $O(x,y)$.

Stacking a series of dithered (jittered) object frames with rejection produces an estimate of the terms

$$\hat{I}(x,y) = f(x,y) \left[S(x,y) + F(x,y) + T(x,y) \right] + d(x,y) \quad (2)$$

therefore,

$$D(x,y) - \hat{I}(x,y) = f(x,y)\, O(x,y) \quad (3)$$

In principle this obviates the need for dark/bias-correcting and fringe removal as both separate data gathering requirements and as separate data processing steps; and minimises the effect of systematic and random errors in the flatfield function by removing the largest potential error terms (i.e. those due to sky).

The caveats here, of course, are that this method may well remove parts of large extended objects, large area nebulosity, large low surface brightness objects and more seriously, the photometry can be compromised by the presence of faint objects not rejected properly during the stacking and subsequent "sky" subtraction phase due to the data not being dark-corrected and flatfielded first.

The alternative we use is to treat the dark/bias correction $d(x,y)$, flatfield $f(x,y)$, and fringe pattern $F(x,y)$, if present, as accurately known master calibration frames, in which case applying the master frames leads to

$$D'(x,y) = S(x,y) + O(x,y) + T(x,y) \qquad (4)$$

reducing the problem to one of sky-correction, including residual instrumental effects such as reset anomaly, and then of detecting astronomical objects on an additive slowly spatially varying background. In this case stacking sequences of processed frames $D'(x,y)$ with rejection works much better, because the majority of the "noise" in the frames has already been removed prior to the rejection operation. Rejection of faint objects can be even further improved by iterating on this process.

5 Object Cataloging

Catalogue generation involves the following series of operations e.g. [2]

- the local sky background is estimated over the field and any variations are tracked at adequate resolution (typically ≈30 arcsec) and temporarily removed – this is invariably a compromise between following detailed background variations, not compromising measurements of larger objects, and maintaining adequate statistical robustness in the local sky estimates;
- objects/blends of objects are detected using a matched filter and a list of pixels belonging to each blend is kept for further analysis – this includes a deblending stage to attempt to disentangle overlapping objects;
- the detected objects are parameterised with the descriptors summarising derived position, shape and intensity information – these measures include assorted aperture flux estimates, intensity-weighted centroid estimates, and shape information, such as intensity-weighted 2nd moments to encode the equivalent elliptical Gaussian light distribution.

The image catalogues are then further processed to yield morphological classification for detected objects (i.e. stellar, non-stellar, noise-like), aperture corrections (i.e. essentially a curve-of-growth for stellar images), and are also used to directly generate relevant astrometric and photometric calibration and other quality control information. An example of how robust an automatic pipeline using aperture photometry can be, even in complex crowded fields, is shown in Fig. 2.

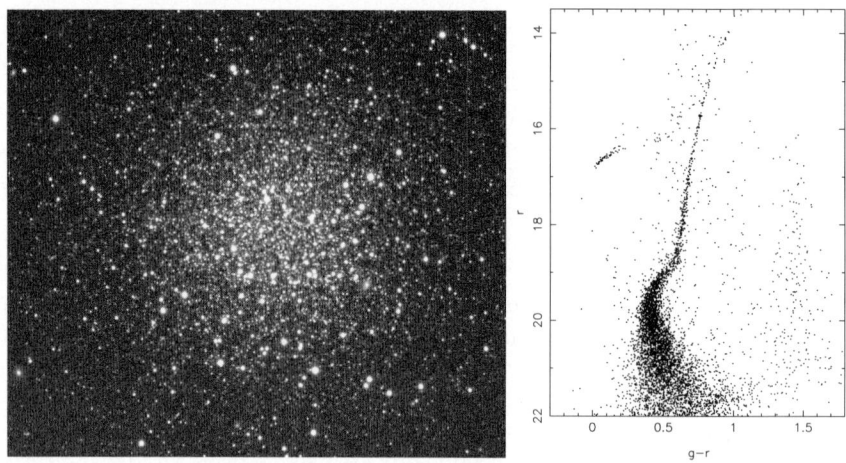

Fig. 2. The central 10'×10' of the globular cluster NGC 5466 and the pipeline generated colour-magnitude diagram for the complete INT WFC pointing

6 Astrometric Calibration

All the mosaic camera telescope systems we have so far encountered admit to a relatively simple astrometric distortion model where the distortion is well described by a radially symmetric polynomial distortion. For most of these the effective radial scale, r', is related to the true scale r by

$$r' = r + k_3 r^3 + k_5 r^5 + \ldots \tag{5}$$

where the 5th order term is usually negligible. This is characterised by a WCS using the ZPN projection [1] with the coefficients for this encoded in the secondary FITS headers using keywords like PV2_1 and PV2_3. The only other WCS system we encounter, or use for mosaicing, is the standard tangent plane projection TAN.

To refine the astrometric calibration, the 2MASS all-sky catalogue [6] is interrogated to define an astrometric reference grid projected into "distorted" standard coordinates centred on the tangent point of the telescope pointing [7]. A binary search over a small grid of offsets centred on the expected location is used to find the best starting point for the refinement with the search radius for a possible match adaptively computed from the 2MASS and object catalogue source densities. Given a list of potential matched sources a bilinear 6 constant solution is found between pixel coordinates and (radially distorted) standard coordinates with respect to the optical axis. This is done on a per-detector basis using

$$\xi' = ax' + by' + c \qquad (6)$$
$$\eta' = dx' + ey' + f \qquad (7)$$

where a, b, d, e encode for scale(s), rotation and shear, c, f are offsets and x', y' denote pixel coordinates shifted to the centre of the matched star distribution for numerical stability in the solution. The refinement is iteratively computed and applied using sequential k-sigma clipping of outlying matched sources until convergence is reached, usually after a few iterations. Typical rms residuals per star are better than 100 mas and are mainly dominated by the 2MASS coordinate accuracy.

We generally investigate more subtle astrometric distortion residuals by stacking the 2MASS astrometric residuals from sets of many nights of data, taken at large numbers of sky positions to minimise effects of residual systematics in 2MASS. Figure 3 shows an example of the stacked residuals from 0.5 million K-band objects from the first two weeks of 05B WFCAM observations. There are still residual low level systematics present but the figure, in addition to being a diagnostic of the problem, demonstrates a possible further refinement. The impact of residual systematics could be reduced still further by using a look-up table to interpolate a correction for the repeatable spatially-dependent component.

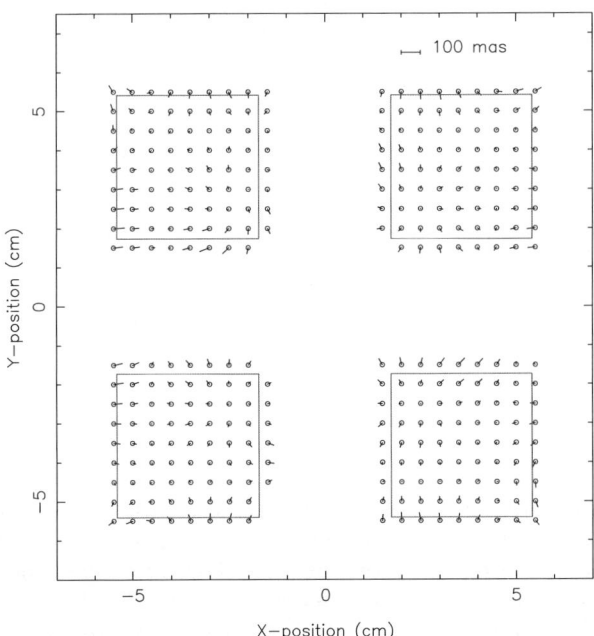

Fig. 3. Stacked average astrometric residuals for WFCAM with respect to 2MASS with the idealised positions of the detectors overlaid

7 Photometric Calibration

The internal gain-correction, applied at the flatfielding stage, places all the detectors on a common zeropoint (ZP) system (at least to first order i.e. ignoring colour equation variations between the detectors). Given a stable instrumental setup, the apparent variation of ZP then directly measures the change in "extinction" without the need to rely on extensive standard field coverage over a range in airmass. Longer term trends in ZPs due to, for example, accumulation of dust on the optical surfaces can be readily decoupled from shorter term (nightly) variations.

All generated object catalogues provide the basis for the photometric calibration. In the NIR, 2MASS can be used to generate photometric calibration for every single pointing and in effect monitor the extinction on a frame-by-frame basis. In the optical, standard fields (e.g. [3]) can be observed several times during the night to provide an equivalent nightly calibration. The distribution of ZPs in each passband per night can also be used to make an estimate of the overall photometric quality of the night and is illustrated in Fig. 4.

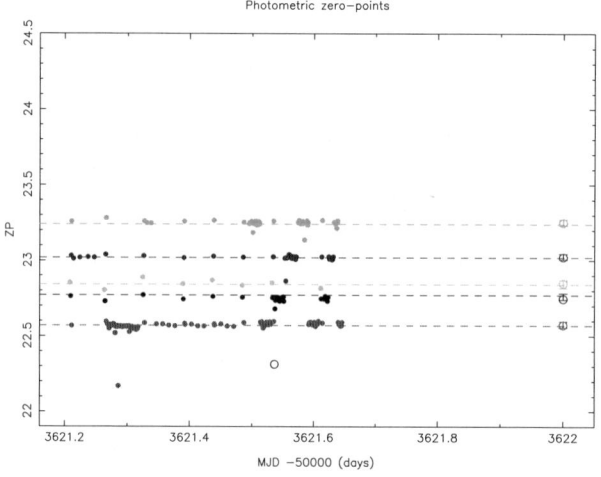

Fig. 4. An example of the WFCAM photometric calibration for the night of 20050908. Each *filled circle* (bottom set K-band, top H-band, 2nd from top J-band, middle Y-band, 2nd from bottom Z-band) is a catalog-based individual ZP. *Open circles* denote either badly trailed or poor seeing data. The *dashed lines* are nominal ZPs for each passband. *Open circles, with error bars*, to the *right* of the figure show the median ZPs for the night with error bars equal to the average robust scatter, i.e. a measure of the photometric quality of the night. The outlying red point is from an incomplete observation group

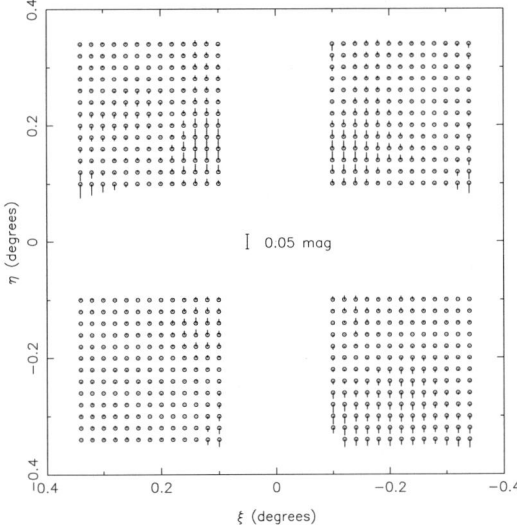

Fig. 5. Spatial variation of J-band photometric residuals within 2 arcsec diameter apertures for June 2005 WFCAM data. The *rms* residual in the binned data is 0.015 mag

If sufficient standards have been observed then it is straightforward to use the stacked residuals from a large series of photometric solutions to directly investigate spatial variations in the photometric calibration. Such effects may be due to a combination of scattered light causing illumination effects in flat-fielding, PSF distortions across the focal plane, or effective scale change due to astrometric distortion. An example of a derived illumination correction for WFCAM is shown in Fig. 5. Note that a correction for astrometric radial distortion is automatically applied during the 2MASS calibration and amounts to ≈1% at the outermost corners.

For most optical telescopes we find negligible illumination correction, though a notable exception is the ESO 2.2 m WFI which requires corrections up to ≈10%.

8 Quality Control Measures

The object catalogues provide an efficient way of determining various quality control measures relevant to assessing both the data integrity and the quality of the science products. These on-sky measures supplement routine daytime operational checks that monitor the general health of the system.

Typical parameters that can readily be automatically derived at the individual detector level include measures of: sky brightness – saturated? too

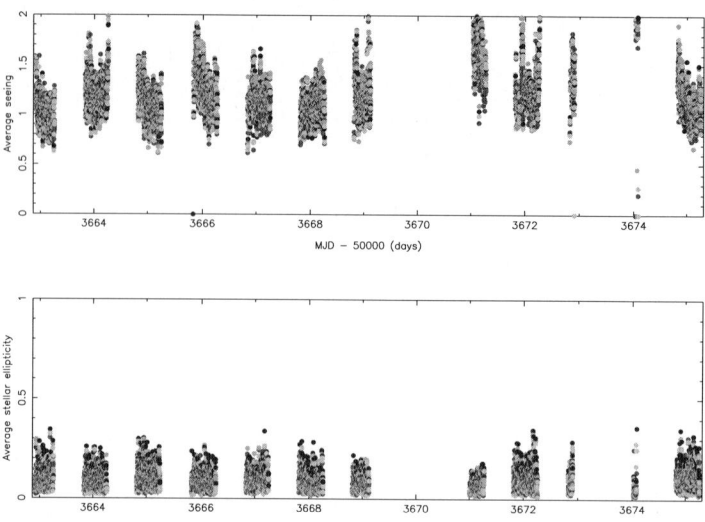

Fig. 6. Data quality measures showing the variation of seeing, *upper panel*, and average stellar ellipticity, *lower panel*, for a two week run on the INT WFC camera in October 2005. Each detector is colour coded separately. Outlying points highlight a data quality problem

bright? no signal?; sky noise – *rms* acceptable? pattern noise?; average stellar ellipticity – trailed images? focus?; average FWHM seeing measure – science requirements? in focus?; stellar aperture corrections – weird PSF?; number of spurious images – corrupted data?; astrometric errors – wrong RA, Dec?; photometric magnitude zeropoint – monitors system throughput?; and on a nightly basis, derived local and global photometricity flags, limiting magnitude estimates, nightly zeropoints and extinction measures.

Figure 6 shows an example use of these measures for monitoring received data frame seeing (FWHM) and image trailing via the average ellipticity of stellar images.

References

1. M.R. Calabretta, E.W. Greisen: A&A **395**, 1077 (2002)
2. M.J. Irwin: MNRAS **214**, 575 (1985)
3. A. Landolt: AJ **104**, 340 (1992)
4. W.D. Pence: Proc. SPIE, 4847, 444 (2002)
5. C.N. Sabbey: PASP **112**, 867 (2000)
6. M.F. Skrutskie, et al.: AJ **131**, 1163 (2006)
7. W.M. Smart: Spherical Astronomy, Cambridge University Press (1965)

Discussion

B. Wolff: On which kind of data do you perform a visual inspection?

M. Irwin: We always visually inspect master calibration frames but only a subset of processed science frames. For WFCAM surveys, there are already around 3 million $2\,\text{k} \times 2\,\text{k}$ images which means it is only practical to visually inspect a small sample of the science output.

The Garching-Bonn Deep Survey (GaBoDS) Wide-Field-Imaging Reduction Pipeline

H. Hildebrandt[1], T. Erben[1], M. Schirmer[2], and J. P. Dietrich[3], and P. Schneider[1]

[1] Argelander-Institut für Astronomie, Universität Bonn, Auf dem Hügel 71, 53121 Bonn, Germany; hendrik@astro.uni-bonn.de
[2] Isaac Newton Group of Telescopes, Apartado de correos 321, 38700 Santa Cruz de La Palma, Tenerife, Spain
[3] ESO, Karl-Schwarzschild-Strasse 2, 85748 Garching, Germany

Abstract. We introduce our publicly available Wide-Field-Imaging reduction pipeline THELI. The procedures applied for the efficient pre-reduction and astrometric calibration are presented. A special emphasis is put on the methods applied to the photometric calibration. As a test case the reduction of optical data from the ESO Deep Public Survey including the WFI-GOODS data is described. The end-products of this project are now available via the ESO archive Advanced Data Products section.

1 Introduction

Over the past years our group has developed a wide-field-imaging reduction pipeline called THELI which is now publicly available.[1] Since data reduction for most optical and near-infrared cameras is very similar the pipeline was designed in an instrument-independent way from the beginning. It is easily adaptable to new cameras by use of instrument configuration files. Nearly fully automatic processing with little need for interaction quickly leads to scientifically exploitable results making the reduction of one night of wide-field-imaging data (e.g. from WFI@ESO/MPG2.2 m) possible in a few hours. Large parts of the pipeline are parallelised increasing speed in presence of a multi-CPU machine. A graphical user interface (GUI) is available which facilitates the handling of reduction and configuration of THELI. A detailed description of the pipeline can be found in Erben et al. [1]. The modules for absolute photometric calibration which are not part of the public release at the moment are described in Hildebrandt et al. [5].

2 Structure

The pipeline is based on a number of excellent existing open source software packages like the *LDAC* tools, different *TERAPIX* packages (*SExtractor*, *SWarp*, etc.), *Eclipse*, *Astrometrix*, and *IMCAT*, besides others. Bash

---
[1] ftp://ftp.ing.iac.es/mischa/THELI/

scripts are wrapped around these packages in order to handle the communication between the different tasks, to control configuration parameters, and to produce some plots for quality control. Due to this modular structure the pipeline is easily extensible and modules can be exchanged if better ones become available. This was done several times in the past; e.g., for resampling and coaddition *drizzle* was replaced by *SWarp*.

Building up a pipeline from many different software packages naturally has some disadvantages when compared to a homogeneous system which is developed from scratch. The data flow is not as transparent and error handling becomes more complicated. For very large projects like the major upcoming imaging surveys it would be desirable to be able to track the history of each of the many thousand reduced images back to the raw images. This can only be done with a sophisticated database system which is at the moment not implemented in THELI.

3 Processing

3.1 Pre-Processing

The pre-processing, i.e. the removal of instrumental signatures from the data, is done on a single chip basis and does not differ from well established procedures applied for single-chip cameras. A standard debiasing and flatfielding is combined with a superflat and, if necessary, a fringe-removal. See Fig. 1 for a visual impression of some WFI data at different reduction steps.

3.2 Astrometric Calibration

THELI was designed with weak-lensing applications in mind. Therefore, a highly accurate astrometric calibration was mandatory from the beginning to minimise the impact of coaddition on the shape of the PSF. *Astrometrix* is used in combination with overlapping astrometry between all chips entering a coaddition to achieve an internal astrometric accuracy of a tenth of a pixel. The external accuracy is obviously limited by the accuracy of today's astrometric standard star catalogues.

3.3 Photometric Calibration

The different chips of a camera are brought to the same photometric zero-point by division with the appropriately rescaled superflat. Since some residuals may be left after this procedure we perform an internal photometric calibration from overlap objects between different chips of different exposures to derive relative zero-points, $ZP_{\text{rel},i}$, which satisfy the additional condition $\sum_i ZP_{\text{rel},i} = 0$.

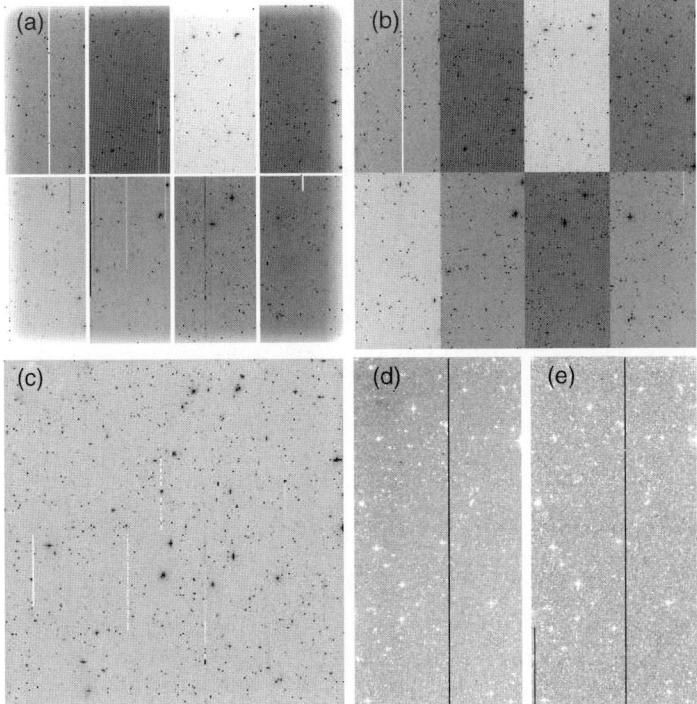

Fig. 1. Pre-reduction steps for WFI V-band data: (**a**) raw data, (**b**) and (**d**) debiased and flatfielded data (**c**) and (**e**) superflatted data. Taken from [1]

All available Landolt/Stetson standard-star exposures of the considered nights are reduced in the same way as the science exposures. Absolute photometric zero-points, colour terms, and extinction coefficients are estimated from these exposures. From those photometric parameters and the relative zero-points we calculate corrected zero-points, $ZP_{corr,i}$, for all images belonging to calibrated nights:

$$ZP_{corr,i} = ZP + \text{Airmass} \cdot \text{EXT} + ZP_{rel,i}$$

Theoretically these corrected zero-points should all coincide. A plot showing the distribution of all corrected zero-points is therefore an excellent tool to identify non-photometric nights that should not be used for absolute calibration (see Fig. 2).

Moreover, we check the absolute photometric calibration by means of galaxy number counts and, if images in more than one filter are available, by means of stellar colour-colour diagrams (see Fig. 3).

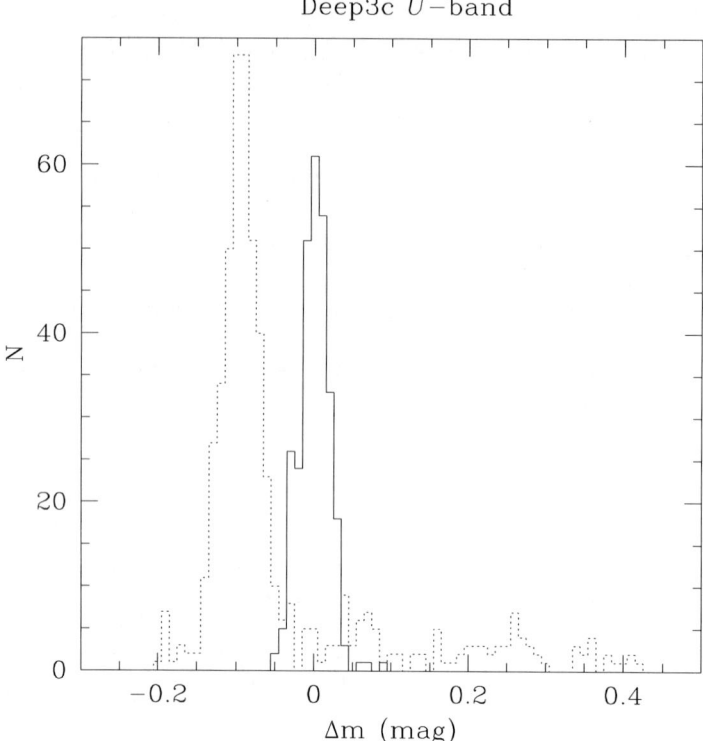

Fig. 2. Distribution of the corrected zero-points before (*dashed line*) and after (*solid line*) rejection of apparently non-photometric nights for the calibration. From [5]

4 Reduction of the ESO Deep Public Survey

The ESO Deep Public Survey (DPS) is a deep multi-colour imaging survey overlapping with GOODS-South and carried out with WFI@ESO/MPG2.2 m and SOFI@NTT. Here, we concentrate on the optical part in the filters $UBVRI$. The survey covers three square degrees in total of which approximately two square degrees have full five-colour coverage to considerable depth. The reduction, quality control, and a comparison to a different reduction by the ESO Imaging Survey team is covered in [5].

4.1 Survey Overview

In Table 1 the main characteristics of the DPS fields are summarised. The 5σ limits in circular apertures of $2''$ diameter are: $\mathrm{mag}_{\mathrm{lim},U} = 25.3$, $\mathrm{mag}_{\mathrm{lim},B} = 26.2$, $\mathrm{mag}_{\mathrm{lim},V} = 25.8$, $\mathrm{mag}_{\mathrm{lim},R} = 25.3$, $\mathrm{mag}_{\mathrm{lim},I} = 24.3$.

The reduction of the DPS was the first application of our pipeline to a large dataset consisting of more than 3000 raw science frames. During this

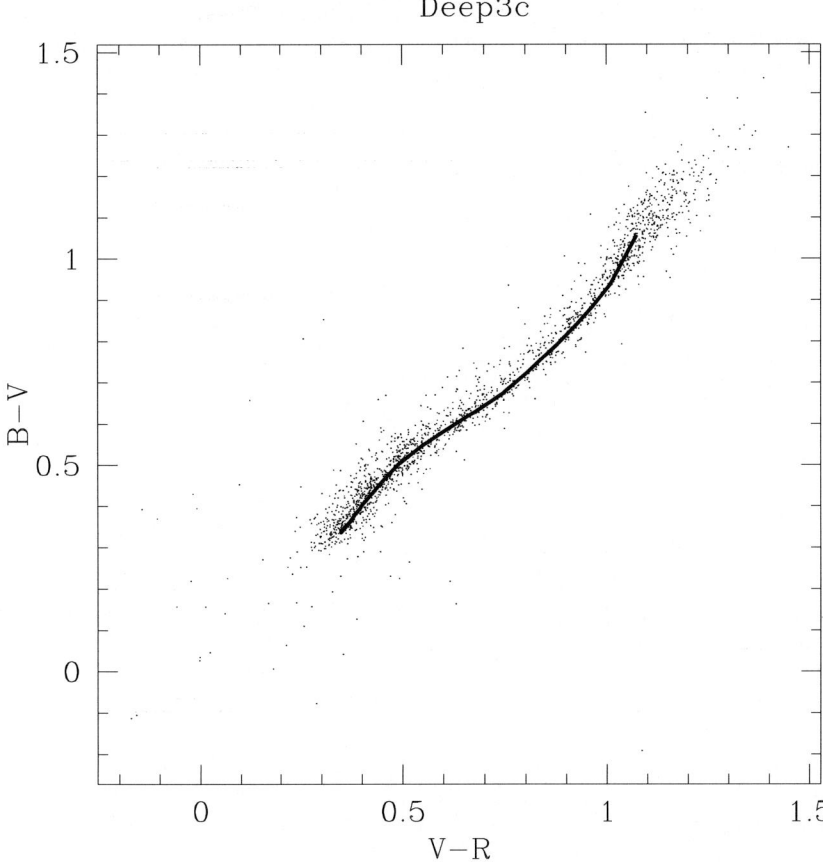

Fig. 3. Colour-colour diagram for stars in the field Deep3c in comparison to theoretical isochrones from [2]. Taken from [5]

reduction the pipeline was extensively debugged and optimised so that it is now in a very stable state.

4.2 Data Release

The data were released to the scientific community after reduction and careful checking. The images can be retrieved from our server[2] or via the ESO archive Advanced Data Products section[3]. The DPS optical data are used by our group for studies of Lyman-break galaxies at $z \sim 3$ [4, 6] and for weak-lensing studies supported by photometric redshifts [3, 7].

---

[2] http://marvin.astro.uni-bonn.de/DPS/
[3] http://archive.eso.org/archive/adp/GaBoDS/DPS_stacked_images_v1.0/

Table 1. Positions and available colours of the twelve DPS fields and the two mispointings (Deep1e and Deep1f). Taken from [5]

Field	RA [h m s] J2000.0	DEC [d m s] J2000.0	Avail. Colours	Comments
Deep1a	22:55:00.0	−40:13:00	$UBVRI$	
b	22:52:07.1	−40:13:00	$UBVRI$	
c	22:49:14.3	−40:13:00	VRI	
d	22:46:21.4	−40:13:00	—	
Deep2a	03:37:27.5	−27:48:46	R	
b	03:34:58.2	−27:48:46	$UBVRI$	
c	03:32:29.0	−27:48:46	$UBVRI$	centred on GOODS-S
d	03:29:59.8	−27:48:46	R	
Deep3a	11:24:50.0	−21:42:00	$UBVRI$	
b	11:22:27.9	−21:42:00	$UBVRI$	
c	11:20:05.9	−21:42:00	$UBVRI$	
d	11:17:43.8	−21:42:00	$BVRI$	
Deep1e	22:47:47.9	−39:31:06	URI	
f	22:44:58.4	−39:31:54	I	

References

1. T. Erben, M. Schirmer, J. P. Dietrich, et al.: Astronomische Nachrichten **326**, 432 (2005)
2. L. Girardi, G. Bertelli, A. Bressan, et al.: A&A **391**, 195 (2002)
3. M. Hetterscheidt, P. Simon, M. Schirmer, et al.: A&A **468**, 859 (2007)
4. H. Hildebrandt, D. J. Bomans, T. Erben, et al.: A&A **441**, 905 (2005)
5. H. Hildebrandt, T. Erben, J. P., Dietrich, et al.: A&A, **452**, 1121 (2006)
6. H. Hildebrandt, J. Pielorz, T. Erben, et al.: A&A, **462**, 865 (2007)
7. H. Hildebrandt, C. Wolf, N. Benitez: arXiv: 0801.2975 A&A in press (2008)

Discussion

A. Grado: What is the SW insert to perform the astrometric calibration?
H. Hildebrandt: ASTROMETRIX.
F. Selman: The illumination correction appears to be similar in BVR but there are reports of color dependencies, such as going redder as the correction increases, reaching 30% peak-to-valley in I.
H. Hildebrandt: This might be true. You may still be left with color errors of a couple of percent.

Calibrating VISTA Data

J. Emerson[1], S. Hodgkin[2], P. Bunclark[2], M. Irwin[2], and J. Lewis[2]

[1] Astronomy Unit, Queen Mary University of London, Mile End Road, London, E1 4NS, UK; j.p.emerson@qmul.ac.uk
[2] Cambridge Astronomical Survey Unit, Institute of Astronomy, University of Cambridge, Madingley Road, Cambridge CB3 0HA, UK

Abstract. When VISTA is operating with its 16 detector 2048×2048 pixel camera it will produce some 300 GB of data each operational night. Calibration plans are outlined, making much use of the 2MASS catalogue, and build on experience with WFCAM on UKIRT.

1 Introduction

VISTA, the Visible & Infrared Survey Telescope for Astronomy [5], is a 4 m Survey Telescope equipped with a single instrument, a 1.65 degree diameter field of view near-infrared camera with only one moving part, the filter wheel, which is equipped with broad-band filters for Z, Y, J, H, & K_s (0.9–2.5 µm), and two spaces for other filter sets. VISTA is located on a peak $\sim$ 1500 m from the VLT at the Cerro Paranal Observatory in Chile, and is being provided to ESO by the UK.

The VISTA camera's focal plane consists of 16 Raytheon 2048 × 2048 VIRGO near-IR detectors with 0.34 arcsec pixels arranged in a sparse 4 × 4 array with spacings of 90% & 42% of detector. A single exposure with VISTA produces a sparsely filled 0.6 sq degree detector 'pawprint'. These pawprints can be most efficiently turned into filled 'tiles' by taking a 3 × 2 set of suitably offset pawprints which when combined provide an area of 1.5 sq degrees covered by (at least) two exposures. VISTA has a wide field of view, so particular attention must be paid to variations across the field, and VISTA's calibration observations are designed to characterize the transfer function of the end-to-end system (image in, data number out) and in particular:

1. the instrument characteristics
2. the astrometric distortions of the images
3. the photometric zero points, distortion (illumination correction) and extinction coefficients.

Using calibration observations and the data itself the VISTA calibration pipeline [6, 2] removes instrumental artefacts, combines the component exposures offset by small jitters into a pawprint, calibrates each pawprint photometrically and astrometrically, and finally provides Quality Control measures [4]. With its sixteen detectors and the relatively short integration times typical of infrared detectors, VISTA will on average (i.e. including weather factors) produce ~ 250 GB of data each night which far exceeds the data volumes produced by the instruments on all the VLTs. The need to handle a continuous high volume data stream implies that VISTA calibration must be a highly automated process, and here we describe our plans to cope with this.

2 Strategy: Prepare for VISTA with WFCAM

Given that we will not have real images with VISTA until it is commissioned later in 2007, we wanted to mitigate the risks in learning to properly handle VISTA data (and its large volume), to ensure data processing will keep up with the incoming data flood. The strategy we adopted has been to first design a data flow system (running in the UK) to handle data from UKIRT's wide-field infrared camera WFCAM, with upgrading to handle VISTA data and volumes in mind, and then to build on this and the experience gained to produce the system for VISTA. The comparison between WFCAM and VISTA is shown in Table 1.

Of course the two systems differ in many details, but setting up a system (the VISTA Data Flow System - VDFS) that is demonstrated to handle WFCAM data well provides the best way to prepare for VISTA data, and we believe that debugging of VISTA processing, including calibration, will be speeded up and de-risked through the WFCAM experience. The examples of data shown in this paper are based on WFCAM data. There are three other talks at this workshop which touch on the VDFS work [2, 4, 1].

3 Instrument Characteristics

3.1 Cross-Talk & Persistence

The WFCAM camera on UKIRT suffers from cross-talk from saturated images. The cross-talk gives 'bumps' symmetrically above and below saturated stars (left hand side of Fig. 1). For WFCAM these are mostly (but not perfectly) correctable (right hand side of Fig. 1). Whilst there is no evidence

Table 1. Similarities between WFCAM and VISTA

	WFCAM	VISTA
Telescope	4 m (UKIRT)	4 m
Camera pixel size	0.4 arcsec	0.34 arcsec
2k×2k Detectors	4 × Hawaii (Rockwell)	16 × VIRGO (Raytheon)
Time camera on	$\sim 50\%$	100%

Fig. 1. Cross Talk from saturated stars in WFCAM observations (of the open cluster M67) before (*left*) and after (*right*) artefact cleaning

so far that VISTA will suffer from any cross-talk, if it does then a similar strategy to that used for WFCAM should largely remove it.

WFCAM also suffers from persistence/remanence effects after observing bright objects. Figure 2 shows the persistence artefacts from a previous 9-point jitter sequence (9 closely spaced decaying images) of the Open Cluster field shown in Fig. 1.

Persistence has proved much harder to completely correct, presumably because the characteristics of the persistence are not constant enough with time. It is not known if VISTA will show similar effects (the detectors and controllers are different) but we plan to take calibrations in case this effect needs to be corrected.

To characterize any persistence effects for VISTA we will choose a fairly empty field containing a star which saturates the VISTA detector. On a

Fig. 2. Persistence effect (pattern of 9 decaying artefacts in the middle and upper middle) for a WFCAM frame following on after frames on M 67

sequence of (monthly) dates we will take an exposure of the saturated star and then a sequence of dark frames to measure the characteristic decay time, which we hope will prove stable.

3.2 Linearity, Flat Fielding & Illumination Correction

Although VISTA has a Dome 'Flat screen' this will not be used for flatfielding but rather for monitoring instrument performance, image structure and generating confidence maps, and measuring system linearity.

To determine any non-linearity we will take a series of differently timed dome flat screen observations under constant illumination, which, with the pixel timing, will give the true linear value for each pixel & bad-pixel maps for each detector.

Twilight (flat) fields, will be used to remove multiplicative instrumental signatures and determine: pixel-to-pixel gain variations; the instrumental vignetting profile; the gain correction between the 16 detectors; and the gain correction between the 16 read out channels of each detector.

The Flat-field should remove all pixel-to-pixel gain differences as well as any large-scale variations due (generally) to vignetting within the focal plane, but any scattered light within the camera may lead to large scale background variations (requiring illumination correction) which cannot easily be modeled and removed, as the level depends critically on the ambient flux.

The illumination correction will be measured in three ways:

1. Observe secondary photometric standard fields (100–200 objects per detector) and the variation of zero-point across each detector gives a map of the spatial systematics across each detector
2. Carry out 'mesostep' sequence of exposures of a sparse field of relatively bright stars on a regular grid of offsets, to completely sample across the face of the detectors in medium-sized steps to monitor residual systematics in photometry
3. Use the stacked zero point differences from 2MASS objects in each pawprint (see Sect. 5).

4 Astrometric Calibration

Astrometric calibration is based on the 2MASS point source catalog which is calibrated globally to ~ 100 milliarcsec and internally to $<\sim 50$ milliarcsec and tied to the TYCHO-2 catalogue and is in International Coordinate Reference System (ICRS).

The wide field of view means that the effects of distortion across the field of view need to be handled particularly carefully. The strongest term in the optical-distortion model is a cubic radial term and the actual distortion will be determined for VISTA from on-sky observations of 2MASS stars.

This radial scale distortion also has an impact on photometric measurements, inducing an error up to 3.5% in the corners of the field, compared to the centre, if uncorrected.

5 Photometric Calibration

To achieve the 2% accuracy goal in the photometric calibration two independent methods are used to derive zero-points for each image.

Method 1: from the 2MASS all-sky point source catalogue

Method 2: from routine observations of standard star fields

Together these allow monitoring of effective zero-points at the ∼ few % level. Subsequent inter-detector comparisons enable residual errors in the gain correction to be detected and calibrated. Off-line analysis gives a measure of median zero-point for the night, and associated error (and scatter) which is indicative of photometric quality.

5.1 2MASS Magnitudes

The initial photometric calibration for all filters will be based on the 2MASS photometric system which is globally consistent to ∼ 1% [3]. We will use colour equations (with some colour and signal-to-noise cuts) to convert 2MASS magnitudes to the VISTA instrumental system which enables each detector image to be calibrated directly from 2MASS stars that fall within field of view. Analysis of WFCAM data with respect to UKIRT standards shows that using 2MASS stars for WFCAM calibration delivers good frame-by-frame photometric zero-points at the ± 2% level (with factored-in extinction tracking). VISTA should be calibrated at least as well.

5.2 Network of Standard Star Fields

We will also setup a network of Secondary Standard photometric fields, spaced every ∼ 2 h in RA. These will be 2MASS 'Touchstone' fields and/or UKIRT faint standard fields which will provide ∼ 100 stars per detector J < 18, K_s < 16 to avoid long exposures which will characterize systematic position dependent photometric effects. These will encompass a broad spread in colour to derive colour terms robustly, and will be observed every two hours elapsed time throughout each night enabling an independent calibration to be made on a nightly basis. The Touchstone fields will also provide information on the stability and can be used to measure the illumination correction.

5.3 Extinction Coefficients

The extinction will be monitored through each (photometric) night assuming a fixed zero point and measuring the touchstone fields over a range of airmass. We then use the 2MASS stars in each pawprint to determine the extinction.

During nights which are less photometric we will assume a fixed zero point and use the 2MASS stars in each pawprint to determine the extinction coefficients as a function of time.

6 Conclusion

We have prepared our calibration plan for VISTA with the benefit of experience of handling and calibrating data from UKIRT's WFCAM, the closest operational analogue to VISTA. The astrometric and photometric calibrations will benefit greatly from the availability of the well-calibrated 2MASS survey. When VISTA is operational and we have characterized its performance we expect to be able to get well calibrated without much delay.

Acknowledgement. We thank the UK's Science and Technology Facilities Council (STFC) for their support of this work.

References

1. M. Irwin: this proceedings (2008)
2. J. Lewis, M. Irwin, P. Bunclark, S. Hodgkin: this proceedings (2008)
3. S. Nikolaev, M.D. Weinberg, M.F. Skrutskie, R.M. Cutri, S.L. Wheelock, J.E. Giziz: Astrophys. J **120**, 3340 (2000)
4. M. Riello: this proceedings (2008)
5. http://www.vista.ac.uk
6. VISTA Infra Red Camera Calibration Plan, VIS-SPE-IOA-20000-0002 v1.4 (2007)

Discussion

U. Hopp: How large is the persistence effect (the remnant) and its decay time for WFCAM and for VISTA?
M. Irwin: The magnitude of the effect is a large fraction of a percent and only happens for saturated images. The time of full decay is around one minute.
J. Emerson: The effect for VISTA has not yet been characterized in detail. This awaits getting VISTA on sky. The detectors are not from the same manufacturer, so there is no reason to suppose VISTA will show the same persistence as WFCAM.
A. Grado: Is ESO planning to transfer VST/VISTA data through network?
A. Kaufer: No, due to the data volume a transfer of the data from the observatory to the ESO science archive and the data centers is no option – at least today. The data will be shipped on hard disks.

The VISTA Data Flow System

J. Lewis, M. Irwin, P. Bunclark, and S. Hodgkin

Cambridge Astronomy Survey Unit, Institute of Astronomy, Madingley Road, Cambridge CB3 0HA, UK; jrl@ast.cam.ac.uk

Abstract. The Visible and Infrared Survey Telescope for Astronomy (VISTA) is a new ESO telescope that will be commissioned in late 2007. The telescope has been specifically designed for survey work in both the visible and near infrared, but to start with will only have an infrared camera (VIRCAM). The focal plane of VIRCAM will consist of 16 2k by 2k non-buttable detectors with a pawprint of 0.6 square degrees. With an expected nightly data rate of 200–500 Gb, automated pipeline processing and data management requirements are paramount.

Pipeline processing of IR data is far more technically challenging than for optical data. IR detectors are inherently more unstable and the sky emission is over 100 times brighter than most objects of interest and varies in a complex spatial and temporal manner. In this presentation we describe the pipeline architecture developed to deal with the IR imaging data from VISTA. We discuss the issues involving robustly removing instrumental signatures, sky correction, astrometric and photometric calibration. We also describe some of the checks that have been put in place to monitor data quality and system integrity.

1 Introduction to CASU and VDFS

The Cambridge Astronomy Survey Unit (CASU) is a small group within the Institute of Astronomy at Cambridge which specialises in survey astronomy. Amongst the group's many specialities are the design, implementation and running of reduction pipelines for imaging data in the optical and the infrared. These activities began in the early 1980s with the scanning of Schmidt plates on the Automatic Plate Measuring machine (APM) and continues to the present day with pipelines that cover many optical and infrared imaging instruments.

The Particle Physics and Astronomy Research Council in the UK have funded the Visible and Infrared Survey Telescope for Astronomy (VISTA) as an in-kind contribution to offset part of the cost of the UK's fee for joining ESO. A major part of the VISTA project is an end-to-end data flow system (the VISTA Data Flow System or VDFS). Although this was specifically targeted at VISTA, the project, as it was conceived, included a data flow system for the WFCAM instrument on UKIRT, which was to be used as a testbed for VISTA. The WFCAM part of the VDFS is currently in operation and consists of a data reduction pipeline in Cambridge and a science archive run by the Wide Field Astronomy Unit (WFAU) in Edinburgh. The VDFS for VISTA is to contain the following:

Quality Control and Calibration Pipelines These will run at Paranal and Garching to extract data quality and calibration information. Although they will use exactly the same software, their results will probably differ as the pipeline in Garching will have access to more up-to-date calibration information.

Science Pipeline This will be run in Cambridge and will be responsible for the full science calibration of the data. There will be some differences between this pipeline and the above mentioned QC pipelines. The major reason for this is that the QC pipelines will work serially by reducing one OB at a time without reference to other OBs taken during the night. The science pipeline will have access to the whole night's data and can optimise its reduction schedule accordingly.

Science Archive This will be run by the Wide Field Astronomy Unit (WFAU) in Edinburgh and will be the single point of contact for the reduced VISTA data for users.

The actual data flow will start with the raw data being ingested into the summit data archive and then assessed by the summit pipeline. This will be used to measure the QC1 parameters required by ESO to assess the health of the system and the observing conditions. The data will then be shipped to Garching on removable discs where it will be ingested into the Garching archive and then assessed by the Garching pipeline for data quality purposes. Finally the discs will be shipped to CASU for the full science data reduction. It is hoped that some time in the near future these raw data transfers can be done via the Internet. The reduced data products will be sent via the Internet to WFAU for ingestion into the science archive.

2 IR Data Reduction

Although there is a great deal of overlap between the techniques used to process optical and infrared imaging data, the latter can be a much more challenging problem [1]. The odd electronic effects (some of which we will discuss later) caused by the fact that IR detectors are inherently more unstable than optical CCDs mean that a lot of extra work needs to be done to remove artifacts from the data. In addition the sky emission is typically a factor of 100 times brighter than most of the objects of interest and it varies both spatially and temporally. To get around this problem of a variable sky, it is standard practice to keep exposure times low. The consequence of this, however is a very high nightly data rate. WFCAM with its 4 detectors can generate between 100–300 Gb per night. The expected data rate for the VISTA public surveys will be 200–500 Gb per night. Fortunately Rice tile compression [2] can be used losslessly on 32 bit integer data and save factors of 3–4 in the total required disc space.

A typical observing sequence for VISTA will entail jittering the telescope around a central location with a series of offsets. This is used to help rid the reduced data of bad pixels. The resultant image is known as a 'pawprint'. It may also be necessary during times of the very best seeing to microstep the telescope. This is done by taking four observations with each exposure being shifted relative to the first one by $N + 0.5$ pixels in the x axis, the y axis and then both, (where N is an exact integer number of pixels and is possibly zero). The resulting four images have their pixels shifted by N and then interleaved onto an output grid of roughly twice the number of pixels as the input images. This is done to recover some of the spatial resolution that may be lost due to undersampling during periods of excellent seeing conditions.

A bias image must be taken with each individual exposure but is subtracted in the data acquisition system. This differencing operation is called 'Correlated Double Sampling' and has the effect of halving the data rate. These difference images are the 'raw' data that will enter the pipeline.

The first steps in the pipeline are to remove the dark current and the non-linearity in the data and to correct for the 'reset anomaly'. The latter effect often shows up at the readout boundary of a channel in an infrared array. The bright parts of Fig. 1 show an example of reset anomaly for WFCAM.

The first laboratory tests of the VIRCAM detectors show very small or no reset anomaly, although experience has shown that it can depend on the level of ambient flux.

Dividing the frame by a twilight flat field will correct for the variation in quantum efficiency between individual pixels and will also remove any large scale variations due to vignetting. Subtracting a 2d background estimate will correct for any large scale additive features, such as the sky emission, fringes and any thermal emission which might result from dust in the optical trail.

Fig. 1. The reset anomaly from a dark current image on WFCAM. Note the bright outside edges on each of the readout quadrants

Three additive electronic effects must now be removed.

stripes Laboratory tests have shown a low level striping effect in the background of VIRCAM frames. These exist as both narrow and broad features as shown by Figs. 2 and 3. Figure 3 shows that both the narrow and broad stripe patterns repeat over all four of the detectors on a single controller. This means there will be a great deal of data redundancy when it comes to modelling and subtracting the effect in the Cambridge pipeline.

Crosstalk Electronic crosstalk is a common problem on digital detectors. Very often images of bright objects will generate ghost images at some location in the focal plane offset from the original. With WFCAM the problem is slightly more complicated by the fact that the ghosts of the bright images appear to be the first derivative of the original (taken along the readout axis). A good example of this is shown in Fig. 4. At the time of writing we have no information about crosstalk in the VIRCAM detectors and this will have to wait until commissioning to assess its total effect.

Persistence Persistence occurs when a ghost of a bright object is visible even after the source has been moved away from the affected pixels. Like the crosstalk, characterising this effect in VIRCAM will have to wait until commission for an assessment of the effect it has on the quality of the output data products.

Once these electronic effects have been removed, what remains to be done to the images is to interleave any images that have been taken as part of a microstep sequence. The final pawprint image is formed by shifting and adding all images that form part of the observed jitter sequence. From this pawprint image the astronomical objects can be extracted and catalogued. Using this object catalogue and the 2MASS catalogue it is possible to calibrate the world coordinate system of the image to better than 100 milli-arcseconds and to calibrate the photometric zeropoint to about 0.02 magnitudes.

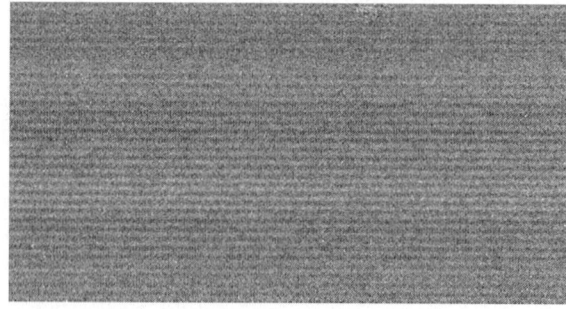

Fig. 2. A closeup of a section of detector 5 showing the high frequency stripes

The VISTA Data Flow System 569

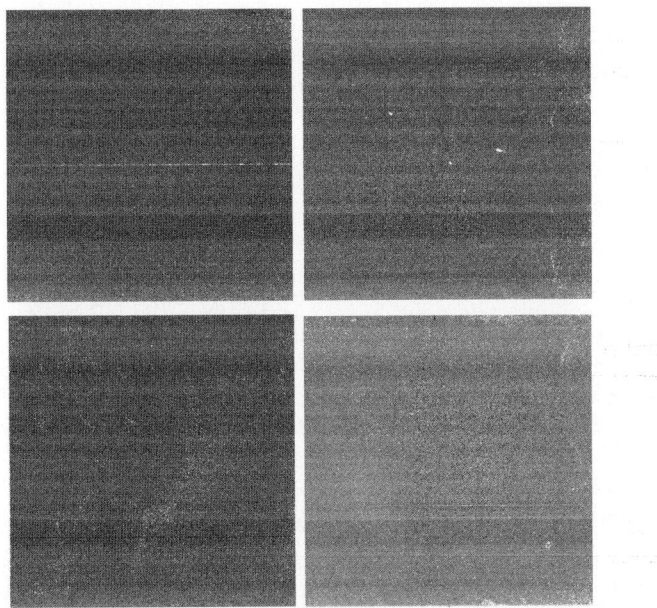

Fig. 3. Detectors 5–8 all showing the same faint broad bands in the background

Fig. 4. A closeup of a bright object on a WFCAM image showing two crosstalk ghosts near the *right* and *left* edges

At this point the data are fully calibrated. However, because the detectors in VIRCAM are not buttable it is necessary to observe six pawprints in order to fill in the gaps properly and create a single contiguous image. Creating a tile from these six pawprints will only be done as a final stage of the Cambridge pipeline.

3 Pipelines and Project Status

The VDFS pipelines will contain recipes to cover the following areas:

- Create master calibration frames (dark, twilight flats, etc)
- Do linearity analysis
- Assess detector noise and dark current properties
- Analyse crosstalk and persistence test data
- Do illumination correction analysis
- Do full reduction of science and standard star fields

The summit and Garching pipelines exist primarily to generate quality control parameters such as photometric zero points and astrometric fit quality. They are written using ESO's qfits/CPL based infrastructure and both will use the same software modules. These reduce data at the pawprint level as they will have no knowledge of how to link pawprints together into a tile. The amount of reduction can be tuned either up or down depending upon time constraints. Calibration images (flats, etc) will come from master calibration libraries.

The Cambridge pipeline will have considerably more flexibility to treat a whole night's data as a single entity. This will allow for better sky correction algorithms and for full tiling of the final results. Catalogue generation will be done for both the pawprints and the tiles. Much of the CPL based software that is used in the QC pipelines will be reused here, but these will be supplemented by new modules to add the flexibility we want.

At the time of writing version 0.5 of the summit and Garching pipeline software has been released to ESO as part of the Preliminary Acceptance for Europe milestone. As this contains nearly a full compliment of recipes (the crosstalk and persistence recipes must wait until commissioning) subsequent releases prior to first light are expected to consist mainly of bug fixes.

References

1. M.J. Irwin: 2007, these proceedings
2. C.N. Sabbey: PASP **112**, 867 (2000)

Discussion

U. Hopp: Does your pipeline include error propagation?

J. Lewis: Yes, the error propagation is done through the "confidence map". A confidence map folds bad pixel map, exposure map and weight map concepts into a single image. The pipeline creates a confidence map for each science image. These are used in subsequent processing such as weighting of pixels during catalogue extraction.

S. Mieske: What is the reason for having two pipelines for VISTA, namely the ESO one and the Cambridge one?

J. Emerson: The one at ESO is to deliver the standard calibration functionality required by ESO for its instruments. The VISTA Data Flow System pipeline at Cambridge is to assist PIs in converting their VISTA data into appropriate science products for their surveys. ESO itself decided some years ago that it could not provide such a facility for VISTA (or VST).

M. Schirmer: What observing strategies are implemented for VISTA surveys in order to ensure proper sky background modeling in the case of extended targets?

J. Lewis: We are currently in discussions with all the survey PIs to ensure that they are fully aware of the sky estimation issues and that they take the time to design their OBs around this.

SkyMapper and the Southern Sky Survey a Resource for the Southern Sky

S. C. Keller, B. P. Schmidt, and M. S. Bessell

Research School of Astronomy and Astrophysics, Cotter Rd., Canberra, ACT 2611, Australia; stefan@mso.anu.edu.au

Abstract. SkyMapper is amongst the first of a new generation of dedicated, wide-field survey telescopes. The 1.3 m SkyMapper telescope features a 5.7 square degree field-of-view Cassegrain imager and will see first light in late 2007. The primary goal of the facility is to conduct the Southern Sky Survey a six colour, six epoch survey of the southern sky. The survey will provide photometry for objects between 8th and 23rd magnitude with global photometric accuracy of 0.03 magnitudes and astrometry to 50 mas. This will represent a valuable scientific resource for the southern sky and in addition provide a basis for photometric and astrometric calibration of imaging data.

1 The SkyMapper Telescope

The SkyMapper telescope is a 1.3 m telescope currently under construction by the Australian National University's Research School of Astronomy and Astrophysics in conjunction with Electro Optic Systems of Canberra, Australia. The telescope will reside at Siding Spring Observatory in central New South Wales, Australia.

The telescope is a modified Cassegrain design with a 1.35 m primary and a 0.7 m secondary. Corrector optics are of fused silica construction for maximum UV throughput and a set of six interchangeable filters can be placed in the optical path. The facility will operate in an automated matter with minimal operator support. Further details on all aspects of our programme can be found in [4].

2 Detectors and Filters

The focal plane is comprised of 32 2k×4k CCDs from E2V, UK. Each CCD has 2048 × 4096, 15 µm square pixels. The devices are deep depleted, backside illuminated and 3-side buttable. They possess excellent quantum efficiency from 350–950 nm (see Fig. 1), low read noise and near perfect cosmetics.

The SkyMapper imager will utilise the recently developed STARGRASP controllers developed for the Pan-STARRS project by Onaka and Tonry et al.

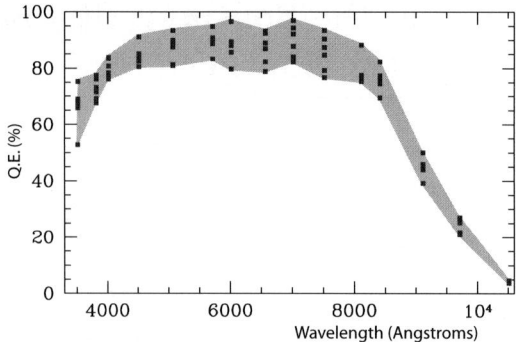

Fig. 1. Spectral response of SkyMapper CCDs as measured in the laboratory. The *shaded* area encloses the range in response exhibited

of the University of Hawaii [5]. Twin 16-channel controllers enable us to read out the array in 12 s with ∼4−5 electron read noise.

Figure 2 shows the expected normalised throughput of our system. The filter set is based upon the Sloan Digital Sky Survey filter set with three important modifications: the movement of the red edge of the u filter to the blue, the blue edge of the g filter to the red, and the introduction of an intermediate band v filter (essentially a DDO38 filter). At this time coloured glass fabrication of filters of these bandpasses offers the best solution for spatial uniformity compared to the competing interference film technology. Our filters are sourced from MacroOptica of Russia.

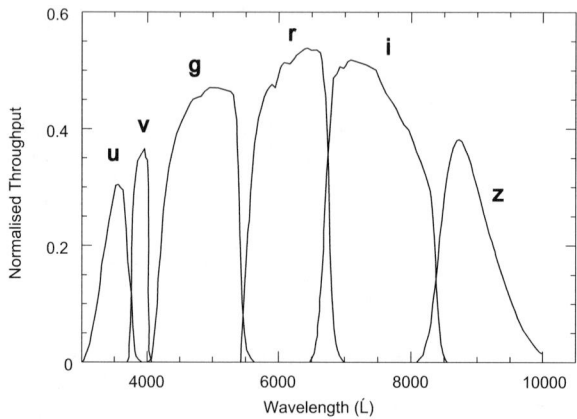

Fig. 2. The predicted throughput of the Southern Sky Survey filter set, excluding atmospheric absorption

Table 1. Southern Sky Survey limits (5 sigma) in AB magnitudes from multiple 110 second exposures

	u	v	g	r	i	z
1 epoch	21.5	21.3	21.9	21.6	21.0	20.6
6 epochs	22.9	22.7	22.9	22.6	22.0	21.5

3 The Southern Sky Survey

Performing the Southern Sky Survey is the primary preoccupation of the SkyMapper telescope. The survey will cover the 2π steradians of the southern hemisphere reaching $g=23$ at a signal-to-noise of 5 sigma. For stars brighter than $g=18$ we require global accuracy of 0.03 magnitudes and astrometry to better than 50 mas.

The survey's six epochs are designed to capture variability on the time scales of days, weeks, months and years over the five year expected lifetime of the survey. The 5 sigma limits attained after one 110 second epoch and after the full six epochs are given in Table 1. In all bands we attain limits slightly deeper (~ 0.5 mag) than the Sloan Digital Sky Survey.

4 Global Photometric Calibration

The greatest impediment to deriving accurate photometry from wide field imaging cameras is the accurate description of the illumination correction. The illumination correction corrects for geometry of the optics and inclusion of scattered light in the system (see Patat and Freudling these proceedings).

During commissioning we will develop an illumination correction for each filter via dithered observations of a field. We will then rotate the instrument and repeat the dithered observations to ensure we rigourously understand the illumination correction for the system. We will establish six such reference fields at declinations of around $-25°$ and spaced in right ascension. Each field will be 4.6 degrees square following the dither pattern.

During the first year of operation we will perform the Five-Second Survey, a rapid survey in photometric conditions to provide all-sky standards between 8–16th magnitude. The Five-Second Survey will consist of a set of at least three images of a field in all filters.

During Five-Second observing we will observe the two highest of our six reference fields every ninety minutes. This will ensure photometry is obtained on a highly accurate standard instrumental system. The Five-Second Survey will provide a network of photometric and astrometric standards to anchor the deeper main survey images. Furthermore, it enables the main survey to proceed in non-photometric conditions.

We will establish the six reference fields to include stars with photometry in the Walraven system [6]. As demonstrated by Pel and Lub (ibid), the Walraven system zeropoint is highly accurate: the closure solution over 2π in right ascension has rms of less than 1 m mag. In addition, the Walraven stars we have selected are spectro-photometric standards from [3]. The use of these standards will provide absolute flux calibration for our system.

5 A Filter Set for Stellar Astrophysics

The majority of science goals identified for SkyMapper are based on the identification of stellar populations. It was therefore fundamental to the science output of the telescope that we choose a filter set that offers optimal diagnostic power for the important stellar characteristics of effective temperature, surface gravity and metallicity. Below I will discuss some specific examples.

Through an exploration of colour parameter space derived from model stellar atmospheres and filter bandpasses we arrived at the filter set shown in Fig. 2. The filter set possesses two filters, u and v, distinctly either side of the Balmer Jump feature at 3646 Å.

5.1 Blue Horizontal Branch Stars

Figure 3 shows the uncertainty in the derived stellar surface gravity as a function of temperature for a range of surface gravities with photometric uncertainties of 0.03 mag per filter. In the case of A-type stars we expect to determine surface gravity to ~10%. The sensitivity to surface gravity arises from the $u-v$ colour which measure the Balmer Jump and the effect of H$^-$ opacity, both of which increase with surface gravity. It is at these temperatures that we find blue horizontal branch stars (BHBs). Due to their characteristic absolute magnitude BHBs are standard candles for the Galactic halo.

A line of sight through the halo inevitably contains a mixture of local main-sequence A-type and blue straggler stars. However as is shown in Fig. 3 the SkyMapper filter set enables us to clearly distinguish the BHBs of interest on the basis of their lower surface gravity. Simulations show that we will be able to derive a sample of BHBs to 130 kpc with less than 5% contamination.

5.2 Extremely Metal-Poor Stars

In the case of cooler stars (F0 and cooler) the u and v filters indicate the level of metal line blanketing blueward of $\sim$ 4000 Å. Figure 4 shows the $v-g$, $g-i$ colour-colour diagram for a range of metallicities and surface gravities. The $v-g$ colour has a strong dependency on the metallicity and little dependency on the surface gravity hotter than K0 ($g-i \sim 1.7$).

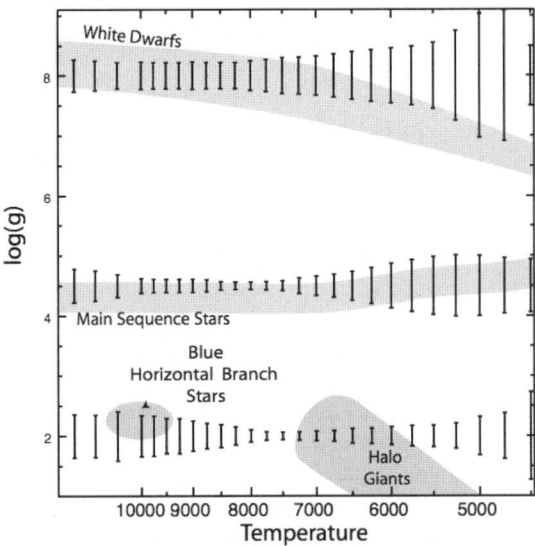

Fig. 3. Precision of determined surface gravity from our filter set as a function of temperature and surface gravity (error bars show estimated uncertainties at each point for photometric uncertainties of 0.03 mag in each filter)

This enables us to cleanly separate the extremely metal-poor stars in the halo from the vast bulk of the halo at [Fe/H]< −2. Our simulations show we should find of order 100 stars with [Fe/H]< −5.

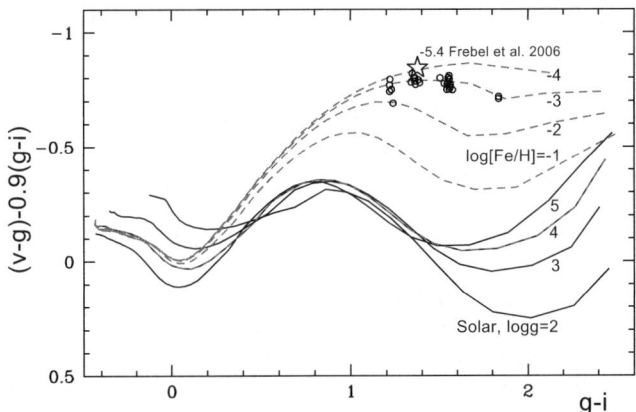

Fig. 4. $v-g$ vs. $g-i$ for stars of solar metallicity (*dashed lines*) and for a range of surface gravity (*solid lines*). Open circles are stars from the sample of [1] and the star symbol is HE1327-2326 from [2]

6 Data Products and Their Possible Application to ESO Calibration

The first SkyMapper data product will be the Five-Second Survey of 8–16th magnitude stars in the southern hemisphere. Two main survey data releases will follow. The first data release will occur when three images in each filter have been reduced for a field and the second (reaching 23rd magnitude in g) when the full set of six have been obtained and undergone quality control.

The survey will provide sufficient density and spectral sampling of standard stars to enable photometric calibration of any field imaged in any broadband filter in the southern hemisphere. The largest source of dispersion in transformations between photometric systems is due to the lack of knowledge of the surface gravity and metallicity of the sample. The SkyMapper photometric system provides a prior on both these points of uncertainty. Consequently we will be able to provide improved transformations from our photometric system to any other system. Scheduled observations on, for instance VLT or VST, may then dispense with photometric standards and also proceed under non-photometric conditions.

References

1. R. Cayrel, et al.: A&A **416**, 1117 (2004)
2. A. Frebel, et al.: Nature **434**, 871 (2005)
3. M. Gregg: Next Generation Spectral Library. In: `http://lifshitz.ucdavis.edu/~mgregg/gregg/ngsl/ngsl.html`
4. S. Keller, et al.: PASA **24**, 1 (2007)
5. P. Onaka, J. Tonry: StarGrasp – Detector Controllers for Science and Astronomy. In: `http://www.stargrasp.org/`
6. J.W. Pel, J. Lub: The Walraven System. In: *The Future of Photometric, Spectrophotometric and Polarmetric Standardization*, ASP Conference Series 364, PASP ed by C. Sterken, pp. 63–80 (2007)

Discussion

U. Hopp: You mentioned that your glass filters are more homogeneous than interferences filters. Did you measure their spatial stability?
S. Keller: Yes. This has been measured in the laboratory. The center to edge variation in peak throughput is $\sim 0.5\%$. Shifts in bandpass edges are < 10 Å.
U. Hopp: You intend to do the Five-Second Survey under photometry conditions. From where do you know that the night is photometric?
S. Keller: We rely on the repeated observations of our standard star fields. Additional information on conditions can be extracted from stars with 2MASS photometry (in our r,I,z filters) and Tycho (in g). If doubt exists, we will reject the entire night as non-photometric.

W. Freudling: Did you consider to use drift scans instead of pointed observations for your survey? What is the advantage of the pointed observations? What is the overhead per exposure for the 5 s exposures?

S. Keller: Drift scan observations would not provide sufficient depth as our aperture is modest. The overhead for exposures is of order 15 s.

Quality Control Monitoring for WFCAM

M. Riello and M. Irwin

Cambridge Astronomy Survey Unit, Institute of Astronomy, Cambridge University, Madingley Road, Cambridge CB3 0HA, UK; mriello@ast.cam.ac.uk

Abstract. We present the data Quality Control (QC) infrastructure that has been put in place at the Cambridge Astronomical Survey Unit (CASU) to deal with the large data volume produced by WFCAM. QC measures are produced during pipeline processing and saved in the FITS headers and, afterwards, ingested into a relational database that serves a number of report-generating tools. QC measures includes: sky brightness and noise, average stellar ellipticity and seeing, astrometric calibration errors, per-image and nightly-averaged photometric zero points and errors and the number of detected sources per chip. The QC system has been used also to investigate the near-infrared sky brightness at Mauna Kea Observatory over a period of 3 semesters.

1 WFCAM Data Quality Control

The United Kingdom Infra-Red Telescope (UKIRT) Wide Field Camera (WFCAM) on Mauna Kea began operations in the first quarter of 2005 and is currently the most capable near-infrared (NIR) imaging survey instrument in the world [1]. The camera is equipped with four Rockwell-Hawaii-II $2\,\text{k} \times 2\,\text{k}$ detectors, each covering about 13.7×13.7 arcmin of the sky, with a separation of $\approx 95\%$ of the detector size and is capable of providing $\approx 0.8\,\text{deg}^2$ with a 4-pointings filled tile, see e.g. [2]. A large fraction of WFCAM time is dedicated to the implementation of the UKIRT Deep Infrared Sky Survey (UKIDSS), a set of five sky surveys with different filter sets, area coverage and scientific objectives [4]. Since WFCAM was commissioned in early 2005, all the raw data have been shipped to Cambridge to be processed by CASU as part of its involvement in the VISTA Data Flow System (VDFS) [5]. During the first three semesters of operations, over 30 Tb of raw data have been pipeline-processed by CASU producing $\approx 60\,\text{Tb}$ of reduced images ($\approx 3.5 \times 10^{13}$ pixels), catalogues ($\approx 10^9$ detected objects) and confidence maps.

To deal with such a large data volume in an effective way, we developed a quality control database based on PostgreSQL that stores all the data quality information for every data product and input science frame. In addition several other relevant pieces of information are extracted from the FITS headers and ingested in the database: complete WCS calibration, weather conditions, temperatures, fractional lunar illumination (FLI), Moon angular distance and elevation, dusk and dawn twilight start times, etc. The ingestion of a pipeline-processed night into the QC database has also proved to be very effective in spotting a number of problems that may occur during the reduction. The QC database has a web-based front-end that is used internally by CASU members to rapidly search for images at a specific position on the sky and, optionally, satisfying user-specified constraints for any QC parameter. The front-end is also complemented with an image cut-out service that is used to create on-the-fly object postage stamps and full-chip previews.

The QC database is also used to keep track of the night status within the processing data flow, to flag missing files when the raw data tapes are ingested, to record every single frame that was transferred to ESO (raw data obtained in the framework of the UKIDSS surveys are public to all ESO member countries) to be ingested in their archive (more than 400000 FITS files have been transferred so far). Such information is used to generate user-friendly web pages (Fig. 1) publicly accessible from the CASU web site[1]. For each semester a list of the available nights is kept up-to-date via a simple daemon-like script that is synchronising the database with the current status of the data every hour. The processing status web pages show, for each night: (1) the date of raw data ingestion at CASU; (2) the current status of the night; (3) number of raw images (including calibration frames); (4) the date when the data was flagged by CASU as ready to be transferred by Wide Field Astronomy Unit (WFAU) in Edinburgh for ingestion into the WFCAM Science Archive; (5) the date when the data were actually transferred by WFAU; (6) three QC summary plots (Fig. 2) showing, for each stacked frame, the photometric zero points, average seeing (FWHM), average stellar ellipticity, the sky brightness and the $5-\sigma$ magnitude limit deduced from the sky noise; (7) the total size in Gb of the raw and reduced data (compressed); (8) the total number of stacked images, the total number of stacked images for the UKIDSS surveys altogether and by survey (once the night has been fully processed and ingested in the QC database). The QC database is also feeding a web-based survey monitoring tool developed by E. Gonzalez-Solares

This page displays the reduction progress of WFCAM data. Information is automatically updated every hour (you need to reload the page).

« Back to WFCAM start page.

Night	Status	N_{raw}	N_{ESO}	Checked	Transferred by WFAU	Version	Summary Plots	Photometry Plots	Summary Info
2006/04/28	reduced	476	361	23 Aug 2006	06 Sep 2006	1	GIF1 GIF2	GIF	summary.list
2006/04/29	reduced	411	366	23 Aug 2006	06 Sep 2006	1	GIF1 GIF2	GIF	summary.list
2006/04/30	reduced	145	133	23 Aug 2006	06 Sep 2006	1	GIF1 GIF2	GIF	summary.list
2006/05/01	reduced	1055	794	23 Aug 2006	06 Sep 2006	1	GIF1 GIF2	GIF	summary.list
2006/05/02	reduced	1290	139	23 Aug 2006	06 Sep 2006	1	GIF1 GIF2	GIF	summary.list
2006/05/03	reduced	2402	640	23 Aug 2006	06 Sep 2006	1	GIF1 GIF2	GIF	summary.list
2006/05/04	reduced	1564	182	23 Aug 2006	06 Sep 2006	1	GIF1 GIF2	GIF	summary.list
2006/05/05	reduced	1639	116	23 Aug 2006	06 Sep 2006	1	GIF1 GIF2	GIF	summary.list
2006/05/06	nodata	66	55						
2006/05/07	nodata	0							
2006/05/08	nodata	554							
2006/05/09	nodata	768	84						
2006/05/10	reduced	2516	1918	25 Aug 2006	27 Sep 2006	1	GIF1 GIF2	GIF	summary.list
2006/05/11	reduced	2603	2553	25 Aug 2006	23 Sep 2006	1	GIF1 GIF2	GIF	summary.list
2006/05/12	reduced	1841	1823	25 Aug 2006	27 Sep 2006	1	GIF1 GIF2	GIF	summary.list
2006/05/13	reduced	1129	563	19 Sep 2006	23 Sep 2006	1	GIF1 GIF2	GIF	summary.list
2006/05/14	reduced	2402	1424	19 Sep 2006	23 Sep 2006	1	GIF1 GIF2	GIF	summary.list
2006/05/15	reduced	2674	1540	19 Sep 2006	23 Sep 2006	1	GIF1 GIF2	GIF	summary.list
2006/05/16	reduced	148	145	19 Sep 2006	23 Sep 2006	1	GIF1 GIF2	GIF	summary.list

Night	Observation Log	Size raw [Gb]	Size red [Gb]	N stacks All	N stacks UKIDSS	N stacks LAS	N stacks DXS	N stacks UDS	N stacks GPS	N stacks GCS
2006/04/28	obs_index	8.86	16.13	39	30	0	0	0	30	0
2006/04/29	obs_index	8.60	2.58	22	0	0	0	0	0	0
2006/04/30	obs_index	2.01	1.68	10	0	0	0	0	0	0
2006/05/01	obs_index	20.46	35.09	126	69	42	0	0	27	0
2006/05/02	obs_index	24.99	50.84	47	0	0	0	0	0	0
2006/05/03	obs_index	32.26	47.46	44	0	0	0	0	0	0
2006/05/04	obs_index	29.42	60.45	60	0	0	0	0	0	0
2006/05/05	obs_index	25.64	44.36	37	0	0	0	0	0	0
2006/05/06	obs_index	0.69								
2006/05/07										
2006/05/08	obs_index	5.57								
2006/05/09	obs_index	7.09								
2006/05/10	obs_index	42.30	76.66	257	226	0	0	0	0	226
2006/05/11	obs_index	49.60	128.29	357	330	2	0	0	107	221
2006/05/12	obs_index	35.96	57.71	229	178	89	0	0	52	37
2006/05/13	obs_index	18.18	60.40	60	0	0	0	0	0	0
2006/05/14	obs_index	46.56	95.81	111	30	1	17	0	12	0
2006/05/15	obs_index	53.59	105.73	103	20	0	20	0	0	0
2006/05/16	obs_index	1.85	1.93	15	0	0	0	0	0	0

Fig. 1. A snapshot of the reduction progress page for semester 06A. See the text for a description of the columns

[1]http://casu.ast.cam.ac.uk/surveys-projects/wfcam/data-processing/.

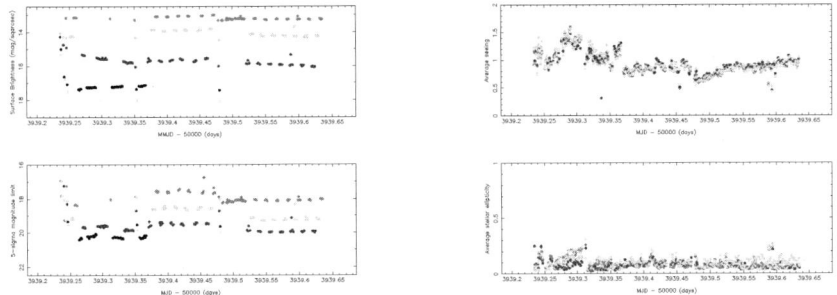

Fig. 2. An example of the QC plots showing the average sky brightness, magnitude limit, seeing (FWHM) and stellar object ellipticity for each stacked frame of a night

that is used to track the progress of the UKIDSS surveys and, for each survey, to visualise the distribution and time evolution of different QC parameters[2].

2 NIR Sky Brightness

In the last few months we have been using the quality control database to investigate the near infrared (NIR) sky brightness (SB, hereafter) properties using public UKIDSS data. The sky level estimation is one of the quality control measures automatically produced by the pipeline during the reduction and is available in the image header of each chip extension. Having all the information stored in the database allows us to easily isolate different samples according to several criteria: e.g. photometric nights, reasonable seeing and stellar ellipticity, fractional lunar illumination (FLI), Moon elevation (h_{moon}) and angular distance from the observed field, time distance from the closest twilight. In Fig. 3 we show the photometric zero points (ZP) as a function of the modified Julian date (MJD) for the three semesters considered in our work. WFCAM photometric calibration is based on 2MASS and has been shown to deliver frame-by-frame photometric ZPs at the ±1–2% level in the J, H, K bands including the majority of non-photometric nights [2, 3].

The primary property of interest is the intrinsic SB and its possible dependence on the Moon position and phase. The dark-time SB was estimated using only data from photometric nights, with a minimum distance from the closest twilight of 1 hr and with an FLI ≤ 0.4 and h_{moon} ≤ −18 deg. The bright-time selection criteria is similar with the only difference being FLI ≥ 0.8, h_{moon} ≥ 30 deg and a minimum Moon-field angular separation of 30 deg. Our results, summarised in Table 1, show some interesting properties of the NIR sky: (1) the sky seems to be much more stable in the K-band compared to the other passbands, as suggested by the much smaller scatter.

---

[2] http://casu.ast.cam.ac.uk/survey-progress/wfcam/.

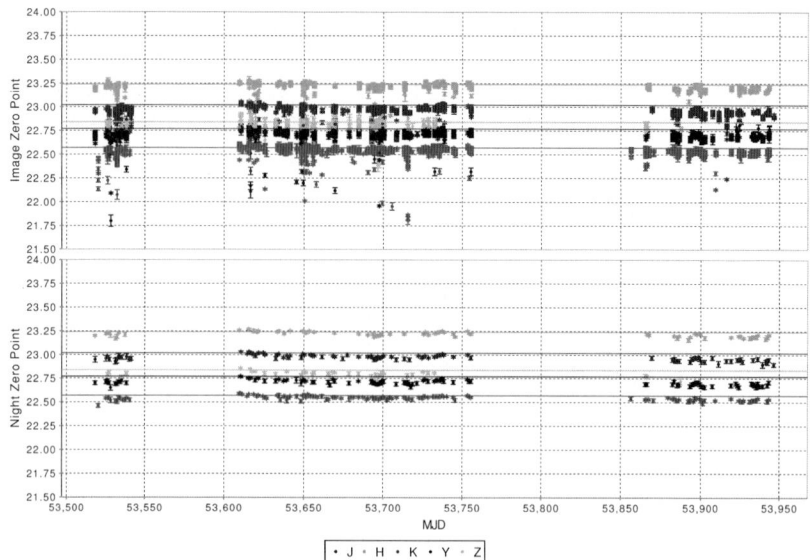

Fig. 3. Photometric ZPs based on 2MASS for the five WFCAM broad-and filters for the photometric nights in 05A, 05B and 06A. The *top panel* shows the image ZPs whereas the *bottom* one shows the nightly derived ones. The horizontal lines show the nominal ZPs in each passband

Also the Z-band appears to be very stable, albeit, with the result based on a much smaller sample; (2) for all the passbands except Z, the average sky level seems not to vary between dark and bright time indicating that the average Moon effect is negligible. The only exception is the Z filter where the sky is ≈0.4 mag brighter in bright time. Looking at the SB as a function of the angular distance from the Moon we found no effect at all for separations larger than 15–20 deg in J, H and K. In Y and Z the effect of scattered Moon light is much stronger but appears to be significant only when observing closer than 25–30 deg from the Moon.

Table 1. Zenith corrected sky brightness during dark (D) and bright (B) time. N is the number of images used, μ is the median sky surface brightness (mag arcsec^{-2}), σ^{mad} is the standard deviation based on the M.A.D. (median absolute deviation from the median) and $\Delta\mu$ is the variation between dark and bright-time

Filter	Dark-Time			Bright-Time			
	N_D	μ_D	σ_D^{mad}	N_B	μ_B	σ_B^{mad}	$\Delta\mu_{D-B}$
Z	170	18.44	0.14	64	18.05	0.15	0.39
Y	917	17.38	0.33	243	17.30	0.29	0.08
J	1072	15.99	0.34	529	16.03	0.50	−0.04
H	768	14.19	0.32	412	14.16	0.19	0.03
K	1322	13.60	0.16	833	13.68	0.10	−0.08

At NIR wavelengths the background is mostly due to the OH airglow in the J and H passbands, and to a lesser extent in K where the dominant component is rather the thermal emission from telescope, instrument and sky. Indeed, WFCAM K-band SB correlates nicely with air temperature with a slope of about -0.05 mag per degree. This result is consistent with the variation expected from a black body integrated over the WFCAM K passband. A similar analysis on the other passbands showed no correlation with the temperature thus confirming the negligible thermal contribution.

In Fig. 4 we show the SB evolution as a function of the elapsed time from the dusk twilight. The data shown in the plots include observations obtained before the formal beginning of the astronomical night to illustrate the sky behavior during twilight. Observations taken during the dawn twilight have been excluded to avoid systematic effects due to the variable length of the night. Both Z and Y appear to be consistent with no evolution of the SB within the night, with the obvious exception of observations obtained in twilight. In the J band the same effect is still visible although on a much smaller scale. After the beginning of the astronomical night, the J-band sky seems to get darker in the first $\sim$2–3 h before settling to a constant level until about 8 h from the dusk twilight after which it seems to darken again slightly in the subsequent 2–3 h. On single nights, the time evolution of the J-band SB shows a variety of cases, therefore the trend described here is an ensemble average. The H-band is consistent with no evolution of the sky brightness along the night and it does not seem to brighten significantly during the twilight. Also in this band a variety of trends can be seen when looking at single nights. Finally, the K-band data show a clear trend with the sky getting darker along the night with a slope of -0.027 mag/h. Given the thermal component contributing to the K-band sky level, we tested whether the trend shown in Fig. 4 could be explained simply by temperature variations. We therefore normalised the SB to a temperature of $0°C$ using the 0.05 mag/degree slope from our fit (see above) to remove the effect of varying temperature. The new fit provides a slope of -0.024 mag/h which is consistent with the previous value. This rules out a possible thermal origin of the effect. This behaviour is explained by the observed properties of OH-Meinel bands [6]. Short-period variations occur in the nightglow due to the passage of perturbation waves through the ionosphere causing periodic variations in the temperature and density of the atmospheric gases, which in turn affects the intensity of the OH emission. In particular [6] found evidence for wave-like variations superimposed on a global decrease of the emission during the beginning of the night. The global decrease during the night is indeed predicted by many models of the night-glow emission. The large scatter seen in Y, J and H bands (see Table 1. and Fig. 4) can be also explained in terms of the stronger effect that the air-glow lines play in these passbands. For more information we refer the interested reader to a paper, by the same authors, that will appear in MNRAS later this year.

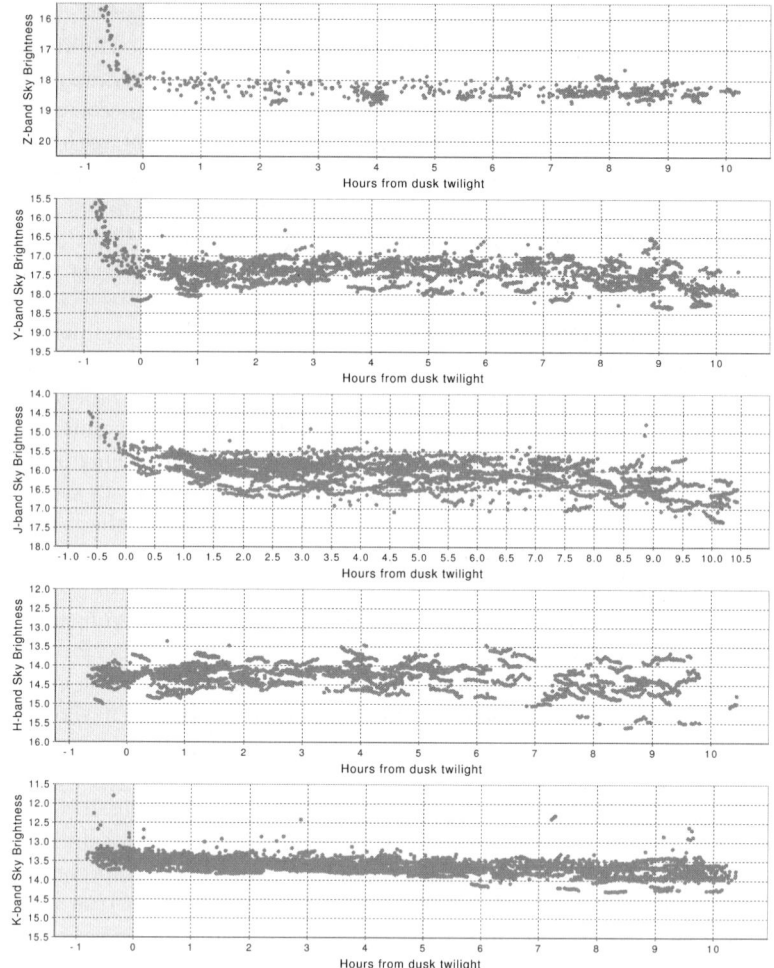

Fig. 4. Sky brightness evolution versus elapsed time (hours) from the dusk twilight for ZYKHK. Shaded regions indicate evening twighlight

References

1. M. Casali, et al.: A&A **467**, 777 (2007)
2. S. Dye, et al.: MNRAS **372**, 1227 (2006)
3. S. Hodgkin, et al.: MNRAS, in preparation (2007)
4. A. Lawrence: MNRAS **379**, 1599 (2007)
5. J. Lewis, et al.: these proceedings (2008)
6. S.K. Ramsey, C.M. Mountain, T.R. Geballe: MNRAS **259**, 751 (1992)

Discussion

M. Rejkuba: In your slides you have shown that the K-band sky gets darker on average towards the end of the night. Also there is a trend with temperature. Have you tried to see how the temperature changes during the night? In other words, if you take out the temperature changes over the night, it there still a residual trend of sky brightness decreasing in K-band?

M. Riello: This is indeed one of the first things I would like to check out. I just did not have time to do that for the talk. This is certainly an important thing to check.

W. Hummel: The QC you presented is mostly on the pipeline products. Is there also some QC implemented related to instrumental health check?

M. Riello: Yes, we are also making some QC health check measurements on a regular basis, but they have not yet been integrated in the infrastructure (database, web pages, etc.) because at the beginning there was a stronger request for science/level QC. It will be implemented as soon as possible anyway.

R. Hanuschik: How large is the approximate fraction of calibration data in the daily volume of 2100 raw files?

M. Riello: There are, on average, $\sim$ 80–100 dark frames per night. Flats are taken every $\sim$20 days because they are quite stable. Standard fields are now at a frequency of 5/night but they used to be at a level of 1/hour at the beginning when the instrument was being characterized.

Relative and Absolute Calibration for Multi-Band Data Collected with the 2.2 m ESO/MPI and 1.54 m Danish Telescopes

A. Calamida[1], C. E. Corsi[1], G. Bono[1], P. B. Stetson[2], L. M. Freyhammer[3], and R. Buonanno[4]

[1] INAF – Osservatorio Astronomico di Roma, Via Frascati 33, 00040, Monte Porzio Catone, Rome, Italy; calamida@mporzio.astro.it
[2] DAO, HIA-NRC, 5071 W. Saanich Road, Victoria, BC V9E 2E7, Canada
[3] Univ. of Central Lancashire, Preston PR1 2HE, UK
[4] Univ. di Roma Tor Vergata, Via della Ricerca Scientifica 1, 00133 Rome, Italy

Abstract. We present the strategies adopted in the relative and absolute calibration of two different data sets: U, B, V, I-band images collected with the Wide Field Imager (WFI) mosaic camera mounted on the 2.2 m ESO/MPI Telescope and u, v, b, y Strömgren images collected with the 1.54 m Danish Telescope (ESO, La Silla). In the case of the WFI camera we adopted two methods for the calibration, one for images collected before 2002, with the ESO filters $U/38_{ESO841}$ and $B/99_{ESO842}$, and a different one for data secured after 2002, with the filters $U/50_{ESO877}$ and $B/123_{ESO878}$. The positional and color effects turned out to be stronger for images collected with the old filters. The eight WFI chips of these images were corrected one by one, while in the case of images secured with the new filters, we corrected the entire mosaic in a single step. In the case of the Danish data set, we compared point-spread function (PSF) and aperture photometry for each frame, finding a trend in both the X and Y directions of the chip. The corrections resulted in a set of first and second order polynomials to be applied to the instrumental magnitudes of each individual frame as a function of the star position.

1 WFI Data Set

1.1 Observations and Data Reduction

Data have been retrieved from the ESO archive and include 8 U, 39 B, 51 V, and 26 I images of ω Cen . Data include both shallow and relatively deep images, with exposures times ranging from 1 to 300 s for the B, V, I bands, and from 300 to 2400 s for the U band, and were collected in several observing runs ranging from 1999 to 2003. During this period two filters were changed: data secured before 2002 were collected with the filters $U/38_{ESO841}$ and $B/99_{ESO842}$, while later ones with the filters $U/50_{ESO877}$ and $B/123_{ESO878}$. These data were obtained in good seeing conditions, and indeed the mean seeing ranges from 0.6" for the I band to 1.1" for the U band. We accurately selected the best PSF stars uniformly distributed across each chip. A moffat analytical function linearly variable on the chip was assumed for the PSF. The data were reduced with DAOPHOT IV/ALLFRAME [5].

1.2 Relative and Absolute Calibration

We are interested in providing accurate relative calibration of individual chips of the WFI mosaic camera because the occurrence of positional effects might cause a spurious color broadening of key evolutionary features such as the Red-Giant Branch (RGB) and the Main-Sequence Turn-Off. Moreover, we need to provide an accurate absolute calibration of our data in order to compare theory with observations. The plausibility of this comparison relies on the accuracy of the absolute zero-point of individual bands. This is a fundamental requirement for the distance modulus, and in turn for the absolute ages. Recent findings [1, 2] based on photometric data collected with the WFI indicate that the eight CCD chips might be affected by positional effects involving zero-point errors of the order of several hundredths of magnitude. This subtle effect could be due to scattered light. It seems that telescopes equipped with mosaic cameras and focal reducers may present this problem [3, 4].

In order to correct the positional effects of the WFI mosaic camera and to perform an accurate absolute calibration of ω Cen data set, we followed these steps:

- obtain a set of local standard stars for ω Cen ;
- identify all the trends of PSF magnitudes compared to standard magnitudes versus position on the frame and correct them;
- estimate the calibration curves.

We thus made use of a set of new multi-band (U, B, V, I) local standard stars for ω Cen [6]. This star list has been selected in photometric accuracy ($\sigma \leq 0.03$ mag) and in 'separation index' ($sep \geq 2.5$). We ended up with a catalog of $\sim 3 \times 10^4$ local standard stars. The sky area covered by these stars is $\sim 37' \times 40'$, and includes a substantial fraction of our WFI data. We thus applied two methods: in the case of images collected before 2002 ($U/38_{ESO841}$ and $B/99_{ESO842}$ filters), the eight chips were corrected one by one, while for images secured starting from 2002 ($U/50_{ESO877}$ and $B/123_{ESO878}$ filters) we corrected the entire mosaic in a single step. The color terms for these filters have a very steep slope, therefore, we decided to estimate a first color curve and apply it. We then studied the residuals, $\Delta mag = mag_S - mag_{PSF}$, where S stands for Standard, as a function of the X and Y position on the chip (or mosaic, see Fig. 1). Once corrected for the positional effects, we estimated the color term once again and applied it for the absolute zero-point calibration. This strategy relies on the assumption that the two corrections are independent, and there is no reason why this should not be the case. Figure 1 shows the magnitude residuals in the V band plotted versus the X and Y positions on the frame, before and after the corrections were applied to an image collected in 2002. The trend with the position is clear, nonlinear, and stronger in the four outermost chips. A small residual trend on

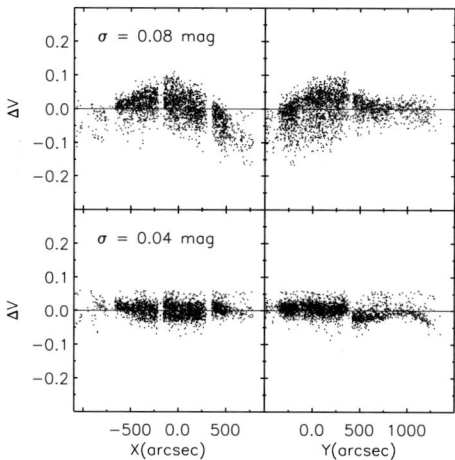

Fig. 1. Residuals, $\Delta mag = mag_S - mag_{PSF}$, in the V band, plotted versus the X, Y position on the chip, before (*top*) and after (*bottom*) the positional corrections were applied, for an image collected in 2002

the position is still present in the external regions, where the lack of standard stars makes it difficult to estimate the corrections.

We thus corrected the positional effects by fitting the Δmag vs X/Y with second, third or fourth order polynomials, and we then estimated the calibration curves for each set. We adopted a first order polynomial to calibrate the V and the I bands as a function of the instrumental $V - I$ color. In the case of the B band, we used a first order polynomial to calibrate the new filter, $B/123_{ESO878}$, and a third order one for the old filter, $B/99_{ESO842}$, as a function of the instrumental $B-V$ color. Particular attention has been paid to the U band calibration, either in the case of the old filter, $U/38_{ESO841}$, as well in the case of the new one, $U/50_{ESO877}$. Having applied the positional corrections to each set, we estimated a first color curve, a fourth order polynomial for U vs $U - I$. After applying this calibration curve, the U magnitudes still showed a residual trend as a function of the U magnitude and the $U - I$ color. We thus corrected these trends with first and sixth order polynomials, respectively. The reason of these very complicated trends with colors is probably due to the shape of the old B and U filters, which is quite different from the shape of the standard Johnson filters.

2 Danish Data Set

A set of 110 u, v, b, y Strömgren images centered on ω Cen were collected in 1999, with the 2048×2048 pixel Ford-Loral CCD mounted on the Danish

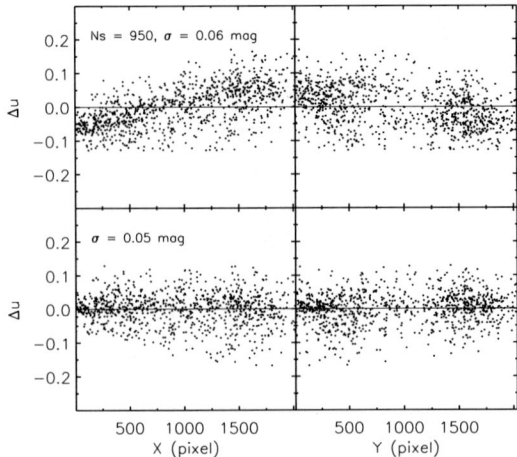

Fig. 2. Residuals, $\Delta mag = mag_{AP} - mag_{PSF}$, in the u band, plotted versus the X, Y position on the chip, before (*top*) and after (*bottom*) the positional corrections were applied

Telescope. Data have been reduced with DAOPHOT IV/ALLFRAME, using an IDL procedure to accurately select the best PSF stars across each frame. We found that the b and the y frames present faint spurious reflected images of bright stars. This effect causes a systematic decrease in the flux measured by the PSF of bright stars and a systematic increase in the flux of the faint neighbor stars. The effect strongly affects the photometry in crowded fields, resulting in a larger dispersion on the color-magnitude diagrams (CMDs), in particular in b versus $b-y$. In order to correct this effect, we first investigated the presence of magnitude trends with X, Y position on the chip, comparing PSF and aperture photometry for each frame. The resulting effect changes from frame to frame, depending on the photometric band, and the magnitude difference from one edge to the other of the image can be as high as 0.15 mag. The result is a set of first and second order polynomials that correct the magnitudes of each individual frame as a function of the X and Y coordinates. Figure 2 shows the magnitude residuals, $\Delta mag = mag_{AP} - mag_{PSF}$, for the u band, before and after applying the corrections.

3 Conclusions

We have presented the relative and absolute calibration strategies applied to extended photometric catalogues obtained with the 2.2 m MPI (Fig. 3) and the 1.54 m Danish ESO Telescopes. The photometry was performed with DAOPHOT IV/ALLFRAME in both cases, using a linearly variable PSF

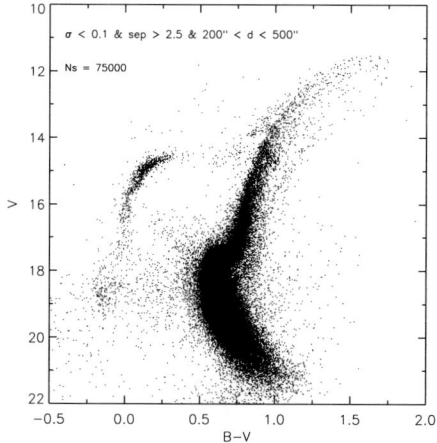

Fig. 3. $V, B - V$ CMD for the entire WFI corrected catalog of ω Cen

across the images. In spite of the attention paid in selecting PSF stars uniformly distributed in the frame, we found a systematic error depending on the position of the stars in the image, when comparing instrumental to standard magnitudes. This effect might be due to problems in adapting the analytical PSF to variations across the image or to intrinsic problems of the optical system. Therefore, in order to perform an accurate relative and absolute calibration it is useful to check for the presence of a positional dependence of photometric errors. This is particularly recommended when dealing with telescopes equipped with focal reducers and mosaic cameras.

References

1. C.E. Corsi, et al.: Mem.SAIT **74**, 884 (2003)
2. A. Koch, E. Grebel, M. Odenkirchen, J.A. Caldwell: The Messenger **115**, 37 (2004)
3. J. Manfroid, P. Royer, G. Rauw, E. Gosset: ASPC **238**, 373 (2001)
4. J. Manfroid, F. Selman: The Messenger **104**, 16 (2001)
5. P.B Stetson: PASP **106**, 250 (1994)
6. P.B. Stetson, et al.: in preparation (2008)

ALBUM: A Tool for the Analysis of Slitless Spectra and its Application to ESO WFI Data

C. Martayan[1,2], D. Baade[2], A.-M. Hubert[1], M. Floquet[1], and J. Fabregat[3], and E. Bertin[4]

[1] GEPI-Observatoire de Paris, 5 place Jules Janssen, 92195 Meudon cedex, France; christophe.martayan@obspm.fr
[2] ESO, Karl-Schwarzschild-Strasse 2, 85748 Garching, Germany
[3] Observatori Astronòmic de la Universitat de València, Edifici Instituts d'Investigació, Polígon La Coma, 46980 Paterna València, Spain
[4] Institut d'Astrophysique de Paris, 98bis boulevard Arago, 75014 Paris, France

Abstract. ALBUM is a general-purpose tool to visualize and screen large amounts of slitless spectra. It was developed for a search for emission-line stars in SMC and LMC clusters. The observations were obtained with ESO's Wide Field Imager (WFI) and comprise ∼8 million low-resolution spectra. The tool as well as the results of its application to the SMC part of the database are presented. The inferred frequency of Be stars is compared to the one in the higher-metallicity environment of the Milky Way.

1 Data Reduction

Observations (see Fig. 1) covering much of the Small Magellanic Cloud (SMC) have been obtained in September 2002 with the WFI attached to the 2.2 m MPG Telescope at La Silla. The instrument was used in its slitless spectroscopic mode. To reduce crowding, the length of the spectra was limited by means of a filter with a bandpass of 7.4 nm centered on Hα. Unfortunately, a large part of the fields suffers from substantial non-homogeneous defocusing, which severely reduces the contrast between stars with and without line emission at Hα.

The basic reduction of the CCD images was performed with the MSCRED IRAF tasks except for the astrometry, for which the ASTROM package [8] was applied to the extracted 1st-order spectra. The achieved accuracy was 0.5-1" rms. The extraction in 2-D of the spectra was accomplished by means of the SExtractor software [1]. All in all, about 1 million of the 3 million spectra available in the SMC part of the survey proved usable.

To recognize and distinguish emission-line stars (Em**) from other objects, we created the ALBUM package in IDL. Its strategy is based on the assumption that the 2-D point-spread function (PSF) is only slowly varying with position in the frame and only insignificantly falsified by the inclusion of emission-line objects in the calculation of the mean local PSF. Typically, 50–250 spectra were aligned (by cross correlation), co-added, and normalized.

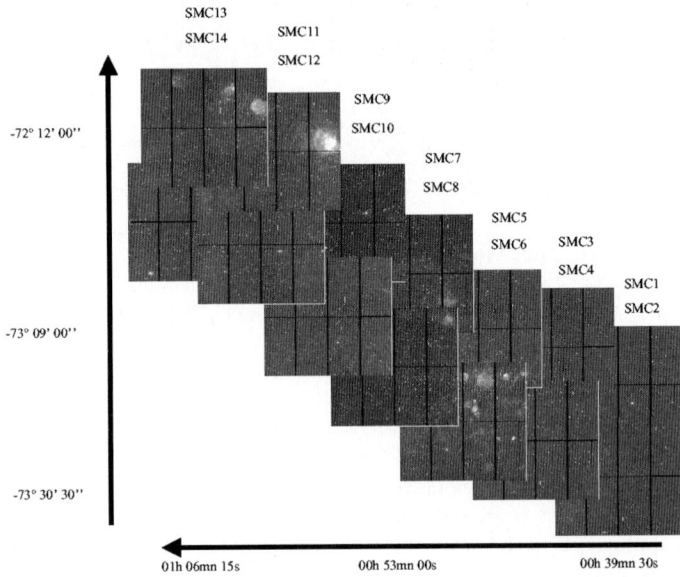

Fig. 1. The Small Magellanic Cloud as projected on the WFI frames used in this study

This local template spectrum was subtracted (after cross correlation and shift) from each normalized 2-D spectrum (see Fig. 2) to be checked for Hα line emission. In the case of Em*, the 2-D spectra show a secondary peak (see Fig. 3). But after subtraction of the mean PSF the resulting difference images display a more characteristic and conspicuous ring-like structure, which is due to the large defocus (see Fig. 3). Since this peculiar structure is more readily and reliably recognized by the human eye than by software, the identification of the Em* was done by visual inspection of the album of PSF-subtracted 2-D spectra.

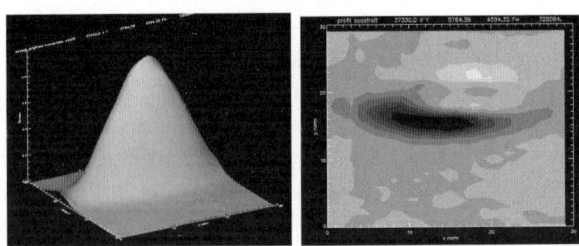

Fig. 2. Non Emission line star. *Left panel*: original source. *Right panel*: projection of the residual of the subtraction of the mean profile

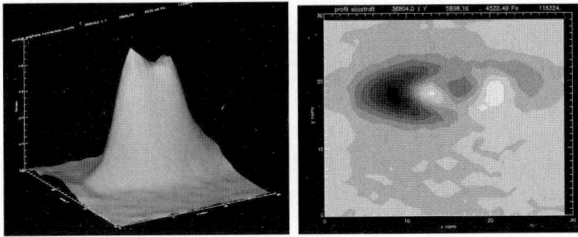

Fig. 3. Emission line star. *Left panel*: original source. *Right panel*: projection of the residual of the subtraction of the mean profile, due to the emission in Hα and the defocus a ring structure with two peaks is observed

2 Results: Frequency of Be Stars vs. Metallicity

We have investigated 85 clusters in the SMC with log(age) between 7 and 9 and E[B-V] available from the OGLE survey [6]. For a total of 7741 stars, V, B, and I magnitudes were obtained from the OGLE database [6, 7]. Figure 4 displays the combined HR diagram of all clusters with the Em** marked. The

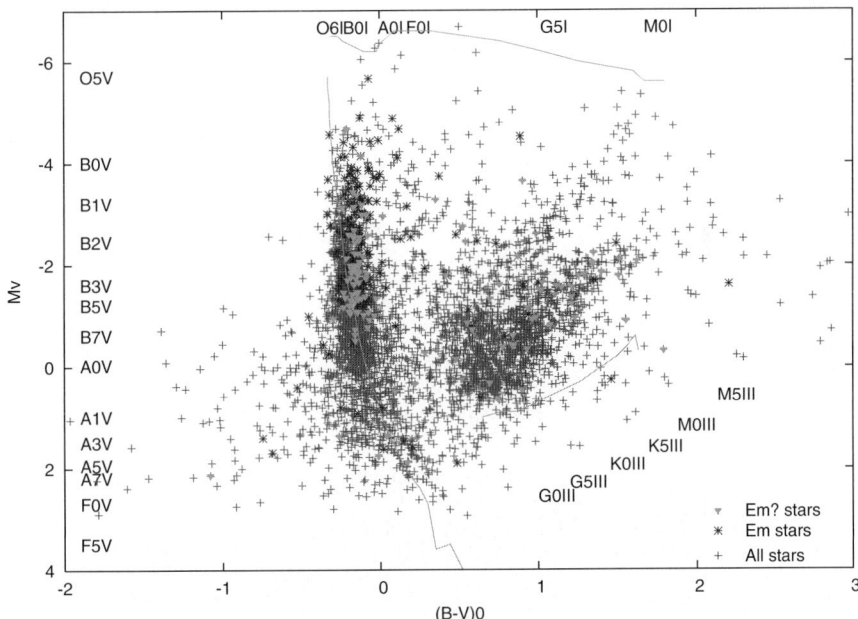

Fig. 4. (Mv,(B-V)$_O$) diagram for the stars cross-correlated in OGLE database. The calibration in spectral types comes from [2]. '+' (red) are for the non-Emission line stars, '*' (blue) for the Emission line stars, and the *triangles* (green) are for the candidate Emission line stars

results can be compared to the relative frequencies of Be stars (Be/(B+Be)) in Milky Way clusters [5] in order to search for any effect of the metallicity on the proportion of Be stars and on the still unknown reason for the development of disks around these extremely rapidly rotating stars. The fractions range from 0 to 46% in the SMC and from 0 to 24% in the Milky Way (MW), depending on the parent cluster. There seems to be a trend in that the lower the metallicity, the higher the proportion of Be stars is. This could be explained by higher rotational velocities in the SMC than in the MW [3, 4].

3 Conclusions

– A new method and software package for the reduction and analysis of slitless spectra was developed.
– It was applied to WFI data. Catalogues of Be stars in 85 SMC clusters were obtained.
– Metallicity seems to influence the relative proportion of Be stars among all B-type stars: The fraction of Be stars is larger in the SMC than in the MW.
– The equivalent study of the LMC is in progress (additional 5 million spectra).
– The combination of a large field and slitless spectroscopy is very powerful for surveys for objects with distinct spectral properties.

Acknowledgement. C.M. gratefully acknowledges support from ESO's DGDF 2006.

References

1. E. Bertin, S. Arnouts: A&AS **117**, 393 (1996)
2. K.R. Lang: *Astrophysical data: Planets and Stars*, Eds Springer-Verlag, New York (1992)
3. A. Maeder, G. Meynet: A&A **373**, 555 (2001)
4. C. Martayan, Y. Frémat, A.-M. Hubert, et al.: A&A **462**, 683 (2007)
5. M.V. McSwain, D. Gies: ApJS **161**, 118 (2005)
6. G. Pietrzyński, A. Udalski, M. Kubiak, et al.: AcA **49**, 521 (1999)
7. M. Szymanski: AcA **55**, 43 (2005)
8. P.T. Wallace, N. Gray: *User's guide of ASTROM* (2003)

Part XIII

Workshop Summary

The 2007 ESO Instrument Calibration Workshop – A Personal Summary

D. Baade

ESO, Karl-Schwarzschild-Strasse 2, 85748 Garching, Germany; dbaade@eso.org

1 Introduction

Nearly 50% of the top 50 hits returned by a *Google* search performed at the end of 2006 for 'calibration workshop' concerned astronomy. At a first glance, this is surprising, given the relatively small share of astronomical research in science in general. On the other hand, it is only careful calibration, which transforms observations into quantitative measurements and permits the non-tangible objects of astronomical research to be projected onto the human world (and on the one of microscopic physics): Without calibration, astronomy would not be a science.

In the electromagnetic spectrum, astronomy only disposes of 8 observables: 2 angular coordinates, time, wavelength, flux, degree and angle of polarization, and phase (phase differences). Accordingly, only 3 basic types of scientific calibrations are needed: radiometry, geometry, and chronometry.

While there is consensus that radiometry is the most delicate (thanks to atrocities such as atmospheric and interstellar extinction, sky and instrument backgrounds, scattered light, slit losses, uneven slit and aperture illumination, variable point spread functions, detector non-linearity, imperfect standards, incompatible passbands, variability, and many more), geometric calibrations have the reputation of being easy in comparison. However, flexure necessitates differential parametric corrections or, in the case of hysteresis, a lot of daytime efforts to repeat the operations of the night. And vibrations, which go entirely unnoticed with single-telescope instruments, can make interferometric observations fully impossible. Perhaps the most important geometry-related question that came up in various contributions is (Gaspare Lo Curto et al.) whether we have been settling for too little. High-precision results obtained with HARPS (Christophe Lovis & Francesco Pepe) and NACO (Andreas Seifahrt et al.) strongly suggest a positive answer.

Amazingly, not one mention was made of challenges associated with time measurements. The advent of GPS has certainly made them much simpler. But it is difficult to believe that all of a sudden they have become trivial.

2 Calibrations at ESO

At ESO, calibrations form the core of the VLT quality control process the depth as well as the breadth of which are without precedence at earlier ground-based observatories. Unfortunately, these important efforts, which were mentioned by many speakers, were not presented in a dedicated contribution. Daily updates of the impressive and comprehensive information about the performance of the VLT instruments can be found at http://www.eso.org/qc.

In 1995, when systematic calibration plans were demanded for all VLT instruments to be built, this requirement met quite a bit of skepticism. But, as we learned from Gianni Marconi, calibration plans seem to be functioning reasonably smoothly in practice although a user satisfaction survey has indicated that some users are perhaps stunned by the level of complexity.

Gianni Marconi also raised the question whether calibrations plans have been optimized for minimal effort at a given level of requirements. Marco Scodeggio's impressions led him to a straight 'No - too much, too complicated.' Reinhard Hanuschik agreed that confusion can arise if people look at the Science Archive and its unstructured manifold of calibration data. By contrast, ESO's Quality Control Department only delivers the calibration data used with the pipeline reduction. However, one needs to remember that pipeline reductions are foreseen for service-mode users only.

David Osip asked whether instrument consortia are given enough time to calibrate their instruments during the commissioning periods. Andreas Kaufer and Ulli Käufl confirmed this, with the latter adding that the success of the commissioning also depends critically on all pre-shipment laboratory activities having been properly completed. Evidently, it is much more difficult to calibrate an instrument, which has problems that still need fixing.

In answering the complementary question whether calibration plans need to be adjusted to meet evolving scientific requirements, Andreas Kaufer admitted that such revisions are not yet part of the standard operations paradigm. The only such science-driven suggestion came from Piero Rosati, who advocated that on-the-sky (i.e., mainly radiometric) calibrations should be given higher weight. Interestingly, for none of the calibration plans discussed in some detail their origin was traced back to specific scientific needs.

Physical modelling introduces a new quality to calibration. After first applications to HST instruments, the scope is being extended to ground-based instruments as well, and we have seen a number of impressive examples (Michael Rosa; Paul Bristow & Florian Kerber; Yves Jung & Paul Bristow; Jeremy Walsh et al.). Physical models are not only useful for performance predictions and the elimination of possible problems. They are at least as useful during the operations phase, and Andreas Kaufer mentioned the case of UVES: Without a physical model, the re-alignment of UVES after an earth quake could take weeks. But with the diagnostic capabilities of a physical model, it becomes clear much faster, where the problems lies.

3 Calibration Challenges from VLT to ELT

It may be useful to project this workshop onto a compilation of accomplishments that have come true already, others that should still be realized with the VLT, and still others that will have to await the advent of the ELT. As it turns out, this leaves behind rather few blank (unmatched) spots. But this does not mean that all other areas are deeply colored through and through. The main body of this summary is randomly interspersed with selected quotes ('Merksätze') from various participants but rephrased by the summarizer.

3.1 Pipelines

Originally, VLT pipelines would only be a secondary spin-off resulting from automated quality-control procedures. But thanks to the success of, and the lessons learnt from, the first-generation pipelines, the initial rather unflexible concept of monolithic pipelines needs to be upgraded. Michèle Péron's talk has shown that ESO has understood the need for more explicit modularity.

> *Calibration is a life-long learning experience.*
> Gianni Marconi

The Finnish Sampo Team under the leadership of Richard Hook, has demonstrated how, with the help of a public-domain tool, existing software can be transformed to a configurable pipeline. Arguably, ESO Reflex is the most significant development in the field since the advent of general-purpose astronomical data reduction systems such as IDL, IRAF or MIDAS. This may enable the structured captivation of a much larger part of the community expertise with data reduction and calibration. One of the first steps could be the identification and implementation of standard modules for standard tasks (flat fielding, wavelength calibration, etc.).

Carlo Izzo et al. (see also Sabine Möhler's talk) have shown us a beautiful example of how a very robust pipeline can result from the consistent adherence to simple morphological principles with minimal assumptions and common sense. Although this is at the extreme opposite of physical models, it is not their negation, and both approaches can be similarly robust. (However, not having put in any physics, one can not expect any physical insights to come out of purely parametric tools.) The additional need for robustness to scale with the data volume was emphasized by Eric Ensellem.

> *Calibration cannot make up for poorly prepared observations.*
> Piercarlo Bonifacio

Any pipeline without error propagation is of little value. At a future calibration workshop, a paper dedicated to this issue could be very beneficial, which in 2007 was at least mentioned by Aniello Grado et al., Matthew Horrobin et al, and Martin Roth et al. An important detail is to not lose track of the uncertainties associated also with independent variables.

While pipelines will often be very useful already if they merely give a rough idea of what might be extractable from the data, the ultimate goal must be publication-ready results. Jonas Møller-Larsen et al. have addressed this challenge, and in an impressive, very careful analysis of UVES spectra of cool stars, Poul Nissen reminded, how many challenges a pipeline is facing to produce as good results as the instrument was intended to supply.

3.2 Science Archive

Poul Nissen also made a plea that, in spite of all possible limitations, the results of the pipeline processing should always be offered to Archive users. Martino Romaniello explained that ESO is working towards this goal. In fact, Piero Rosati described advanced data products, which have been derived for some selected data sets, are available from the Archive, and illustrate more general objectives. Hendrik Hildebrandt presented a pipeline set up for the reduction of archival data from the ESO Deep Public Survey.

> *Calibration maps are a strong alternative to re-sampling.*
> Richard Davies, Carlo Izzo
> *Artifacts are removed most effectively by multiple re-sampling.*
> Eric Emsellem

Most other areas of calibration-related activity for the Science Archive remained unmentioned. For instance, the Science Archive could provide highly valuable services by supporting *in situ* operations on images, extracted parameters, and header values, by permitting the nearly inexhaustable VLT log files to be projected onto individual exposures, and by the provision of pro-active filters for the selection of candidate calibration data. This leaves much room for future workshops.

3.3 Atmospheric Extinction and Background

There was unanimous agreement among the participants that the Earth's atmosphere poses a larger challenge than any reasonably solid instrument. Based on many years of experience with the FORSes, Ferdinando Patat managed to let the optical night-sky lines dance in harmony with a tune of his choice, Mario Riello extended the description of the symptoms to wide-field imaging in the IR, and David Burke told us about the efforts of the LSST project to cope with them. The latter speaker and Piercarlo Bonifacio emphasized that atmospheric monitoring does not require a 42 m telescope. In fact, a dedicated, albeit small, facility will be much better. Although the ever-erratic nature of the variability on all time scales remains a burden for all ground-based observatories, its calibration should be more strongly based on physical understanding as Michael Rosa suggested.

Ferdinando Patat and Wolfram Freudling et al. pointed out that the ASM/LOSSAM data could be quite profitably integrated into photometric data reduction pipelines, another potential future task for the Science Archive. However, line of sight ('LOS') means a viewing direction, which is not in any way related to the line of sight towards the target of interest. The latter may be much better characterized by the autoguider data, which could also be contributed through the Science Archive.

> *A suite of software can be a pipeline if it can reduce data from several instruments of the same category.*
> Carlo Izzo, Sabine Möhler

Elena Mason gave an example of how the analysis of repeated observations of standard star fields can give a good view of the severity of the problem. The transparency of the IR sky can with good accuracy be deduced from telluric lines the strength of which scales with the precipitable water vapor, thereby reducing the need for additional calibration observations, as was found by Alain Smette et al. and Joanna Thomas-Osip et al. Participants were reminded by Danuta Dobrzycka et al. and Eric Pantin et al. that good IR standards have become available mainly thanks to space-borne projects and have reached a quality that is not easy to fully exploit from the ground.

The majority of observations requiring large telescopes benefit from good seeing. But in the case of interferometry, the matter of quality becomes much more quickly one of feasibility as Christian Hummel illustrated.

Adaptive optics observations have special radiometric properties but did not feature on the workshop agenda. Only Yann Clénet as well as Richard Davies came into the vicinity of a detailed discussion of this topic, which for the ELT will gain substantially in importance.

> *Involve the end users. Set up an agency/forum for coordination as well as management for prioritized allocation of resources.*
> David Lindler, Eric Emsellem

MOS with and without IFUs has vastly amplified the operational throughput of spectrographs. However, if the term 'quantitative results' is not to mean mere mass production, the calibration of the flux throughput of such devices requires considerable care as Piercarlo Bonifacio and Matteo Monelli showed for FLAMES (fibers) and FORS2 (masks), respectively.

3.4 Illumination Corrections

A particular nasty problem is the uneven illumination of instrument apertures. Illustrations of its occurrence in various disguises in imagers were given by Jim Emerson et al., Wolfram Freudling et al., Piero Rosati et al., and Fernando Selman et al. Dome flats are basically hopeless, sky flats can even at the antisolar point still feature significant natural gradients, focal reducers often suffer from so-called sky concentration due to background light reflected between the detector and the field lens, and, finally, there may be vignetting. As the FORS Absolute Photometry project (Freudling et al.) has shown, an instrument may be stable in time to much better than 1%, but the spatial variations of the illumination can still limit the accuracy to 3%.

> *False matches can confirm expectations most beautifully.*
> Carlo Izzo
> *If it's wrong, a polynomial will fit it.*
> Michael Rosa
> 'Calibration Manifesto': *A combination of meritocracy (physical models) and democracy (plain data) is the optimal constitutional basis.*
> Carlo Izzo & Michael Rosa

A physically entirely unrelated effect is the contamination of detectors or optics or both. The Wide Field Imager had developed such a problem. But, as Fernando Selman et al. reported, it went unnoticed in spite of regular quality monitoring of the calibration data because the degradation was masked by the poorly reproducible illumination of both dome and sky flats.

The spectroscopic (slit) variant of illumination problems results in a continuum shape not intrinsic to the object. At high signal-to-noise ratios, this effect eventually dominates any non-local high-precision analysis. Obviously, echelle spectrographs suffer particularly strongly because there will always be a residual echelle ripple function in the extracted spectra. Poul Nissen has exemplified this from a number of very accurate applications. Contrary to imaging, spectroscopy has one additional calibration option, namely the observation of metal poor stars. But even the spectra of the most extreme such stars are not feature free, and in particular they share at least the strong hydrogen lines, which are often needed as diagnostics such as of T_{eff}.

3.5 Detector Fringing

Interference fringes are one of the most vicious calibration problems because they compromise multiplicative calibration data (flat fields) but require subtractive correction. Instrument-internal calibration tools are of no use because they do not achieve an accuracy of better than a couple of percent, depending on the amplitude of the fringes in monochromatic light and the nature of the instrument. Moreover, the night sky lines, which give rise to the problem in imaging, are not only rapidly variable but are so also in an inhomologous fashion, as Ferdinando Patat illustrated in his movie (and sound track). It will be useful, though, to analyze the variability separately for each passband. If the line-to-line differences in variability are not too bad, or average out with time, physical modelling of the thickness of the silicon layer, as done for the FORS2 CCDs by Jeremy Walsh et al., provides a fringe map, which after proper scaling can be profitably subtracted from the data.

> *Near-IR polarimetry at ESO is a last-minute add-on for enthusiasts.*
> *Polarimetry would be more popular with better support.*
> Nancy Ageorges, Hans Martin Schmid

Since it is better to avoid a problem than to correct it, Poul Nissen asked about ESO's plans to deploy CCDs without fringing. Deeply or even fully depleted CCDs have a larger physical thickness for enhanced sensitivity to red photons, which are only poorly absorbed by silicon. Because infra-substrate path differences become rather different from the wavelength, a very welcome side effect is a substantial reduction of fringing. In fact, ESO has ordered such devices. If the laboratory tests are successful, candidate instruments for the installation are Giraffe, the red arm of UVES, and MUSE.

3.6 Instrument Design

Calibration begins long before first light: An instrument can only be well calibrated if it is well designed. Reproducible illumination (equal for internal and celestial sources) and the elimination of parasitic light are the most critical for the radiometric calibration. At a seismically active site, this also includes a quasi-dynamic component in that it must be easy to check and correct the optical alignment. As the case of UVES has shown, a good physical model can be of invaluable help, especially if it is developed in close interaction with the design of the instrument. Considerable and careful efforts for pre-construction simulations were presented by Nuria Lorente et al. (IFU's) and by Suzanne Ramsay Howat et al. (KMOS), and Yann Clenet et al. talked about quality control with atmospheric parameters in the NACO FITS headers.

> *If you know how to improve your calibrations, do it.*
> Tycho Brahe, quoted by Michael Rosa

Instrumental stiffness facilitates both radiometric and geometric calibrations (cf. Marco Scodeggio's talk and others). But stiffness requires additional mass, which requires more mass elsewhere to maintain the overall stiffness and, therefore, is a signifcant cost driver. Soon after the success of ESO's NTT as the first active telescope, suggestions mushroomed to develop also active instruments. The pace of actual progress has been slower than that but Hans Ulrich Käufl mentioned the cryogenic piezo-electric actuator in the pre-disperser collimator of CRIRES used to position-stabilize the spectrum on the detector by reference to some telluric features.

Measurements obtained with broadband filters can only be transformed to the standard system if the actual passbands are close to the orginal ones. This has been a painful experience with the WFI (Fernando Selman et al.).

3.7 Surveys

No other type of observational projects is more closely tied to calibrations than surveys are. Only a carefully and densely woven net of calibrations covering the full range in both sky and time leads to a homogeneous database, the quality of which is independent of year and airmass of the observation. The scope of ESO's standard pipelines ends with the removal of the instrumental signature. But in surveys, the extraction and subsequent quantitative characterization of sources provides for a much deeper quality control.

> *Thou shalt not have parallel pipelines.*
> Eric Emsellem, Christian Hummel

The community interest in surveys manifested itself in a considerable number of contributions (e.g., David Burke, Hendrik Hildebrand, Mike Irwin, Piero Rosati, Marco Riello, and Fernando Selman et al.). Stefan Keller et al. announced the all-digital all-sky survey with the Australian SkyMapper, which will consist of a Main Survey and a Five-Second Survey as a precursor and calibration anchor, which could be a useful reference also for VST surveys. The imminent coming into operation of VISTA was also well echoed (Jim Emerson et al. and Jim Lewis et al.).

3.8 Standard Stars and Laboratory Calibration Tools and Data

The vast majority of astronomical measurements are calibrated relative to some on-sky and/or laboratory reference data. Arguably, astronomy has become a science when it incorporated the creation of such standards. Their permanent improvement has ever since been a most vital part of astronomy.

> *The sky is the limit.*
> Many - implying the Earth's atmosphere

The large number of contributions reporting on new or improved standard stars already, or still to be, established gives evidence of this fact. Some of them are conceived as standards in a particular wavelength regime (Danuta Dobrzycka et al., Eva Bauwens et al.), partly extending the wavelength beyond traditional boundaries (Joel Vernet et al.) or form integral part of wide-area imaging surveys, which will also yield large numbers of tertiary standards (David Burke, Stefan Keller et al.), still others specialize on particular observing techniques such as interferometry (Sebastien Morel et al.), or polarimetry (Rafael Barrena et al.) or address the problem encountered by 10 m telescopes with standards established with 50-cm telescopes (José Miguel Rodriguez Espinosa et al.). For spectrophotomteric standards, there is an increasing contribution from stellar-atmosphere models (Danuta Dobrzycka, Eva Bauwens, and Joel Vernet with their respective collaborators).

Theoretical models are employed also in the area of terrestrial atmospheric extinction (David Burke). Telluric features provide useful, since simultaneous, fiducial marks for wavelength calibration (Christina Papadaki & Linda Schmidtobreik). But at higher spectral resolution, highly accurate atomic data (Florian Kerber et al.) are essential to start with but require very careful selection (Christophe Lovis & Francesco Pepe, Francesco Saitta et al.) to achieve their potential in real spectra. Finally, a calibration source not hitherto used in astronomy are laser combs, which promise to establish a new meaning of precision (Constanza Araujo-Hauck et al., Piet Schmidt et al.).

3.9 The Role of the Telescope

The telescope does not just feed the instruments with light but is also the dynamical platform for their operation. At our workshop, the VLT as such was not explicitly mentioned as a limiting factor for the calibration of instruments. This may partly be a side effect of the somewhat restrictive title to this workshop. But it probably also testifies to the quality of the optomechanics and software of the VLT. However, the routine delivery of this quality requires numerous auxiliary subsystems, each of which poses a potential risk in other areas, e.g., due to vibrations (VLTI: Florentin Millour et al., Sebastien Morel et al., Andrea Richichi) or electromagnetic emissivity (Florentin Millour et al., Eric Pantin et al., Fernando Selman et al.). For the ELT, the associated challenges will only be larger and more numerous.

3.10 Performance Monitoring and Optimization of the Tools

The need to control the numerical performance of the calibration tools and procedures was hardly touched upon during the workshop. For instance, one could think of public test data as well as the results of their processing. Of interest are also means to find the optimal set of parameters for automated procedures, e.g., high-level looping or low-level interactivity. This could be a useful topic for future ESO calibration workshops.

3.11 Closing the Loops

Nine talks by members of ESO Instrument Operations Teams (Nancy Ageorges et al. [2x], Paola Amico et al., Gaspare Lo Curto et al., Elena Mason et al., Sebastien Morel et al., Ferdinando Patat et al., Linda Schmidtobreick et al., Fernando Selman et al.) have formed the backbone of this workshop's program. This is also a tribute to the success of the underlying concept of giving each instrument into the custody of multidisciplinary teams in order to have available the broadest possible professional expertise at any time. Although some of the presentations focussed perhaps a bit too strongly on the description of the instruments rather than the actual calibration challenges and strategies, the resulting ability to quickly recognize and effectively correct problems has become apparent. Another very strong concept is the feedback of the results of the pipeline processing into engineering calibrations and instrumental health checks (Nancy Ageorges et al., Gianni Marconi, Sabine Möhler, Fernando Selman et al.), which one might call preventive calibration.

The most important of all feedback loops is the one from the users, for which ESO provides many channels. In the given context of calibration, workshops such as the present one are very promising as the extremely valuable critical appraisals of the performance of various instruments by Piercarlo Bonifacio, Eric Emsellem, Poul Nissen, Hans-Martin Schmid, and others have amply shown - we should try to get more of them.

> *The best quality check is a logarithmically scaled 3-color composite.*
> Mike Irwin

Very much of the community experience with instrument calibration is also encapsulated in software. Currently, ESO captivates this expertise at the time of the instrument construction, of which the development of calibration and data reduction software is an integral part. Modular and configurable open-source pipelines assembled from this input will permit this process to continue throughout the life cycle of instrument and with the broadest possible community involvement as both providers and recipients.

4 Conclusions

The title to this workshop includes the year, suggesting that the organizers are considering to hold other ESO instrument calibration workshops in future. They should feel strongly encouraged by the quality and amount of (calibrated!) feedback ESO received from its users (which might have been larger still had the Astronet symposium *Science Vision for European Astronomy* not taken place the same week). The Community and ESO have exchanged numerous critical encouragements as well as new inspirations although one would have wished the discussions to have been a bit more extensive.

> *Pre-natal modelling is better than post-mortem calibration.*
> Michael Rosa

In the domain of public surveys, ESO and its users are entering an era of qualitatively new cooperations. Their format is not dissimilar to the very successful paradigm of the joint Community/ESO development of VLT instruments, and their structural backbone could look like ESO Reflex. Such a 'Public Private Partnership' could be a promising *ansatz* to further enhance within European astronomy the Linux principle of projects under central coordination and control but with vital community participation.

Maybe, the most tangible result of the workshop is the resolution proposed by Eric Emsellem and adopted by the large majority of the attendants that ESO should accept observing proposals aiming for improved calibrations.

Index

κ-matrix, 458, 475
2MASS, 546, 562, 583
47 Tuc, 273

AB magnitude, 418
adaptive optics, 133, 239, 249, 259, 272
advanced data products, 170, 553, 604
ALMA, 434
AMBER, 453, 461, 479, 483
APEX, 434
arc lamp, 101, 149, 195, 207, 307, 323, 348
astrometric calibration, 546, 554
astrometry, 573
atmospheric extinction, 417, 439
atmospheric models, 419, 447
atmospheric remote sensing, 391
atmospheric window, 154
aXe, 186

background, 604
Beer's law, 391
beta-light, 13, 533
Blue horizontal branch stars, 576
bottom-up approach, 193
Bouguer curve, 439
broad-band photometry, 416

calibration plan, 6, 27, 110, 602
calibration proposals, 610
calibration reference data, 181
calibration star, 146
CalVin, 457, 485
Cassegrain focus, 501
Cataclysmic Variables, 523
CCD, 63
CES, 355
cesium clock, 388

chopping, 111, 120
circular polarimetry, 499, 523
closure phase, 452, 462
CODEX, 387, 412
Color-Magnitude Diagram, 43
common pipeline library, 12, 141, 169
Continuum rectification, 370
conversion factor, 123, 141
correlated noise, 467
covariance, 222
CPL, 163, 235, 403, 570
CRIRES, 132, 191, 200, 211, 225, 357, 397, 411
cross talk, 560, 568
cryo-coolers, 121
curvelets, 126

dark, 111
data flow, 159, 236, 443, 471, 479, 541, 554, 560, 565, 581
data reduction pipeline, 177
data reduction software, 96
datacube, 249
deep depletion devices, 332
deformable mirror, 239
Dense wave division multiplexing, 516
deployable masks, 301
detector gain, 121
detector monitoring, 11
differential chromatic refraction, 271
diffuse screen, 421
DIMM, 483
distortion, 322, 546, 559

E-ELT, 212
Easy-Z, 102
Efits, 217
EFOSC2, 75, 492, 501

EMMI, 76, 191, 355
error propagation, 221
ESO Reflex, 169, 171
EsoRex, 97, 163, 169
etendue, 416, 532
Euro 3D, 290
exposure time calculator, 160
extinction, 559, 604
extrasolar planets, 271
extremely metal-poor stars, 576

Fabry-Perot, 383
femtosecond, 389, 409
FEROS, 354
Fiducial Acquisition Bundles, 81
field alignment, 81
filter transmission, 55
Finite aperature, 393
FINITO, 467
FITS, 161, 179, 217, 541
FLAMES, 49, 77, 81, 279
flat cube, 419
flat field, 111
flexure, 284, 307
flux calibration, 69, 308
FORS, 35, 87, 191, 214
FORS1, 76, 492, 501, 524
FORS2, 43, 63, 77
forward analysis, 212
Fourier Transform Spectroscopy, 58, 391
FPOSS, 81
Fried parameter, 485
fringing, 63, 343, 366, 606
FTS, 379, 399
FWHM, 348

GAIA, 330
gas cell, 132, 397
Gasgano, 97, 163, 169
GEISA databank, 392
GIRAFFE, 43, 49, 81, 191, 356
GOES, 434, 447
GPS, 388, 409
GTC, 425

H band, 155
H_2O, 369
HARPS, 354, 412

Hipparcos binaries, 273
HITRAN, 392, 434
hollow cathode lamp, 57, 132, 201, 359, 378, 385, 397
HST, 153, 155, 177
Hubble Legacy Archive, 185
hydroxyl radical, 397

IDL, 236, 295
illumination correction, 38, 575, 605
imaging surveys, 541, 554
instrument model, 293
Instrument Operations Team, 8, 29
instrument profile, 369
instrumental line shape, 392
Integral-Field Spectrograph, 280, 289, 311, 444
Integral-Field Unit, 145, 233, 295, 301, 319, 338, 515
inter-order straylight, 136
interaction matrix, 242
iodine, 378
ISAAC, 109, 149, 439, 490, 501
IUE, 177

J band, 155
JPL catalog, 392
JWST, 145, 297

K band, 155
KMOS, 214, 296, 311, 319

laboratory reference data, 608
Large Synoptic Survey Telescope, 415
laser frequency comb, 383, 386, 409, 609
laser guide star, 279, 328
light concentration, 27, 530
limb-darkened disk, 455
line broadening, 392
linear polarization, 499
Long Baseline Interferometry, 451, 471
long-slit spectroscopy, 89, 134
LOSSAM, 28
Lyot Coronography, 337

MACAO, 133, 239, 483
magnetic field, 515, 519, 523
MARCS, 146

MASS-DIMM, 423
massive binaries, 49
meta data, 201
MIDI, 453, 471, 483
MIRI, 145, 297
model based calibration, 182
Molecular spectroscopy, 391
MSX, 456
Mueller matrices, 505
multi-object spectroscopy, 75, 87, 95, 192
MUSE, 233, 292, 325
MXU, 77, 89

N band, 119, 139
N_2O, 397
NaCo, 109, 261, 271, 490, 501
Nasmyth, 312, 319, 501
NIST, 60, 212, 397, 419
nodding, 111, 120
noise map, 217
nuclear spin, 398

observation block, 160
odd-even effect, 116
OH airglow emission, 154, 249, 397, 443, 585
OH spectrum, 296
Optical apodization, see Finite aperture
optical pathlength difference, 452
optimal extraction, 226, 403
OSIRIS, 427
Ozone, 393

P2VM, 459, 479, 483
pattern matching, 87, 191
pawprint, 567
pen ray lamps, 58
persistence, 561, 568
Phase errors, 393
photodiode detectors, 419
photometric accuracy, 35, 590
photometric calibration, 425, 548, 553, 563
photometric standard star, 115, 439, 486
photometry, 573
physical model, 227, 602
pixel-to-pixel gain, 562

polarimetric standard star, 491, 505, 512
polarimetry, 489, 512
polynomial, 208
precipitable water vapor, 122, 433, 440, 447, 605
PSF, 249, 259, 417, 592, 595
PSF reconstruction, 242, 251, 260
public surveys, 566
pupil alignment, 246
Python, 97, 171, 236

Q band, 119, 139
quality control, 7, 14, 27, 75, 91, 116, 140, 170, 221, 283, 471, 534, 545, 554, 559, 566, 582, 602
quantum efficiency, 11

radial velocity, 41, 50, 271, 375
refractive index, 63
repetition frequency, 410
resolving power, 347
response curve, 69
retarder plates, 519, 524
rotator angle, 35, 284

SAMPO, 164, 169
Science Archive, 604
SDSS, 416
SED, 429
Shack-Hartmann, 239, 251, 265
signal-to-noise, 365, 403
Simulated Annealing, 199
SINFONI, 109, 155, 249, 279, 315, 443
sky background, 89, 444
sky brightness, 583
sky emission, 349
sky lines, 286
Sky Model, 297
sky spectra, 123
sky subtraction, 83, 231, 322, 404
SkyMapper telescope, 573
slitless spectra, 185, 595
Sloan Digital Sky Survey, 574
SofI, 109, 490, 501
speckle noise, 337
Specsim, 295
spectral templates, 427
spectro-imager, 109

spectro-photometric, 140, 153
spectro-photometric calibration, 123
spectro-photometric standard star, 53, 69, 153, 282, 576
spectro-polarimetry, 499, 515, 519
SPHERE, 337
standard star, 123, 140, 153, 425, 605
static aberrations, 245, 268
stellar populations, 41
STIS, 199, 211
Stokes parameters, 490, 511, 521
Stokes vectors, 505
Strehl ratio, 242, 260, 328, 485
striping effect, 121
surveys, 607

Taverna, 169
technical downtime, 3
telluric absorption lines, 149, 154, 365
Th–Ar, 359, 397
Th–Ar, 57, 132, 201, 378, 385, 409
THELI, 553
thermal IR camera, 420
top-down approach, 192
transfer function, 245, 473
tritium, 13
tunable monochromatic laser, 421
twilight flat, 562

u,g,r,i,z,y, 415
UBVRI, 418
UKIRT, 581
uniform disk, 455
UVES, 199, 211, 356, 403

vibrational bands, 444
vibrations, 463

Vignetting, 393
VIMOS, 53, 69, 77, 95, 191, 214, 279, 289, 301, 343
VIMOS-IFU, 347
VIPGI, 95
Virtual Observatory, 170, 183, 185
visibilities, 462, 479
VISIR, 109, 119, 139, 357, 434, 485
VISTA, 559, 565
VLT White Book, 6
VLTI, 483
volume phase holographic gratings, 332
VVDS, 78, 95

water vapor, 422, 466
wavefront sensor, 239, 265
wavelength calibration, 89, 207, 227, 307, 349, 377
wavelength reference, 376
wavelets, 126
WFCAM, 560, 581
WFI, 229, 554, 589, 595
white dwarf, 155
Wide Field Imager, 529
wire-grid, 490
Wollaston prism, 490
workflow, 169

X-shooter, 57, 156, 203, 211, 221

zCOSMOS, 95
Zeeman effect, 519
zero-point, 27, 35, 53, 208, 418, 428, 439, 532, 548, 554, 559, 568, 583, 590
ZIMPOL, 337

ESO ASTROPHYSICS SYMPOSIA
European Southern Observatory

Series Editor: Bruno Leibundgut

A.J. Banday, S. Zaroubi, M. Bartelmann (Eds.),
Mining the Sky
Proceedings, 2000. XV, 705 pages. 2001.

E. Costa, F. Frontera, J. Hjorth (Eds.),
Gamma-Ray Bursts in the Afterglow Era
Proceedings, 2000. XIX, 459 pages. 2001.

S. Cristiani, A. Renzini, R.E. Williams (Eds.),
Deep Fields
Proceedings, 2000. XXVI, 379 pages. 2001.

J.F. Alves, M.J. McCaughrean (Eds.),
The Origins of Stars and Planets: The VLT View
Proceedings, 2001. XXVII, 515 pages. 2002.

J. Bergeron, G. Monnet (Eds.),
Scientific Drivers for ESO Future VLT/VLTI Instrumentation
Proceedings, 2001. XVII, 356 pages. 2002.

M. Gilfanov, R. Sunyaev, E. Churazov (Eds.),
Lighthouses of the Universe: The Most Luminous Celestial Objects and Their Use for Cosmology
Proceedings, 2001. XIV, 618 pages. 2002.

R. Bender, A. Renzini (Eds.),
The Mass of Galaxies at Low and High Redshift
Proceedings, 2001. XXII, 363 pages. 2003.

W. Hillebrandt, B. Leibundgut (Eds.),
From Twilight to Highlight: The Physics of Supernovae
Proceedings, 2002. XVII, 414 pages. 2003.

P.A. Shaver, L. DiLella, A. Giménez (Eds.),
Astronomy, Cosmology and Fundamental Physics
Proceedings, 2002. XXI, 501 pages. 2003.

M. Kissler-Patig (Ed.),
Extragalactic Globular Cluster Systems
Proceedings, 2002. XVI, 356 pages. 2003.

P.J. Quinn, K.M. Górski (Eds.),
Toward an International Virtual Observatory
Proceedings, 2002. XXVII, 341 pages. 2004.

W. Brander, M. Kasper (Eds.),
Science with Adaptive Optics
Proceedings, 2003. XX, 387 pages. 2005.

A. Merloni, S. Nayakshin, R.A. Sunyaev (Eds.),
Growing Black Holes: Accretion in a Cosmological Context
Proceedings, 2004. XIV, 506 pages. 2005.

L. Stanghellini, J.R. Walsh, N.G. Douglas (Eds.)
Planetary Nebulae Beyond the Milky Way
Proceedings, 2004. XVI, 372 pages. 2006.

S. Randich, L. Pasquini (Eds.)
Chemical Abundances and Mixing in Stars in the Milky Way and its Satellites
Proceedings, 2004. XXIV, 411 pages. 2006.

A.P. Lobanov, J.A. Zensus, C. Cesarsky, P.J. Diamond (Eds.)
Exploring the Cosmic Frontier
Astrophysical Instruments for the 21st Century.
XXVI, 270 pages. 2006.

I. Saviane, V.D. Ivanov, J. Borissova (Eds.)
Groups of Galaxies in the Nearby Universe
Proceedings, 2005. XX, 390 pages. 2007.

G.W. Pratt, P. Schuecker, H. Boehringer (Eds.)
Heating versus Cooling in Galaxies and Clusters of Galaxies
Proceedings, 2006. XIV, 468 pages. 2007.

M. Kissler-Patig, R.J. Walsh, M.M. Roth (Eds.)
Science Perspectives for 3D Spectroscopy
Proceedings of the ESO Workshop held in Garching, Germany, 10-14 October 2005
Proceedings, 2006. XVIII, 402 pages. 2007.

S. Hubrig, M. Petr-Gotzens, A. Tokovinin (Eds.)
Multiple Stars Across the H-R Diagram
Proceedings of the ESO Workshop held in Garching, Germany, 12-15 July 2005
Proceedings, 2006. XIV, 304 pages. 2007.

N.C. Santos, L. Pasquini, A.C.M. Correia, M. Romaniello (Eds.)
Precision Spectroscopy in Astrophysics
Proceedings of the ESO/Lisbon/Aveiro Conference held in Aveiro, Portugal, 11-15 September 2006
Proceedings, 2006. XXII, 330 pages. 2008.

A. Kaufer, F. Kerber (Eds.),
The 2007 ESO Instrument Calibration Workshop
Proceedings of the ESO Workshop held in Garching, Germany, 23-26 January 2007
Proceedings, 2007. I, 644 pages. 2008.

Printing: Krips bv, Meppel, The Netherlands
Binding: Stürtz, Würzburg, Germany